SAP PRESS e-books

Print or e-book, Kindle or iPad, workplace or airplane: Choose where and how to read your SAP PRESS books! You can now get all our titles as e-books, too:

- By download and online access
- For all popular devices
- And, of course, DRM-free

Convinced? Then go to www.sap-press.com and get your e-book today.

Configuring SAP S/4HANA® Finance

SAP PRESS is a joint initiative of SAP and Rheinwerk Publishing. The know-how offered by SAP specialists combined with the expertise of Rheinwerk Publishing offers the reader expert books in the field. SAP PRESS features first-hand information and expert advice, and provides useful skills for professional decision-making.

SAP PRESS offers a variety of books on technical and business-related topics for the SAP user. For further information, please visit our website: *www.sap-press.com*.

Janet Salmon, Michel Haesendonckx
SAP S/4HANA Finance: The Reference Guide to What's New
2019, 505 pages, hardcover and e-book
www.sap-press.com/4838

Paul Ovigele
Material Ledger in SAP S/4HANA
2019, 540 pages, hardcover and e-book
www.sap-press.com/4863

Kathrin Schmalzing
CO-PA in SAP S/4HANA Finance: Business Processes, Functionality, and Configuration
2018, 337 pages, hardcover and e-book
www.sap-press.com/4383

Dirk Neumann, Lawrence Liang
Cash Management with SAP S/4HANA: Functionality and Implementation
2018, 477 pages, hardcover and e-book
www.sap-press.com/4479

Stoil Jotev

Configuring SAP S/4HANA® Finance

Editor Meagan White
Acquisitions Editor Emily Nicholls
Copyeditor Melinda Rankin
Cover Design Graham Geary
Photo Credit iStockphoto.com/172685774/© nycshooter
Layout Design Vera Brauner
Production Hannah Lane
Typesetting III-satz, Husby (Germany)
Printed and bound in the United States of America, on paper from sustainable sources

ISBN 978-1-4932-1815-8

© 2019 by Rheinwerk Publishing, Inc., Boston (MA)
1st edition 2019

Library of Congress Cataloging-in-Publication Data
Names: Jotev, Stoil, author.
Title: Configuring SAP S/4HANA finance / Stoil Jotev.
Description: 1st edition. | Bonn ; Boston : [2019] | Includes index.
Identifiers: LCCN 2019020421 (print) | LCCN 2019981058 (ebook) | ISBN 9781493218158 | ISBN 9781493218165 (ebook)
Subjects: LCSH: SAP HANA (Electronic resource) | Accounting--Data processing. | Financial statements--Data processing.
Classification: LCC HF5679 J6358 2019 (print) | LCC HF5679 (ebook) | DDC 657.0285/53--dc23
LC record available at https://lccn.loc.gov/2019020421
LC ebook record available at https://lccn.loc.gov/2019981058

All rights reserved. Neither this publication nor any part of it may be copied or reproduced in any form or by any means or translated into another language, without the prior consent of Rheinwerk Publishing, 2 Heritage Drive, Suite 305, Quincy, MA 02171.

Rheinwerk Publishing makes no warranties or representations with respect to the content hereof and specifically disclaims any implied warranties of merchantability or fitness for any particular purpose. Rheinwerk Publishing assumes no responsibility for any errors that may appear in this publication.

"Rheinwerk Publishing" and the Rheinwerk Publishing logo are registered trademarks of Rheinwerk Verlag GmbH, Bonn, Germany. SAP PRESS is an imprint of Rheinwerk Verlag GmbH and Rheinwerk Publishing, Inc.

All of the screenshots and graphics reproduced in this book are subject to copyright © SAP SE, Dietmar-Hopp-Allee 16, 69190 Walldorf, Germany.

SAP, the SAP logo, ABAP, Ariba, ASAP, Concur, Concur ExpenseIt, Concur TripIt, Duet, SAP Adaptive Server Enterprise, SAP Advantage Database Server, SAP Afaria, SAP ArchiveLink, SAP Ariba, SAP Business ByDesign, SAP Business Explorer, SAP BusinessObjects, SAP BusinessObjects Explorer, SAP BusinessObjects Lumira, SAP BusinessObjects Roambi, SAP BusinessObjects Web Intelligence, SAP Business One, SAP Business Workflow, SAP Crystal Reports, SAP EarlyWatch, SAP Exchange Media (SAP XM), SAP Fieldglass, SAP Fiori, SAP Global Trade Services (SAP GTS), SAP GoingLive, SAP HANA, SAP HANA Vora, SAP Hybris, SAP Jam, SAP MaxAttention, SAP MaxDB, SAP NetWeaver, SAP PartnerEdge, SAPPHIRE NOW, SAP PowerBuilder, SAP PowerDesigner, SAP R/2, SAP R/3, SAP Replication Server, SAP S/4HANA, SAP SQL Anywhere, SAP Strategic Enterprise Management (SAP SEM), SAP SuccessFactors, The Best-Run Businesses Run SAP, TwoGo are registered or unregistered trademarks of SAP SE, Walldorf, Germany.

All other products mentioned in this book are registered or unregistered trademarks of their respective companies.

I would like to thank my lovely wife, Mina, and my wonderful children, Petar and Dimitar, for their support and for enduring all the long nights and weekends without me, which made possible the completion of this book.

Contents at a Glance

1	Project Preparation	47
2	Requirements Analysis	59
3	Financial Accounting Global Settings	73
4	General Ledger	113
5	Accounts Payable	157
6	Accounts Receivable	197
7	Fixed Assets	239
8	Bank Accounting	299
9	General Controlling and Cost Element Accounting	341
10	Cost Center Accounting	365
11	Internal Orders	425
12	Profit Center Accounting	477
13	Profitability Analysis	509
14	Product Costing	567
15	Group Reporting	631
16	Data Migration	663
17	Testing	699
18	Go-Live and Support	719

Dear Reader,

Adding a brand-new writer to our pool of talented SAP PRESS authors is always an exciting, if sometimes apprehensive, process.

Will their writing style, pace, and habits be a good fit for our processes? Should I expect regular updates—or will they appear after six weeks of silence with three new, beautiful chapters? What kind of individualized coaching will help their book reach its potential? Am *I* the best editor for them, or would one of my colleagues be a better match?

With Stoil, answering all of these questions was a breeze. His very first chapter was remarkable, particularly for a first-time author getting accustomed to our template and specs. Since our starting point was so good, we were able to elevate the manuscript to the next level by really focusing on the small details that can set a book apart. From our first call to his last edits, Stoil was an excellent partner, and I look forward to working with him for many editions to come!

What did you think about *Configuring SAP S/4HANA Finance*? Your comments and suggestions are the most useful tools to help us make our books the best they can be. Please feel free to contact me and share any praise or criticism you may have.

Thank you for purchasing a book from SAP PRESS!

Meagan White
Editor, SAP PRESS

meaganw@rheinwerk-publishing.com
www.sap-press.com
Rheinwerk Publishing · Boston, MA

Contents

Preface 21
Introduction 29

1   Project Preparation   47

1.1	Defining Your Project Objectives	48
1.2	Comparing Greenfield versus Brownfield Implementations	49
1.3	Defining the Project Scope	51
1.4	Defining the Project Timeline	53
1.5	Assembling the Project Team	54
1.6	Summary	57

2   Requirements Analysis   59

2.1	Template Requirements Analysis	60
	2.1.1 Financial Accounting	60
	2.1.2 Controlling	65
	2.1.3 Integration with Logistics	65
2.2	Localization Fit/Gap Analysis	66
	2.2.1 Localization Overview	67
	2.2.2 Local Accounting Standards	68
	2.2.3 Local Tax Requirements	69
	2.2.4 Other Local Requirements	70
2.3	Summary	71

9

Contents

3 Financial Accounting Global Settings — 73

3.1 The New Finance Data Model in SAP S/4HANA — 73
- 3.1.1 The Universal Journal — 74
- 3.1.2 Real-Time Integration with Controlling — 76

3.2 Organizational Structure — 77
- 3.2.1 Company — 77
- 3.2.2 Company Code — 78
- 3.2.3 Controlling Area — 84
- 3.2.4 Operating Concern — 88

3.3 Ledgers — 90

3.4 Document Types — 94
- 3.4.1 Document Type Settings — 95
- 3.4.2 Number Ranges — 97
- 3.4.3 Document Types for Entry View in a Ledger — 98

3.5 Currencies — 100
- 3.5.1 Currency Types — 100
- 3.5.2 Exchange Rate Type — 101
- 3.5.3 Exchange Rates — 102

3.6 Taxes — 103
- 3.6.1 Tax Procedure — 103
- 3.6.2 Tax Codes — 109

3.7 Summary — 111

4 General Ledger — 113

4.1 Master Data — 113
- 4.1.1 Chart of Accounts — 114
- 4.1.2 Account Groups — 117
- 4.1.3 General Ledger Accounts — 118

4.2 Document Splitting — 123
- 4.2.1 Document Splitting Characteristics — 124

	4.2.2	Classification of General Ledger Accounts for Document Splitting	125
	4.2.3	Classification of Document Types for Document Splitting	127
	4.2.4	Define Document Splitting Characteristics for General Ledger Accounting	128
	4.2.5	Define Zero-Balance Clearing Account	129
4.3	**Automatic Postings and Account Determination**		131
	4.3.1	Purchasing Flows	132
	4.3.2	Sales Flows	133
	4.3.3	Automatic Postings in Financial Accounting	134
4.4	**Periodic Processing and Financial Closing**		137
	4.4.1	Posting Periods	137
	4.4.2	Intercompany Reconciliation	140
	4.4.3	Foreign Currency Valuation	141
	4.4.4	Account Clearing	144
	4.4.5	Balance Carry-Forward	145
4.5	**Information System**		147
	4.5.1	Balance Reports	147
	4.5.2	Financial Statements	148
	4.5.3	Tax Reports	150
	4.5.4	Drilldown Reporting	151
4.6	**SAP Fiori Applications**		153
	4.6.1	Post General Journal Entries App	154
	4.6.2	Upload General Journal Entries	155
4.7	**Summary**		155

5 Accounts Payable   157

5.1	**Business Partner**		157
	5.1.1	General Data	158
	5.1.2	Company Code Data	159
	5.1.3	Business Partner Configuration	160
5.2	**Business Transactions**		166
	5.2.1	Incoming Invoices/Credit Memos in Financial Accounting	166

	5.2.2	Posting with Alternative Reconciliation Account	168
	5.2.3	Incoming Invoices/Credit Memos from Materials Management	170
	5.2.4	Tax Determination in the Purchasing Process	172
	5.2.5	GR/IR Clearing	176
	5.2.6	Payment Terms and Outgoing Payments	179
	5.2.7	Integration with Vendor Invoice Management	184
5.3	**Information System**		185
	5.3.1	Master Data Reports	186
	5.3.2	Balance Reports	187
	5.3.3	Line Item Reports	189
5.4	**Summary**		195

6   Accounts Receivable   197

6.1	**Business Partner**		197
	6.1.1	General Data	198
	6.1.2	Company Code Data	200
	6.1.3	Sales Data	202
	6.1.4	Business Partner Configuration	204
6.2	**Business Transactions**		213
	6.2.1	Outgoing Invoices/Credit Memos in Financial Accounting	213
	6.2.2	Outgoing Invoices/Credit Memos in Sales and Distribution	218
	6.2.3	Pricing Procedure in Sales	220
	6.2.4	Incoming Payments and Payment Terms	221
6.3	**Taxes**		223
	6.3.1	Taxes in Financial Accounting Invoices	223
	6.3.2	Tax Determination in the Sales Process	224
6.4	**Information System**		231
	6.4.1	Master Data Reports	232
	6.4.2	Balance Reports	233
	6.4.3	Line Item Reports	236
6.5	**Summary**		238

7 Fixed Assets 239

7.1 Organizational Structures 240
7.1.1 Chart of Depreciation 240
7.1.2 Depreciation Area 244
7.1.3 Asset Class 248

7.2 Master Data 251
7.2.1 Screen Layout 251
7.2.2 Account Determination 255
7.2.3 Number Ranges 260
7.2.4 User Fields and Asset Supernumbers 262
7.2.5 Asset Numbers, Subnumbers, and Group Numbers 265

7.3 Business Transactions 268
7.3.1 Acquisitions 268
7.3.2 Transfers 270
7.3.3 Retirement 274

7.4 Valuation and Closing 276
7.4.1 New Asset Accounting Concept 276
7.4.2 Multiple Valuation Principles 277
7.4.3 Depreciation Key 279
7.4.4 Depreciation Run 281
7.4.5 Revaluation 283
7.4.6 Manual Value Correction 285
7.4.7 Year-End Closing Activities 286

7.5 Information System 288
7.5.1 Asset Explorer 288
7.5.2 Asset Balance Reports 290
7.5.3 Asset History Sheet 292

7.6 Summary 297

8 Bank Accounting 299

8.1 Master Data 299
8.1.1 Bank Keys 300

	8.1.2	House Banks	303
	8.1.3	Bank Accounts and IBANs	306
8.2	**Automatic Payment Program**		308
	8.2.1	Automatic Payment Program Parameters	309
	8.2.2	Automatic Payment Program Global Settings	311
	8.2.3	Payment Method	314
	8.2.4	Bank Determination	320
	8.2.5	Common Issues with the Payment Program	323
8.3	**Payment Files**		326
	8.3.1	SEPA Payment Files	327
	8.3.2	Other Common Formats	329
8.4	**Electronic Bank Statements**		331
	8.4.1	Overview	332
	8.4.2	Account Symbols	333
	8.4.3	Posting Rules	335
	8.4.4	Transaction Types	336
8.5	**Summary**		338

9   General Controlling and Cost Element Accounting   341

9.1	**General Controlling Settings**		341
	9.1.1	Maintain Controlling Area	342
	9.1.2	Number Ranges	345
	9.1.3	Versions	348
9.2	**Master Data**		352
	9.2.1	Cost Elements	352
	9.2.2	Cost Element Groups	355
9.3	**Actual Postings**		357
	9.3.1	Manual Reposting	357
	9.3.2	Activity Allocation	360
9.4	**Summary**		363

10 Cost Center Accounting — 365

10.1 Master Data — 365
- 10.1.1 Cost Centers — 366
- 10.1.2 Cost Center Groups — 376
- 10.1.3 Activity Types — 379
- 10.1.4 Statistical Key Figures — 383

10.2 Actual Postings — 384
- 10.2.1 Automatic Account Assignment — 385
- 10.2.2 Substitutions for Account Assignment — 387

10.3 Periodic Allocations — 391
- 10.3.1 Accrual Calculation — 391
- 10.3.2 Distribution — 392
- 10.3.3 Assessment — 400
- 10.3.4 Activity Allocation — 406

10.4 Planning — 408
- 10.4.1 Basic Settings for Planning — 409
- 10.4.2 Manual Planning — 410

10.5 Information System — 415
- 10.5.1 Standard Reports — 416
- 10.5.2 User-Defined Reports — 418

10.6 Summary — 423

11 Internal Orders — 425

11.1 Master Data — 426
- 11.1.1 Order Types — 427
- 11.1.2 Screen Layouts — 432
- 11.1.3 Number Ranges — 434
- 11.1.4 Create Internal Order — 438

11.2 Budgeting — 443

11.3	Actual Postings and Periodic Allocations		449
	11.3.1	Settlement	449
	11.3.2	Periodic Reposting	457
11.4	Planning		461
	11.4.1	Basic Settings	462
	11.4.2	Statistical Key Figures	464
	11.4.3	Allocations	466
11.5	Information System		468
	11.5.1	Standard Reports	468
	11.5.2	Report Painter Reports	472
11.6	Summary		476

12 Profit Center Accounting — 477

12.1	Master Data		478
	12.1.1	Profit Center	478
	12.1.2	Profit Center Group	484
	12.1.3	Standard Hierarchy	486
12.2	Profit Center Derivation		487
	12.2.1	Account Assignment Objects	488
	12.2.2	Document Splitting	490
	12.2.3	Profit Center Substitution	491
12.3	Information System		496
	12.3.1	Standard Reporting	496
	12.3.2	Drilldown Reporting	500
12.4	Summary		507

13 Profitability Analysis — 509

13.1	Overview of Profitability Analysis		509
	13.1.1	Costing-Based Profitability Analysis	510
	13.1.2	Account-Based Profitability Analysis	511

	13.2	**Master Data**	512
	13.2.1	Operating Concern	512
	13.2.2	Data Structure	513
	13.2.3	Operating Concern Attributes	523
	13.2.4	Characteristics Hierarchy	524
	13.2.5	Characteristic Derivation	527
13.3		**Data Flow**	529
	13.3.1	Invoice Value Flow	530
	13.3.2	Overhead Costs Flow	537
	13.3.3	Production Costs Flow	545
13.4		**Integrated Planning**	550
	13.4.1	Planning Framework	550
	13.4.2	Planning Elements	551
13.5		**Information System**	558
	13.5.1	Line Item Lists	559
	13.5.2	Drilldown Reporting	561
13.6		**Summary**	565

14 Product Costing   567

14.1		**Master Data**	567
	14.1.1	Material Master	568
	14.1.2	Bill of Materials	574
	14.1.3	Work Center	575
	14.1.4	Routing	576
14.2		**Product Cost Planning**	578
	14.2.1	Costing Variant Components	579
	14.2.2	Creating the Costing Variant	593
	14.2.3	Cost Component Structure	598
	14.2.4	Costing Sheet	603
	14.2.5	Material Cost Estimate	610
14.3		**Actual Costing and Material Ledger**	613
	14.3.1	Overview and Material Ledger Activation	614
	14.3.2	Multiple Currencies and Valuations	615

	14.3.3	Material Ledger Update	616
	14.3.4	Actual Costing	621
	14.3.5	Actual Costing Cockpit	622
14.4	**Information System**		**623**
	14.4.1	Product Cost Planning	624
	14.4.2	Actual Costing and Material Ledger	626
	14.4.3	Drilldown Reporting	627
14.5	**Summary**		**629**

15 Group Reporting — 631

15.1	**Group Reporting Basics**		**631**
	15.1.1	What Is Group Reporting?	632
	15.1.2	Historical Group Reporting in SAP	632
	15.1.3	Key Benefits	633
15.2	**Global Settings**		**634**
	15.2.1	Prerequisites	635
	15.2.2	Consolidation Ledger	635
	15.2.3	Consolidation Version	637
	15.2.4	Dimensions	640
15.3	**Data Collection and Consolidation Configuration**		**641**
	15.3.1	Financial Statement Items	642
	15.3.2	Subitem Categories and Subitems	643
	15.3.3	Document Type	647
	15.3.4	Number Ranges	652
	15.3.5	Data Collection Tasks	653
	15.3.6	Consolidation of Investments Methods	655
	15.3.7	Task Group	658
15.4	**Summary**		**661**

16 Data Migration — 663

16.1	**Brownfield Implementation Migration**	663
	16.1.1 Check Programs for SAP S/4HANA Readiness	664
	16.1.2 Migration to SAP S/4HANA	668
16.2	**Greenfield Implementation Migration**	676
	16.2.1 Migration Options	676
	16.2.2 Migration Cockpit and Migration Object Modeler	677
	16.2.3 Legacy Data Load	688
16.3	**Financial Migration Objects**	690
	16.3.1 General Ledger Data	690
	16.3.2 Accounts Payable and Accounts Receivable Data	692
	16.3.3 Fixed Assets Data	692
	16.3.4 Controlling-Related Data	695
16.4	**Summary**	696

17 Testing — 699

17.1	**The Testing Process**	699
	17.1.1 Test Plan	700
	17.1.2 Testing Tools	702
	17.1.3 Testing Documentation	704
17.2	**Unit Testing**	705
	17.2.1 Sandbox Client Testing	706
	17.2.2 Unit Testing Client Testing	707
17.3	**Integration Testing**	708
	17.3.1 Planning	708
	17.3.2 Phases	709
	17.3.3 Documentation	711
17.4	**User Acceptance Testing**	712
	17.4.1 Planning	713

	17.4.2	Execution	713
	17.4.3	Documentation	715
17.5	Summary		717

18 Go-Live and Support — 719

18.1 Preparation for the Go-Live — 719
- 18.1.1 Choosing a Go-Live Date — 719
- 18.1.2 Defining a Cutover Plan and Responsibilities — 720
- 18.1.3 Preparing Back-Up Plan — 721

18.2 Activities during the Go-Live — 722
- 18.2.1 Technical Activities — 722
- 18.2.2 Financial Accounting Activities — 723
- 18.2.3 Controlling Activities — 726

18.3 Validation of the Go-Live — 730
- 18.3.1 Project Team Validation — 730
- 18.3.2 Subject Matter Expert Validation — 730

18.4 Hypercare Production Support — 731
- 18.4.1 The First Day — 732
- 18.4.2 Background Jobs — 732
- 18.4.3 Managing Critical Support Incidents — 734
- 18.4.4 Organizing Long-Term Support — 735

18.5 Summary — 736

Appendices — 737

A	Obsolete and New Transaction Codes and Tables in SAP S/4HANA	737
B	The Author	743

Index — 745

Preface

SAP S/4HANA is the biggest innovation in the world of business software applications in more than a quarter of a century. With SAP S/4HANA, SAP introduced a fundamentally improved cloud-based solution that takes full advantage of the revolutionary in-memory database SAP HANA. SAP S/4HANA is a game changer and will set the trend in business software computing for many years to come.

The number of companies on SAP S/4HANA is growing very quickly, and many SAP customers are still planning to move to SAP S/4HANA. Also, many companies that currently do not run SAP are planning SAP implementations to take advantage of the new in-memory cloud solution. The interest in SAP S/4HANA is enormous and dwarfs all previous major release upgrade plans. The projection for the next few years is that so many companies will start or continue major SAP S/4HANA projects that the shortage of SAP consultants with experience in SAP S/4HANA will be enormous. There will be a real fight for talent among companies because an experienced consulting team is the number one factor for the success of an SAP implementation.

Objective of This Book

The objective of this book is to provide a comprehensive guide to customizing SAP S/4HANA Finance. The goal is to cover all main financial and controlling functionalities so that this book can serve as a single source of truth for anyone who is looking to configure SAP S/4HANA Finance—much like the new Universal Journal, which will be discussed in detail later and which is referred to as the single source of truth for financial accounting in SAP S/4HANA. We will cover both the financial accounting (FI) and controlling (CO) areas of the system. The idea of the book is to be equally helpful for new greenfield implementations and for system conversions from older SAP systems.

This book aims not only to cover all main functionalities of the system from the finance point of view, but also to provide deep insight into the whole implementation process, drawing on the 20 years of practical experience of the author in numerous challenging SAP projects. This book will discuss in detail not only the technicalities and the configuration options available, but also methods to gather business requirements and to create system solutions that can fully meet those requirements. The objective of the book will be fulfilled if after reading it you have been transformed into a real SAP S/4HANA financial solution architect, capable of creating real state-of-the-art system solutions.

Target Audience

The target audience of this book includes FI/CO consultants, SAP S/4HANA project managers, finance managers, and business process owners. This is not a beginner's book. We will not cover SAP basics such as how to log on or post a financial document. Some SAP experience is required, but this book should be very useful not only for senior consultants, but also for junior consultants and financial users alike who look to gain more technical experience under the hood of SAP S/4HANA Finance.

This book will cover the whole configuration process in SAP S/4HANA Finance, so some information will not be new for experienced FI/CO consultants coming from an SAP ERP background. However, we'll emphasize the changes between SAP ERP and SAP S/4HANA and will explain the fundamental improvements of SAP S/4HANA.

This book is not only technical. It will give you the methods to deliver successful projects and solutions using the latest techniques from SAP, so it should be also very beneficial for project managers and team leads who are looking to learn best practices for how to lead their projects and at the same time to become more familiar with the underlying technical architecture.

Last but not least, this book should also benefit FI/CO experts who already have SAP S/4HANA experience because its scope is vast and covers all the main financial and controlling areas. Usually SAP consultants tend to focus on specific areas, and this book can be a very useful guide when looking to expand into other areas or just to check your current level of knowledge.

Organization of This Book

This book teaches you how to implement SAP S/4HANA Finance, covering all the phases of the project. It explains in detail not only how to configure all financial accounting and controlling functional areas, but also how to organize the project, perform fit-gap analysis, test the system, perform migration, and support the productive system.

The book is organized to follow the phases of the project, starting with project preparation and finishing with production support, as described in the following chapters:

- **Chapter 1: Project Preparation**
  In this chapter, we discuss how to effectively prepare the SAP S/4HANA Finance implementation. You will learn how to choose a greenfield or brownfield implementation approach, how to define the project scope and objectives, and how to assemble the project team.

- **Chapter 2: Requirements Analysis**
  This chapter explains how to perform analysis of the business requirements for the new SAP S/4HANA system, how to identify gaps with the standard functionalities provided by SAP, and how to manage localization requirements.

- **Chapter 3: Financial Accounting Global Settings**
  In this chapter, we cover the financial accounting global settings, which need to be configured first and which are the basis for all the other financial functional areas.

- **Chapter 4: General Ledger**
  In this chapter, you'll learn how to configure the general ledger in the new SAP S/4HANA environment, placing special emphasis on the new Universal Journal and the real-time integration with controlling and other functional areas.

- **Chapter 5: Accounts Payable**
  In this chapter, we configure accounts payable and discuss the new integrated business partner concept and how to set up various business transactions with vendors.

- **Chapter 6: Accounts Receivable**
  In this chapter, we configure accounts receivable, configuring the business partners as customers and the various business transactions with customers.

- **Chapter 7: Fixed Assets**
  This chapter provides an extensive configuration guide for the fixed assets functional area, which has been completely redesigned to take advantage of the new SAP HANA database and tremendously improved, providing real-time integration in all processes with the general ledger.

- **Chapter 8: Bank Accounting**
  In this chapter, you'll learn how to set up house banks and bank accounts, using the new SAP Fiori app, as well as how to configure the payment program, payment file formats, and the electronic bank statement.

- **Chapter 9: General Controlling and Cost Element Accounting**
  In this chapter, we start the configuration of controlling with the general controlling settings and the configuration for cost elements, which are now completely integrated as general ledger accounts. We pay special attention to the new integration between financial accounting and controlling.

- **Chapter 10: Cost Center Accounting**
  In this chapter, we start the overhead costing configuration with the settings for cost center accounting. You'll learn how to set up the required master data, actual postings, periodic allocations, and planning.

- **Chapter 11: Internal Orders**
  This chapter provides a detailed guide to customizing internal orders in SAP S/4HANA. Covered are the required master data, the settings for internal order budgeting and planning, and actual postings.

- **Chapter 12: Profit Center Accounting**
  In this chapter, we'll configure profit centers, which in SAP S/4HANA are tightly integrated with the general ledger. You'll learn how to set up the relevant master data and how to configure profit center derivation and substitution.

- **Chapter 13: Profitability Analysis**
  This chapter provides a guide to configuring profitability analysis. Special emphasis is placed on account-based profitability analysis, which is the recommended type of profitability analysis in SAP S/4HANA, fully benefiting from the revolutionary SAP HANA database. We also cover the required configuration for costing-based profitability analysis, which is especially important for brownfield SAP S/4HANA implementations, in which often both forms of profitability analysis should be used.

- **Chapter 14: Product Costing**
  This chapter explains how to configure product costing. As with profitability analysis, this functional area has undergone major redesign and improvement in SAP S/4HANA, and special emphasis is placed on new and changed features such as in the material ledger and actual costing.

- **Chapter 15: Group Reporting**
  This chapter provides an overview configuration guide to the new SAP solution for consolidation available from SAP S/4HANA 1809. This is a best-of-breed application that dramatically improves the consolidation process in SAP.

- **Chapter 16: Data Migration**
  After configuring the various financial functional areas in the previous chapters,

in this chapter we'll focus on the data migration process in the area of finance. Separately, we'll discuss the migration tools and processes for brownfield and greenfield implementations. You'll also learn best practices and find advice for various financial migration objects.

- **Chapter 17: Testing**
  This chapter teaches you how to perform various testing cycles during the implementation projects. Specially emphasized are the SAP S/4HANA financial processes and objects, but the best practices and tools discussed should benefit testing efforts in other application areas and systems as well.

- **Chapter 18: Go-Live and Support**
  This chapter finally takes you to the most exciting phase of the project: go-live and ensuing support. We'll first discuss how to prepare for and execute go-live. Then you'll learn how to perform and manage the hypercare initial support of the new SAP S/4HANA system and how to hand over the long-term support to assigned resources.

How to Read This Book

This book is a guide to configuring SAP S/4HANA Finance, both during new implementations and system conversions from SAP ERP. New S/4HANA implementations are referred to as *greenfield implementations*, whereas an upgrade from a previous SAP release is referred to as a *brownfield implementation*.

New (Greenfield) Implementations

Greenfield implementations fall into two categories: implementations for completely new SAP customers and implementations for customers who already run SAP ERP or an older SAP release but decide not to upgrade the current system and instead to implement SAP S/4HANA from scratch. There are many reasons for such an approach. Conversion from an existing SAP system to SAP S/4HANA isn't merely an upgrade. SAP S/4HANA is the biggest SAP release and most fundamental change in almost 25 years. There are many challenges to migrate existing SAP systems, especially heavily customized systems with a lot of custom developments. In addition, many SAP systems are already quite old, with obsolete business processes, which are cumbersome to change—so you can see that often the choice of a greenfield implementation, even when SAP ERP may be already in place, is quite logical.

So how should readers interested in a greenfield implementation read this book? After the preface and introduction, we'll begin at the start of the implementation project with project strategy definition and requirements analysis. These chapters should be very important for both greenfield and brownfield implementations, but especially for greenfield ones as this may very well be your first SAP project. In these chapters, you'll learn the main concepts to properly set up your project and how to go about gathering the important business requirements.

Chapter 3 through Chapter 15 are the technical chapters, which will teach you in detail how to configure all the different financial and controlling areas of SAP S/4HANA Finance, including the newly redesigned SAP S/4HANA Finance for group reporting, which came in release 1809. These chapters will include all the settings required for new implementations, pointing out the required steps for system conversion, which can be skipped for new implementations.

Chapter 16 focuses on migration and focuses more on brownfield implementations: there are some specific migration activities to be performed in the system to be upgraded. However, greenfield readers also will find it useful because we'll discuss the various migration objects needed from legacy systems and the best practices for how to obtain and prepare the data and execute the migration.

Chapter 17 is equally very important for greenfield and brownfield readers: profound and extensive testing is of crucial importance for the success of every SAP S/4HANA project. Greenfield readers will find it especially interesting because they may not have much experience in testing in SAP projects. These readers in particular will learn very important concepts that are the backbone of successful testing.

Chapter 18 will teach you how to organize and execute a successful go-live and hypercare support, which again will benefit both greenfield and brownfield users, but it will also teach important fundamental concepts for readers that are new to implementation projects. In brownfield implementations, there are some specific steps to be executed post go-live, which will be explained here.

System Conversion (Brownfield) Implementations

For existing SAP customers, this is the most desired approach. Building and maintaining an SAP ERP system is a huge effort and expense, and most companies want to keep this investment while still benefiting from the new revolutionary technology brought by SAP S/4HANA. Each existing SAP customer normally will perform extensive and deep analysis, which involves all IT and business counterparts relevant to the SAP ERP system to assess whether it makes sense to upgrade the existing system. Normally, with a newer system that serves the business well without obsolete and hard-to-change processes, the prudent choice is to go with a brownfield approach, even if it will involve a lot of technical effort. These users will benefit tremendously from this book, which will provide them with all the technical and project management knowledge needed to execute the system upgrade.

Chapter 1 and Chapter 2 teach you how to set up your project definition and perform the requirements analysis. Normally, there is less effort in these stages of the project during system conversion than in a new implementation, but it still will be very important to establish a high-performing project structure and to gather detailed, correct business requirements. A system upgrade to SAP S/4HANA is the perfect time to revisit the core business processes and make improvements as necessary. In Chapter 2, we'll point out specific areas where SAP S/4HANA offers improvements and cover how to discuss these with the business to define the best possible solution.

The functional chapters, Chapters 3 through 15, will cover a lot of settings that are already set up in an existing SAP ERP system and may not need to change. Specific areas that are required to change and others that are optional to change in SAP S/4HANA will be pointed out and emphasized. Pay specific attention to Chapter 3 and Chapter 4, in which we'll discuss the general ledger architecture. Very important in Chapter 5 and Chapter 6 is the business partner concept, which is new to SAP S/4HANA. Chapter 7 is very important because SAP S/4HANA offers so-called new asset accounting, which is fundamentally improved compared with the classic asset accounting in SAP ERP and previous releases. From a controlling point of view, especially important are Chapter 13—because in SAP S/4HANA, account-based profitability analysis is required—and Chapter 14—because now the material ledger is required and actual costing is completely redesigned. Chapter 15 will also offer a lot of value for any brownfield reader because group reporting is redesigned and improved in SAP S/4HANA 1809.

Chapter 16 focuses on the data migration process during a brownfield system conversion, so it is crucial to read it carefully. Chapter 17 and Chapter 18, as discussed earlier, will prove equally important for brownfield and greenfield readers alike because comprehensive testing is key to any successful SAP project, as is a well-organized and executed go-live and continuing production support.

Introduction

In this Introduction, we'll discuss the core functional and technical components of an SAP S/4HANA implementation and explain how SAP S/4HANA offers process and user-oriented improvements for financial departments.

Congratulations on starting your journey with the most fascinating and revolutionary new technology in the ERP world—SAP S/4HANA!

SAP S/4HANA is the new ERP solution from SAP. SAP calls SAP S/4HANA its biggest innovation in the last 25 years. We'll examine in detail why it is just that and what is so different and improved in SAP S/4HANA compared with previous SAP ERP systems.

This introduction will start by showing you what's new with your SAP S/4HANA system, then dive more deeply into what's new specifically in SAP S/4HANA Finance. We'll also introduce the configuration interface you'll be using throughout the book.

Your New SAP S/4HANA System

Let's now look under the hood of the SAP S/4HANA system. We'll discuss in detail the new SAP HANA database, which empowers SAP S/4HANA with its great performance and data simplification.

SAP HANA Database

Let's first start with some history. It all started with the development of a new database by SAP called SAP HANA in 2011. *SAP HANA* is SAP's *h*igh-performance *an*alytic *a*ppliance and is a fundamentally new type of database that uses in-memory database technology, which enables the processing of vast amounts of data in real time.

The SAP ERP software runs on a relational database management system (RDMS). Before SAP S/4HANA, SAP didn't offer these database systems itself; instead SAP customers would have to buy them from other vendors such as Oracle, IBM, or Microsoft. This model proved to be extremely successful, with SAP becoming the de facto standard for financial, sales, purchasing, and manufacturing applications all across the globe. These applications formed the SAP ERP system, which is an online transaction processing (OLTP) system, because its main function is to process transactions. Analytical and reporting systems that use data for performing complex and resource-intensive analysis are known as online analytical processing (OLAP) systems. SAP's goal had always been to provide a single system that covers both needs, but the core SAP ERP solution focused on OLTP applications; OLAP applications were provided as separate data warehouse systems. In order to solve this, SAP developed the SAP HANA database, which takes advantage of advancements in processing power and the lower cost of memory, and unites the two types of systems.

SAP HANA is fundamentally different than previous databases used to run SAP ERP first because it is an in-memory database. An in-memory database stores its data online in its memory. This means the data is not stored in hard disc storage devices like in traditional databases, but in the random access memory (RAM). RAM access is much faster, and that enables SAP HANA to achieve tremendous speed compared to other databases.

The other key difference between SAP HANA and older databases is the column store. Traditionally databases store their data in rows. This means that each data record is presented as a row, with each field from this record in a separate column. The data is stored in a table form. To access these records, the database searches by specific fields, but it has to check all the rows to obtain the relevant data. In the column store, each column in the table performs as a separate table and is stored individually. This allows the database to index and compress each of these columns separately, containing only unique records, without duplicate values. This allows for efficient processing of the data.

Let's see how this looks schematically. Figure 1 shows sample data related to vendors.

ID	Vendor Number	Last Name	First Name	Payable
1	1001	Jones	Joe	1000
2	1002	Smith	Mary	2000
3	1003	Connor	Cathy	5000
4	1004	Peterson	Bob	8000
5	1005	Connor	Steve	1400
6	1006	Reagan	Anton	750
7	1007	Bishop	Tom	800
8	1008	Jameson	Stewart	25000
9	1009	Smith	Harry	1300

Figure 1 Vendor Data

Classical row representation of that data is shown in Figure 2, in which each data record is represented as a row consisting of multiple fields.

```
1,1001,Jones,Joe,1000
2,1002,Smith,Mary,2000
3,1003,Connor,Cathy,5000
4,1004,Peterson,Bob,8000
5,1005,Connor,Steve,1400
6,1006,Reagan,Anton,750
7,1007,Bishop,Tom,800
8,1008,Jameson,Stewart,25000
9,1009,Smith,Harry,1300
```

Figure 2 Classic Row-Oriented Database

But how this is represented in a new column-store database, such as SAP HANA? In this case, all the values of a field are serialized in one column, then followed by the values of the next column, and so on, as shown in Figure 3.

Vendor Number	ID	Last Name	ID	First Name	ID	Payable	ID
1001	1	Jones	1	Joe	1	1000	1
1002	2	Smith	2	Mary	2	2000	2
1003	3	Connor	3	Cathy	3	5000	3
1004	4	Peterson	4	Bob	4	8000	4
1005	5	Connor	5	Steve	5	1400	5
1006	6	Reagan	6	Anton	6	750	6
1007	7	Bishop	7	Tom	7	800	7
1008	8	Jameson	8	Stewart	8	25000	8
1009	9	Smith	9	Harry	9	1300	9

Figure 3 Column Store

Now when the system needs to find all records pertaining to, for example, vendor number 1002, it will only search in the vendor number column. It will identify the ID for that vendor (1002) and it will retrieve the rest of the data from the other columns. Therefore, it won't have to go over all the data records as in a traditional row-based database.

As you can see, the new in-memory, column-based SAP HANA database offers significant advantages and not surprisingly is much faster than the traditional database systems used until now. Combine this with the fact that now SAP customers can have both ERP software and a database system from the same vendor, SAP, and it's not surprising that SAP S/4HANA is the biggest improvement in the SAP world in the last 25 years.

> **Note**
>
> Anyone interested to learn more about how SAP HANA evolved from an idea to the revolutionary product that it is today and how it benefits companies looking to combine the power of OLTP and OLAP systems can read the excellent article "A Common Database Approach for OLTP and OLAP Using an In-Memory Column Database" from the inventor of SAP HANA, SAP founder Hasso Plattner himself, which is available at *http://s-prs.co/v485708*.

SAP S/4HANA

SAP S/4HANA has had a long journey. Starting in 2011, SAP transitioned more and more SAP transactions from the SAP ERP system to be based on the HANA database. This led to the SAP Business Suite powered by SAP HANA, available from 2013. This means that the traditional SAP ERP system could work entirely on the SAP HANA database, instead of another database provided by a third-party provider. That benefited SAP customers, who could take advantage of the faster in-memory column-based database. However, from a functionality point of view there was not much change visible to the end users.

Then in 2014 SAP released the first completely rewritten application for the SAP HANA database: SAP Simple Finance. Finance had always been the backbone of the SAP ERP solution, so not surprisingly SAP chose finance to be the first area to be completely redesigned to benefit fully from the new SAP HANA database. SAP Simple Finance was an add-on that could be installed on an existing SAP ERP system, which provided a redesigned table structure and integration at the general ledger level of the various financial components.

Logically, this trend continued: in 2015 SAP, released SAP S/4HANA, a fully functional, fully integrated product entirely based on the SAP HANA database. SAP S/4HANA is the replacement for SAP ERP. With SAP S/4HANA, the use of the SAP HANA database is mandatory. All components are redesigned and rewritten to take full advantage of the enormous capabilities of SAP HANA.

Another tremendous improvement in the SAP S/4HANA system is the new SAP Fiori user experience. SAP Fiori provides a web-based interface to users that makes working with the SAP transactions and reports more approachable.

SAP S/4HANA comes with two product offerings: SAP S/4HANA (by default, this refers to the on-premise version) and SAP S/4HANA Cloud. SAP S/4HANA is the main product, which covers all functionalities provided previously by SAP Business Suite/SAP ERP. SAP S/4HANA Cloud is a software-as-a-service (SaaS) offering, which has somewhat limited functionalities and is suitable for customers looking for a solution that is easier, quicker, and cheaper to deploy, benefiting from standardized provided processes. It is important to note, however, that SAP S/4HANA also can be hosted in the cloud by another provider. So companies can benefit fully from the on-premise version and develop the very complex business processes they need without having to physically host the system themselves.

The release strategy for SAP S/4HANA involves annual updates, which use a naming convention that includes the name and year of the release. As of the printing of this book, the following releases are available for SAP S/4HANA:

- SAP S/4HANA Finance 1503: March 2015
- SAP S/4HANA 1511: November 2015
- SAP S/4HANA Finance 1605: May 2016
- SAP S/4HANA 1610: October 2016
- SAP S/4HANA 1709: September 2017
- SAP S/4HANA 1809: September 2018

This book is based on the SAP S/4HANA 1809 release. As we proceed, we'll point out functionalities that become available only at a certain release and aren't available in prior releases.

Your New Finance Solution

SAP S/4HANA Finance is the most mature solution in SAP S/4HANA. Originally it was released as the SAP Simple Finance add-on back in 2014, which means that it already has five years of continuous improvements, along with a solid customer base, which keeps increasing very fast. Every new SAP S/4HANA release brings new fundamental improvements.

In this section, we'll discuss the most important advancements in S/4HANA in the finance area, including the new SAP Fiori user interface, which provides a beautiful and streamlined user experience.

Advances in Finance

Let's look at the key advancements in the finance area in SAP S/4HANA. There are so many, but the most fundamental are as follows:

- Universal Journal
- Material ledger
- Account-based profitability analysis
- New asset accounting
- Group reporting

Universal Journal

The Universal Journal is the most fundamental advancement in SAP S/4HANA Finance. It combines all finance relevant data in one single table, table ACDOCA, which is often referred to as the *single source of truth*. People working for a long time with finance applications can fully appreciate what an amazing, revolutionary improvement that is. Now all financial and controlling fields, such fixed assets, cost centers, or internal orders, are available together in the same table with pure general ledger information such as the general ledger account, company code, and amount.

SAP's financial solutions traveled a long way to get to this point. Another major improvement in the past was the new general ledger, available from 2005. Before that in the classic general ledger, it was possible to have only one general ledger, but from accounting point of view companies often had to report based on different accounting frameworks. The early solution for that involved the so-called account approach, in which all accounts were duplicated in another range (often starting with the letter *Z* followed by the same account number) to portray another accounting principle's postings. This was a cumbersome solution, both to maintain the additional master data and for the users to make all these redundant postings. A more advanced solution for the time was to use a component called the special purpose ledger, which provided additional ledgers, which could be used to make postings for additional accounting frameworks using the same general ledger accounts. However, these ledgers were not integrated with the general ledger, and the solution involved a lot of reconciliation and custom developments.

Your New Finance Solution

Based on the special purpose ledger fundamentals, SAP delivered the new general ledger, which provided multiple ledgers integrated into the general ledger component. This was a huge step forward, but still not perfect; more often than not the postings to the nonleading ledgers were not happening in real time. Also, additional reconciliation between financial accounting and controlling often was required, using the so-called reconciliation ledger.

With the Universal Journal in SAP S/4HANA Finance, SAP delivered the perfect solution, fully integrated across all the modules and fully in real time. Now every ledger can post in real time in the Universal Journal, and all controlling, fixed asset, and other financial fields are available in the same table ACDOCA. Let's look at the structure of table ACDOCA, as shown in Figure 4.

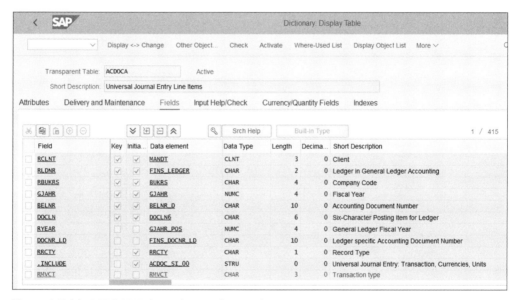

Figure 4 Table ACDOCA Universal Journal Entry Line Items

In Figure 4, you can see the structure of table ACDOCA, the Universal Journal entry line items table, as seen in the data dictionary. This table contains all finance-related line-item data, and this makes many index and summarization tables obsolete. They are not removed altogether, for compatibility reasons, but continue to exist as core data services (CDS) views.

The Universal Journal includes several includes, which contain the fields from controlling, fixed assets, and so on. For example, include ACDOC_SI_GL_ACCAS contains additional account assignments, as shown in Figure 5.

35

Introduction

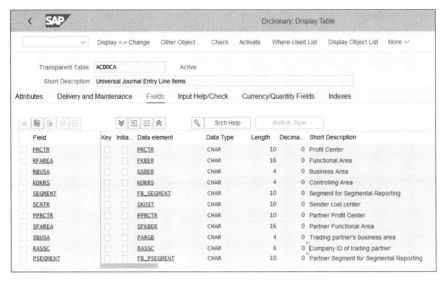

Figure 5 Additional Account Assignemnts in Universal Journal

Now with the Universal Journal, there is only one financial document. There are no more separate controlling, asset accounting, and material ledger documents posted by the system. Figure 6 shows the content of the Universal Journal, with the cost center and a financial document number, in the **DocumentNo** column.

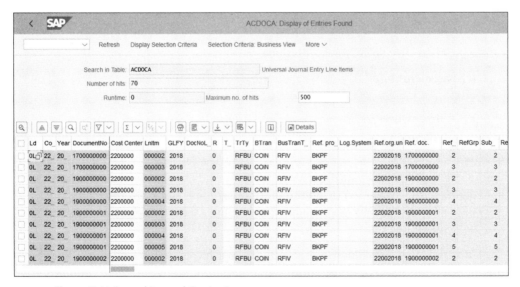

Figure 6 Universal Journal Content

Material Ledger

The material ledger has two main functions:

- **Actual costing**

  The system stores price differences during any material movement. At the end of the month the actual costing run calculates the actual prices for inventory on stock and consumptions.

- **Parallel currencies and parallel valuations**

  The material ledger provides valuations in multiple currencies and valuation principles.

Starting with SAP S/4HANA 1511, the activation of the material ledger becomes mandatory. Previously it had always been an optional component. However, it's important to understand that only the parallel currencies/valuations part is mandatory; actual costing remains an optional component.

Under SAP S/4HANA, the tables of the material ledger are also aggregated to table ACDOCA, and the old material ledger tables become obsolete. Also, there will be no more material ledger documents posted separately because everything is integrated into the Universal Journal. As such, the material ledger is very much streamlined and optimized in SAP S/4HANA.

The actual costing run also is completely redesigned and improved. It includes a lot fewer steps than the old transaction and runs much faster. So even if it's only an optional component, we highly recommend that it be implemented for any company that can benefit from the calculation of actual material prices.

Account-Based Profitability Analysis

Profitability analysis had been for a very long time one of the most powerful tools in finance. It provides a deep, sophisticated analysis of the profitability of a company, which naturally is one of the most important analyses for any business. However, it always caused some confusion. There were two types of profitability analysis: account-based and costing-based. The account-based one was easy to reconcile with the general ledger because it was based on cost elements, but it wasn't flexible enough to analyze vital sales data in every possible way needed. Therefore, SAP provided the costing-based option, which was based on value fields that grouped together various postings based on complex configuration rules. The costing based profitability analysis didn't please accountants, though, because it was virtually impossible to reconcile with FI. Many companies chose to implement both, which added to the complexity of the implementation, support, and data volumes—and profitability analysis reports were always some of the slowest-running SAP transactions due to the underlying data architecture.

Then SAP S/4HANA came along. Profitability analysis is one of the areas that benefits the most from the SAP HANA database and the new integrated data architecture of the financial modules. In SAP S/4HANA, account-based profitability analysis is mandatory, whereas costing-based version is optional. The account-based option is fully integrated with the Universal Journal so it should be able to fulfil both profitability analysis needs and accounting reconciliation needs. One of the main reasons that account-based profitability analysis now should fulfil all needs is that it provides cost-of-goods-sold split functionality.

New Asset Accounting

Asset accounting (commonly referred to as fixed assets) is another finance area that is completely redesigned to take full advantage of the SAP HANA database. Available as an add-on in SAP ERP, now in SAP S/4HANA the new asset accounting is mandatory. It offers real-time integration with the Universal Journal, as shown in Figure 7.

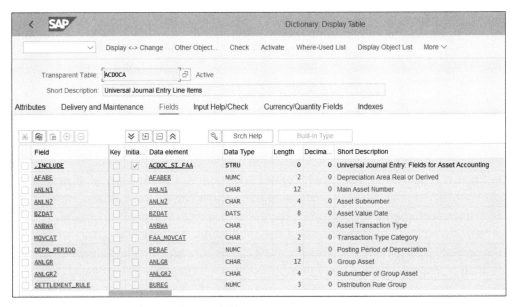

Figure 7 Universal Journal Asset Fields

Include `ACDOC_SI_FAA` provides the fields for fixed assets. Every asset posting is fully integrated with the Universal Journal, and there is no separate asset document generated.

A key benefit of the new asset accounting is that now valuations according to different valuation frameworks, such as IFRS and US GAAP, for example, are fully integrated and can post in real time to the general ledger. In classic asset accounting, this was done using periodic programs and delta postings with the leading ledger, but now separate documents can be posted in real time to the leading and nonleading ledgers.

Group Reporting

Group reporting is an area in which SAP changed its solution a few times. Among the more recent solutions are SAP Business Planning and Consolidation (SAP BPC) and SAP Strategic Enterprise Management Business Consolidation (SEMC-BCS). Now with SAP S/4HANA 1809, SAP provides the new SAP S/4HANA Finance for group reporting. It's fully based on SAP HANA and provides the following benefits:

- Complete package of consolidation functions, such as interunit eliminations and intercompany profit elimination
- Integrated planning, budgeting, and data analysis within the cloud
- SAP Fiori user experience
- Full integration with Microsoft Excel through an Excel add-in
- Open architecture for cloud and on-premise systems
- Single application for local closing and group closing procedures

Thus with the latest SAP S/4HANA release, group reporting in SAP also is fundamentally redesigned and offers many key benefits.

SAP Fiori User Experience

From a user's point of view, one of the main benefits of SAP S/4HANA is the SAP Fiori user experience (UX). In a world of connected devices in which users would like to access their ERP systems increasingly using mobile devices such as tablets and mobile phones, SAP provides the new SAP Fiori UX, which can be used in addition to the classical SAP GUI.

SAP Fiori is entirely web-based and very intuitive and easy to use. It continues to evolve as more and more SAP transactions become available as SAP Fiori apps. The focus is now on the user transactions, as user reports especially are quite convenient to run using SAP Fiori. However, many configuration transactions also are becoming available on SAP Fiori, and some are now even only possible to run as SAP Fiori apps, with the old SAP GUI transaction codes made obsolete. For example, Figure 8 shows the Maintain Banks SAP Fiori app, which replaces Transaction FI12.

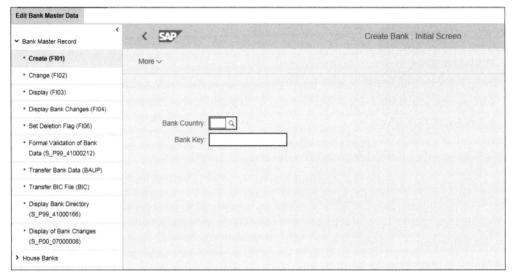

Figure 8 SAP Fiori App Maintain Banks

You can access SAP Fiori apps using the SAP Fiori launchpad. You can start the launchpad from SAP GUI by entering Transaction /UI2/FLP from your SAP S/4HANA system (normally there is a dedicated system in your SAP GUI in which SAP Fiori is enabled because it needs the SAP Fiori server). You can also use a web link provided by your system administrator, which you can enter in a web browser.

If you enter the link directly in a web browser, you'll see a beautiful logon screen, as shown in Figure 9. Enter your **User** name and **Password** (matching those from your SAP S/4HANA system) and click the **Log On** button.

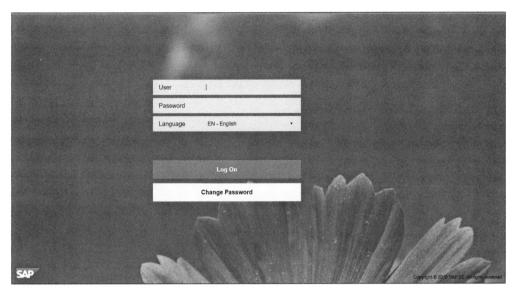

Figure 9 SAP Fiori Logon Screen

Figure 10 shows the home screen of the SAP Fiori launchpad, where you can set your most commonly used apps as tiles. You can easily configure this screen by creating various sections and placing tiles with SAP Fiori apps in them for quick execution.

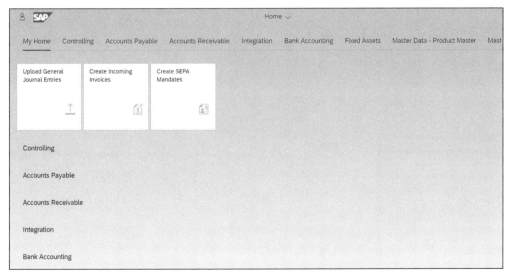

Figure 10 SAP Fiori Home Screen

Introduction

If you click the 👤 (<your name>) button in the top-left corner, you can modify the settings of the SAP Fiori launchpad as shown in Figure 11. With the ✏️ (**Edit Home Page**) button, you can add and remove apps as tiles on the home page. The ⚙️ (**Settings**) button allows you to modify general settings, such as appearance, and language and region.

You can click the 🔍 (**App Finder**) button to search for SAP Fiori apps.

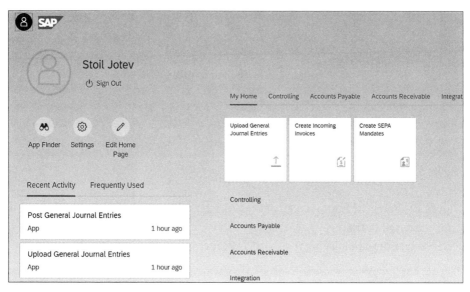

Figure 11 SAP Fiori Settings

As shown in Figure 12, in the App Finder, you can search for apps by process area.

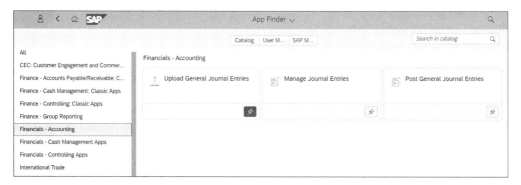

Figure 12 Find SAP Fiori Apps

There are many SAP Fiori apps available, but they need to be configured by your SAP Basis team so you can use them in your system. Normally the finance consultants on the project together with the client users will discuss which apps will be beneficial to use and prepare an inventory of them, and the SAP Basis team then will configure those apps.

A list of all available SAP Fiori apps can be found in the SAP Fiori apps reference library, available at *https://s-prs.co/v485709*.

Configuration Interface

Before we start the first chapter, let's walk through a short introduction to the configuration interface of the SAP S/4HANA system for those that are just starting to configure the SAP system. Experienced readers can skip this: the design of the SAP Reference IMG in SAP S/4HANA hasn't changed compared with previous releases.

Most of the configuration activities are performed in the SAP Reference IMG, which can be accessed with Transaction SPRO, or from the SAP main application menu under menu path **Tools • Customizing • IMG • SPRO • Execute Project**.

Going forward in this book, we'll omit the SPRO portion of the menu path (and anything before it) to reduce redundancy.

One you follow this menu path or enter Transaction SPRO, you're presented with a list of implementation projects, if defined. If not, you can just select **SAP Reference IMG** from the top menu, as shown in Figure 13.

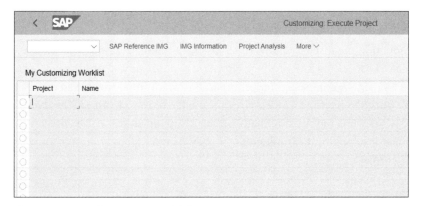

Figure 13 SAP Reference IMG Initial Screen

Then you can see the whole tree-like configuration menu (see Figure 14), which is separated into sections such as **Enterprise Structure**, where we'll define objects, **Financial Accounting**, **Controlling**, and so on.

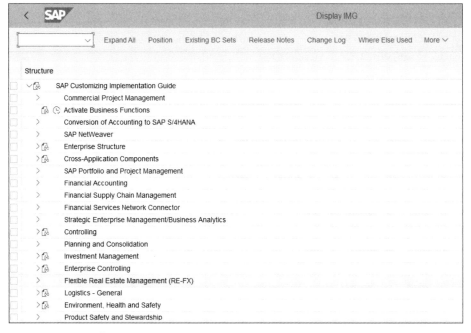

Figure 14 SAP Configuration Menu

Summary

In this Introduction, we discussed how the SAP HANA database came into existence, from an idea to best-selling SAP product. Now you should know about the key benefits that the SAP HANA database offers compared to traditional row-based database systems. We outlined how the SAP HANA–based ERP solution from SAP evolved to become today's mature, highly sought after, state-of-the-art solution that virtually all SAP customers are looking to implement, if they haven't already started or completed such a project.

We discussed the key benefits in the finance area that SAP S/4HANA brings. We touched on the most important ones; in coming chapters, we'll go into deep detail about these and other key advancements in SAP S/4HANA Finance.

The biggest benefit of SAP S/4HANA from a finance point of view is the simplicity it provides, both in terms of processes and in terms of database design. It isn't surprising that the solution initially was called SAP Simple Finance because that highlights exactly what it provides: simplification. This enables faster data processing, huge improvements in reporting capabilities, and improved business processes. Also, the new SAP Fiori UX is a dramatic improvement over SAP GUI from the user's point of view. It provides fast, convenient, and beautifully designed access to the SAP ERP system from any device, from anywhere with a mobile connection.

We also introduced the configuration interface, which enables you to configure the SAP S/4HANA system according to your business requirements. You're now familiar with how to access the configuration of the various functional areas and how are they organized in a tree-like structure.

Let's now start our journey deep under the hood of SAP S/4HANA Finance, examining how to configure all financial modules in detail in the following chapters.

Chapter 1
Project Preparation

This chapter discusses how to choose the right implementation approach for your SAP S/4HANA project (greenfield or brownfield) and outlines best practices for setting the project scope, timeline, and implementation team.

Careful and well-thought-out project preparation is required to ensure a successful ERP project, especially in such an important and challenging project as an SAP S/4HANA implementation. Even if your company is already running SAP ERP, you should consider SAP S/4HANA not just as the next technical upgrade, but as a great opportunity to rethink and improve business processes, taking advantage of the breakthrough in simplification offered by SAP S/4HANA's technology.

There are many key elements of successful project preparation, which we'll now explain in detail.

We'll start with the definition of the project objectives because every successful implementation is based on clearly defined and realistic objectives. We'll then discuss the differences between greenfield and brownfield SAP S/4HANA implementations and the benefits and drawbacks of each approach. This is one of the most important decisions for existing SAP customers, so we'll provide a solid basis on which to base this key decision. We'll guide you through how to define the project scope, which has to be carefully thought out so that the project will add maximum value for the company, but at the same time be realistic and within the agreed-upon budget. We'll also discuss the project timeline, which is dependent mainly on the project scope. It's very important to set a project timeline that's agreed upon by the project management and the customer management and then adhere strictly to it.

Finally, we'll advise you on how to assemble a project team that can perform effectively and deliver the SAP S/4HANA project successfully, adding value to the business.

1.1 Defining Your Project Objectives

First, it's very important to define clearly your project objectives. Carefully established project objectives that are accepted by all levels of the organization ensure that the project will be successful.

For new SAP customers, clearly the objective is to implement the leading and most advanced ERP system in the world: SAP S/4HANA. However, there are different options. Companies that have a well-defined budget at their disposal and with diverse regional and product footprints should look at SAP S/4HANA (the on-premise version) to maximize the value and the benefits they can get from the new software system. They have to be fully aware that such an implementation takes time and effort and should be able to assemble a project team that's prepared for challenging tasks, both technically and from a business process point of view.

New SAP customers who are looking for a quicker, lower-cost implementation could consider SAP S/4HANA Cloud. It comes delivered with predefined business scenarios and content and is more suitable for companies with not as complicated business requirements that can benefit from predefined template processes.

Most SAP S/4HANA implementations will be done by existing SAP customers because SAP already has an enormous customer base. SAP is by far the world market leader in ERP systems, and most big international companies already run one of its ERP business suite, most commonly SAP ERP. They need to set their project objectives and expectations carefully. SAP S/4HANA is the perfect opportunity to rethink your business processes and get rid of old, obsolete ones while implementing new, streamlined processes that can fully benefit from the simplified data architecture of SAP S/4HANA. Especially for customers that implemented SAP ten years or longer ago, their current ERP systems probably aren't in line with the latest business global trends and requirements, and the project objectives should include not only moving to SAP S/4HANA but also redesigning their core processes.

However, there are also many companies that have implemented SAP ERP relatively recently. Their systems likely are functioning well and management and business users are satisfied. Still, those companies should not postpone the migration to SAP S/4HANA because the benefits are enormous, such as those offered by the Universal Journal, the new asset accounting, and the account-based profitability analysis, just to name a few in the finance area. For those companies, the main project objectives will be not so much redesign of processes, but getting the latest functionalities the SAP HANA technology can bring.

Defining the project scope is vital because it determines how much value the implementation of SAP S/4HANA will bring to the customer. SAP S/4HANA delivers cutting-edge technology, but it's how you employ the technology to streamline your business and enhance its processes that could make the implementation a tremendous success. It's a well-defined project scope that will help you utilize SAP S/4HANA for the best benefit to your company.

To define the scope properly, it will help to answer some questions:

- How will the new system help to optimize business processes?
- How will it increase the efficiency of the operations?
- How will it decrease operation costs?
- How will it increase the return on investment (ROI)?
- How will it motivate our business users?

If you can't find reasonable answers to some of these questions, perhaps you have to rethink the scope and what the system will be used for. The implementation should not be a goal in itself. The business case for implementing SAP S/4HANA should answer these questions, and then it will be easier to define the project scope because you'll select functionalities that will increase the ROI, decrease the operation costs, optimize the business processes, and overall add value to the business.

Regardless of when the current ERP system was implemented, any existing SAP customer will inevitably need to have a big discussion and go through a decision-making process to determine whether to undertake a greenfield or brownfield approach to SAP S/4HANA implementation.

1.2 Comparing Greenfield versus Brownfield Implementations

A greenfield SAP S/4HANA implementation is implementing a completely new system from scratch, similar to implementation at a new SAP customer. In this case, the existing SAP system is treated as a legacy system and used as a source for legacy data migration.

A brownfield SAP S/4HANA implementation, on the other hand, is conversion of an existing SAP system to SAP S/4HANA without reimplementation. It does require checking and modifying some of the existing customizing and existing custom programs.

Companies that are already running SAP Business Suite face a difficult dilemma. As already mentioned, the migration to SAP S/4HANA can't be compared with previous SAP upgrades from one SAP ERP enhancement pack to another, or even from the old SAP R/3 system to SAP ERP. This is because SAP S/4HANA is fundamentally different. It comes with a new database and a much enhanced and simplified data model. Therefore, upgrading an existing SAP ERP system to SAP S/4HANA is a huge effort, especially for heavily customized systems with a lot of custom developments. A lot of custom programs have to be changed. All the replaced tables still exist as core data services (CDS) views, and in general programs that only read from them should work without modification because the SELECT statement shouldn't be impacted. However, programs that write to those tables have to be modified.

Therefore, it's compelling to choose the greenfield approach: in this case companies don't have to worry about modifying their old custom code programs and can benefit from a system designed for the new SAP HANA simplified data architecture. But another element to consider is the cost factor. Companies have already invested a huge amount of money, time, and effort into their existing SAP systems, and implementing a new system may seem too much, especially if the current system is relatively new.

There is no clear recommendation for which option in better. It needs to be assessed very carefully on a case-by-case basis, comparing the costs and benefits of both the greenfield and brownfield approaches. Some key factors to consider are as follows:

- **Data volume**
  SAP S/4HANA enormously increases speed and performance. Therefore companies that have huge data volumes can benefit dramatically from conversion of an existing system to SAP S/4HANA.

- **Business processes**
  If the company is satisfied with how the business processes are portrayed in the current system, this is a major reason to consider the brownfield approach. If, on the contrary, there is a lot of room for improvement and there are a lot of obsolete business processes in the current system, the greenfield approach should be considered.

- **Custom code**
  A system with fewer modifications and custom enhancements is a better candidate for brownfield system conversion. If you're running a highly customized SAP system, at the end it may turn out that implementing SAP S/4HANA greenfield is more cost-effective.

- **Technical readiness**

  For an existing system to be converted to SAP S/4HANA, it needs to fulfill several technical prerequisites. SAP S/4HANA supports only Unicode. If the existing system is on multiple code pages, it has to be converted to Unicode first. Also, the existing system should be running a minimum of SAP ERP 6.0 EHP 7, so older systems will have to be upgraded first. All these factors will add time, effort, and cost to the conversion and may be a good reason to go for a greenfield approach.

To summarize, Table 1.1 shows the pros and cons of the greenfield and brownfield approaches.

	Brownfield Approach	Greenfield Approach
Pros	<ul><li>Lower cost</li><li>Shorter implementation time</li><li>Keeping current custom developments and modifications</li><li>Keeping existing system processes</li></ul>	<ul><li>Opportunity to implement new and improve current processes</li><li>New configuration designed for SAP S/4HANA</li><li>Opportunity for data cleaning and harmonization</li><li>Limiting number of custom developments</li></ul>
Cons	<ul><li>Higher system complexity</li><li>Existing custom developments need to be checked and modified</li><li>Using older processes</li></ul>	<ul><li>Higher implementation cost</li><li>Longer implementation time</li><li>Need to develop custom developments from scratch</li></ul>

Table 1.1 Pros and Cons of Brownfield and Greenfield Approaches

1.3 Defining the Project Scope

The project scope is key element of the project preparation phase. It involves clearly defining which modules, components, and functionalities will need to be delivered as part of the SAP S/4HANA implementation project.

This is also a topic that usually causes a lot of discussions at all levels of the organization. SAP provides myriad modules and functionalities, and there is no SAP implementation that uses all of them. Some of them require additional licensing fees, but most are included in the standard SAP S/4HANA license. However, even those come with high implementation and support costs for consulting services and subject matter expert know-how. Therefore, it's important to analyze carefully and select the most needed and beneficial functionalities.

It's good to analyze requirements as mandatory or nice to have. Mandatory requirements consist of processes that are needed to run the business and local legal requirements. These need to be implemented. Typical examples include VAT tax reporting and whatever other reports are mandatory according to local tax authorities.

There are also many requirements that can benefit the business but can be classified as nice to have only. For these, you should perform a cost-benefit analysis and assess potential manual workarounds. Usually there are many competing requirements, and a task for the project management, usually represented by the steering committee of the project, is to assess all options and decide which to include in the project scope. Some functionalities may be left for the next project phase; others may be deemed too expensive or too complex relative to their presumed benefit.

More specifically, in terms of the financial areas in SAP S/4HANA, typically core areas such as the general ledger, accounts payable, accounts receivable, and fixed assets will be included in the project scope. On the controlling side, overhead costing is almost always included, as well as profitability analysis. Product costing should be included at least for production companies. Other finance-related functional areas, such as a project system and financial supply chain management solutions, are more often optional and included in the project scope only with good reason; companies in specific industries will benefit from them more than others.

There are also numerous functionalities that are needed for specific countries or business processes, which are not included in the project scope by default and are often delivered as add-ons to the SAP system. For example, this includes the SAP Revenue Accounting and Reporting tool, which is the SAP solution for the requirements of IFRS 15 (Revenue from Contracts with Customers). This obviously provides an important solution for companies that have a lot of customer contracts, but there are also possible manual workarounds to satisfy reporting requirements, so it's part of the cost-benefit assessment to decide whether to include it in the project scope or not. Also consider the SII (Suministro Inmediato de Información del IVA) tax requirement in Spain, which stipulates that all invoices should be reported to tax authorities in a specific electronic format. For this, SAP provides a solution based on the eDocument add-on, which involves additional licensing. Still, because it's required by law, all companies that implement SAP S/4HANA in Spain will have to include this add-on in the project scope.

The project scope is defined not only by the functionalities implemented, but also geographically. SAP is used by many global companies that have diverse footprints across the globe. These companies need to decide which regions and countries will be included in which implementation waves. Many companies take the cluster wave approach, in which countries that are similar in terms of geographical location and local requirements go live together in the same wave. This may include, for example, Spain and Portugal, or Belgium, the Netherlands, and Luxembourg. Big countries such as the United States usually will be in implementation waves by themselves.

Clearly defining the functional scope and the geographical scope of the project is an important prerequisite for the successful SAP S/4HANA implementation. Of course, as the project goes along, small changes to the scope are possible, but having a well-defined scope from the very beginning is crucial for keeping the budget under control and having a well-motivated project team.

1.4 Defining the Project Timeline

The project timeline is the next key element in your project planning. You have to plan realistically how much time it will take to implement the project scope. This directly impacts the project budget and expectations of the high-level management and the other interested parties.

Sometimes SAP projects don't meet the established timeline, sometimes due to unrealistic goals or to deficiencies in the project execution. SAP S/4HANA projects are not immune to such problems. The fact that a cutting-edge technology solution is being implemented doesn't guarantee that it will be delivered on time and in line with the planned budget. The human factor is as important in SAP S/4HANA implementation as it has been in traditional SAP implementations in the past decades. Therefore, it's imperative to set the timeline expectations ambitiously, but also realistically. Some time ago, we were involved in an implementation in which the implementation partner promised to implement SAP in 70 countries within four years. For anyone well experienced in SAP implementations, that would seem a selling point far from the reality. And in fact, in four years only one country went live. You can imagine the disappointment at all levels of the organization.

In general, the project timeline should be planned aggressively, as SAP implementation projects take a lot of expensive resources from both technical side and the business. However, significant delays and postponements in the project plan should be avoided; they're not only bad for the high-level management and the project sponsors, but also demotivate the whole project team. Therefore, there should be a cushion planned that can counter some unforeseen obstacles.

1   Project Preparation

It's a good idea to choose some of the countries and companies with less complex requirements to go live first and prove the system template is working fine, and then to add complexity into the system landscape. The project timeline should take into account the complexity of the business processes and the local legal requirements in the project scope. This project timeline should be developed and communicated as soon as possible to all interested parties of the project and be discussed in detail before it's officially announced to the outside world. Every project is different and the timeline will vary from project to project, but Figure 1.1 shows a sample project plan timeline with the major phases that are typically part of an SAP S/4HANA implementation project.

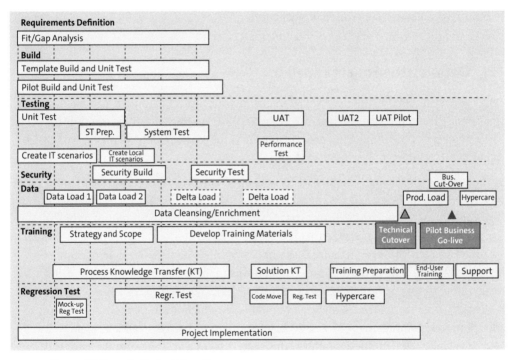

Figure 1.1 Sample Project Timeline

1.5   Assembling the Project Team

The human factor is perhaps the most important element of a successful SAP S/4HANA implementation. You need a well-motivated and technically competent team of project managers, SAP consultants, subject matter experts, and business process owners to navigate through such a complex implementation.

Most companies should engage an implementation partner to provide most of the consulting services and to manage the SAP S/4HANA project. Some companies opt out and assemble an entirely internal team, but that requires a lot of effort and energy, and currently it's very difficult to attract and retain SAP talent, especially with strong SAP S/4HANA experience. Therefore, the default choice is to contract with an experienced IT solution provider that has strong experience in SAP S/4HANA implementations.

A key factor for the success of the implementation project is the commitment of the senior management to the success of the project and the engagement of the client key resources early on. Even the best consultants will fail if there is no good motivation and engagement on client side. It is very important that knowledgeable subject matter experts from client side are gathered as part of the project team. These experts should be very familiar with the business specifics and the key requirements that the system should fulfill.

The project team should be structured by process area—for example, record to report (which corresponds to financial accounting and controlling in SAP S/4HANA), purchase to pay, order to cash, and so on. Each process team should have work stream leads both from the consulting side and from the customer side. Only strong and joint leadership from both implementation partner and client sides can ensure good project planning, execution, and delivery.

It's also a good idea to have an overall team of solution architects to oversee the end-to-end processes and the integration overall in the system. Often, requirements are well defined and good solutions are being delivered in the separate modules, but the end-to-end integration is where problems occur. Therefore, it's good to be proactive as early as possible and have dedicated resources that look after the integration and overall solution architecture.

The technical architecture of the system should be managed by dedicated team(s) also, responsible for custom development and system basis support. Although SAP S/4HANA offers simplified processes and data architectures, it's still expected that custom development will be needed, although probably not as much as in the older SAP releases.

Another very important area is testing. Even the best technical solutions need to undergo extensive testing. The testing usually is divided into a few waves. In unit testing, the responsible consultants test their solutions on their own. During integration testing, end-to-end (or E2E) processes are tested in a joint effort by the whole project team. In user-acceptance testing (UAT), users test the processes and functionalities and sign off after successful testing. The testing effort should be managed by a dedicated test lead/manager, and there should be a team of testers that can help with test execution and documentation.

Once the SAP S/4HANA template is implemented and the project goes into the country rollout implementation phase, for each country/cluster wave there should be a similar, albeit smaller team responsible for the country implementation, including the local specific requirements. It's a very good idea to engage some local experts in the finance area. The local tax requirements vary greatly from country to country, and taxation is an important topic in most countries.

Finally, there also should be a dedicated team for the organization of the go-live and for the hypercare production support. The initial support after go-live—usually one month, but sometimes more—is called *hypercare* and should be provided by a team formed from the original implementation team. This team should perform knowledge transfer to the team that will be responsible for ongoing support and maintenance, which often more or less offshore-based.

The project team should be well structured, with clear areas of responsibility, as presented in Figure 1.2.

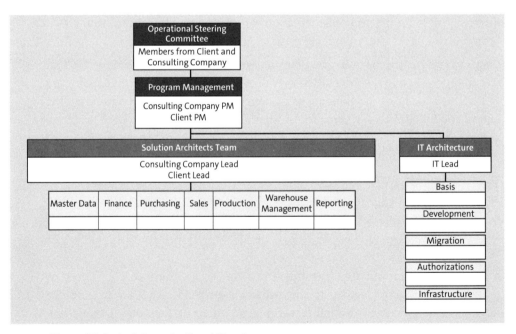

Figure 1.2 Project Organizational Structure

In terms of size, the teams we discussed really vary a lot from project to project. Some projects are huge, multiyear implementations, which form project teams consisting of hundreds of experts. But it's also possible to implement SAP S/4HANA in much smaller companies, with much smaller budgets and project teams. Whatever the size of the business and the size of the budget, the principles for successful project preparation remain the same. There should be clearly defined teams (smaller or bigger), with clearly defined responsibilities and strong leadership from both the consulting side and the client side.

1.6 Summary

In this chapter, we discussed best practices for how to structure your SAP S/4HANA implementation project. There is no single formula that can fit all projects, but the principles discussed in this chapter should be a good guide for how to choose the proper implementation type for your company: greenfield or brownfield.

You should be now well prepared to define your project objectives and the scope of your implementation. There are many potential pitfalls, which can be avoided with careful project preparation, and we discussed how to make good project decisions from the onset of the project.

We discussed what project timelines make sense in various cases and how to avoid costly delays that could demotivate the project sponsors, high-level management, and project team members.

Last but not least, you learned how to assemble a strong and well-performing project team and what types of managers and experts are needed at the various stages of the project. As discussed, acquiring and retaining smart and knowledgeable talent for your project is the most important factor for the success of the implementation. Arguably one of the biggest challenges currently is the lack of strong, senior SAP consultants with SAP S/4HANA experience because the market is booming and many companies are implementing or planning soon to start an SAP S/4HANA implementation. Hopefully this book will help resolve this situation to some extent, by helping more and more SAP FI/CO consultants learn the secrets of SAP S/4HANA.

Chapter 2
Requirements Analysis

This chapter explains how to conduct and document the business requirement analysis that provides the foundation of an SAP S/4HANA implementation. It discusses the best practices for collecting vital information from the business, both on the template level and the localization-specific level, and how to manage multiple requirements from various countries within budget.

After the project preparation is complete, the official kickoff of the project follows, and it starts with the requirements analysis phase. This extremely important phase lays down in document form what business processes the system should perform, how they should be executed, and what legal and business requirements should be met. The requirements analysis provides the backbone for the SAP S/4HANA implementation. It consists of multiple meetings, workshops, and requirements-gathering sessions, performed at various levels of the organization, and at the end should produce a full set of documents that clearly define the business requirements for the new SAP S/4HANA system.

The form and naming of these documents vary from project to project. Sometimes a single, very big document with all the requirements should be produced, which often is called a *business blueprint*. Sometimes you instead create a set of documents per process area, or even per process, which could be called *business requirement specifications*, *fit/gap analysis*, *business process definition*, or some other name used exclusively by a particular implementation partner. Whatever the name, these documents should contain very well-defined business processes and their expected design in the system and are the result of months of hard work by both the project team members and the relevant business resources.

The requirements analysis is performed at the template level and the localization level. In the beginning of the project, requirements for the system template are defined, and after its successful build and testing, the country implementation phase continues with the localization requirements definition for the pilot countries. We'll examine these phases separately, starting with the template requirements analysis.

2 Requirements Analysis

2.1 Template Requirements Analysis

In the old days of SAP implementations, many companies were implementing separate SAP systems in their major markets, and sometimes even in countries with smaller market representation. Now in the current highly globalized business environment, this approach is long gone, and companies want to have one central system, either globally or at least per major region, such as Americas, Europe and Middle East (EMEA), Asia Pacific (APAC), and so on.

The implementation of such a centralized system starts with defining and building the so-called system template. The template is a system that fulfills all the global requirements of the business and is the foundation for future country implementations. After the build is complete, rollout starts to the various markets where the company operates. Clearly, the template is the foundation for a successful SAP S/4HANA system that meets the business requirements.

We will discuss in detail how the template requirements analysis should be performed and what to expect from it in the financials area, starting with financial accounting (FI).

2.1.1 Financial Accounting

Financial accounting is the very foundation of the SAP ERP system: all other areas post into it. It is the only area that isn't optional; it has to be implanted in every SAP system. Sometimes early on in implementations, its importance may be underestimated by some not-finance-related managers or users ("Finance? It's just numbers"; "Sales is what matters most," you may even hear), but the proper implementation and the careful definition of business requirements in the area of financial accounting is of paramount importance for every project. This goes especially true for SAP S/4HANA, in which finance completely integrates all information into one table, the single source of truth: the Universal Journal. To fully benefit from the great simplification in finance that comes with SAP S/4HANA, you need to define its requirements in great detail.

There are different ways to structure the requirement documents and different methods for how to proceed with the discussions with the business and perform the review and approval process. The following is an example structure and process based on our 20 years of experience, both in SAP S/4HANA and older SAP implementations. It's to be used as best practices guidelines, but of course each company and project team may implement its own specifics to enhance the process.

This is a sample list of process areas for FI, under which the business requirements documents may be divided:

- Manage financial organizational structures
- Manage financial global setting
- Manage general ledger accounting
- Manage accounting subledgers
- Manage bank accounting

Within each of these areas, several documents will be defined, which correspond to particular process areas. From our point of view, this approach is better than having a single business blueprint document for financial accounting. This way, the requirements are better structured and easier to read and approve. Also, this approach is in line with the latest SAP project methodology, which involves using the SAP-specific project implementation tool SAP Solution Manager to manage the requirements definition.

Each of these documents should be well structured. It should include administrative information such as who created the document, when, and which versions have been created with what major changes to the document. This will give the reviewer important information about how the requirements-gathering process evolved. It should also include information about who verified and approved the document and when. It should have an overview section that describes the process in general business terms. Then it should have all the different process steps covered by the requirement document. It's convenient to assign each of these process steps an identifier for better tracking. For each of these process steps, what requirements the system should perform should be clearly written.

Where relevant, the documents should note for which process areas you need to develop reports, interfaces, conversions, enhancements, forms, and workflows (RICEFW) objects. These are the objects to be custom developed for each project, rather than standardly delivered by SAP. *Reports* are executable programs that retrieve data from the database and meet specific customer reporting requirements. *Interfaces* provide a link between the SAP S/4HANA system and other SAP or non-SAP systems. *Conversions* are programs that convert data to meet specific requirements. *Enhancements* provide custom code to be triggered in specific areas of the system, such as using user exits and BAdIs. *Forms* provide custom layouts for printed or electronically sent documents from the system. *Workflows* provide custom workflow functionality to trigger some approval processes within the system (e.g., the purchase order approval process).

The requirements definition document also should contain information about dependencies with other configuration documents and test case information.

Each requirements definition document should be uploaded as a separate business requirement definition item in SAP Solution Manager, in which the whole review and approval process is organized, as shown in Figure 2.1.

Figure 2.1 Manage Fixed Assets Requirements Document in SAP Solution Manager

Figure 2.1 shows a manage fixed assets accounting requirements document, which is part of the manage accounting subledgers process as defined in SAP Solution Manager. In this section of SAP Solution Manager, you can track the testing requirements, the approval process, the changes to the document, and the performed configuration in the system related to these requirements.

SAP Solution Manager is a convenient repository for everything related to the configuration of the system, and as a SAP-delivered tool it's fully integrated with the SAP S/4HANA system. You can assign a configuration unit to each configuration document, which contains the various configuration activities to be performed. This configuration unit in turn contains several configuration transactions. From within SAP Solution Manager, if properly configured, you can double-click a configuration element to go to the relevant configuration transaction in the linked SAP S/4HANA system. Figure 2.2 shows the list of configuration elements for definition of financial global settings.

2.1 Template Requirements Analysis

Figure 2.2 Configuration Elements in SAP Solution Manager

Double-clicking any of these elements would take you directly to the relevant configuration path in the SAP Reference IMG. For example, double-clicking **Define Accounting Principles** would open a new SAP GUI window within the linked SAP S/4HANA system and take you to the configuration transaction, as shown in Figure 2.3.

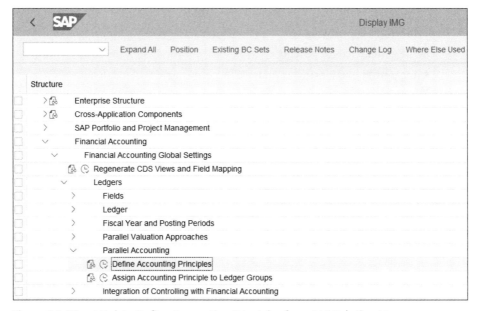

Figure 2.3 Direct Link to Define Accounting Principles from SAP Solution Manager

This is a convenient function for anyone reviewing, approving, or checking the requirements gathered and how they are fulfilled in the system.

The structuring of specific financial accounting configuration documents should be logical, so in one configuration document the complete configuration elements for a particular function are included. They should enable the business reviewer to be able to check and approve all the steps within his realm of knowledge.

The following is an example structure for financial accounting requirement documents:

- **Financial organizational structures**
  - Define accounting structures
  - Define fixed asset structures
  - Define controlling structures
- **Financial global settings**
  - Define currencies
  - Define ledgers
  - Define tax codes and tax determination
  - Define account determination
  - Define output forms
- **General ledger accounting**
  - Define general ledger postings
  - Define period-end closing in financial accounting
- **Accounting subledgers**
  - Define accounts payable
  - Define accounts receivable
  - Define fixed assets
- **Bank accounting**
  - Define bank accounting
  - Define cash journals
  - Define electronic bank statement

This is, of course, only an example. Use it as a guideline, but structure the requirement documents in the most clear and easy to review and approve fashion based on your project specific requirements.

2.1.2 Controlling

Similarly, in controlling, the requirements definition documents should be organized in logical, coherent process areas. This also varies from project to project, especially given that only certain subareas of controlling are used in some projects.

The following is an example of possible structuring of the controlling requirement definition documents for a company using all the main controlling functionalities, including product costing:

- **Overhead costing**
    - Define general controlling settings
    - Define master data for overhead costing
    - Manage actual costs in overhead costing
    - Define planning in overhead costing
- **Profitability**
    - Manage profit center accounting
    - Manage profitability analysis
    - Manage period-end closing in controlling
- **Inventory valuation and product costing**
    - Define inventory valuation
    - Perform actual costing
    - Define material ledger
    - Perform inventory controlling and reporting

As with financial accounting, this structure is just a guideline; most companies and industries have some specifics that need to be taken into account.

In terms of content of the controlling documents, they should have the same structure that we already outlined for the financial accounting documents. In general, all requirement definition documents for all areas should follow the same structure and be consistent, even if they're being prepared by different consultants and reviewed by different business users.

2.1.3 Integration with Logistics

There are many functions that are integration points between finance and logistics. The requirement definition in these areas should be a joint effort of both the financial and logistic teams. SAP is a highly integrated system, which is where its main strength lies. To define efficient processes that meet the business requirements of all departments, you need to have many integration meetings in which both the financial and logistic teams discuss the end-to-end processes, not only separate processes within a process area.

Typical requirements definition areas that should be a joint effort between finance and logistics are inventory valuation, account determination, and tax code determination. Indeed, the configuration transactions for those can be found in both the finance and logistics areas of the SAP Reference IMG.

For example, tax codes are defined by the finance department. But the automatic determination of the proper tax code in the different purchase and sales processes comes from the so-called condition records and condition tables, which are typically maintained by the sales and purchasing consultants. The discussion of these logistic processes should involve users from both finance and logistics to determine the proper tax treatment for each process.

Account determination is typically a finance task, but many of the logistic processes post automatically in financial accounting and need to have correct accounts assigned to work properly. Normally, the requirements definition document for this topic should be owned by finance, but it should be worked on in conjunction with the logistics team. Similarly, inventory valuation usually is owned by finance, but its nature makes it required to be developed in close integration with logistics.

Also, there are many requirement definition documents that are typically owned by logistics, but which finance needs to be consulted on. Just to name a few, these include material master data, credit management, and pricing.

So it's very clear that requirements analysis is a complex stage of the project, which needs to be worked on in a highly integrated manner between financial and logistic consultants and subject matter experts.

2.2 Localization Fit/Gap Analysis

The template is the foundation of the SAP S/4HANA implementation, but the real deployments are done at a country level. Typically, once the template is configured, tested, and signed off, the country deployments start in earnest. Most companies opt for a wave approach, in which one or a few similar countries are selected as pilot countries, and then over the next couple of years all the other countries are deployed in carefully selected waves. In very big companies operating in many markets, these deployments can last for 5 or 10 years, or even longer (in very rare cases).

In the following sections, we'll discuss in detail how to localize your SAP S/4HANA system. We'll start with an overview of the localization process, then we'll deep-dive into the implementation of local accounting standards, tax requirements, and other localization topics for the various countries in scope.

2.2.1 Localization Overview

The method to roll out the template to different countries is the same, whether it's for the pilot country or a few years down the implementation, and whether it's one of the major markets for the company or some small market with very limited presence. The template needs to be localized, which essentially means performing a fit/gap analysis for the local business and legal requirements.

Fit/gap analysis is a process in which the baseline solution (the template solution) is analyzed in all its process areas and discussed with the local business and consulting resources. Any identified gaps should be recorded—either in the original requirements definition documents, which are being "localized" and enhanced, or in a new set of localized requirements definition documents on the country level.

Then all the gaps should be assessed in terms of their importance. Obviously, fulfilling each gap comes at a certain cost, and it should be compared with the benefit it provides. This is called cost-benefit analysis. The gaps should be classified as mandatory to fulfill, essential, and nice to have. Accordingly, the project management will decide which ones to be implemented based on the available budget and timing.

Because each country has its own specifics, and some countries' solutions in SAP are sciences of their own due to the very high complexity of the tax and statutory requirements (notably, Russia), it makes sense to enlist the help of local consultants. Of course, there is no SAP implementation team that is knowledgeable about all the specifics in all the countries. Therefore, a good approach is to have lead consultants from the template project team manage the rollout effort, but also hire one or more local SAP financial consultants that know about the local taxation, VAT, and other relevant topics. Of course, another key factor to properly gather the local requirements is to engage very deeply with the local subject matter experts and discuss all the local topics with them.

Still, the core implementation team should educate itself about the local specifics as much as possible prior to starting the requirements analysis. A very good source for this is the SAP Globalization Services website, which provides detailed information about country and language versions for various countries: *https://support.sap.com/ en/product/globalization.html*. There you can browse by region and country and find important country localization information. Another good source for country localization information is found in SAP Notes, available at *https://support.sap.com*.

An SAP ID is required to access SAP Notes, which usually can be provided by your project manager. SAP Notes provide up-to-date information as SAP constantly monitors changes in the legal requirements for various countries and provides updates in these SAP Notes, which can be implemented before the change becomes part of the standard system. It's important to gather all relevant SAP Notes for the countries for which you're performing requirements analysis.

In the areas of financial accounting and controlling, we can find localization gaps in few key areas: local accounting standards requirements, local tax requirements, and other business-related local requirements, which we'll analyze next.

2.2.2 Local Accounting Standards

In the current globalized world, there is a huge movement toward international harmonization in the area of accounting. International Financial Reporting Standards (IFRS) has gone a long way toward defining common accounting rules for most business transactions and situations. It's being adopted in many countries in the world, including the European Union, Russia, many Asian countries—such as South Korea, India, Singapore, and Hong Kong—Australia, South Africa, Turkey, and many others. However, it's not universally accepted everywhere, and notably is still not adopted in the United States, where US Generally Accepted Accounting Principles (US GAAP) is the main accounting standard.

Managing different accounting standards is a major requirement for most SAP S/4HANA systems. In fact, the new general ledger architecture, powered by the simplified SAP HANA database is perfect to manage these requirements. This will be discussed in detail in Chapter 3 and Chapter 7, but for now we can say that using the leading ledger to portray the leading accounting principle from a group point of view and nonleading ledgers for the other accounting principles is the perfect technical solution.

From a requirements definition point of view, consultants need to discuss with each country which accounting standards need to be portrayed in the system. Even in some countries in which IFRS is adopted, companies may need to follow another local accounting standard in parallel. Fortunately, with the new SAP S/4HANA Finance solution this is much easier than before. In the past, sometimes a complete set of duplicated general ledger accounts was deployed for this task.

After confirming the required accounting standard framework, the requirements definition documents should be enhanced and signed off to reflect the desired setup. Typically, this will include setting up nonleading ledgers for each accounting framework that needs to be reported on. For example, if from the template point of view a leading ledger for US GAAP and one nonleading ledger for IFRS are set up, as part of the localization fit/gap phase you could set up an additional nonleading ledger to be activated to cover the requirements of the local accounting principles.

2.2.3 Local Tax Requirements

In many countries, there are specific tax rules that differ from IFRS and from the locally accepted accounting principles. This means that from an accounting point of view, companies can follow certain rules and guidelines, but then when preparing their tax declarations and calculating their taxes due they need to follow rules specifically postulated by the local tax authorities. Therefore, in such countries requirements analysis to include these tax rules and define how the system should meet them is key. These are mandatory requirements, and not meeting them in the SAP S/4HANA system could result in severe penalties.

Some countries have more stringent tax requirements than others. In most countries these local tax requirements revolve around the value-added tax (VAT), which is a tax added on most sales and purchases. As it's a main source of taxes for tax administrations and is difficult to track, with a lot of fraud around it, in many countries there are very tough reporting requirements related to VAT. For example, in Spain it's mandatory to use the SII reporting, which requires that all incoming and outgoing invoices should be reported to the tax office no later than four days after their issue. This is the type of requirement that's very important to capture during the fit/gap localization phase. Also, many other countries have a specific required form for the VAT reporting to be produced by the accounting system.

SAP aims to cover all these legally required specifics in different countries. It constantly monitors the tax requirements around the world (which often change and create greater demands on information systems) and provides updates and new functionalities to meet those requirements. For most countries and most requirements, this is part of the standard system. But for some of the requirements, you need to install some additional add-ons to the system, which may require additional licensing costs. For example, for the SII requirement in Spain, as well as for similar real-time reporting requirements to tax offices in other countries, you need the eDocument processing add-on from SAP. In some countries, there are also third-party add-ons to the SAP system developed by other companies that have locality-specific experience. Choosing the right solution is part of the requirements analysis phase and is a joint effort of the consulting and business teams on the project.

In countries with sophisticated local tax requirements, usually a local tax ledger also is activated. This is another nonleading ledger, similar to what we discussed for the local accounting standards, which is activated only in countries with very special tax needs. Using this tax ledger, it's possible to make ledger-specific postings that allow different tax treatments for some transactions. For example, in fixed assets often local tax authorities stipulate different depreciation rules from a tax point of view. This is easy to accomplish using the tax ledger. Local tax depreciation areas are set up that reflect those rules, and these areas are mapped to the tax nonleading ledger. Now in SAP S/4HANA this process is much enhanced compared to older SAP systems because these postings happen in real time. In the past, these differences were tracked in delta depreciation areas and posted to the nonleading ledgers offline. The nonleading ledger concept is very advanced in SAP S/4HANA, as we'll discuss in detail in Chapter 3 and Chapter 7.

As with the local accounting standards, the local tax requirements need to be written in deep detail in the relevant requirements definition documents, as well as how the solution should work to meet these requirements. This is a key deliverable of the requirements analysis phase.

2.2.4 Other Local Requirements

There are also other local requirements that don't fall precisely into the category of local accounting standards or local tax code. They stem from local business practices or local ways of accounting for certain transactions, even though they may not be written explicitly in accounting laws. They range from nice to have to mandatory, depending on the significance of the requirements.

These requirements also need to be discussed in detail with the local subject matter experts, and then you need to carefully analyze the costs and benefits of each requirement. It's easy to get carried away and promise to fulfill countless local requirements, but it's important to remember that it's the template that it's being rolled out. Normally this template should be able to function properly in each country with only limited local changes, so most of the nice-to-have elements should be scrutinized very carefully to decide whether it makes sense to implement them.

Let's look at few examples for local business requirements. Inventory valuation is a topic that deserves a lot of attention on the local level. Some countries have specific requirements for inventory valuation, which are indeed required to be implemented. Countries such as Brazil, Russia, and Turkey require inventory to be valued at actual cost. Therefore, in such countries you must implement actual costing. It's important to understand the local requirements properly. In this case, the requirement isn't to use parallel currencies and valuation, but to use actual costing. This is a point that sometimes raises confusion in SAP S/4HANA. Indeed, in SAP S/4HANA the material ledger is required, but it doesn't provide actual costing by default. Actual costing is different functionality, which is activated separately from the material ledger. Therefore, the material ledger is always a template requirement in SAP S/4HANA, whereas actual costing could be a local requirement activated only for specific countries. (The material ledger will be discussed in detail in Chapter 14.)

Other key local requirements that usually should be implemented are the need for a local chart of accounts, for translation into the local language of various forms and reports, and any required interfaces with tax and other government offices.

Other requirements—such as when a country executes particular business processes in its own way, which is not explicitly required by local law and is not in line with group policies and template settings—are prime targets for optimization. In this situation, the SAP S/4HANA implementation could be viewed as an opportunity to help improve obsolete processes and optimize the business at a local level.

2.3 Summary

In this chapter, we covered requirements gathering: a very important, yet sometimes underestimated phase of the project. It's hard to emphasize strongly enough that proper definition of the business requirements, both on the template level and the localization level, is crucial for delivering a good SAP S/4HANA solution that meets all key requirements.

In this chapter, you learned a method for how to approach requirements analysis, what the key elements are that need to be discussed in detail with the business, and how to document them. You learned one possible way to structure the requirements definition documents in financial accounting and controlling. This is a guideline you could use and adapt to your project- and business-specific requirements.

The key to success for the requirements analysis project phase is to cover all the business requirements and discuss them with the business, but also to be able to assess properly the costs and benefits of them. In most projects, one of the main challenges is the infinite desire of various business users to include more and more requirements. The success of the project depends on the ability of the senior project management, the subject matter experts, and the consultants to select only the key, important requirements and deliver those within the agreed-upon budget and timeframe.

Hopefully, with the methods and guidelines discussed in this chapter you'll be able to play this important role in your SAP S/4HANA implementation and deliver good, clear, and successful requirements analysis.

Chapter 3
Financial Accounting Global Settings

This chapter introduces the new data model in SAP S/4HANA and how it improves financial processes and reporting. It describes configuring global finance settings in SAP S/4HANA, such as organizational structure, ledgers, document types, and other settings.

After completing the requirements gathering phase of the project, which produces signed-off business requirements definition documents, it's time to start configuring SAP S/4HANA Finance. The configuration process starts with configuring the financial accounting global settings, which provide the organizational structure and basic configuration elements such as ledgers, document types, currencies, and tax codes.

As briefly discussed in the introduction, SAP S/4HANA offers a new simplified data model, which greatly increases the speed and performance of the finance processes. It's of paramount importance to understand this new data model, the real-time integration of financial accounting and controlling, and how the new Universal Journal functions. So we'll start with detailed explanations of the new data model in SAP S/4HANA Finance.

3.1 The New Finance Data Model in SAP S/4HANA

SAP is a highly integrated system, which manages data from various areas of the business, such as accounting, sales, purchasing, production, and so on. This integration comes with a certain degree of complexity, which results in the data being stored in many different tables, and sometimes even for experienced consultants it's a challenge to pick the best way to find and retrieve the relevant data.

In the area of finance, traditionally financial accounting and controlling (management accounting) were separate applications in SAP, which was a design mainly driven from traditions in the German-language world. However, in today's globalized world there is a strong case for integration of processes and applications and simplification of systems. SAP's answer to and excellent solution for this is the SAP S/4HANA Finance solution, which provides full integration of the financial accounting and controlling applications, both from a process point of view and a database point of view.

We will discuss in detail the two key elements of the new finance data model in SAP S/4HANA: the Universal Journal and the real-time integration between financial accounting and controlling.

3.1.1 The Universal Journal

The Universal Journal provides a solution for a seemingly simple but until SAP S/4HANA elusive goal: bringing together and fully integrating all financial information in one single line-item table that has all financial accounting, controlling, and material valuation information. There were many reasons that there were multiple financial accounting and controlling tables until SAP S/4HANA that were storing data that now is available in the Universal Journal. Some were business-process based, on the presumption that financial accounting and controlling should be separate applications, which is not the case in the current business world. Some were technical reasons: only now with the amazing speed and columnar design of SAP S/4HANA is it technically feasible to have such a vast amount of data in a single table.

The Universal Journal is a new table in SAP S/4HANA, called table ACDOCA. It's a line-item table that brings together information from the general ledger, controlling, asset accounting, and the material ledger, as shown in Figure 3.1.

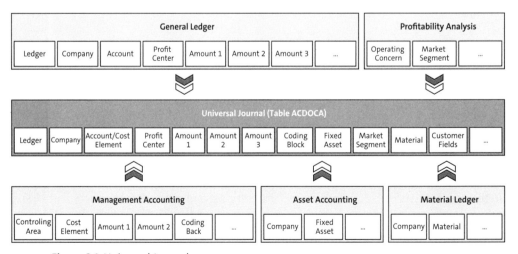

Figure 3.1 Universal Journal

As shown, table ACDOCA, the Universal Journal table, combines fields that previously were stored in the tables of various financial accounting and controlling modules. This means that once a financial document is posted in table ACDOCA, fields such as cost center, asset number, profitability segment fields, and so on are also recorded. This makes a whole lot of tables from controlling, fixed assets, and the material ledger redundant because the information is now integrated in the Universal Journal. For compatibility reasons, they exist as core data services (CDS) views so that they can still be referenced in the custom programs of companies doing brownfield implementations of SAP S/4HANA.

Table 3.1 shows the main financial accounting tables that are now obsolete in SAP S/4HANA because their data is part of table ACDOCA.

Table	Description
BSIS	Accounting: Secondary Index for G/L Accounts
BSAS	Accounting: Secondary Index for G/L Accounts (Clearing Postings)
BSID	Accounting: Secondary Index for Customers
BSAD	Accounting: Secondary Index for Customers (Clearing Postings)
BSIK	Accounting: Secondary Index for Vendors
BSAK	Accounting: Secondary Index for Vendors (Clearing Postings)
GLT0	G/L Account Master Record Transaction Figures (Totals Table)
FAGLFLEXT	General Ledger: Totals (New GL totals table)

Table 3.1 Obsolete Tables in Financial Accounting

Tables BSIS, BSAS, BSID, BSAD, BSIK, and BSAK are index tables containing open and cleared items for general ledger accounts, customers, and vendors, which are now all in table ACDOCA. Tables GLT0 and FAGLFLEXT are totals tables (FAGLFLEXT was introduced with the new general ledger), which are now also obsolete because SAP S/4HANA calculates totals on the fly. Table 3.2 shows other important controlling, fixed assets, and material ledger tables which are now obsolete due to the Universal Journal.

Table	Description
COEP	CO Object: Line Items (by Period)
COBK	CO Object: Document Header

Table 3.2 Obsolete Tables in Controlling, Fixed Assets, and Material Ledger

Table	Description
ANEP	Asset Line Items
ANEA	Asset Line Items for Proportional Values
ANLP	Asset Line Items
MLHD	Material Ledger Document: Header
MLIT	Material Ledger Document: Items

Table 3.2 Obsolete Tables in Controlling, Fixed Assets, and Material Ledger (Cont.)

As you can see, now in SAP S/4HANA the Universal Journal combines the key tables of all the financial applications in a single table, which is commonly referred to as the single source of truth. Now you have all the information needed to present the financials of the company in one place. This is an enormous advantage compared to previous SAP releases and to other ERP systems.

3.1.2 Real-Time Integration with Controlling

The real-time integration of financial accounting with controlling follows logically from the integration design of the Universal Journal that we discussed previously. Indeed, because the controlling-relevant data now is brought together with the financial accounting data in the Universal Journal, there are no technical obstacles preventing the system from providing real-time integration between any financial accounting and controlling documents.

In the past, the reconciliation ledger had to be configured to ensure that financial accounting and controlling were always in sync. This is no longer required because with the real-time integration with financial accounting, such reconciliation is obsolete. Also, secondary cost elements are created as general ledger accounts to ensure this integration.

In SAP S/4HANA, controlling documents are still generated along with FI document numbers. However, even internal controlling movements, such as reallocation of costs from one controlling object to another generate financial accounting document numbers, which ensures real-time integration; this wasn't the case in SAP ERP. In terms of configuration, document types that are used for posting in controlling are defined to post to general ledger accounts as well. These document types are linked to the controlling internal business transactions and generate financial accounting postings as well as controlling postings.

3.2 Organizational Structure

We'll start configuring the SAP S/4HANA Finance system by defining the organizational structure. The organizational structure in SAP is defined to represent the business organizational structure of the enterprise, and it consists of various configuration objects in finance, controlling, sales, purchasing, production, and so on. So it's the foundation of any further system setup and is extremely important that it be designed and defined in a proper, flexible way.

We will examine in detail how to configure the organizational structures in finance and controlling in SAP S/4HANA, such as company, company code, controlling area, and operating concern.

3.2.1 Company

A *company* in SAP is an organizational unit that represents a business from a commercial point of view. It can consist of multiple legal entities and is used to perform consolidation in SAP.

If there is no need for a consolidation process, it's possible not to set up companies in SAP. It's an optional organizational object, and it could be set up later. However, this would require significant effort, so it's better to set it up in the beginning even if consolidation won't be performed until later.

To create a company, follow menu path **Enterprise Structure • Definition • Financial Accounting • Define company**. As you'll recall from the Introduction, this and all other menu paths are accessed via Transaction SPRO.

Then you can create a new company using the **New Entries** option from the top menu. In Figure 3.2, we create a new company for the United States and give it code 1000. The naming conventions of companies, company codes, controlling areas, and so on vary greatly from project to project. A good idea is to use simple, easy to remember numbering. You should make a well-defined proposal and confirm it with the business.

In this configuration transaction, you enter the name and address of the company, the country, language key, and currency, and then you can save using the **Save** button in the lower-right corner. If you are configuring in a development system, you'll be prompted with a customizing request; this stores the changed configuration settings, which need to be transported to other test and productive systems.

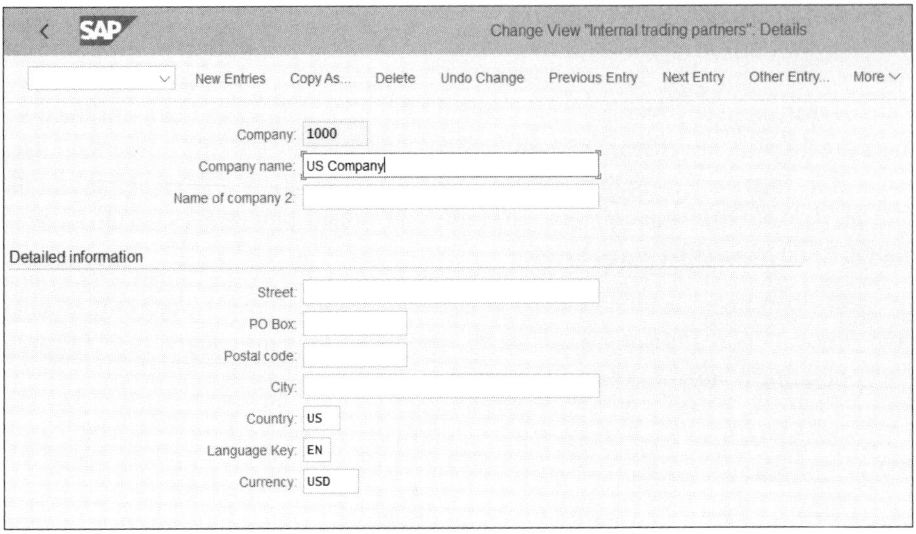

3 Financial Accounting Global Settings

Figure 3.2 Create Company

Configuration changes in SAP S/4HANA, as in previous SAP releases, are essentially changes to configuration tables. Normally you would do a first round of configuration in a so-called sandbox system, which doesn't record the changes in customizing transports. After initial testing there, you would make the configuration settings in the development "golden" client, which should have the settings to be transported to other clients and no data. Then these transports are transported to test systems for unit testing, integration testing, and user acceptance testing, and finally to the production system. This concept will be discussed in detail in Chapter 17.

Now you've created your first company. Your enterprise may decide to set up one company for each country in which it operates and then assign the various legal entities in this country to that company. Then in the consolidation process it will be able to view the financial statements from the group point of view on the level of the company, eliminating intercompany profit and transactions between the different legal entities. We'll come back to this point after you create your first company codes.

3.2.2 Company Code

The *company code* is the main organizational unit in financial accounting. Usually it represents a separate legal entity. For example, a global pharmaceutical company may have a few different legal entities in the United States, which are registered as legally independent companies: perhaps one that manufactures generic drugs, one that is developing biotechnology medications, and one that is performing testing for the pharmaceutical industry. It makes sense that each of these companies is set up as separate company code in SAP S/4HANA. Then if those companies have a common parent company, it can be set up as a company in SAP. So normally there will be as many company codes in the system as the organization has legal entities.

The company code is the main unit for which a complete set of financial statements can be generated. Every financial accounting document is posted per company code. Therefore, the company code is the most fundamental organizational object in financial accounting and is very important to set up correctly to begin with.

We highly recommend copying existing company codes, either standard SAP-provided company codes or already created ones, when creating new company codes. This is because there a lot of configuration settings that are maintained at the company code level, and if creating all configuration manually from scratch it's possible to miss some important settings.

To create a company code, follow menu path **Enterprise Structure • Definition • Financial Accounting • Edit, Copy, Delete, Check Company Code**, then select the **Copy, Delete, Check Company Code** activity—or you can enter Transaction EC01 directly.

For those that are just starting to configure SAP, *transaction codes* are helpful shortcuts to enter into user or configuration transactions without having to navigate through the application or configuration menu. They are entered in the command field in the top-left section of the main SAP application screen, as shown in Figure 3.3.

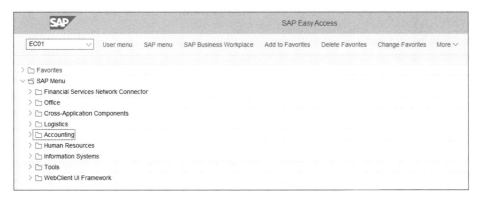

Figure 3.3 Command Field in SAP

You can enter also "/N" before a transaction code from within any transaction, which will end the current transaction and start the new transaction. Or you can enter "/O" before the transaction, which will open it in a new SAP GUI window.

Back to our example, select **Copy Org. Object** from the top menu and select the source and target company codes to be copied, as shown in Figure 3.4.

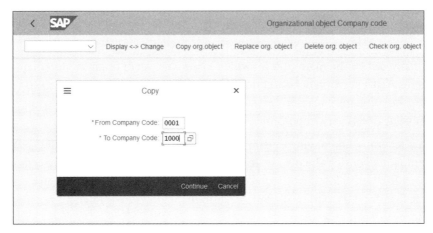

Figure 3.4 Copy Company Code

In the **From Company Code** field, enter "0001" as the source company code, which is a standard SAP-provided company code. In the **To Company Code** field, enter "1000" to copy to new company code 1000, which we'll use to represent a US-based legal entity.

The system will issue the message shown in Figure 3.5.

Figure 3.5 Copy General Ledger Accounts

This provides you with an option to copy all the general ledger accounts from the source to the target company code, which makes sense if they use the same chart of accounts. General ledger accounts are maintained at the chart of accounts level and at the company code level, and confirming this option allows you to automatically extend all the accounts also to the new company code.

After that, the system issues another message regarding the assignment of the controlling area to the company code, as shown in Figure 3.6.

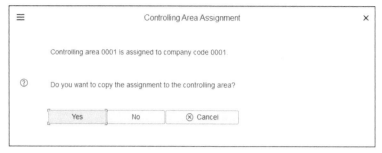

Figure 3.6 Assignment of CO Area

You have the opportunity here to copy the assignment of the same controlling area. If you are going to create a new controlling area, you can reject that option and then assign the new controlling area to the new company code. Next, the system shows confirmation of the copying of the company code, as shown in Figure 3.7.

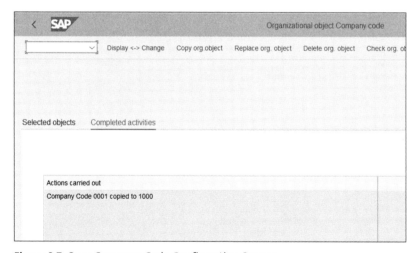

Figure 3.7 Copy Company Code Confirmation Screen

Now go back and select the **Edit Company Code Data** activity to display the list of company codes in the system. Double-click the new company code **1000** to change its basic data, as shown in Figure 3.8.

3 Financial Accounting Global Settings

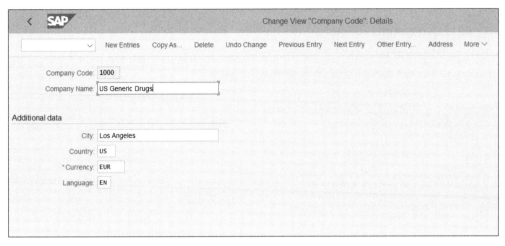

Figure 3.8 Company Code Details

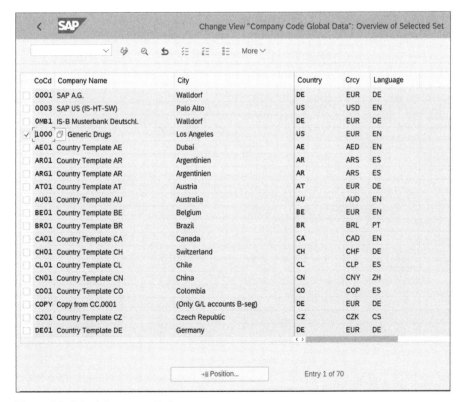

Figure 3.9 Select Company Code

Here you can change the name, city, country, currency, and language of the company code. Enter the required details, including currency; enter "EUR" in this case because this example's US company has a European group parent.

Next, you'll enter the global company code settings. Follow menu path **Financial Accounting • Financial Accounting Global Settings • Global Parameters for Company Code • Enter Global Parameters**, then double-click the company code you want to check or modify from the screen shown in Figure 3.9.

Figure 3.10 shows the settings for new company code 1000.

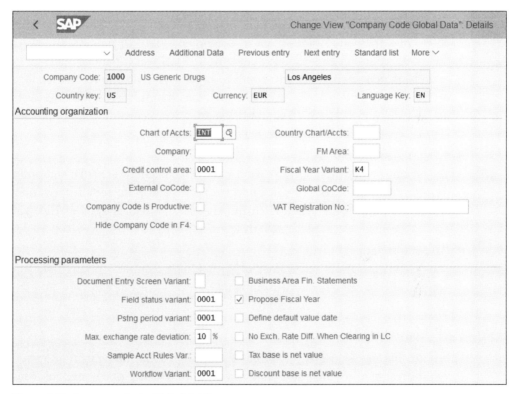

Figure 3.10 Company Code Global Settings

Here are the important fields that need to be configured:

- **Chart of Accts**
  The chart of accounts defines the general ledger accounts used and is maintained at a central (valid for all company codes) level and a company code level. We'll

examine the chart of accounts in detail in Chapter 4. Here you can configured the chart of accounts to be used by the company code.

- **Company**
  Here you can enter the company to which the company code is assigned. The company represents the parent legal entity for the company code.

- **Credit Control Area**
  This is used to perform credit management for the company code. It manages the available credit limits for customers for the company code.

- **Fiscal Year Variant**
  This is the fiscal year variant used for this company code. The fiscal year variant determines the periods and calendar assignments used to post documents in financial accounting. For example, standard SAP fiscal year variant K4 matches the calendar periods: period 01 corresponds to January, period 02 to February, and so on. However, it's possible to use other fiscal year variants, such as the 4-4-5 calendar popular in the United States, in which each quarter consists of three periods, consisting of four weeks, four weeks, and five weeks.

- **Pstng. period Variant**
  The posting period variant in SAP determines which periods are open and closed for postings. It provides a separate option to open and close periods for various types of accounts (general ledger, customer, vendor, assets, and so on). Here you specify the posting variant used for the company code.

- **Field Status Variant**
  The field status variant determines which fields are required, optional, and suppressed when posting financial documents.

3.2.3 Controlling Area

The *controlling area* is the main organizational unit in the controlling area and it structures the organization from a cost point of view. It can include one or multiple company codes and defines which components of controlling are active. In SAP S/4HANA, financial accounting and controlling are integrated, but the controlling area is still the core configuration object, which determines the global controlling settings.

To create a controlling area, follow menu path **Enterprise Structure • Definition • Controlling • Maintain Controlling Area**, then select activity **Copy, Delete, Check Controlling**

Area. As with company codes, we highly recommend copying an existing controlling area to copy all the important settings that are linked to it. SAP standard controlling areas such as US01 (which is designed for the United States) or 0001 (which uses EUR currency) are very good source candidates if you are about to create your first controlling area because they are delivered standard by SAP with all the standard setup needed. Copying a controlling area is very similar to copying a company code. The configuration settings that go along with the controlling area are copied, and then you can adapt them in the next steps.

First, select the **Maintain Controlling Area** activity, as shown in Figure 3.11.

Figure 3.11 Maintain Controlling Area

Double-click controlling area **US01** from the list shown in Figure 3.12 to examine the settings, which will open the screen shown in Figure 3.13.

3   Financial Accounting Global Settings

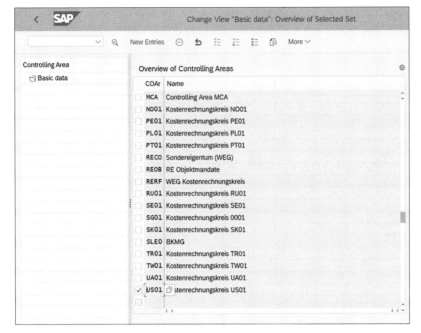

Figure 3.12 Select Controlling Area

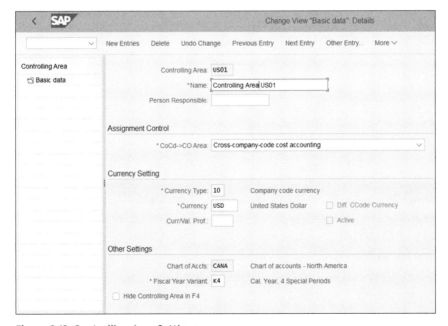

Figure 3.13 Controlling Area Settings

The following important fields need to be configured:

- **CoCd->Co Area**
  This field controls whether multiple company codes are managed for this controlling area (cross-company-code cost accounting) or just one (controlling area same as company code). Most companies choose **Cross-Company-Code Cost Accounting** because usually in today's highly interconnected business world cost responsibilities cross legal entities.

- **Currency Type**
  Currency types in SAP determine the currency based on its purpose, such as company code currency (main currency of the legal entity), group currency (the main currency from business group point of view), hard currency (used in inflation environments), and so on. Here on the controlling area level most commonly currency type 30 (group currency) is used because controlling is managed from a group point of view, but of course other options are possible too.

- **Currency**
  This is the currency of the controlling area itself and is driven by the currency type.

- **Chart of Accounts**
  The chart of accounts defines the general ledger accounts used and is maintained at a central (valid for all company codes) level and a company code level. We'll examine the chart of accounts in detail in Chapter 4. Here the chart of accounts of the controlling area should match the chart of accounts of the company code.

- **Fiscal Year Variant**
  The fiscal year variant of the controlling area is configured here.

We'll configure the assignment of active controlling components and other general controlling area settings in Chapter 9. For now, let's check the assignment of company codes to the controlling area.

To assign company codes to a controlling area, follow menu path **Enterprise Structure • Assignment • Controlling • Assign Company Code to Controlling Area**, as shown in Figure 3.14.

Here, select the controlling area by selecting the checkbox to its left, then click **Assignment of Company Code(s)** in the left pane of the configuration screen, as shown in Figure 3.15.

This shows the company codes have been successfully assigned to the controlling area.

3 Financial Accounting Global Settings

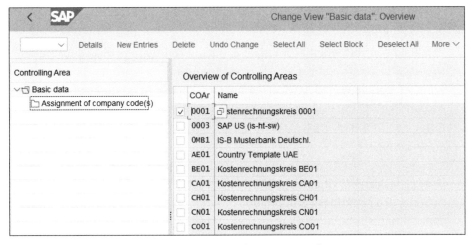

Figure 3.14 Assignment of Controlling Area and Company Codes

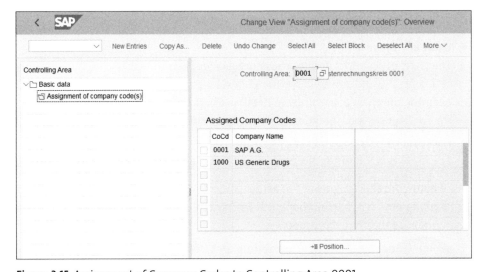

Figure 3.15 Assignment of Company Codes to Controlling Area 0001

3.2.4 Operating Concern

The operating concern is the main organizational unit from a profitability analysis point of view. Profitability analysis is part of controlling, which analyzes the costs against the revenues per various market characteristics and therefore provides invaluable profitability analysis on various levels of the organization.

3.2 Organizational Structure

Here in the organizational structure, you just need to define the operating concern as an organizational object and assign it to controlling area. Follow menu path **Enterprise Structure • Definition • Controlling • Create Operating Concern**, which takes you to a table with the existing operating concerns, as shown in Figure 3.16.

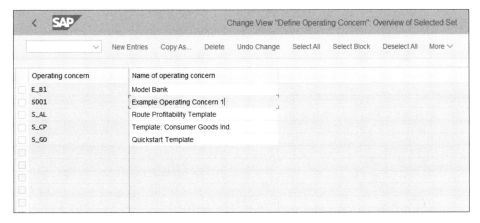

Figure 3.16 Define Operating Concern

As with company codes and operating concerns, here SAP provides sample organizational objects that you can use as references. You can select one of them by selecting the checkbox to its left and then selecting **Copy As...** from the top menu. Create a new operating concern US01 in this way and name it "US Operating Concern", as shown in Figure 3.17.

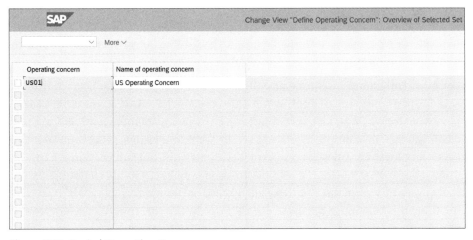

Figure 3.17 Copied Operating Concern

89

The next step for the operating concern is to define its data structure before it can be assigned to a controlling area, but this will be covered in Chapter 13.

Now that you've defined the main organizational structures, let's discuss the main general settings that need to be configured in the system, starting with ledgers.

3.3 Ledgers

Ledgers is an area in the general ledger application that stores accounting documents based on different accounting principles. You are required to have at minimum a leading ledger, which always is called 0L and which represents the main accounting principle from a group point of view. Then you can set up as many nonleading ledgers as required, to represent, for example, local accounting principles, local taxation rules, and so on.

In the old days, there was a separate financial module called Special Purpose Ledger, which used that concept of separate ledgers to store postings and data related to different accounting principles or purposes. For example, different special purpose ledgers were used to handle profit center accounting, consolidation, and funds management.

With SAP S/4HANA, nonleading ledgers are fully integrated and post in real time across all applications. So, let's examine how you need to configure ledgers in SAP S/4HANA.

Most importantly, you need to define which ledgers are required in your organization from the very beginning; subsequent introduction of ledgers is complicated and requires additional effort. The accounting and taxation reporting requirements have to be discussed in detail with the business. The leading ledger should represent the main accounting framework used by the group.

For most companies in Europe and other regions, that would be IFRS—but in the United States the main accounting rules are based on US GAAP. So most big US companies opt for US GAAP for the leading ledger, then many of them have IFRS in a nonleading ledger. In addition, it's wise to set up nonleading ledgers that represent local GAAP and local tax rules for companies with a significant international footprint. Companies that will be doing rollouts to various markets would undoubtedly find that at least in some countries these ledgers will be required, so it is good to set them up from the beginning and activate them only for the countries where they're needed. Some countries are known to have complex local tax requirements, such as Russia and Brazil, among others, and for them local tax ledgers are a must.

Now let's delve into the configuration for ledgers. Follow menu path **Financial Accounting** • **Financial Accounting Global Settings** • **Ledgers** • **Ledger** • **Define Settings for Ledgers and Currency Types**, which shows the list of ledgers in the system, as shown in Figure 3.18.

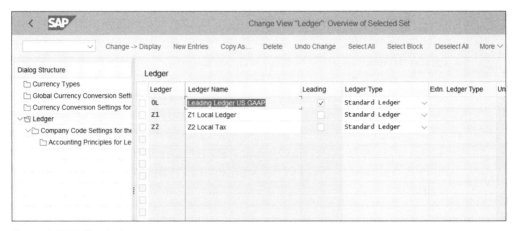

Figure 3.18 Define Ledgers

Here you can define new ledgers by selecting either **New Entries** or **Copy As...** from the top menu. In this example, we have the leading ledger, which is always called 0L, to represent US GAAP valuation, and we've created two nonleading ledgers: Z1 to represent local GAAP and Z2 to represent local tax. The checkmark in the **Leading** column indicates that 0L is the leading ledger; only one ledger can be marked as leading. The **Ledger Type** column determines whether the ledger is standard or extension. Most ledgers are defined as standard. The extension ledger extends a standard ledger and contains the postings of its linked standard ledger. It's used to make additional manual entries, such as adjustments needed for a specific accounting principle.

Now you should make the company code and currency settings for each ledger. Select each ledger individually and click the **Company Code Settings for the Ledger** option on the left side of the screen. Then, using the **New Entries** command from the top menu, you can add the required company codes.

Figure 3.19 shows the following important settings:

- **Fiscal Year Variant**
  This is the fiscal year variant used for this ledger. Different ledgers can have different fiscal year variants, which is normal; different valuation principles may require different fiscal years. For example, the 4-4-5 variant used often in the

3   Financial Accounting Global Settings

United States doesn't correspond with the calendar year, which is used most often throughout the world.

- **Pstng. period Variant**
  The posting period variant in SAP determines which periods are open and closed for postings. It provides separate options to open and close periods for various types of accounts (general ledger, customer, vendor, assets, and so on). Here on the ledger level you specify the variant.

- **Parallel Accounting Using Additional G/L Accounts**
  This checkbox indicates that for this ledger parallel general ledger accounts will be used instead of different ledgers to portray parallel accounting principles. This is used rarely, when one ledger needs to portray parallel accounting principles.

- **Local Currency**
  Here you specify the currency type of the local currency of the ledger. Local currency is the main currency of the company and is stored in each posting and is maintained at the company code level, but here also you can have different local currencies per ledger.

- **Global Currency**
  Here you specify the currency type of the global currency of the ledger. Global currency is the group currency of the company and is stored in parallel to the local currency for each posting.

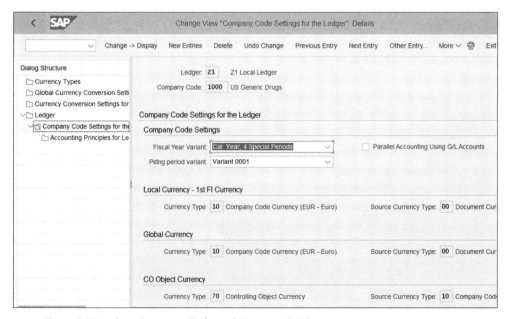

Figure 3.19 Ledger Company Code and Currency Settings

3.3 Ledgers

- **CO Object Currency**
  Here you specify the currency type of the controlling object currency of the ledger. This is the currency used in the controlling objects master and may differ from the transaction currency.

The next step is to define the accounting principles for the ledgers. The accounting principle is a new configuration object in SAP S/4HANA (the ACC_PRINCIPLE field). It maps the ledger with the relevant accounting framework that it needs to portray. To view the accounting principle for the ledger, click the **Accounting Principles for Ledger and Company Code** activity on the left side of the same configuration screen.

In Figure 3.20, you can see that accounting principle LOCL, which portrays local accounting standards, is mapped to ledger Z1.

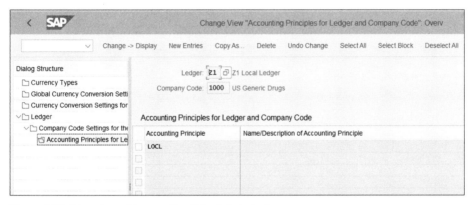

Figure 3.20 Mapping of Accounting Principle to Ledger

The actual creation of accounting principles is done under menu path **Financial Accounting • Financial Accounting Global Settings • Ledgers • Parallel Accounting • Define Accounting Principles**, where you define the accounting principles as shown in Figure 3.21.

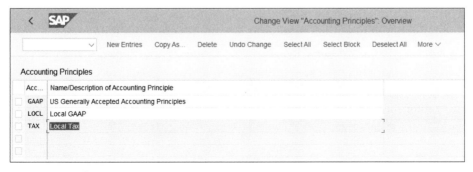

Figure 3.21 Define Accounting Principles

3　Financial Accounting Global Settings

In this example, we defined three accounting principles to portray US GAAP, local GAAP, and local tax rules.

In the next step, you assign these accounting principles to ledger groups. A ledger group normally contains one ledger (the system automatically creates a ledger group for each ledger you define), but it's also possible to have multiple ledgers in one ledger group. The assignment of accounting principles is at the ledger group level. Follow menu path **Financial Accounting • Financial Accounting Global Settings • Ledgers • Parallel Accounting • Assign Accounting Principle to Ledger Groups**, where you can assign the accounting principles to ledger groups as shown in Figure 3.22.

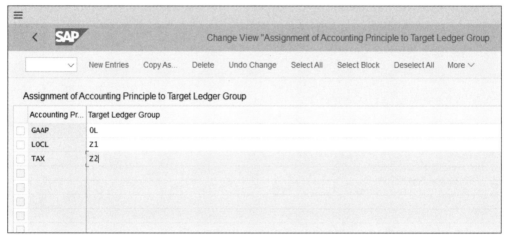

Figure 3.22 Assign Accounting Principles to Ledger Groups

This is where the link between the ledger and the accounting principle, which you saw in Figure 3.20, comes from.

3.4　Document Types

Document types in SAP serve to classify the various transactions posted in financial accounting. Each financial accounting document is assigned a document type, such as vendor invoice, customer invoice, asset posting, and so on. The document types determine the numbers assigned to the documents, as well as many other important configuration parameters, which we'll now examine in detail.

3.4.1 Document Type Settings

To configure the document types, follow menu path **Financial Accounting • Financial Accounting Global Settings • Document • Document Types • Define Document Types for Entry View**. The entry view represents the entry of financial documents in the system, whereas the ledger view shows the posted document in each individual ledger posted. You define the document types for the entry view and also have to define the document types for posting to nonleading ledgers.

Figure 3.23 shows the listing of the defined document types. Most of them are standard SAP-delivered document types, which should suffice for most business needs. Of course, you can copy them into custom specific document types. Sometimes this is needed for local reporting needs or to meet some specific business process.

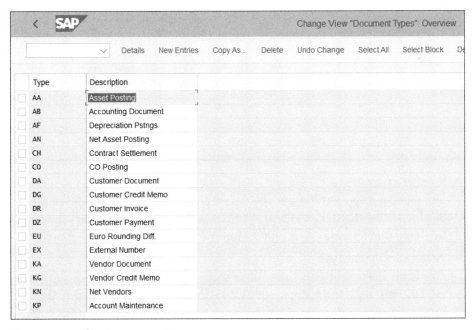

Figure 3.23 Define Document Types

Double-click document type **KR—Vendor Invoice** to examine the relevant settings, as shown in Figure 3.24.

3 Financial Accounting Global Settings

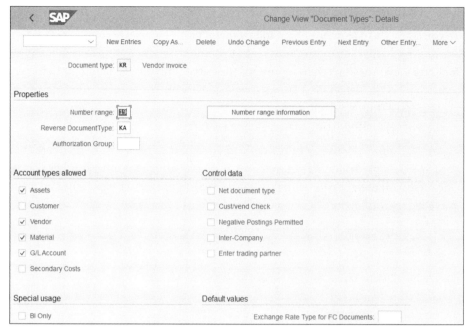

Figure 3.24 Document Type Settings

Figure 3.24 shows the following important settings:

- **Number Range**
  The number range determines the document numbers assigned when posting documents for this document type. You'll create the actual number range in the following section; here you assign the number range object to the document type.

- **Reverse Document Type**
  This is the document type that will be used when making reversals of postings with the selected document type. If a value isn't maintained here, the reversal will be done with the same document type as the original document.

- **Authorization Group**
  This allows you to set up an authorization check on this document type level.

- **Account Types Allowed**
  Here you select what types of accounts are allowed to be posted using this document type. For example, for document type KR, assets, vendors, materials, and general ledger accounts are checked, which means that only accounts of these types can be posted in documents of document type KR; customers and secondary costs are not allowed.

- **Negative Postings Permitted**
  This indicator allows reversal of documents for this document type to be done as negative postings. A negative posting means that the items will be posted on the same side as the original document but with a minus sign. So when reversing, a debit item will remain on the debit side, but as a negative posting.

- **Required during Document Entry**
  Here you can specify that the reference field and/or the document header text field are required during posting of documents with this document type.

These are the most important control parameters of a document type. Now let's look at its number ranges.

3.4.2 Number Ranges

Number ranges are used throughout the system to assign numbers for various transactions and master data objects. Accordingly, every document type in SAP needs to have an assigned number range, which will control the document numbers assigned and whether they are internally generated or have to be entered externally.

To configure number ranges, follow menu path **Financial Accounting • Financial Accounting Global Settings • Document • Document Number Ranges • Define Document Number Ranges for Entry View**. Enter the company code 1000 and click the *Intervals* button to modify the number ranges for the company code, as shown in Figure 3.25.

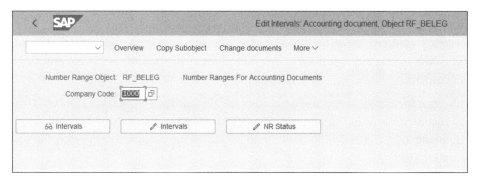

Figure 3.25 Define Number Ranges

The number ranges for your new company code 1000 were copied along with other parameters when you created the company code. You can see the ranges defined in Figure 3.26.

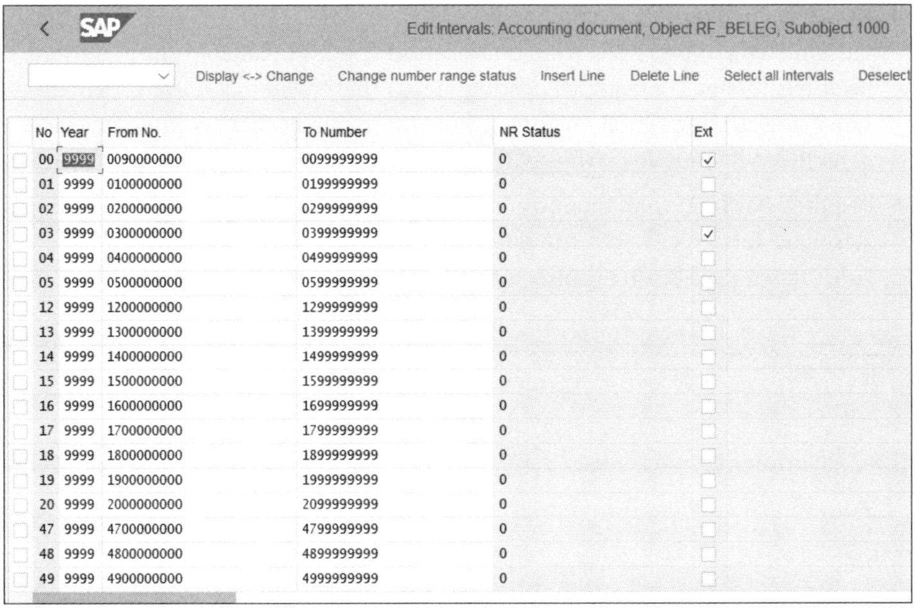

3 Financial Accounting Global Settings

Figure 3.26 Number Range Intervals

Each interval is identified with its number in the first column from the left (01, 02, 03, and so on), and this is the number to be assigned in the document type. Then there is the validity year; good practice is to set this to 9999, which means there is no limitation. Then you enter the **From No.** and **To Number** of the interval, within which the system will assign the document numbers consecutively (if they're to be internally assigned). In the **NR Status** column, you can see the current number (which is 0 in a development system without data). In the last column, **Ext**, a checkmark means that numbers in this interval need to be entered manually by the user when entering a document.

Changes to number ranges are not automatically transported because this could lead to inconsistencies in the target clients. It's good practice to set the number ranges manually in each client, and this should be part of the cutover activities during production start.

3.4.3 Document Types for Entry View in a Ledger

Documents types that should be posted to nonleading ledgers only should be separately configured. By default, when you post to the leading ledger, the system also posts the same document to all the nonleading ledgers. However, you can make ledger-specific postings, and you need to configure the document types for them with their number ranges here. To do so, follow menu path **Financial Accounting • Financial Accounting Global Settings • Document • Document Types • Define Document Types in a Ledger**.

Enter the nonleading ledger for which to maintain document types, as shown in Figure 3.27. The system will not allow you to enter the leading ledger here.

Figure 3.27 Select Ledger

Next, you're presented with the configuration screen shown in Figure 3.28.

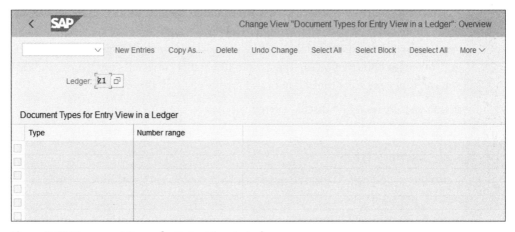

Figure 3.28 Document Types for Entry View in Ledger

On this screen, select **New Entries** from the top menu, then enter the document **Type** and **Number range** (Figure 3.29), which can be then posted in this nonleading ledger. Save your entries.

3 Financial Accounting Global Settings

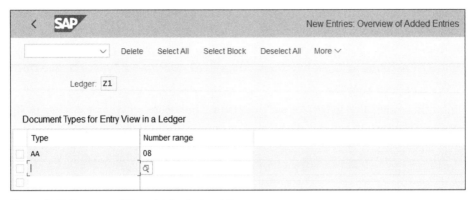

Figure 3.29 Document Type AA for Ledger Z1

Similarly, you can add document types for the entry view in other nonleading ledgers.

3.5 Currencies

SAP provides all the currency codes you will need. In the system, you have to configure which currencies will be used for which purposes. The currency type in SAP defines the purpose of the use for a particular currency, such as local currency or group currency. Then you can have as transaction currency any currency for which the exchange rates are maintained. This enables parallel currency valuation, which is very important in today's globalized business world. In SAP S/4HANA, there is big improvement because you can have up to 10 parallel currencies per ledger, and then you can easily monitor the balances and line items in all these currencies.

In this section, we'll first discuss currency types before moving on to exchange rates.

3.5.1 Currency Types

A *currency type* defines what the purpose of a currency is. The following standard currency types are defined:

- 10: Company code currency
- 30: Group currency
- 40: Hard currency
- 50: Index-based currency
- 60: Global company currency

As you've seen when configuring the ledgers, you can choose from these currency types to select the local and group currency of your company, and you can also use some of the other currency types in special situations, such as when working in a high-inflationary environment.

The configuration of the currency type is done per company code and ledger using the now-familiar menu path **Financial Accounting • Financial Accounting Global Settings • Ledgers • Ledger • Define Settings for Ledgers and Currency Types**. After selecting the ledger, click **Company Code Settings for the Ledger** in the left side of the screen, as shown in Figure 3.30.

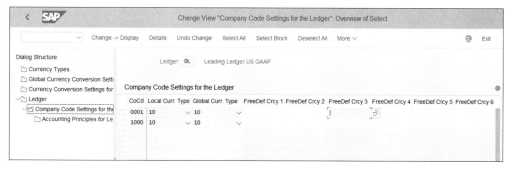

Figure 3.30 Currency Types per Ledger and Company Code

Here you see the settings per company code for the selected ledger. You can maintain the local and global currency type, as well as others free definition currencies. Then in this ledger and company code each transaction will be stored also in these currencies.

3.5.2 Exchange Rate Type

Exchange rates in the system need to be maintained for the currencies in use. These exchange rates are maintained always per exchange rate type. These are keys under which exchange rates of particular types are stored. For example, you can enter specific buy, sell, and average exchange rates under different exchange rate types.

As with other important general settings, SAP provides a list of standard exchange rate types, which usually meet most requirements. To check the exchange rate types, follow menu path **SAP NetWeaver • General Settings • Currencies • Check Exchange Rate Types**, as shown in Figure 3.31.

You can see the list of defined exchange rate types for various purposes. In accounting, the most commonly used standard exchange rate is type **M**, **Standard Translation at Average Rate**.

101

3 Financial Accounting Global Settings

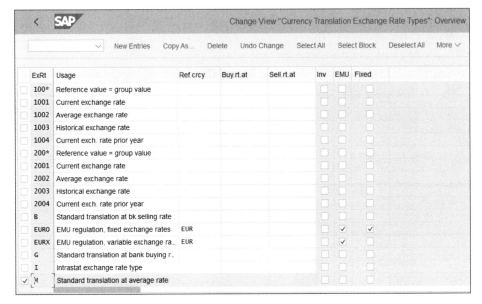

Figure 3.31 Exchange Rate Types

3.5.3 Exchange Rates

Now let's maintain exchange rates between the currencies to be used. Maintaining exchange rates during production use of the system normally is a user task, and many companies also establish some interface to automatically upload exchange rates using a feed from a central bank or other financial institution.

To enter exchange rates, follow menu path **SAP NetWeaver • General Settings • Currencies • Enter Exchange Rates**, as shown in Figure 3.32.

Here you maintain for each exchange rate type (in our example, M), the exchange rates between **From** and **To** currencies, using either direct or indirect quotation. In the direct quotation method, the exchange rate gives the price in the **To** currency that you have to pay for a unit of the **From** currency. In the indirect method, this is reversed. The **Valid From** date determines the date from which the entered exchange rate is valid, and it will remain valid until a rate with a subsequent date is maintained.

Maintain as many exchange rates as required, then save the entries.

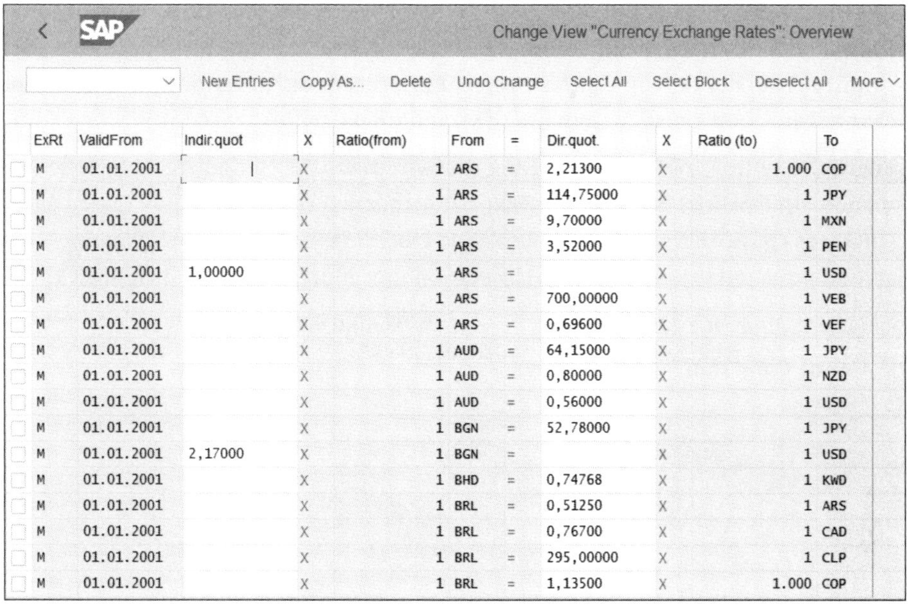

ExRt	ValidFrom	Indir.quot	X	Ratio(from)	From	=	Dir.quot	X	Ratio (to)	To
M	01.01.2001		X		1 ARS	=	2,21300	X	1.000	COP
M	01.01.2001		X		1 ARS	=	114,75000	X	1	JPY
M	01.01.2001		X		1 ARS	=	9,70000	X	1	MXN
M	01.01.2001		X		1 ARS	=	3,52000	X	1	PEN
M	01.01.2001	1,00000	X		1 ARS	=		X	1	USD
M	01.01.2001		X		1 ARS	=	700,00000	X	1	VEB
M	01.01.2001		X		1 ARS	=	0,69600	X	1	VEF
M	01.01.2001		X		1 AUD	=	64,15000	X	1	JPY
M	01.01.2001		X		1 AUD	=	0,80000	X	1	NZD
M	01.01.2001		X		1 AUD	=	0,56000	X	1	USD
M	01.01.2001		X		1 BGN	=	52,78000	X	1	JPY
M	01.01.2001	2,17000	X		1 BGN	=		X	1	USD
M	01.01.2001		X		1 BHD	=	0,74768	X	1	KWD
M	01.01.2001		X		1 BRL	=	0,51250	X	1	ARS
M	01.01.2001		X		1 BRL	=	0,76700	X	1	CAD
M	01.01.2001		X		1 BRL	=	295,00000	X	1	CLP
M	01.01.2001		X		1 BRL	=	1,13500	X	1.000	COP

Figure 3.32 Maintain Exchange Rates

3.6 Taxes

Taxes are big topic in SAP. Most selling and purchasing transactions are affected by taxes, and there are very stringent requirements for tax reporting around the world. Therefore, it's important that the tax setup in your SAP S/4HANA system reflects the tax requirements from both process and reporting points of view.

As part of the financial accounting global settings, you need to set up the tax procedure and assign it to your company codes. Then you need to set up the relevant tax codes that this procedure uses. The tax determination will be discussed in detail in Chapter 5 for the purchasing processes and in Chapter 6 for the sales processes.

3.6.1 Tax Procedure

The tax procedure contains the settings to perform tax calculations in SAP S/4HANA. It's a very complex configuration object, which uses access sequences and condition techniques to determine the proper tax codes, which in turn determine the tax rates, general ledger accounts to be posted to, and other relevant settings.

The tax procedure is maintained at the country level, which means it's valid for all company codes for a given country. SAP supplies sample tax procedures for each country. You should copy those to new tax procedures to modify them, or if no changes are envisioned you can use the standard procedures.

Check the settings of the standard tax procedures for the United States by following menu path **Financial Accounting • Financial Accounting Global Settings • Tax on Sales/Purchases • Basic Settings • Check Calculation Procedure**. You'll find the following three activities related to setting up the calculation procedure, as shown in Figure 3.33:

- Access Sequence
- Define Condition Types
- Define Procedures

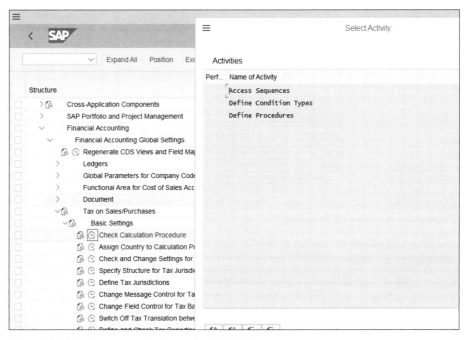

Figure 3.33 Tax Procedure Configuration Activities

The tax procedure is a collection of condition types, which in turn use access sequences to determine tax records based on specific fields, defined in those access sequences.

Double-click the **Define Procedure** activity to examine the tax procedure for the United States. Figure 3.34 shows a list of the tax procedures. The standard tax procedures for the United States provided by SAP are **TAXUS** and **TAXUSJ**, which are based on jurisdiction codes (the tax rates differ by jurisdictions, which are determined with these jurisdiction codes). Select procedure **TAXUS** and click **Control Data** on the left side of the screen.

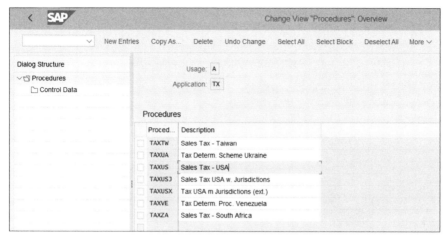

Figure 3.34 Tax Procedures

Figure 3.35 shows how the tax procedure is defined. Condition types are assigned per step numbers. Then, when defining tax codes for this procedure, the tax codes will be assigned at this condition type level.

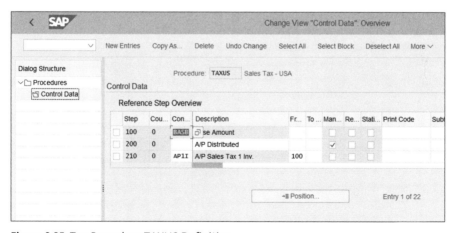

Figure 3.35 Tax Procedure TAXUS Definition

Now go back and select the **Define Condition Types** activity. Figure 3.36 shows the list of condition types, which can be assigned to steps in the tax procedures.

3   Financial Accounting Global Settings

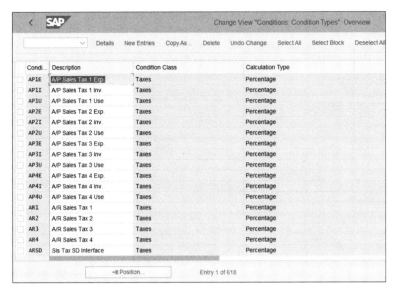

Figure 3.36 Condition Types

Select and double-click **MWAS**, which is the output tax condition, to see its settings, as shown in Figure 3.37.

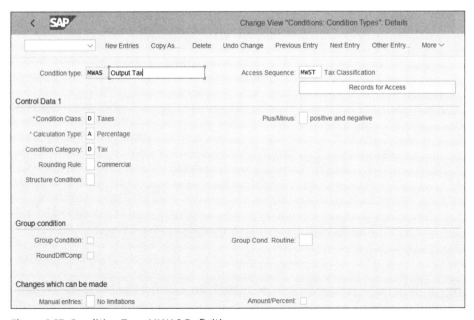

Figure 3.37 Condition Type MWAS Definition

3.6 Taxes

Here, if you click the **Records for Access** button, you can see the condition records, based on the fields defined in access sequence MWST. As you can see in Figure 3.38, **Country** and **Tax Code** are the fields that would determine the taxes in this case.

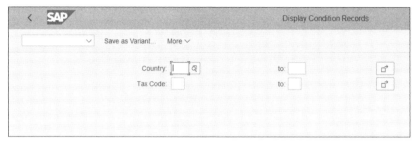

Figure 3.38 Condition Record Fields

Click the **Execute** button to see the existing records, as shown in Figure 3.39.

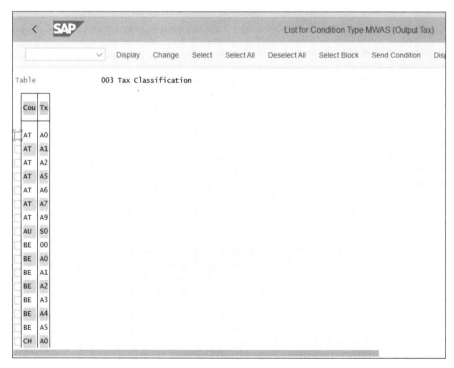

Figure 3.39 Condition Record Values for MWAS

Finally, check the definition of this access sequence. Go back to the screen shown in Figure 3.33 and select **Access Sequences**. The system issues a message that this is a cross-client table. This means that the configuration in this table is very fundamental and affects all clients of the SAP system. Such a configuration is to be maintained only in the golden configuration client, and you have to proceed with caution.

Figure 3.40 shows the list of access sequences defined. Select **MWST** and then click **Accesses** in the left side of the screen. You can have one or more accesses.

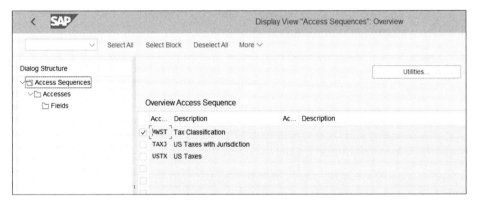

Figure 3.40 Access Sequences

In this case, it's just one, number 10. Select it and click **Fields** on the left side of the screen, and you're presented with the screen shown in Figure 3.41.

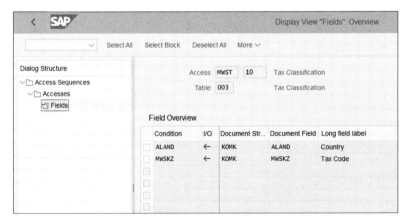

Figure 3.41 Access Fields

So here is how the system determined that has to check the **Country** and **Tax Code** fields for condition type MWAS. It comes from the setup of access sequence MWST, shown earlier and mapped to condition type MWAS.

Let's see how this is used in tax codes to determine the correct tax.

3.6.2 Tax Codes

Tax codes in SAP determine the tax percentage and tax account posted and are assigned at the line item level in documents. They are created per tax procedure and either are determined automatically (usually in logistic documents) or can be entered manually in financial documents.

Let's walk through how to create tax codes for tax procedure TAXUS. First, follow menu path **Financial Accounting** • **Financial Accounting Global Settings** • **Tax on Sales/Purchases** • **Calculation** • **Define Tax Codes for Sales and Purchases**, or enter Transaction FTXP.

The system asks you for which country you need to create a tax code. Enter "US" and continue. Then enter a two-character tax code. The naming of the tax codes should be uniform within the project, and there are different strategies to choose. Some companies opt to have a letter as the first symbol and a number as the second, with the letter representing whether it's an input or output tax code. Whatever the naming convention is, you have to make sure there will be enough space in the naming ranges to accommodate all the tax codes needed. It's normal for a country to use 30–40 tax codes, and sometimes it is required to have even more tax codes.

In this case, name the new tax code "O2" to represent a 10% sales tax, as shown in Figure 3.42. O indicates that this is output tax code, whereas our input tax codes would start with I. The various tax codes, O1, O2, O3, and so on, will represent output tax codes with different rates or purposes.

Once you enter the tax code number, the system opens the properties screen of the new tax code, in which you need to select whether the tax code is input (for purchasing transactions) or output (for sales transactions). You also give a description for the tax code and can define some other optional settings. For example, the **Check ID** indicator can make sure that there will be an error message if the tax amount entered is incorrect. The **EU Code** setting is used for European Union reporting.

After you click **Continue**, you'll see the main configuration screen of the tax code. Figure 3.43 shows the condition types available from the tax procedure for which you created a tax code. They are mapped with account keys (here you see account key NVV), which determine the general ledger accounts to be posted to. You can enter tax rates in one or more of the condition type levels. The system will go through all the levels of the tax code when determining the proper taxes.

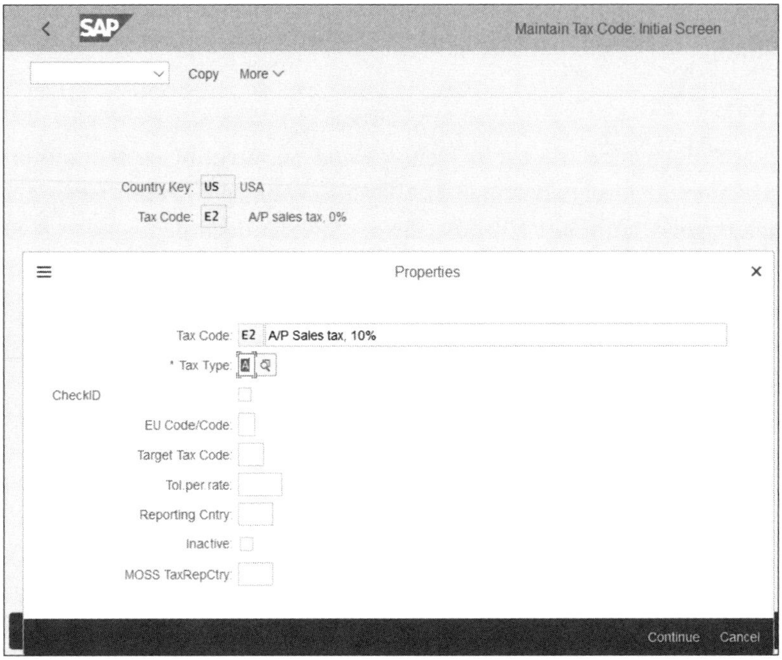

3 Financial Accounting Global Settings

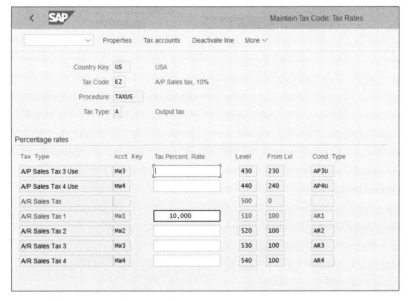

Figure 3.42 Create Tax Code

Figure 3.43 Tax Code Configuration

You can also check the general ledger account assigned by clicking **Tax Accounts** from the top menu. Then the system asks for the chart of accounts and shows the general ledger account, which will be posted to with this tax code, as shown in Figure 3.44.

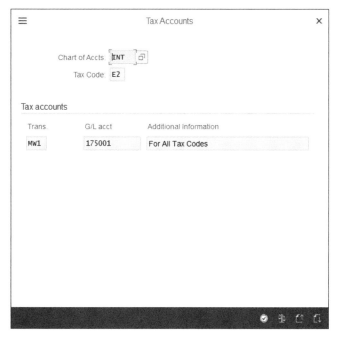

Figure 3.44 Tax Code General Ledger Account Definition

We'll examine the tax code determination in various purchasing and sales flows in detail in Chapter 5 and Chapter 6.

3.7 Summary

In this chapter, we examined the global settings that need to be performed in SAP S/4HANA Finance in detail. We started by explaining the concept of the new finance data model in SAP S/4HANA so that you're in a position to properly define your organizational setup and global settings, taking into consideration the advancements SAP S/4HANA offers in the finance area.

We then explained how to configure the organizational structure of the enterprise, including the company, company code, controlling area, and operating concern. The proper decisions about how to structure your organization in the system provide a good foundation on which the system can be built and meet your business requirements. With the guidelines and practical advice from this chapter, you should be in a position to design your organizational structure well.

Then we covered the main configuration objects that are part of the global settings of the system, such as ledgers, document types, currencies, and taxes. These are used throughout the system and by all modules, so their proper configuration is of paramount importance. We examined the various important settings that can be set for these objects to ensure proper functioning of the SAP S/4HANA system.

With that done, now let's start configuring the various financial accounting and controlling areas of the system, starting with the general ledger.

Chapter 4
General Ledger

This chapter gives step-by-step instructions for configuring the general ledger in SAP S/4HANA. It explains how the general ledger integrates with the rest of the system and how account determination works in the system.

So far, you've configured the financial global settings required for SAP S/4HANA, and now you'll learn how to configure the backbone of all financial processes: the general ledger. The general ledger contains all financial and controlling postings in SAP S/4HANA and provides the full accounting picture of your organization. In SAP S/4HANA, it's in the general ledger that previously separate components such as controlling and fixed assets are fully integrated. Therefore, its proper configuration is of paramount importance.

In the following sections, we'll cover the following general ledger topics in detail:

- Master data
- Document splitting
- Automatic postings and account determination
- Periodic processing and financial closing
- Information system

We'll also look at a few SAP Fiori apps available for the general ledger that should be of use to you.

4.1 Master Data

The master data for the general ledger consists of general ledger accounts. They are used to record accounting transactions and are organized in a chart of accounts, which structures them into main accounting categories such as asset accounts, liabilities, profit and loss accounts, and so on.

At the general ledger accounts level, you can see one of the main benefits of SAP S/4HANA. In SAP S/4HANA, financial accounting and controlling are fully integrated, so now cost elements are also part of general ledger accounts. Cost elements used to be separate master data objects in controlling used to record the controlling transactions. Now they're just a type of general ledger account in the SAP S/4HANA simplified data model.

Let's look at how charts of accounts are configured.

4.1.1 Chart of Accounts

The *chart of accounts* is a classification structure for the general ledger accounts used. There are three types of charts of accounts:

- **Operational chart of accounts**
  The operational chart of accounts is the main chart of accounts, which is assigned to the company code. It's used to make postings in financial accounting.

- **Group chart of accounts**
  The group chart of accounts links one or multiple operational general ledger account numbers to a group account number. It's used in the consolidation process.

- **Country chart of accounts**
  This chart of accounts can be used on the country level to portray local requirements. It's optional and can be used for only some countries. When the country chart of accounts is assigned to the company code, you can assign an alternative account number in the master record of the operational general ledger account.

Now let's examine the settings for the chart of accounts. Follow menu path **Financial Accounting • General Ledger Accounting • Master Data • G/L Accounts • Preparations • Edit Chart of Accounts List**, or enter Transaction OB13.

Figure 4.1 shows a list of defined charts of accounts in the system. As you can see, SAP delivers sample charts of accounts for many countries. It's a good idea to use one of these as a basis for your new chart of accounts, depending on the country.

Double-click the **INT** chart of accounts (**Sample Chart of Accounts**) to see the screen shown in Figure 4.2.

4.1 Master Data

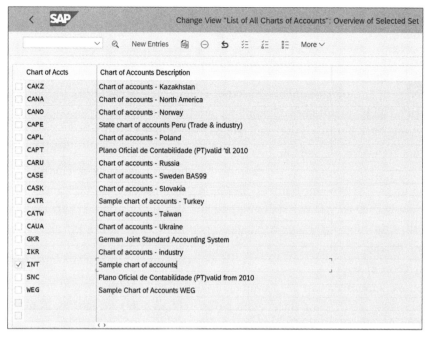

Figure 4.1 Chart of Accounts List

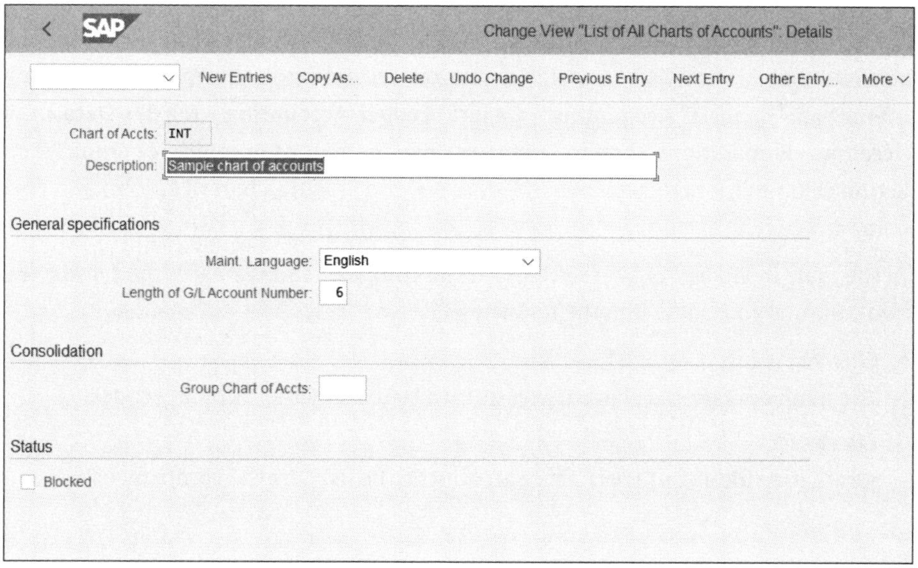

Figure 4.2 Chart of Accounts Settings

Here, the following fields can be configured:

- **Description**
  The description of the chart of accounts, which should indicate its use.

- **Maint. Language**
  This is the main language under which the general ledger accounts for the chart of accounts will be maintained. General ledger accounts can have descriptions in other languages, but this is the main language used in the country.

- **Length of G/L Account Number**
  The length of the general ledger account number. For example, a length of six would mean that account numbers will be 100001, 100002, 100003, and so on. Commonly used lengths are six or eight digits. Just make sure there are enough digits to structure your accounts in logical account number ranges. For most companies, six digits is enough.

- **Group Chart of Accts**
  Here the operational chart of accounts is linked with the group chart of accounts. Then in each account from the operational chart of accounts, an account from the group chart of accounts will be entered.

- **Blocked**
  If this checkbox is selected, no accounts can be created for this chart of accounts.

After you finish checking or changing these settings, save and proceed to the next configuration transaction to assign a chart of accounts to a company code. Follow menu path **Financial Accounting • General Ledger Accounting • Master Data • G/L Accounts • Preparations • Assign Company Code to Chart of Accounts**, or enter Transaction OB62. In the screen shown in Figure 4.3, you assign a chart of accounts to your company code.

In the table in Figure 4.3, you see a list of the company codes created, and in the last two columns you can assign the following:

- **Chrt/Accts**
  The main operational chart of accounts to be used by the company code

- **Cty ch/act**
  Alternative local country chart of accounts to be used by the company code

4.1 Master Data

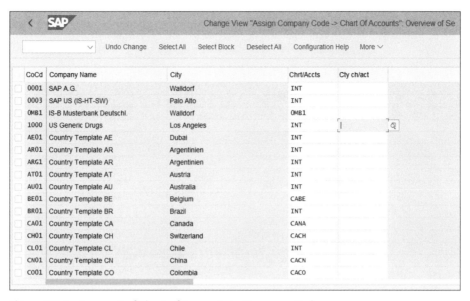

Figure 4.3 Assignment of Chart of Accounts to Company Code

The next step is to define the account groups used for your chart of accounts.

4.1.2 Account Groups

Account groups classify general ledger accounts based on common accounting purposes. Typically, separate account groups are created for fixed asset accounts, material accounts, profit and loss (P&L) accounts, and so on.

Follow menu path **Financial Accounting • General Ledger Accounting • Master Data • G/L Accounts • Preparations • Define Account Group**, or enter Transaction OBD4. In Figure 4.4, you can see that multiple account groups are defined for each chart of accounts.

For example, for chart of accounts INT, the following account groups are defined:

- Fixed assets accounts
- Liquid funds accounts
- General G/L Accounts
- Materials management accounts
- P&L Statement Accounts
- Recon. Accts Ready for Input

117

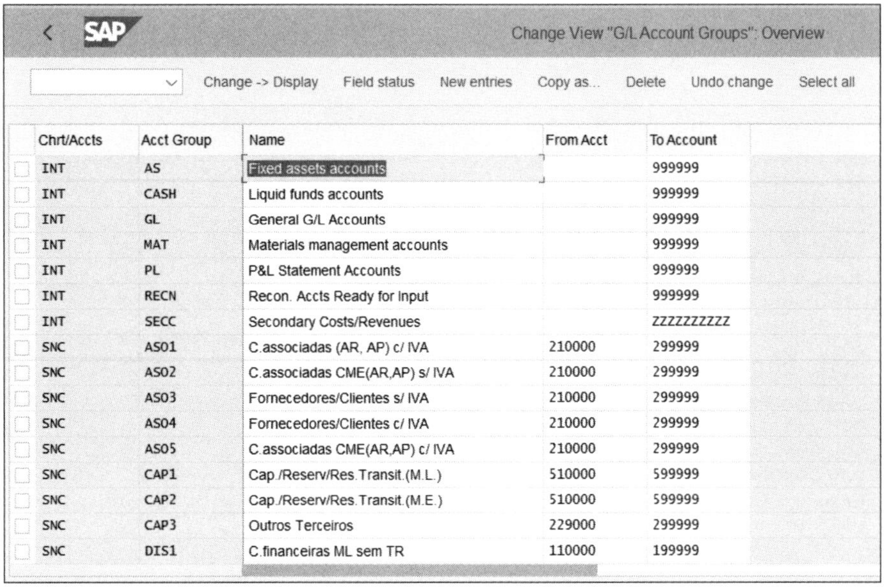

Chrt/Accts	Acct Group	Name	From Acct	To Account
INT	AS	Fixed assets accounts		999999
INT	CASH	Liquid funds accounts		999999
INT	GL	General G/L Accounts		999999
INT	MAT	Materials management accounts		999999
INT	PL	P&L Statement Accounts		999999
INT	RECN	Recon. Accts Ready for Input		999999
INT	SECC	Secondary Costs/Revenues		ZZZZZZZZZZ
SNC	AS01	C.associadas (AR, AP) c/ IVA	210000	299999
SNC	AS02	C.associadas CME(AR,AP) s/ IVA	210000	299999
SNC	AS03	Fornecedores/Clientes s/ IVA	210000	299999
SNC	AS04	Fornecedores/Clientes c/ IVA	210000	299999
SNC	AS05	C.associadas CME(AR,AP) c/ IVA	210000	299999
SNC	CAP1	Cap./Reserv/Res.Transit.(M.L.)	510000	599999
SNC	CAP2	Cap./Reserv/Res.Transit.(M.E.)	510000	599999
SNC	CAP3	Outros Terceiros	229000	299999
SNC	DIS1	C.financeiras ML sem TR	110000	199999

Figure 4.4 Define Account Groups

Each account group is defined with an identifier entered in the **Acct Group** column. In the **From Acct** and **To Account** columns the possible numbers of the accounts are defined. From empty value to 999999 means that accounts have to consist of numbers up to 999999. You can also define more precise ranges, such as 100000 to 199999 for one group, 200000 to 299999 for another, and so on. It's also possible to have accounts consisting of letters, defined as ZZZZZZZZZZ.

4.1.3 General Ledger Accounts

General ledger accounts are master data and are maintained from the SAP S/4HANA application menu. They can be maintained in each client individually, but the more common and better approach is to maintain them centrally in a single master data client and distribute them to all other systems and clients via Application Link Enabling (ALE).

General ledger accounts are maintained at the chart of accounts level and company code level. At the chart of accounts level settings valid for all company codes that use this chart of accounts are maintained, such as the long and short description and the group account number. Then at the company code level most of the control settings of the accounts are maintained, such as the currency, open item management, line item display, and so on.

To maintain general ledger accounts, follow application menu path **Accounting • Financial Accounting • General Ledger • Master Records • G/L Accounts • Individual Processing**. Here you'll see three transactions available to maintain general ledger accounts, as shown in Figure 4.5.

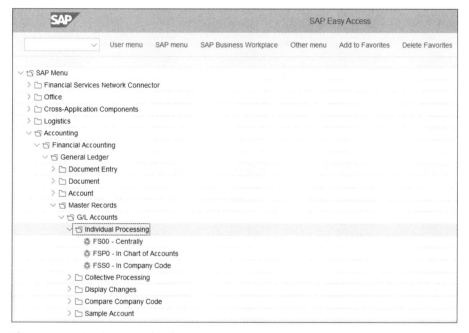

Figure 4.5 Maintain General Ledger Accounts Transactions

These three transactions are as follows:

- **FS00—Centrally**
- **FSP0—In Chart of Accounts**
- **FSS0—In Company Code**

Enter transaction **FS00—Centrally**, which gives access to both the chart of accounts and company codes settings. In the screen shown in Figure 4.6, enter "630200" for the **G/L Account** number and your new **Company Code** 1000 (in which we copied the accounts from the sample company code 0001), then click the 🖉 (**Change**) button, which allows you to change the fields of the general ledger account.

4   General Ledger

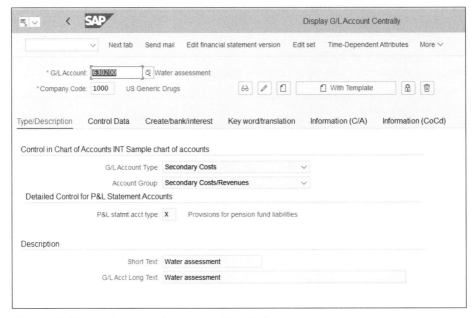

Figure 4.6 Display General Ledger Account Centrally

The fields are organized in different tabs. Which fields appear as optional or required or are hidden is controlled by the field status group of the account. Let's examine the most commonly used fields.

In the **Type/Description** tab, which is maintained at the chart of accounts level, the following fields are available:

- **G/L Account Type**
  Determines the type of account, such as balance sheet account or profit and loss account.

- **Account Group**
  Here you select one of the account groups we discussed in Section 4.1.2.

- **P&L Statement Acct Type**
  This field is relevant for P&L accounts and determines the target retained earnings account to which the result during year-end closing will be transferred to calculate the profit or loss.

- **Short Text**
  The short description of the general ledger account.

- **G/L Acct Long Text**
  The long description of the general ledger account.

If you scroll down, you'll see more fields available, related to consolidation:

- **Trading Partner**
  Used to determine the intercompany partner related to this account.
- **Group Account Number**
  The general ledger account from the group chart of accounts assigned to the operative chart of accounts. The relationship is *1:n*, meaning that the same group account number could be assigned to multiple general ledger accounts from the operative chart of accounts, but only one group account can be assigned to each operative account.

Now select the **Control Data** tab, as shown in Figure 4.7.

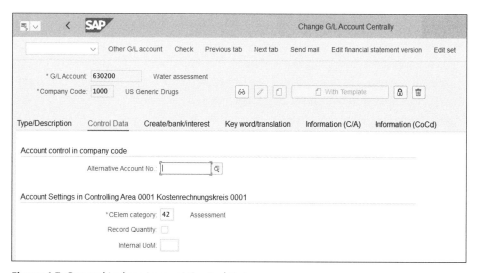

Figure 4.7 General Ledger Account Control Data

Different fields are available, based on the account. The most commonly used ones are as follows:

- **Account Currency**
  The currency of the account, which is stored in every posting.
- **Balances in Local Crcy Only**
  If checked, balances will be updated only in local currency when posting to this account.

- **Tax Category**
  Determines what type of taxes are allowed to be posted to this account.

- **Posting without Tax Allowed**
  If checked, entering a tax code when posting to the account is optional.

- **Recon Account for Acct Type**
  This is used for general ledger accounts, which are reconciliation accounts for assets, customers, or vendor accounts. Each posting to the relevant asset, customer, or vendor also updates the reconciliation general ledger account.

- **Alternative Account No**
  This is the local account number from the country chart of accounts assigned to the operative chart of accounts. Its use is optional in countries where a country chart of accounts is needed and is used to depict local accounting rules.

- **Sort Key**
  Defines a rule for default sorting of the line items in line item reports.

- **CElem. Category**
  As discussed previously, in SAP S/4HANA, financial accounting and controlling are fully integrated, which means that now cost elements are general ledger accounts. With this field, you determine their category and use; before, this was a field available in the cost element master.

- **Record Quantity**
  Indicates that the system will also record quantity for postings to this cost element.

- **Internal UoM**
  Default unit of measure used for postings to this cost element.

Now go to the **Create/Bank/Interest** tab. Here, the following fields are commonly used:

- **Field Status Group**
  This field controls which fields are optional, required, and hidden in the general ledger account master record.

- **Post Automatically Only**
  If this field is checked, then manual postings are not allowed. Typically this is used for accounts posted by automatic processes from logistics, such as sales revenue accounts or material accounts.

Next, in the **Key Word/Translation** tab you can maintain keywords related to the account in different languages, and you can maintain the account description in multiple languages, as shown in Figure 4.8.

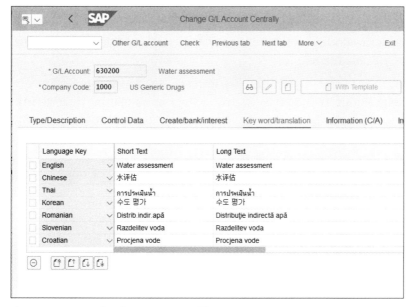

Figure 4.8 General Ledger Account Different Language Descriptions

Go to the **Information (C/A)** tab to find administrative information about who created the account on the chart of accounts level and when, as well as relevant change documents for its changes.

Finally, visit the **Information (CoCd)** tab to find administrative information about who created the account on the company code level and when, as well as relevant change documents for its changes.

After you've finished making changes to the general ledger account, you can save your settings.

4.2　Document Splitting

Document splitting is an extremely powerful concept that was first introduced as a special feature in the special purpose ledger, then was included as standard functionality with the new general ledger, and now in SAP S/4HANA is commonly used to provide a full set of balanced financial statements on various types of characteristic levels, such as profit center or segment.

Essentially, *document splitting* enables you to automatically split financial line items for selected dimensions (such as receivable line items by profit center or payable line items by segment), and therefore to have zero balance in the document per those dimensions. During document splitting, the system creates additional clearing lines to balance these dimensions.

We'll now examine the various configuration elements that you need to configure for the document splitting. We'll start by discussing the various splitting characteristics that can be used to split the documents. We'll start the configuration by classifying the general ledger accounts into different categories, based on which they will be split. Then we'll classify the document types for splitting. Next we'll configure the splitting characteristics, and finally we'll define a zero-balance general ledger account, which is used to offset the splitting entries.

4.2.1 Document Splitting Characteristics

Document splitting's purpose is to create complete financial statements for fields (characteristics) you configure in customizing. The following characteristics are available:

- Segment
- Profit center
- Business area
- Fund (used in public-sector accounting)
- Customer-defined characteristic (customer field)

Let's look in detail at how the system performs document splitting, using as an example a posting of a vendor invoice. A common situation in such a case is to have multiple expense line items, which are assigned to different cost centers and therefore to different profit centers. If you need to have a full, balanced financial statement on the profit center level, you need to split these items per profit center.

Let's assume that one expense line item for 5,000 USD is assigned to profit center A and another one for 10,000 USD is assigned to profit center B. Then the system will perform the splitting as shown in Figure 4.9.

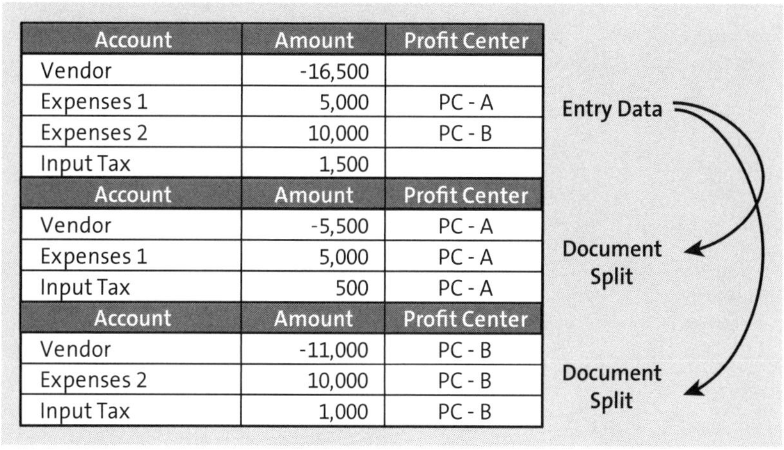

Account	Amount	Profit Center
Vendor	-16,500	
Expenses 1	5,000	PC - A
Expenses 2	10,000	PC - B
Input Tax	1,500	
Account	**Amount**	**Profit Center**
Vendor	-5,500	PC - A
Expenses 1	5,000	PC - A
Input Tax	500	PC - A
Account	**Amount**	**Profit Center**
Vendor	-11,000	PC - B
Expenses 2	10,000	PC - B
Input Tax	1,000	PC - B

Entry Data
Document Split
Document Split

Figure 4.9 Document Splitting Example

As you can see, the system uses the ratio between the expense line items (1:2 in this case, as the first line item is 5000 and the second is 10000) to determine the amounts that need to be allocated to each of the profit centers. To do that, it posts additional line items at the general ledger level for each of the profit centers. It splits the input tax line item into two line items based on this 1:2 ratio. Thus it enhances this input tax item for which originally no profit center could be determined with profit center information based on the expense line items. These line items are available only in the general ledger view, which is different from the entry view in which the document is posted in financial accounting.

Now let's examine how to configure document splitting in SAP S/4HANA.

4.2.2 Classification of General Ledger Accounts for Document Splitting

We need to tell the system how the general ledger accounts should be split. For that, we need to assign the general ledger accounts to item categories. *Item categories* are predefined objects that contain the logic for the splitting based on the type of account.

To classify general ledger accounts for splitting, follow menu path **Financial Accounting • General Ledger Accounting • Business Transactions • Document Splitting • Classify G/L Accounts for Document Splitting**. As shown in Figure 4.10, here you define ranges of general ledger accounts and assign them to item categories.

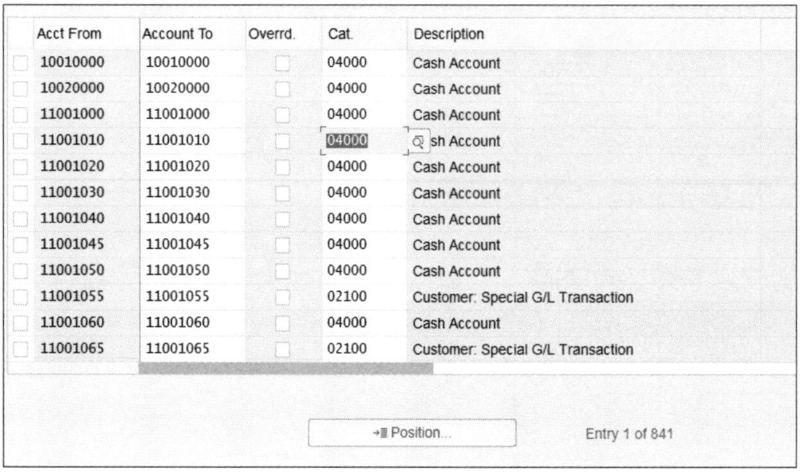

4 General Ledger

Acct From	Account To	Overrd.	Cat.	Description
10010000	10010000	☐	04000	Cash Account
10020000	10020000	☐	04000	Cash Account
11001000	11001000	☐	04000	Cash Account
11001010	11001010	☐	04000	Cash Account
11001020	11001020	☐	04000	Cash Account
11001030	11001030	☐	04000	Cash Account
11001040	11001040	☐	04000	Cash Account
11001045	11001045	☐	04000	Cash Account
11001050	11001050	☐	04000	Cash Account
11001055	11001055	☐	02100	Customer: Special G/L Transaction
11001060	11001060	☐	04000	Cash Account
11001065	11001065	☐	02100	Customer: Special G/L Transaction

Position... Entry 1 of 841

Figure 4.10 Classify General Ledger Accounts for Splitting

The following standard item categories are available:

- 01000: Balance Sheet Account
- 01001: Zero Balance Posting (Free Balancing Units)
- 01100: Company Code Clearing
- 01300: Cash Discount Clearing
- 02000: Customer
- 02100: Customer: Special G/L Transaction
- 03000: Vendor
- 03100: Vendor: Special G/L Transaction
- 04000: Cash Account
- 05100: Taxes on Sales/Purchases
- 05200: Withholding Tax
- 06000: Material
- 07000: Fixed Assets
- 20000: Expense
- 30000: Revenue
- 40100: Cash Discount (Expense/Revenue/Loss)

- 40200: Exchange Rate Difference
- 80000: Customer-Specific Item Category

As you can see, SAP provides item categories for all types of accounts out of the box. Still, if you have special splitting requirements, you can program your own logic using the customer-specific item category.

When you create new general ledger accounts, you need to make sure they are also assigned here to a split item category.

4.2.3 Classification of Document Types for Document Splitting

The next step is to define per document type how the splitting should be performed. For this, you need to assign for each document type a transaction variant. First, follow menu path **Financial Accounting • General Ledger Accounting • Business Transactions • Document Splitting • Classify Document Types for Document Splitting**.

In the screen shown in Figure 4.11, you assign a business transaction and variant for each document type.

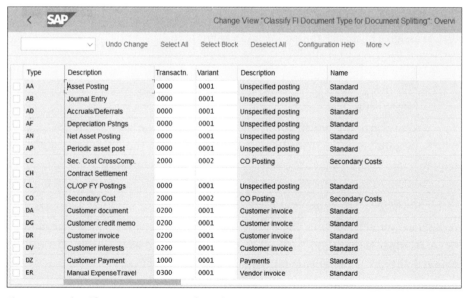

Figure 4.11 Classify Document Types for Splitting

The business transactions are predefined by SAP, and you can choose from the following:

- 0000: Unspecified Posting
- 0100: Transfer Posting from P&L to B/S Account
- 0200: Customer Invoice
- 0300: Vendor Invoice
- 0400: Bank Account Statement
- 0500: Advance Tax Return (Regular Tax Burden)
- 0600: Goods Receipt for Purchase Order
- 1000: Payments
- 1010: Clearing Transactions (Account Maint.)
- 1020: Reset Cleared Items
- 2000: CO Posting

The business transaction variant further enhances the business transaction, which is necessary to have different splitting rules.

When you create a new document type, make sure also to configure it in this transaction to assign a business transaction and variant to it.

4.2.4 Define Document Splitting Characteristics for General Ledger Accounting

You need to also define which document splitting characteristics you'll use. This will tell the system on which fields the document line items should be split and if you require a zero balance to be achieved per these fields. This means that the system will always balance the document per such zero-balance characteristics by creating additional line items.

To define document splitting characteristics, follow menu path **Financial Accounting • General Ledger Accounting • Business Transactions • Document Splitting • Define Document Splitting Characteristics for General Ledger Accounting**.

In Figure 4.12, two fields are defined as splitting characteristics: profit center and segment. You can also add more fields by selecting **New Entries** from the top menu.

4.2 Document Splitting

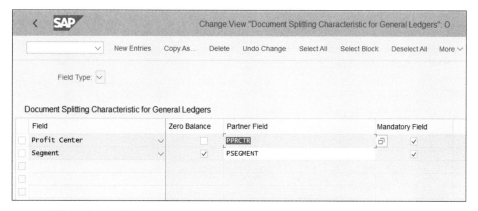

Figure 4.12 Define Splitting Characteristics

The following fields are to be configured on this screen:

- **Field**
  This is the document splitting characteristic you're adding.
- **Zero Balance**
  By checking this flag, the system checks during document posting whether the balance for the characteristic is equal to zero. If it isn't, it generates additional clearing lines in the document.
- **Partner Field**
  A sender/receiver relationship is established in the additionally created clearing lines for this field.
- **Mandatory Field**
  Specifies that this field must be filled with a value after the document splitting.

4.2.5 Define Zero-Balance Clearing Account

You also need to configure a zero-balance clearing account if you need to have balanced financial statements on a certain field's level. For fields for which you want to have a zero-balance setting, the system checks whether the balance for those objects is zero after document splitting. If it isn't, the system generates additional clearing items, which are posted to this clearing account.

To define a zero-balance clearing account, follow menu path **Financial Accounting • General Ledger Accounting • Business Transactions • Document Splitting • Define Zero-Balance Clearing Account**.

Figure 4.13 shows two account keys: one for postings originating from financial accounting and one from controlling. Select **000** and click **Accounts** on the left side of the screen.

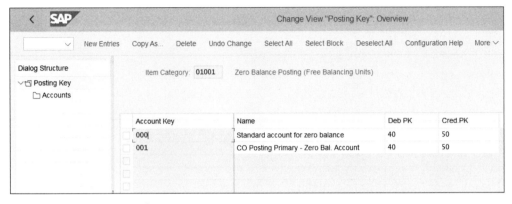

Figure 4.13 Zero Balance Keys

After entering your chart of accounts, you have the opportunity to enter the account used as the zero-balance clearing account, as shown in Figure 4.14.

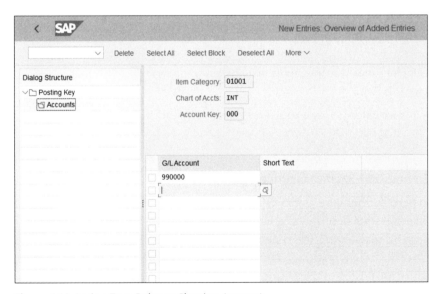

Figure 4.14 Assign Zero-Balance Clearing Account

Then save your entry with the **Save** button.

4.3 Automatic Postings and Account Determination

SAP S/4HANA is a highly integrated system that can manage data from various areas of the business, such as sales, purchases, production, human resources, and so on. At the end, all documents should post to financial accounting. The general ledger is where all the postings are collected, and table ACDOCA has all the postings for the whole organization.

All these integrated processes work well if the automatic postings are properly configured in the system. The automatic postings use various account determination techniques to provide the proper general ledger accounts to be posted to.

To understand well the automatic postings and account determination in financial accounting, let's first look at overall organizational structures and document flow in logistics. Figure 4.15 shows how the logistics organizational structures are integrated with the financial organizational structures.

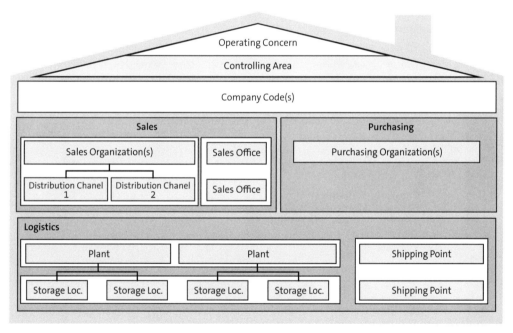

Figure 4.15 Organizational Structures

The company code, controlling area, and operating concern financial organization structures sit on top of the integration flow and receive postings from the logistics processes. Not all logistic transactions result in finance documents; or example, the purchase order doesn't generate a finance document. However, ultimately every logistic flow ends up in finance. For example, after the purchase order is fulfilled, the goods received and invoice received are posted in materials management, which generates financial documents. Similar logic results from the sales flows. Therefore, the logistic organizational objects such as purchasing organization and sales organization are under the financial organizational objects.

Let's examine now in detail the relevant logistic flows, starting with purchasing.

4.3.1 Purchasing Flows

From a purchasing point of view, the purchasing organization is the main organizational unit, in which purchasing is planned and executed. The goods are supplied in storage locations, which belong to plants. These plants are mapped to company codes.

Figure 4.16 shows the basic purchasing flow in SAP S/4HANA. Most commonly, the process starts with a purchase requisition, which establishes the requirement for certain goods or services and is sent to a vendor. Then a purchase order (PO) is created, which among other elements has the account assignment, which determines how the resulting goods receipt and invoice receipt will be posted in accounting. Note that at the time of the purchase requisition and purchase order, no accounting documents are created; these are purely logistic documents.

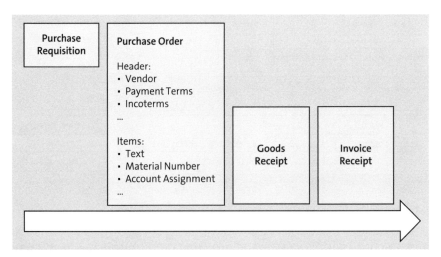

Figure 4.16 Purchasing Flow

Next, in a typical process with goods receipt (GR) and invoice receipt (IR)—which is commonly referred to as a *three-way match* because the system matches the PO, GR, and IR—accounting documents are generated, which usually debit an expense account and credit a vendor account. In between the GR/IR clearing account is posted to, which is credited during GR and debited during IR; therefore, its balance should be zero at the end of the complete process.

It's also possible to do a *two-way match*, in which there is no goods receipt. In this case, the invoice receipt follows the purchase order and there is no GR/IR clearing account used.

4.3.2 Sales Flows

Look again at Figure 4.15. There you can see that the sales organization is the main organizational unit in the sales processes. At lower levels, the sales are being performed at the distribution channel, division, and sales office levels.

Figure 4.17 shows the main sales flow in the system.

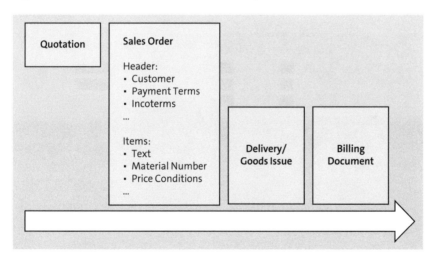

Figure 4.17 Sales Flow

The sales process usually starts with quotation, which is sent to the customer. Then if he decides to proceed, a sales order is created, which contains the elements, which later on determine the automatic account determination during the delivery of the goods, when goods issue is posted, and during the billing document, which generates the financial document for the customer invoice. SAP uses sophisticated techniques involving condition tables and access sequences to determine the correct revenue accounts, in which the material and customer master play key roles, along with other characteristics. We'll examine these in detail in Chapter 6.

Now, let's examine how the automatic postings are configured.

4.3.3 Automatic Postings in Financial Accounting

Because SAP S/4HANA is such an integrated system that is used to manage virtually every area of an organization, there are numerous automatic postings that can be configured. They come from different process areas and modules, so there is no single configuration path in Transaction SPRO that can lead to all the automatic postings configurations. There is, however, a very convenient configuration code that SAP financial experts should know by heart: Transaction FBKP. This transaction provides most account determinations for automatic postings in one place. A notable exception is the sales automatic postings process, which we'll examine in Chapter 6, along with the pricing procedure.

So now, enter Transaction FBKP. You're taken to the screen shown in Figure 4.18.

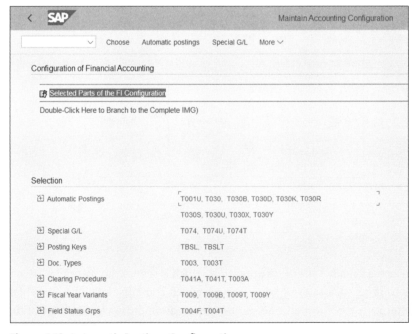

Figure 4.18 Automatic Postings Configuration

4.3 Automatic Postings and Account Determination

As you can see here, you have the link to many financial configuration tasks, such as automatic postings, but also posting keys, document types, fiscal year variants, and so on.

Expand the automatic postings section by clicking it. As shown in Figure 4.19, you'll see the various automatic posting settings, organized by group. Select the **Materials Management postings** (**MM**) group.

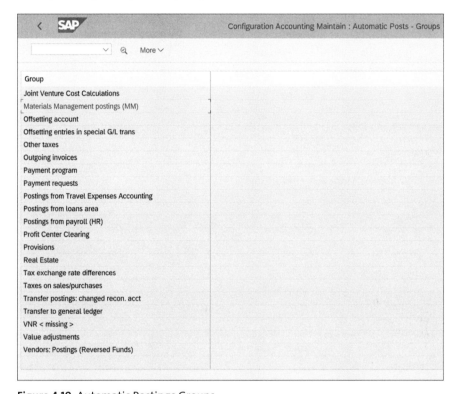

Figure 4.19 Automatic Postings Groups

Figure 4.20 shows the automatic postings related to materials management. You can also directly go into this configuration transaction with Transaction OBYC (which every experienced SAP financial expert also should know by heart).

Here the settings are done by various transactions, and each transaction is defined with its three-character code. For example, select **GBB** for offsetting entry for inventory posting. After entering your chart of accounts, you'll see the screen shown in Figure 4.21.

135

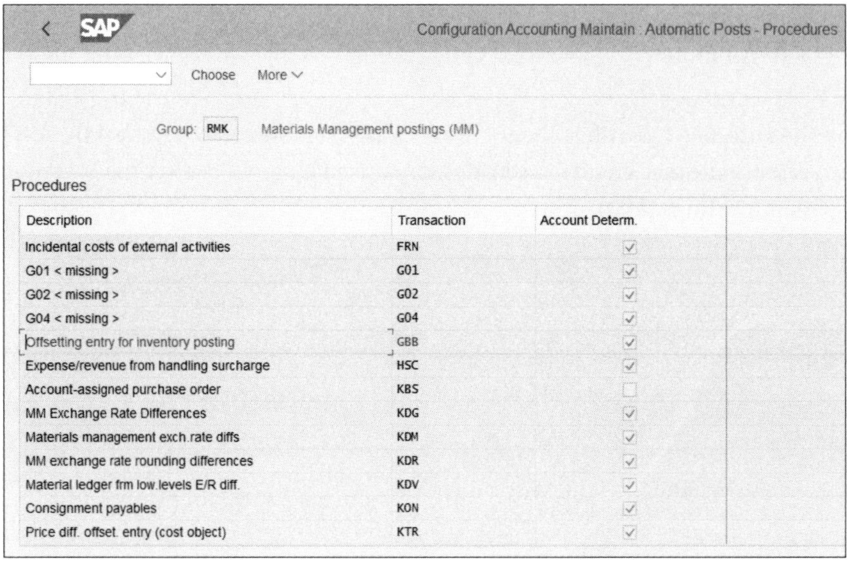

4 General Ledger

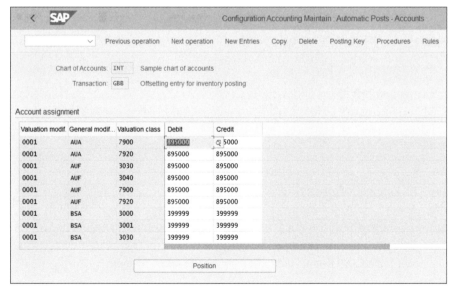

Figure 4.20 Materials Management Automatic Postings

Figure 4.21 Automatic Postings for Offsetting Entry for Inventory Postings

For each transaction, different criteria are possible, based on which you can assign the accounts. Here we see the criteria in the following columns:

- Valuation modifier
- General modifier
- Valuation class
- Debit
- Credit

So there are lots of possibilities to have different accounts determined based on various criteria. The valuation class comes from the material master record. The valuation modifier groups the valuation areas (in most cases, those are the plants). Using this valuation modifier, is possible to assign different general ledger accounts for the same valuation class materials. The general modifier provides groupings that can have different options based on the transaction. For example, it could be dependent on the movement type in materials management.

Using this logic, you also can modify the accounts for other transaction keys and areas in SAP S/4HANA.

4.4 Periodic Processing and Financial Closing

Period-end closing is a vital part of the general ledger processes. After each period (calendar month for most companies, but it can differ based on the fiscal year variant), companies need to perform several procedures, which are usually very well defined in their period-end schedules. At the end of the calendar year, there also are several year-end activities that need to be performed. We'll examine in detail how to configure these procedures and activities in SAP S/4HANA.

4.4.1 Posting Periods

In SAP S/4HANA, you can control which periods are allowed to be posted to. This can be determined by type of account or in general for all accounts and is configured using the posting period variant. The posting period variant groups together the posting periods for different types of accounts and is assigned at the company code level.

To define a posting period variant, follow menu path **Financial Accounting • Financial Accounting Global Settings • Ledgers • Fiscal Year and Posting Periods • Posting Periods • Define Variants for Open Posting Periods**. Here you define the posting period variant as a customizing object. Select **New Entries** from the top menu and create a new variant, 1000, for company code 1000, as shown in Figure 4.22.

4 General Ledger

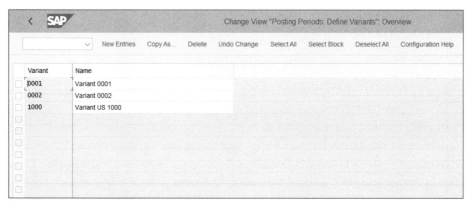

Figure 4.22 Define Posting Period Variant

The next step is to assign the posting period variant to the company code. Follow menu path **Financial Accounting • Financial Accounting Global Settings • Ledgers • Fiscal Year and Posting Periods • Posting Periods • Assign Variants to Company Code** to do so.

As you can see in Figure 4.23, the company code is assigned to variant 0001, which was the variant assigned to the source company code 0001 from which you copied to company code 1000. Variant 0001 is a standard posting period variant provided by SAP. Change it to the new variant you created (1000) and save your entry with the **Save** button.

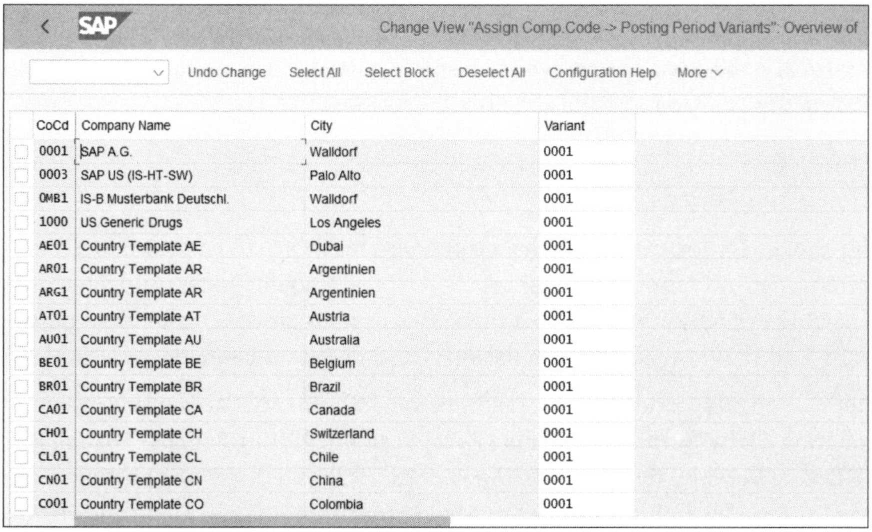

Figure 4.23 Assign Posting Period Variant

138

4.4 Periodic Processing and Financial Closing

Now you need to define the actual periods within this posting period variant by following menu path **Financial Accounting • Financial Accounting Global Settings • Ledgers • Fiscal Year and Posting Periods • Posting Periods • Open and Close Posting Periods**. After entering posting period variant 1000 in the pop-up window to determine the work area, you'll see a blank screen with no periods defined. Enter the posting periods as shown in Figure 4.24.

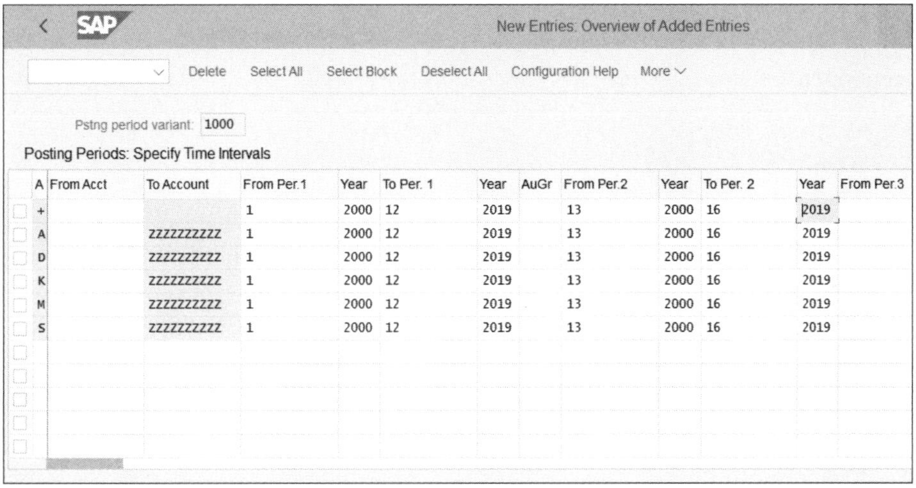

A	From Acct	To Account	From Per.1	Year	To Per. 1	Year	AuGr	From Per.2	Year	To Per. 2	Year	From Per.3
+			1	2000	12	2019		13	2000	16	2019	
A		ZZZZZZZZZZ	1	2000	12	2019		13	2000	16	2019	
D		ZZZZZZZZZZ	1	2000	12	2019		13	2000	16	2019	
K		ZZZZZZZZZZ	1	2000	12	2019		13	2000	16	2019	
M		ZZZZZZZZZZ	1	2000	12	2019		13	2000	16	2019	
S		ZZZZZZZZZZ	1	2000	12	2019		13	2000	16	2019	

Figure 4.24 Define Posting Periods

In the first column, **Account Type**, the different account types for which the posting periods rule in the row is valid are defined. The following account types are possible:

- **+**: Valid for all account types
- **A**: Assets
- **D**: Customers
- **K**: Vendors
- **M**: Materials
- **S**: General ledger accounts
- **V**: Contract accounts

Now define the **From Acct** and **To Account** range for which the rule will be valid. In this example, don't set any restrictions for the account numbers. Next, define the **From** and **To Period/Year** combination. The periods within this range are open for postings; anything outside of this range is closed. You can enter multiple intervals in the three sets of columns available. Enter one range of periods from 1 to 12, which are the normal 12 accounting periods within a fiscal year, then enter special periods from 13 to 16, which are used to make special adjustment postings after the fiscal year is closed.

When you implement SAP S/4HANA, you need to configure these periods initially. Normally, you leave the older periods open, but after the system is productive it will be a user task, which is also available in the application menu, to maintain the posting periods. It's a question of accounting policy whether only the current period will be open, which is the most common case, or if perhaps two or more periods will be kept open.

4.4.2 Intercompany Reconciliation

Intercompany reconciliation (ICR) provides periodic reconciliation of documents between legal entities within a group. Its purpose is to ensure that intercompany documents from different legal entities within the group match each another. It checks that documents have been correctly assigned and allows you to find corresponding intercompany documents and to assign them if needed.

To configure intercompany reconciliation, you need to configure the intercompany relations between the company codes. For all company codes that post in intercompany documents, you need to assign general ledger accounts that are debited and credited when posting between them.

To assign general ledger accounts, follow menu path **Financial Accounting • General Ledger Accounting • Business Transactions • Correspondence: Internal Document • Prepare Cross-Company Code Transactions**.

The system prompts you to enter the two company codes for which you are configuring the relation. Enter the two company codes as shown in Figure 4.25.

Figure 4.25 Cross-Company Code Relations

4.4 Periodic Processing and Financial Closing

In the next screen, you enter the receivables and payables accounts posted in each company code when posting the intercompany code with the other company code. As shown in Figure 4.26, you need to enter the receivables and payables accounts posted in company code 1000 and cleared against 0001 and vice versa.

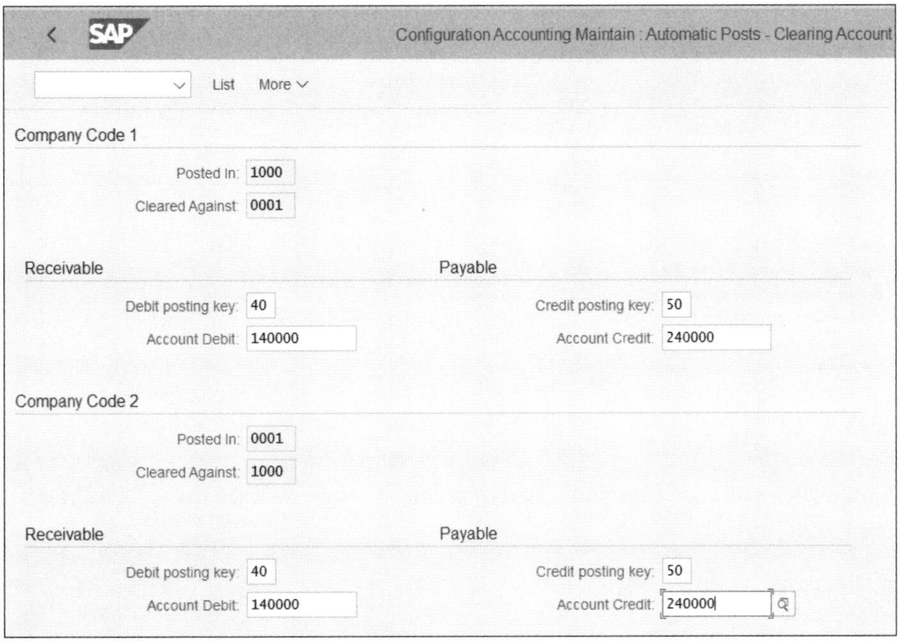

Company Code 1			
Posted In:	1000		
Cleared Against:	0001		
Receivable		**Payable**	
Debit posting key:	40	Credit posting key:	50
Account Debit:	140000	Account Credit:	240000
Company Code 2			
Posted In:	0001		
Cleared Against:	1000		
Receivable		**Payable**	
Debit posting key:	40	Credit posting key:	50
Account Debit:	140000	Account Credit:	240000

Figure 4.26 Intercompany Code Accounts

Update these accounts for your company codes and save your entries.

4.4.3 Foreign Currency Valuation

Foreign currency valuation is the process to revalue open items and balances in a foreign currency to reflect changes in the exchange rates at the end of the period compared to the time of posting the transactions. Thus, foreign currency exchange rate profit or loss can be recognized, which can be posted to configured P&L accounts, depending on the requirements.

You need to configure the accounts to be automatically posted in the process. Follow menu path **Financial Accounting • General Ledger Accounting • Periodic Processing • Valuate • Foreign Currency Valuation • Prepare Automatic Postings for Foreign Currency Valuation** or enter Transaction OBA1 to do so.

As shown in Figure 4.27, here you can configure the various exchange rate difference processes. Most commonly used are transactions **KDB: Exch. Rate Diff. using Exch. Rate Key** used for revaluation of G/L account balances and **KDF: Exchange Rate Dif.: Open Items/GL Acct**, used to revalue open items in foreign currency.

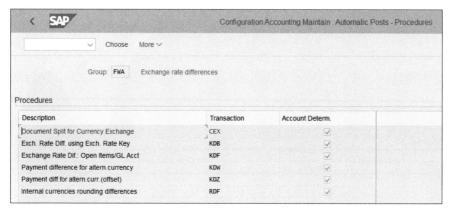

Figure 4.27 Foreign Currency Exchange Rate Differences Automatic Postings

Double-click **KDF** and enter your chart of accounts, and you'll see the screen shown in Figure 4.28.

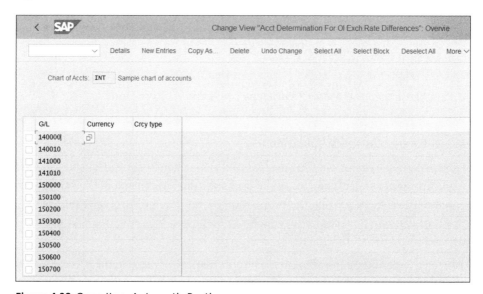

Figure 4.28 Open Item Automatic Postings

4.4 Periodic Processing and Financial Closing

Figure 4.28 shows a list of general ledger open item accounts, which are configured to be revalued for FX rate differences. Double-click the first line to see the screen shown in Figure 4.29.

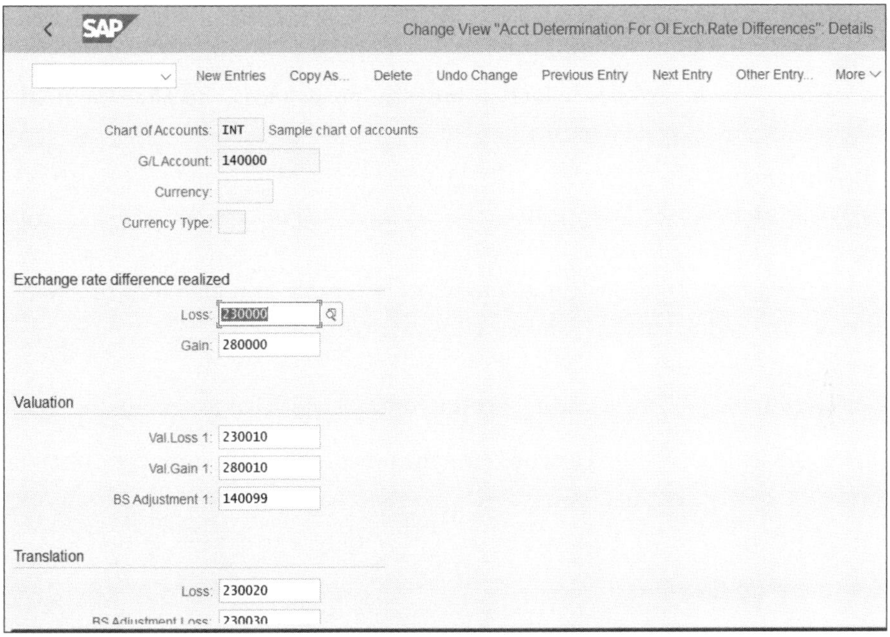

Figure 4.29 Automatic Accounts Assignment for FX Differences

Here you configure the following accounts:

- **Loss**
  Account for realized exchange rate losses. This account is posted to when the loss is realized—for example, during clearing of the open item.

- **Gain**
  Account for realized exchange rate gain. This account is posted to when the gain is realized—for example, during clearing of the open item.

- **Val. Loss 1**
  Account for loss realized during the foreign currency valuation period-end closing program.

- **Val. Gain 1**
  Account for gain realized during the foreign currency valuation period-end closing program.

- **BS Adjustment 1**
  Account used to post the receivables and payables adjustment during the foreign currency valuation of open items.

In the **Translation** section, you can configure the following accounts:

- **Loss**
  After the foreign currency valuation program posts a loss and after the item is cleared, the loss is posted to this account.
- **BS Adjustment Loss**
  This account is used as a clearing account for translation based on a gain.
- **Gain**
  After the foreign currency valuation program posts a gain and after the item is cleared, the gain is posted to this account.
- **BS Adjustment Gain**
  This account is used as a clearing account for translation based on a loss.

In a similar fashion, you also can configure the other exchange rate automatic transactions.

4.4.4 Account Clearing

General ledger accounts can be managed on an open item basis, and in such cases it makes sense to run the account clearing program also as part of the period-end closing, which automatically clears open items that correspond to specific criteria. You need to configure these criteria.

Follow menu path **Financial Accounting • General Ledger Accounting • Business Transactions • Open Item Clearing • Open Item Processing • Prepare Automatic Clearing** to configure the automatic clearing criteria.

Figure 4.30 shows the fields that you can configure:

- **ChAcct**
  Here optionally you can specify that the criteria are valid only for a particular chart of accounts.
- **AccTy**
  Account type, which could be a customer, vendor, or general ledger account. In this example, you're interested in the account type S for general ledger accounts, but it's the same configuration transaction in which you configure the rules for automatic clearing of customers and vendors.

4.4 Periodic Processing and Financial Closing

- **From...To Account**

    You can specify a range of accounts for which the criteria are valid.

- **Criterion 1...5**

    You can specify up to five criteria in these columns. The automatic clearing program will clear only open items that match in terms of all criteria specified here for the selected type and range of accounts. The criteria are entered with the technical names of the fields, such as ZUONR, which is the assignment field, or GSBER, which is the business area.

    Usually at a minimum it's good idea to use the assignment field because very often items that should be cleared contain the same information in this field.

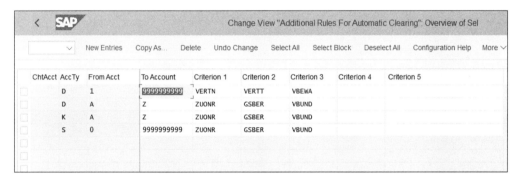

Figure 4.30 Automatic Clearing Configuration

Update the criteria based on your requirements for clearing, then save your entries.

4.4.5 Balance Carry-Forward

Balance carry-forward is a technical step in the year-end closing that transfers the closing balances of the accounts as starting balances in the new year and calculates the retained earnings for the P&L accounts.

In SAP S/4HANA, there's a new transaction for balance carry-forward: Transaction FAGLVTR. In fact, the balance carry-forward procedure in SAP S/4HANA works differently because now the totals tables (table GLT0 in the classic general ledger and table FAGLFLEXT in the new general ledger) are obsolete, and the totals are calculated from table ACDOCA, which is a line-item table. However, table ACDOCA also contains general ledger carry-forward balances, but they're posted in the table as period 0 documents. The balance carry-forward program creates documents in period 0 of the newly opened fiscal year, which represents the cumulative balances as of the beginning of the fiscal year.

For the balance carry-forward program, you need to configure the retained earnings accounts to be posted to by following menu path **Financial Accounting • General Ledger Accounting • Periodic Processing • Carry Forward • Define Retained Earnings Account**. After entering your chart of accounts in the pop-up window, you'll see the screen shown in Figure 4.31.

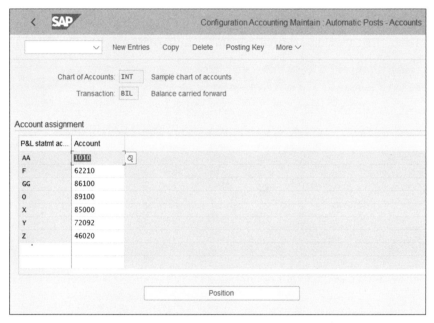

Figure 4.31 Configure Retained Earnings Accounts

Here you assign a retained earnings account to be posted to per P&L statement account type, which is part of the general ledger account master record (**Type/Description** tab). When executing the balance carry-forward program, the system will check the P&L type from the general ledger account master record and will transfer the result from that account to the retained earnings account specified here. In the beginning of the new year, the P&L accounts will start with a zero balance.

This completes our review of the settings for the period-end closing procedures. Next, we'll look at some of the reports provided by the general ledger information system.

4.5 Information System

The general ledger is the main area for providing accounting and taxation reports and unsurprisingly provides a very extensive information system that can suit even the most complicated reporting requirements.

From the myriad reports provided, we'll look at the most important: balance reports, financial statements, tax reports, and drilldown reporting.

4.5.1 Balance Reports

There are several standard reports that provide various balance information on the general ledger account level. They are located in the application menu under **Accounting • Financial Accounting • General Ledger • Information System • General Ledger Reports • Account Balances**. Some of them are general, and many are provided by country; the latter meet most of the local country requirements.

Still, you also have flexibility to configure balance reports, the characteristics of which are available per ledger as selection characteristics when displaying balances. Follow menu path **Financial Accounting • General Ledger Accounting • Information System • Define Balance Display**. Then select **New Entries** from the top menu to specify up to five characteristics per ledger, as shown in Figure 4.32.

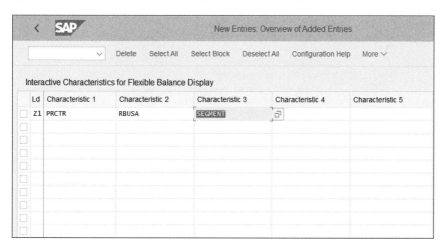

Figure 4.32 Configure Balance Reports

Here you specify which characteristics are available for the ledgers as selection characteristics when displaying balances. You can specify up to five characteristics per

4 General Ledger

ledger. It makes sense to configure the main reporting dimensions here you would use, such as profit center and segment, for example. Configure the desired characteristics per ledger and save your entries.

4.5.2 Financial Statements

Financial statements are legally required financial reports, such as balance sheets and profit and loss statements. They're configured in SAP S/4HANA as a collection of accounts, organized in sections and rows.

These settings are maintained in financial statement versions (FSVs). To configure FSVs, follow menu path **Financial Accounting • General Ledger Accounting • Master Data • G/L Accounts • Define Financial Statement Versions**. The system displays the available financial statement versions, as shown in Figure 4.33.

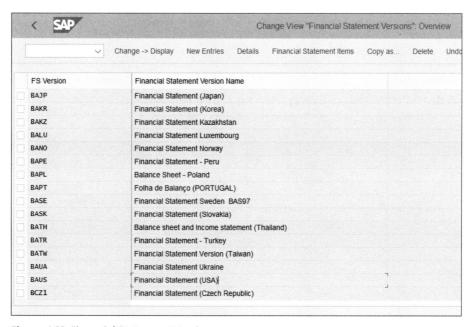

Figure 4.33 Financial Statement Versions

Normally each country requires a separate FSV. Select **BAUS: Financial Statement (USA)**, and copy it to create your new FSV by selecting **Copy as...** from the top menu.

Next, give the new FSV a name and description and map it to your chart of accounts, as shown in Figure 4.34.

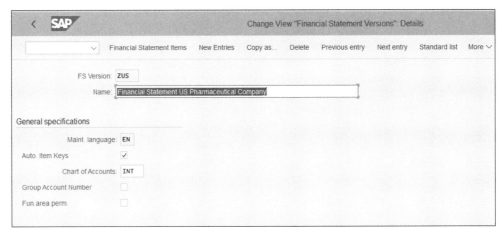

Figure 4.34 Copy Financial Statement Version

Now select **Financial Statement Items** from the top menu, which shows the structure of the financial statement, as shown in Figure 4.35.

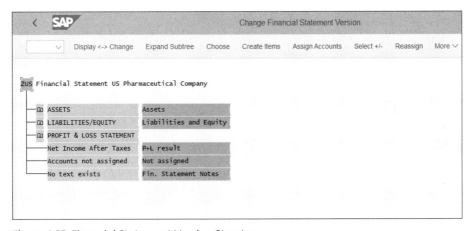

Figure 4.35 Financial Statement Version Structure

You can expand each row by positioning the cursor on it and clicking on **Expand Subtree** from the top menu. This way you can drill-down to the lowest rows, where you assign accounts by selecting **Assign Accounts** from the top menu.

Here you can assign ranges of accounts, as shown in Figure 4.36. Select checkboxes **D** (for debit) and/or **C** (for credit), which tells the system to include the debit and/or the credit balances.

4 General Ledger

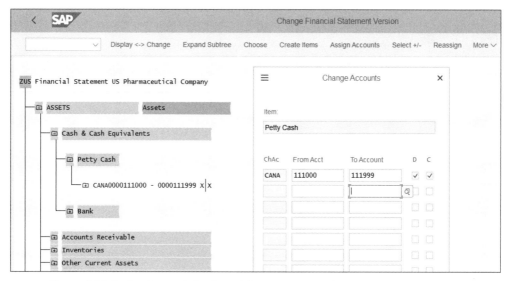

Figure 4.36 Assign Accounts in Financial Statement Version

In this fashion, you can assign accounts to all lines of the financial statement version.

4.5.3 Tax Reports

Tax reports are some of the most important reports provided by the general ledger in SAP S/4HANA because tax requirements are some of the most important legal requirements. They're located in the application menu under **Accounting • Financial Accounting • General Ledger • Reporting • Tax Reports**. Some of them are in the **General** folder, which means they're meant to be used in most countries. There also are numerous country-specific reports.

Most of these reports revolve around requirements for sales/use taxes on transactions, which are knows as value-added tax (VAT) in most countries and use and sales tax in the United States. Because these taxes are a main source of revenue for tax authorities and are difficult to track, there are stringent requirements for ERP systems for reporting them.

The main report for sales and use tax in the United States is located in the application menu under **Accounting • Financial Accounting • General Ledger • Reporting • Tax Reports • USA • S_ALR_87012394 - Record of Use and Sales Taxes (USA)**, as shown in Figure 4.37.

4.5 Information System

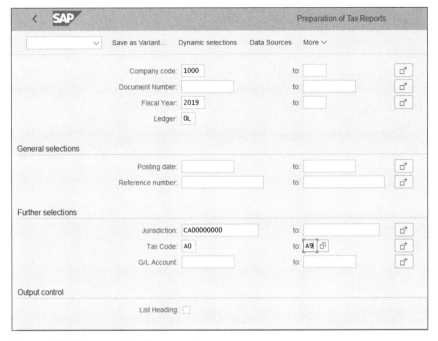

Figure 4.37 Sales and Use Tax Report

This report is based on jurisdiction codes and tax codes. The jurisdiction codes play an important role in determining the local tax rates per jurisdiction (state, county, city) in the United States, and their configuration will be covered in detail in the next chapter.

4.5.4 Drilldown Reporting

Drilldown reporting is a flexible and convenient reporting technique in SAP in which a user can interactively navigate through a report by clicking certain characteristics to see additional details for any combination of these characteristics. Originally developed mostly for the Controlling module in SAP ERP, in SAP S/4HANA they are widely available and useful in the general ledger.

There are several standard drilldown reports available for the general ledger, and you can easily configure additional reports. The configuration activities are located under menu path **Financial Accounting • General Ledger Accounting • Information System • Drilldown Reports (G/L Accounts)**. First you need to define a form, which defines the

4 General Ledger

structure of available characteristics and their values. Then you define reports based on this form, specifying the report layout, selection criteria, and output settings.

Let's see how a form is defined. Follow menu path **Financial Accounting • General Ledger Accounting • Information System • Drilldown Reports (G/L Accounts) • Form • Specify Form**, then select the **Change Form** activity. In the resulting screen, navigate to the 0SAPBLNCE-01 G/L Accounts—Bal. form, as shown in Figure 4.38.

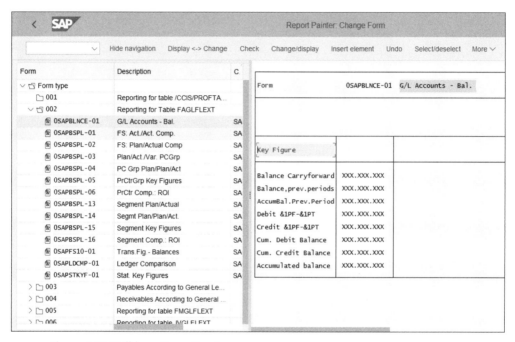

Figure 4.38 Drilldown Reporting Form

This standard form can be used for reporting of general ledger account balances. You can copy it to your form and change it as needed. Each line in the form is defined with its characterstics and values. Double-click on the **Balance Carryforward** line, for example, to open the screen shown in Figure 4.39.

On the right side, you see the list of available characteristics, and on the left side the posting period is selected. The **From** and **To** columns contain the values to be included in the report. They can be entered directly in the columns, or you can use sets (predefined group of values), if you tick the checkbox in the ⛁ column, or variables (parameters to be entered during program executing in the selection screen), if you tick the checkbox in the ▦ column. In Figure 4.39, a variable named 1PT is selected.

152

This means that the posting period will be entered in the selection screen of the drill-down reports based on this form.

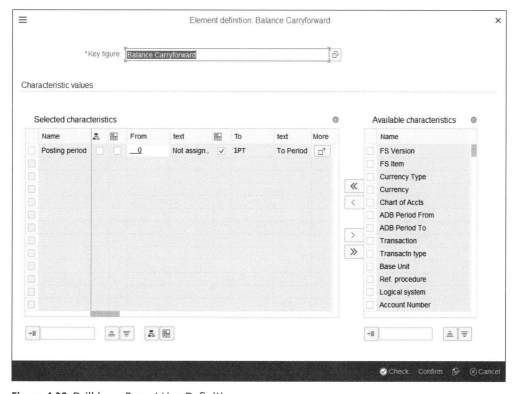

Figure 4.39 Drilldown Report Line Definition

In this fashion, you also can define the other lines and columns of the report form and create your own reports based on it.

4.6 SAP Fiori Applications

As you know by now, SAP Fiori is the new web-based, very efficient user interface, specially designed and optimized for SAP S/4HANA. Although its adoption can take some time because SAP keeps adding functions as SAP Fiori apps and because existing SAP users are very used to using the classic SAP GUI, but SAP Fiori is the future and will become even more important as it matures.

SAP Fiori by itself could be a topic for another large book, but here we'll look at just some of the useful SAP Fiori apps available for the general ledger, such as Post General Journal Entries and Upload Journal Entries.

4.6.1 Post General Journal Entries App

The Post General Journal Entries app is a useful app that enables users to post general ledger entries directly in financial accounting using the SAP Fiori web interface.

As shown in Figure 4.40, the field layout and organization of the screen is very similar to the classic SAP GUI transaction to enter journal entries (Transaction FB50). There is a header section, where information relevant for the document as a whole is entered, such as document and posting date, company code, and so on. Below that, the relevant line items are entered.

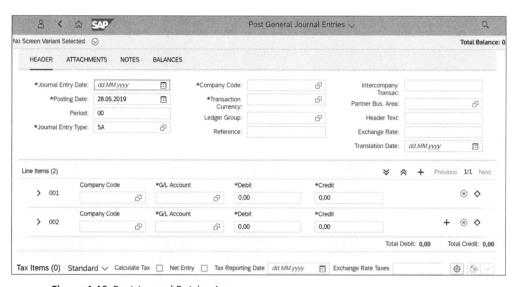

Figure 4.40 Post Journal Entries App

Users of the classic SAP GUI transaction will be very familiar with this app, but it will be easier for them to use the SAP Fiori web-based interface, because the SAP Fiori app offers an intuitive web interface that has responsive design, which automatically adapts based on the type of device that is used.

4.6.2 Upload General Journal Entries

The Upload General Journal Entries app is a particularly useful SAP Fiori app. People who've worked with previous SAP ERP systems are well aware that SAP didn't have a convenient out-of-the-box transaction to mass upload general ledger journals from an Excel file. Because this is an important and often used function in most companies, SAP customers often would buy additional third-party tools to integrate with their SAP system or develop their own custom programs in SAP.

Now in SAP S/4HANA (starting with the 1709 release) there is an SAP Fiori app available to upload journal entries, as shown in Figure 4.41.

Figure 4.41 Upload Journal Entries App

Here you can select a template-based file using the **Browse** button, and then it can be uploaded easily as journal entries in the system. No additional programming is required now for this important task!

4.7 Summary

In this chapter, we covered in depth the configuration to be performed for the general ledger in SAP S/4HANA. The general ledger is the most fundamental functionality in SAP S/4HANA, into which ultimately all postings flow and are recorded in the Universal Journal. As such, the general ledger is the area where we see at their best the great improvements that SAP S/4HANA brings to finance with its simplification and streamlining of processes.

After finishing this chapter, you should understand the master data in the general ledger. You are now in a position to properly organize your chart of accounts and create efficient general ledger accounts that enable your organization to accurately depict its financial processes.

We also covered document splitting, which is a fundamental feature of SAP S/4HANA that enables full reporting on various levels of characteristics.

We discussed account determination and automatic postings, which allow all financial data to be posted in the general ledger. We also discussed the main logistic processes in the system so that you can better understand the flow of values to the general ledger.

We then configured the main periodic processes and financial closing processes, such as managing posting periods, intercompany reconciliation, foreign currency revaluation, account clearing, and balance carry-forward. Month-end and year-end closing are some of the most important activities in the general ledger area because they provide financial results for the company.

Then we explained the various important reports available from the general ledger information system and what configuration is needed for them. Last but not least, we provided an overview of some useful general ledger SAP Fiori apps. With that, we finish the general ledger configuration, and we can move to configuring the subledgers in accounting, starting with accounts payable.

Chapter 5
Accounts Payable

This chapter gives step-by-step instructions for configuring accounts payable in SAP S/4HANA and explains the integration with materials management processes. We also introduce the new business partner concept in SAP S/4HANA, payment processing, and available reporting for accounts payable.

The next area we'll configure is accounts payable, which is used to manage the financial processes related to purchasing, including posting vendor invoices, payment processing, and clearing and reconciliation of open items. We'll provide a detailed explanation of how to configure the master data, business transactions, and information system for accounts payable.

5.1 Business Partner

In SAP S/4HANA, the master data for accounts payable is managed via the business partner master data object. Business partners are not an entirely new concept in SAP S/4HANA. In the SAP ERP system, business partners were used to manage certain partner master data objects, mostly in financial supply change management. However, now in SAP S/4HANA the use of business partner is mandatory in financial accounting, and vendor (used in accounts payable) and customer (used in accounts receivable) master records are managed via the business partner concept.

This enables many benefits, such as the following:

- There is now a fully integrated relationship between vendors and customers when the same company/person serves as both a supplier and customer of the organization.
- General data is shared between different business partner roles.
- Application-specific business partner roles can be created that extend already defined business partners, such as credit management or collections management.
- There is now a harmonized architecture between SAP functional areas.

5 Accounts Payable

There is now a single transaction, Transaction BP, available to manage all business partner master data processes, which makes obsolete the previous create/change/display transaction codes for vendors, such as Transaction FK01/02/03, Transaction XK01/02/03, and so on. In fact, if you execute one of these transaction codes you'll be redirected to Transaction BP to maintain a business partner.

As you'll see in the following sections, the finance data for vendors in the business partner record is divided into general data and company code data. Once we've covered the finance data, we'll quickly walk you through the underlying configuration for business partners.

5.1.1 General Data

Let's see how to navigate through the business partner master record. Enter Transaction BP to see the screen shown in Figure 5.1.

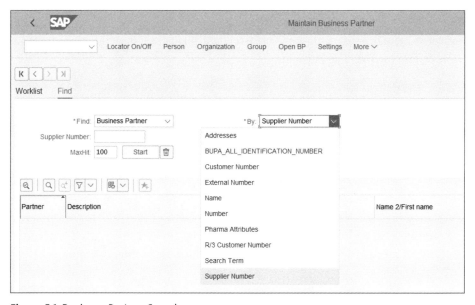

Figure 5.1 Business Partner Search

Here, using the dropdown option on the right side, you can search business partners by various roles. Select **By Supplier Number** to view the business partners maintained as vendors. Remove the value **100** in the **MaxHit** field so that there's no limitation in the search, then click the **Start** button. In the resulting list, double-click any vendor number. This takes you to the general data for the business partner, as shown in Figure 5.2.

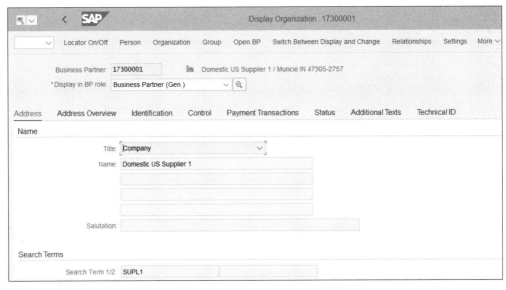

Figure 5.2 Business Partner General Data

This data is valid for all partner functions of the business partner. It's organized into tabs, which group together similar fields, such as address data, control, payment transactions, and so on. When you click the dropdown list in **Display in BP Role**, you'll see the roles maintained for the business partner. The **FI Supplier Role** contains the vendor financials data and **Supplier** holds the purchasing data for the vendor. Select the **FI Supplier Role** also to see FI-related tabs, such as **Tax Data**.

You can update fields by selecting **Switch Between Display and Change** in the top menu.

5.1.2 Company Code Data

To view the company code–level data for the vendor, first select the **Creditor (FI Vendor)** business partner role in the dropdown list. Then select **Company Code** from the top menu. Figure 5.3 shows the company code data of the financial accounting vendor account.

You can change the company code and maintain the vendor for another company code using the **Switch Company Code** button.

5   Accounts Payable

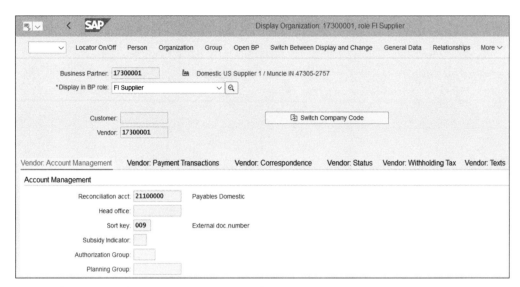

Figure 5.3 Business Partner Company Code Data

All these fields are maintained on the company code level and can be different from company code to company code. In the **Reconciliation acct**. field, you maintain the reconciliation account, which is the general ledger account posted to each time something is posted to this vendor. This is the account that's normally credited for vendor invoices and debited for payments, credit memos, or when otherwise clearing open vendor items.

As you can see, the business partner and vendor have the same account numbers. This is best practice and is controlled with the configuration of the number ranges, assigned to the vendor account groups and to the business partner account groups. They are separate objects and can have different number ranges, but that can cause confusion.

5.1.3 Business Partner Configuration

Now let's look at the underlying configuration for business partners. To define business partner groups and number ranges, follow menu path **Cross-Application Components • SAP Business Partner • Business Partner • Basic Settings • Number Ranges and Groupings • Define Groupings and Assign Number Ranges**.

Figure 5.4 shows a list of business partner groups and their number ranges. When creating a business partner from a specific group, a number will be assigned from the corresponding number range. Note that the system creates separate account numbers for the business partner and the vendor. When posting to the vendor, the vendor account number is used, but it's good to have the same numbers assigned on the business partner and vendor levels.

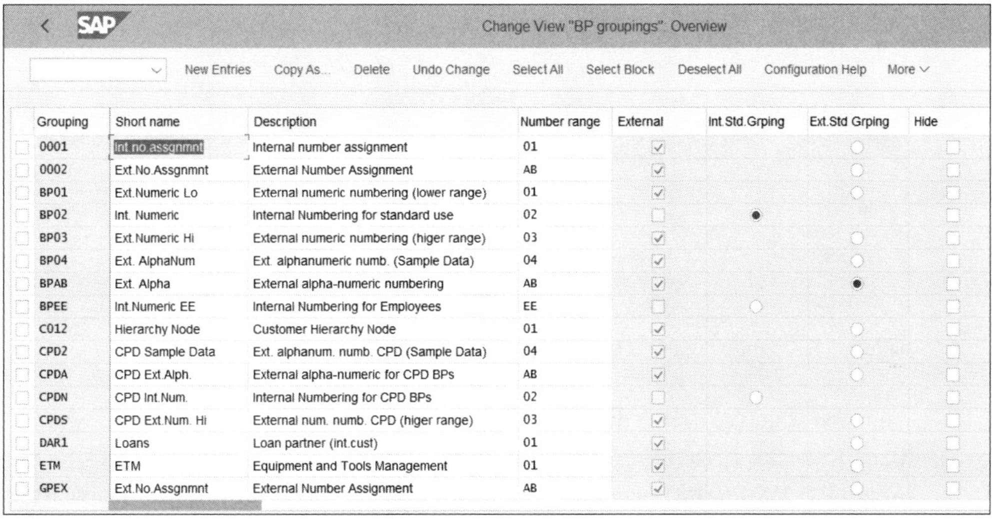

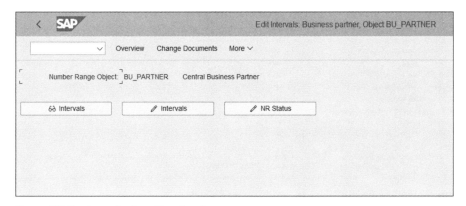

Figure 5.4 Business Partner Groups and Number Ranges

The actual number ranges for business partners are maintained in menu path **Cross-Application Components • SAP Business Partner • Business Partner • Basic Settings • Number Ranges and Groupings • Define Number Ranges**. As shown in Figure 5.5, the number range object is BU_PARTNER.

Figure 5.5 Business Partner Number Ranges

5 Accounts Payable

Click the ✏ Intervals button to change the number ranges. On the screen shown in Figure 5.6, you can modify or create new number ranges to be used in business partner groups.

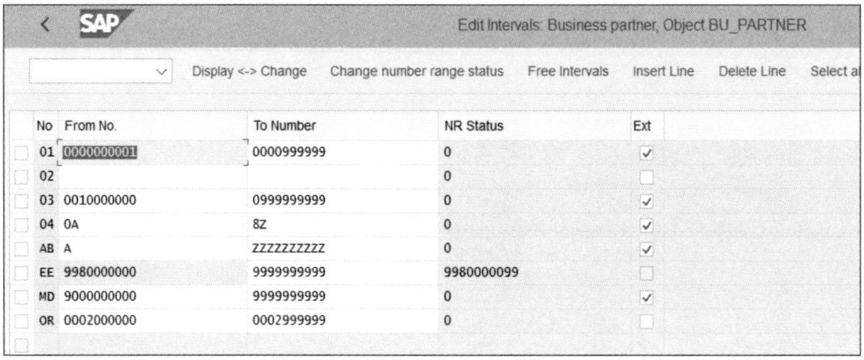

No	From No.	To Number	NR Status	Ext
01	0000000001	0000999999	0	✓
02			0	
03	0010000000	0999999999	0	✓
04	0A	8Z	0	✓
AB	A	ZZZZZZZZZZ	0	✓
EE	9980000000	9999999999	9980000099	
MD	9000000000	9999999999	0	✓
OR	0002000000	0002999999	0	

Figure 5.6 Change Business Partner Number Ranges

The next step is to define the vendor groups with their number ranges. Follow menu path **Financial Accounting • Accounts Receivable and Accounts Payable • Vendor Accounts • Master Data • Preparations for Creating Vendor Master Data • Assign Number Ranges to Vendor Account Groups** to do so. Figure 5.7 shows how the number ranges are assigned to the various vendor groups.

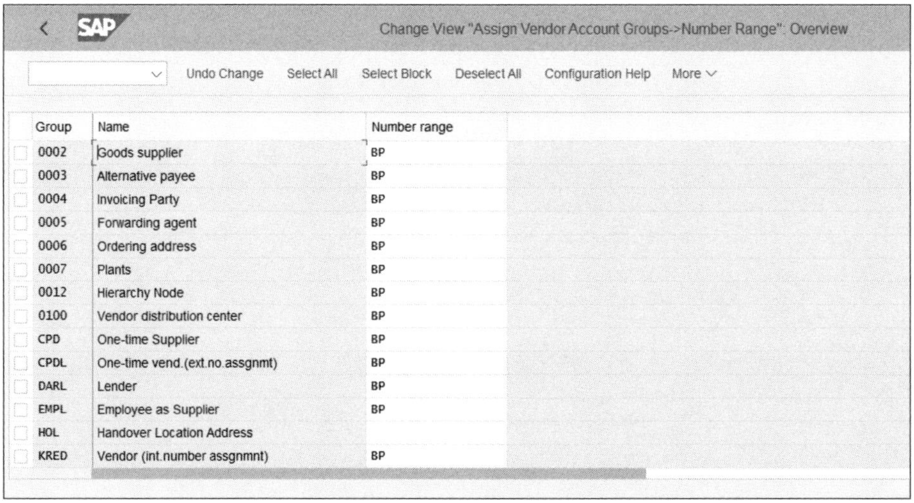

Group	Name	Number range
0002	Goods supplier	BP
0003	Alternative payee	BP
0004	Invoicing Party	BP
0005	Forwarding agent	BP
0006	Ordering address	BP
0007	Plants	BP
0012	Hierarchy Node	BP
0100	Vendor distribution center	BP
CPD	One-time Supplier	BP
CPDL	One-time vend.(ext.no.assgnmt)	BP
DARL	Lender	BP
EMPL	Employee as Supplier	BP
HOL	Handover Location Address	
KRED	Vendor (int.number assgnmnt)	BP

Figure 5.7 Vendor Account Groups and Number Ranges

For this example, assign number range BP for the groups you are going to use.

The actual number ranges are defined in menu path **Financial Accounting • Accounts Receivable and Accounts Payable • Vendor Accounts • Master Data • Preparations for Creating Vendor Master Data • Create Number Ranges for Vendor Accounts**. As shown in Figure 5.8, the number range object is KREDITOR. Click the ✎ Intervals button to change the number ranges.

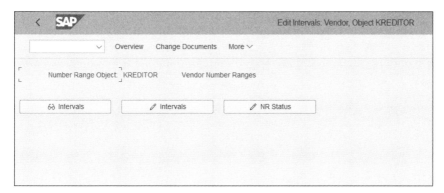

Figure 5.8 Vendor Number Ranges

On the screen shown in Figure 5.9, you can change the numbers behind the vendor number ranges by changing the **From** and **To** numbers of the range.

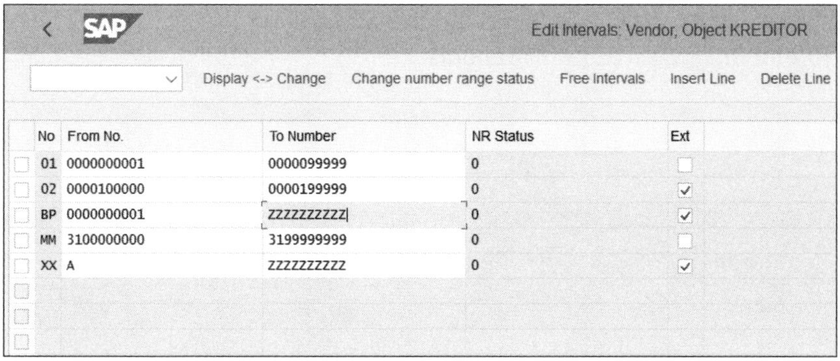

No	From No.	To Number	NR Status	Ext
01	0000000001	0000099999	0	
02	0000100000	0000199999	0	✓
BP	0000000001	ZZZZZZZZZZ	0	✓
MM	3100000000	3199999999	0	
XX	A	ZZZZZZZZZZ	0	✓

Figure 5.9 Define Vendor Number Ranges

The next step is to configure the customer-vendor integration (CVI). CVI is the link between the business partner from one side and the vendor and customer from another.

163

First you need to configure the synchronization control for business partners by following menu path **Cross-Application Components • Master Data Synchronization • Synchronization Control • Synchronization Control • Activate PPO Requests for Platform Objects in the Dialog**. As shown in Figure 5.10, you need to make sure that synchronization object BP is active for postprocessing by selecting the **PPO Active** checkbox.

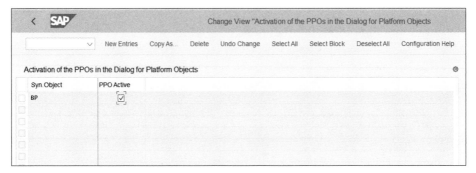

Figure 5.10 Activation of PPO in Business Partner

Now you need to configure the link between the business partner and vendor by following menu path **Cross-Application Components • Master Data Synchronization • Customer/Vendor Integration • Business Partner Settings • Settings for Vendor Integration • Set BP Role Category for Direction BP to Vendor**. Figure 5.11 shows the list of business partner roles related to vendors. Double-click **FLVN00**, which is the business partner role for financial accounting vendors.

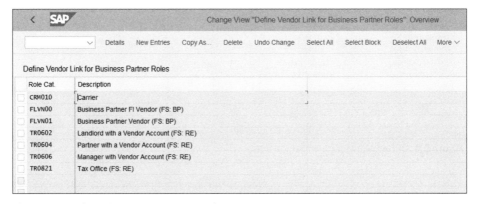

Figure 5.11 Link Business Partner to Vendors

5.1 Business Partner

Here, make sure that **Vendor-Based** is selected, as shown in Figure 5.12. This ensures that it's mandatory to have a vendor for this business partner role. Then when you process a business partner in this role, the system automatically creates or updates a vendor with the relevant data in financial accounting.

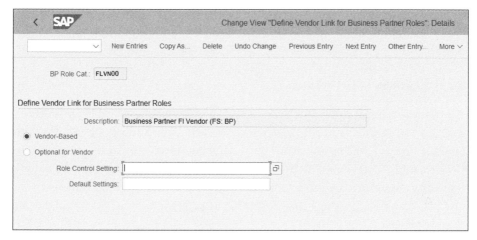

Figure 5.12 Business Partner to Vendor Settings for Finance Vendors

The next step is to configure the integration from vendor to business partner by following menu path **Cross-Application Components • Master Data Synchronization • Customer/Vendor Integration • Business Partner Settings • Settings for Vendor Integration • Define BP Role for Direction Vendor to BP**.

On the screen shown in Figure 5.13, you can map the vendor groups with business partner roles.

Figure 5.13 Link Vendor Groups to Business Partner Roles

165

Select **New Entries** from the top menu and link the relevant vendor groups with the business partner roles. This ensures that when you process a vendor with the relevant account group, the system will create or update a business partner in the BP roles that are assigned to the vendor group.

Now that we've established the settings for business partners, let's examine the configuration of the various business transactions.

5.2 Business Transactions

The main business transactions that are managed in accounts payable are incoming invoices and credit memos from vendors, as well as outgoing payments for these invoices and supporting processes such as goods receipt/invoice receipt (GR/IR) clearing. In this section, we'll explain not only those processes, but also the integration with vendor invoice management (VIM), which scans for vendor invoices and provides automatic posting in SAP S/4HANA, as well as tax determination in the purchasing process.

5.2.1 Incoming Invoices/Credit Memos in Financial Accounting

In SAP S/4HANA, there are two ways to post vendor invoices and credit memos. The most common way is to post them through materials management, where they reference a purchase order and are integrated with the purchasing process. However, in some cases you also need to post invoices and credit memos directly in financial accounting, without reference to a purchase order. This is usually done for smaller purchases that don't require specific approval, such as stationery supplies. They are posted in financial accounting, usually debiting an expense account and cost center or another cost object, and crediting a vendor account.

Not much specific customizing is needed to enable the processing of these transactions in financial accounting. You need to have set up your vendor accounts as business partners, as explained in the previous section. You also need to have defined the document types you're going to use—typically the standard SAP document type KR for vendor invoices and KG for vendor credit memos. We covered the document type configuration in Chapter 3, Section 3.4.

You need to configure the tax codes that can be used when posting invoices and credit memos using Transaction FB60, in which you don't have to specify overly technical objects such as posting keys. This interface is easier for end users to use.

To enable the tax codes to be used when posting invoices and credit memos, follow menu path **Financial Accounting • Accounts Receivable and Accounts Payable • Business Transactions • Incoming Invoices/Credit Memos • Incoming Invoices/Credit Memos—Enjoy • Define Tax Code per Transaction**.

Any new tax codes created for your country need to be configured here so that they can be used in Transaction FB60. Enter "US" in the **Country Key** field in the pop-window, as shown in Figure 5.14.

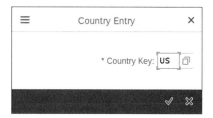

Figure 5.14 Select Country

Initially the following screen is blank, as shown in Figure 5.15.

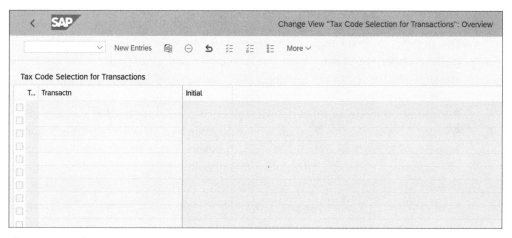

Figure 5.15 Define Tax Codes for Invoices and Credit Memos: Initial Screen

Select **New Entries** from the top menu to configure new tax codes. As shown in Figure 5.16, you'll then enter the input tax codes. In the **Transaction** column, you can select **Relevant for All Transactions** or select **Financial Accounting Invoice Receipt** if you want to limit the tax code to financial accounting document posting only. If you check the **Initial** flag, the tax code will be the default for the transaction.

167

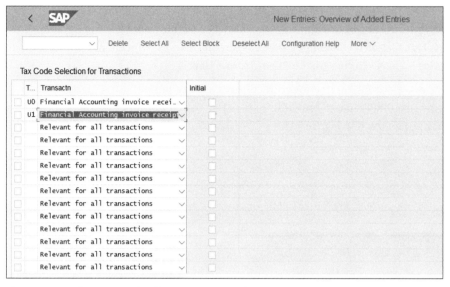

Figure 5.16 Define Tax Codes for Enjoy Invoices/Credit Memos

5.2.2 Posting with Alternative Reconciliation Account

You've seen that in the business partner master record when maintaining the financial accounting vendor data on the company code, you specify a reconciliation account, which is the general ledger account automatically posted with each posting to the vendor account. This is only one account, which you can specify in the master record—but sometimes, for various business cases, you may need to post to another account. This is because from an accounting point of view, certain transactions—such as posting of discounts, guarantees, or payment requests—need to post to specific accounts. Therefore, SAP provides functionality to post to alternative reconciliation accounts.

In this process, the special general ledger indicator is used. This indicator can be entered in documents and will swap the general ledger reconciliation account from the vendor master record with another alternative general ledger account configured for this special general ledger indicator. Let's walk through how to set up this configuration.

To define alternative reconciliation accounts, follow menu path **Financial Accounting • Accounts Receivable and Accounts Payable • Business Transactions • Postings with Alternative Reconciliation Accounts • Other Special G/L Transactions • Define Alternative Reconciliation Account for Vendors**.

As shown in Figure 5.17, you'll see the list of special general ledger indicators provided by SAP. You can also create new indicators if these don't meet your needs. Double-click indicator **D**, which is used to post discounts.

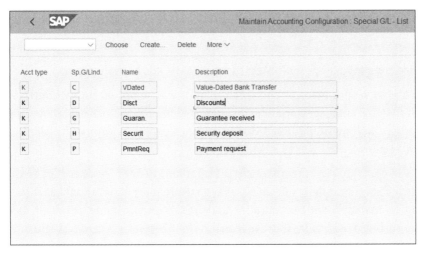

Figure 5.17 Special General Ledger Indicators

After entering your chart of accounts, you're taken to the screen shown in Figure 5.18.

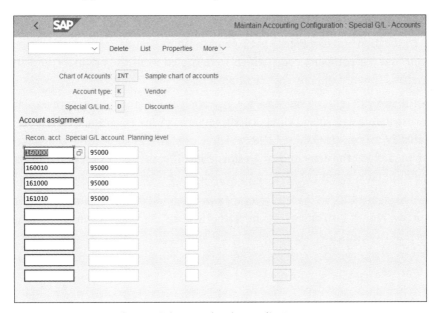

Figure 5.18 Accounts for Special General Ledger Indicators

Here, for each account defined in the **Recon. Acct** column, which are used in the vendor master records, you define an alternative account in the **Special G/L Account** column. Then when posting with the special general ledger indicator for a vendor linked with this reconciliation account, the system will automatically reroute the posting to the alternative reconciliation account.

5.2.3 Incoming Invoices/Credit Memos from Materials Management

Most incoming vendor invoices and credit memos will be posted through integration with materials management. The resulting documents in financial accounting will be posted automatically based on the account determination we covered in Chapter 4, Section 4.3.3.

Most of the configuration for materials management vendor invoices and credit memos (which are commonly referred to as Logistics Invoice Verification) lies within the materials management configuration and area of responsibility. However, as a finance expert, also have to be aware of it and in some projects have to configure it.

One important aspect that affects financial accounting is the invoice automatic block for payment. SAP S/4HANA is a system built to provide very good business controls, and one part of that is that it can be configured to automatically block vendor invoices that match certain criteria. Let's see what options are available.

To configure tolerance limits, follow menu path **Materials Management • Logistics Invoice Verification • Invoice Block • Set Tolerance Limits**. Figure 5.19 shows the tolerance limits set per company code (which were copied from the original company code used as a source company code in Chapter 3).

There are many different tolerance keys, which apply in different business cases, and you can also define new rules. Double-click tolerance key **BD: Form Small Differences Automatically** for company code 1000, which is used to enable posting of invoices that don't match on the debit and credit side if the difference isn't significant and to post that difference automatically to a small differences account.

As shown in Figure 5.20, you set the tolerance amount up to which this is possible. In this case, we enter "2,00" euros in the **Value** field of the **Check Limit** section; all invoices that have mismatches between debits and credits of up to two euros will be posted.

5.2 Business Transactions

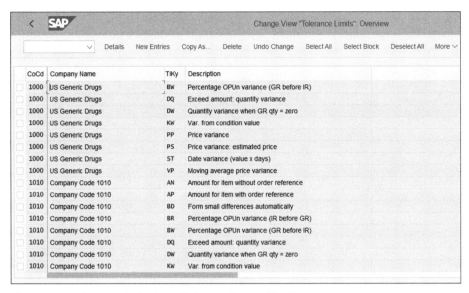

Figure 5.19 Tolerance Limits in Logistics Invoice Verification

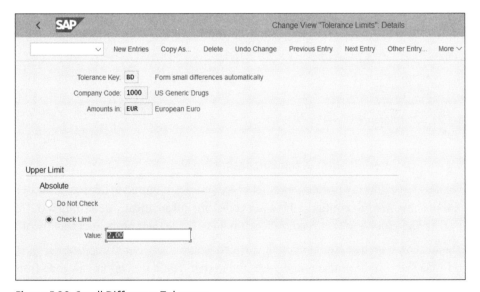

Figure 5.20 Small Difference Tolerance

Now go back to the previous screen and check the tolerance limit for **PP: Price variance** for company code 1000 by double-clicking it.

Figure 5.21 illustrates defining the lower and upper limits. Any invoices outside these limits will be posted but automatically blocked for payment in financial accounting until either the situation is remedied or the block is manually removed by an authorized user.

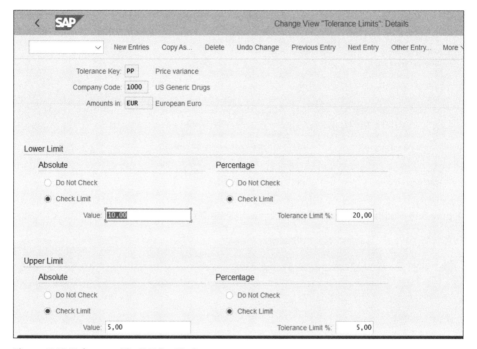

Figure 5.21 Tolerance Limit Price Variance

5.2.4 Tax Determination in the Purchasing Process

We discussed how tax procedures and tax codes are configured in Chapter 3, Section 3.6. We also need to configure how tax codes are automatically determined in the purchasing process so that financial documents receive their proper tax treatment.

The tax code can be entered manually, which would overwrite tax code determination via the condition technique. Also, the system could derive the tax code from another related document (in the following sequence): reference item, contract, request for quotation, and info record. Tax codes are determined from the purchasing orders using the so-called condition technique. Condition records are created based on configured condition tables, which based on certain parameters coming mostly from the vendor master record and the material master record determine the proper tax code for the different business cases.

The purchase order has a pricing procedure that includes a tax condition (the standard SAP condition is MWST input tax), which accesses the relevant condition records and determines the tax code and the tax rate. The configuration of this procedure is a cross-topic between finance and purchasing, but usually the finance consultant should be ultimately responsible for the tax setup. Let's see how these tax condition records are configured.

To define condition types for this process, follow menu path **Materials Management • Purchasing • Conditions • Define Price Determination Process • Define Condition Types**, then select activity **Set Pricing Condition Types - Purchasing**.

Figure 5.22 shows a list of the purchasing condition types. You can also copy a standard condition type as a custom type starting with Z, then make changes as appropriate.

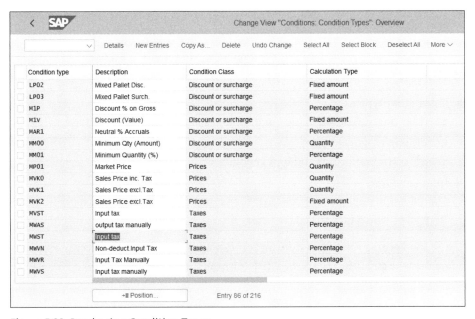

Figure 5.22 Purchasing Condition Types

Double-click condition type **MWST: Input Tax** to see its settings, as shown in Figure 5.23.

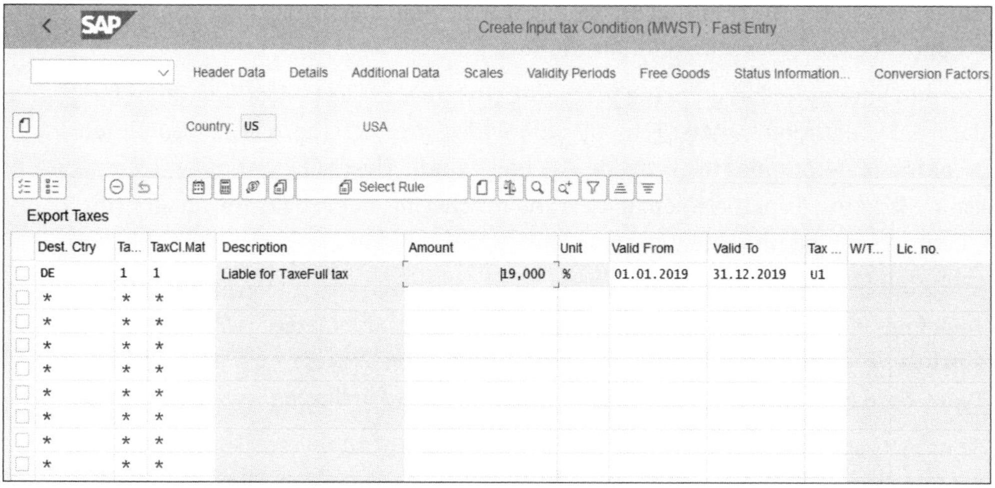

5 Accounts Payable

Dest. Ctry	Ta...	TaxCl.Mat	Description	Amount		Unit	Valid From	Valid To	Tax ...	W/T...	Lic. no.
DE	1	1	Liable for TaxeFull tax	19,000		%	01.01.2019	31.12.2019	U1		
*	*	*									
*	*	*									
*	*	*									
*	*	*									
*	*	*									
*	*	*									
*	*	*									

Figure 5.23 Input Tax Condition Type

The following fields are important:

- **Condition Class**
  This determines the usage of the condition type; for tax conditions, it's always **D Taxes**.

- **Calculation Type**
  Specifies the type of calculation and for taxes is normally **A Percentage**.

- **Condition Category**
  Classifies the conditions and for tax conditions should be **D Tax**.

- **Item Condition**
  Should be checked for taxes because it ensures that the calculation happens at the item level, which is required for taxes.

Using the **Records for Access** button, you can see what records for access are maintained for the condition types, which determine the tax codes.

You can maintain these records using Transaction MEK1. They are considered application data, and you can also navigate to them through the application menu path **Materials Management • Purchasing • Master Data • Conditions • Other • MEK1— Create**. Then enter "MWST" for the **Condition Type** to open a pop-up window asking for the combination of characteristics for which you want to create records, as shown in Figure 5.24. These characteristics are dependent on the configuration of the condition record.

5.2 Business Transactions

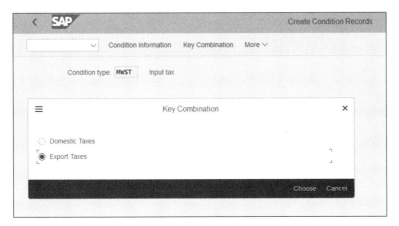

Figure 5.24 Key Combination for Condition Records

Select **Export Taxes** to access the next screen, in which you enter the records, as shown in Figure 5.25.

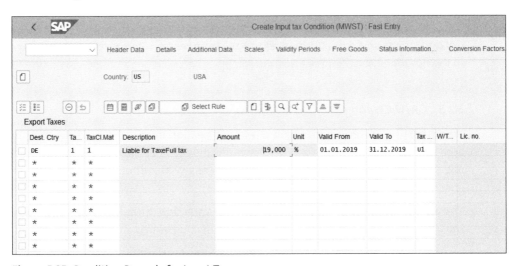

Figure 5.25 Condition Records for Input Tax

Here in the header you enter the source country from which goods or services are sent. In the lines below, you can have multiple records per destination country.

The following fields need to be configured:

- **Destination Country**
  This is the country to which the goods/services are supplied.

175

- **Tax Classification 1**
  Here you select a tax classification indicator, which is maintained in the business partner master records and normally specifies whether this business partner is liable for taxes or exempt.

- **Tax Classification Material**
  Here you select a tax classification indicator, which is maintained in the material master records and normally specifies whether this material is taxable at the full rate, taxable at a reduced rate, or not taxable.

- **Valid From and Valid To**
  Here you specify the validity period of the tax code. Tax rates are often dynamic and change from time to time, so this way you have the flexibility to add more condition records when the tax rates changes.

- **Tax Code**
  Here you specify the tax code, which should be automatically assigned when the criteria in the condition record are met.

Save your entries. Purchasing documents that match these condition criteria now will receive the assigned tax codes automatically, which will then flow into the financial documents and post automatically to the tax general ledger accounts, configured in the tax codes (see Chapter 3, Section 3.6.2).

5.2.5 GR/IR Clearing

When posting materials management-based vendor invoices, the GR/IR clearing account is often used, which in a perfect world should be balanced out if the invoices and the corresponding goods receipts match. In reality, there are almost always differences.

The GR/IR clearing process is executed with Transaction MR11. It's designed to show the GRs that are not yet fully invoiced and the IRs that are yet fully received.

In the normal course of business, there will always be some invoices that have not yet been received and some goods receipts that have not yet been invoiced. But the GR/IR clearing process is meant to be used to clear the old items that probably won't be completed. There isn't much configuration behind Transaction MR11. It uses the tolerances we configured in Section 5.2.3.

You need to also set up document types and number ranges for the GR/IR clearing process by following menu path **Materials Management • Logistics Invoice Verification • Clearing Account Maintenance • Maintain Number Assignments for Accounting**

Documents. On the screen shown in Figure 5.26, click the **Document Types in Invoice Verification** button.

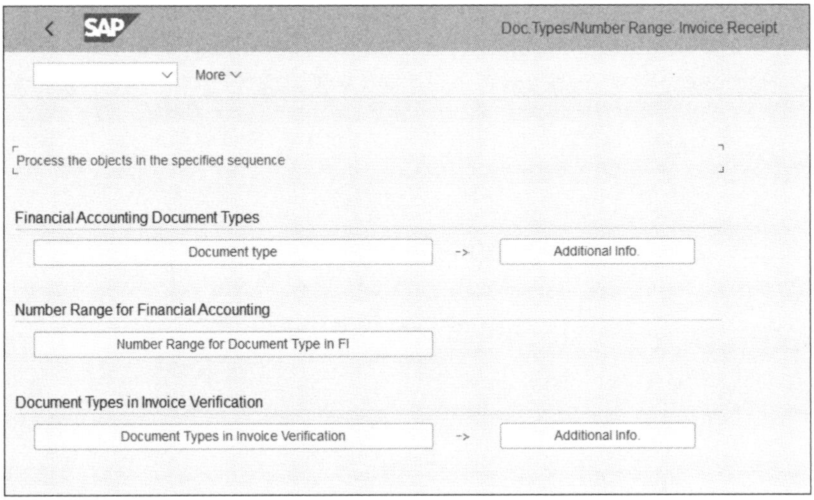

![Figure 5.26 screenshot]

Figure 5.26 Document Types Assignment

Then you'll see the screen shown in Figure 5.27, listing the various transactions used in the invoice verification process. Double-click **MR11: GR/IR account maintenance**.

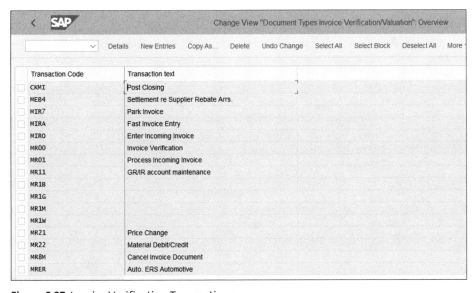

Figure 5.27 Invoice Verification Transactions

As shown in Figure 5.28, here you specify the document type to be used in the GR/IR clearing process. The default is **KP Account Maintenance**. You can change it if you need to define a special document type for the purposes of GR/IR account maintenance.

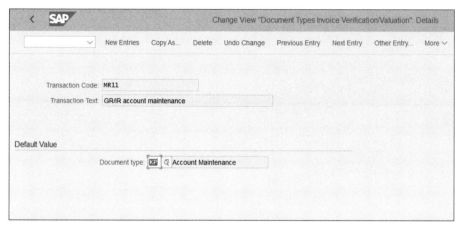

Figure 5.28 GR/IR Clearing Document Type Configuration

In the next step, you need to select the number range to be used for this document type by following menu path **Materials Management • Logistics Invoice Verification • Clearing Account Maintenance • Maintain Number Assignment for Account Maintenance Documents • Maintain Number Range for Account Maintenance Document**. Here you assign the number range to be used with the document type, configured for the GR/IR process. Specify range 04 for document type **KP**, as shown in Figure 5.29.

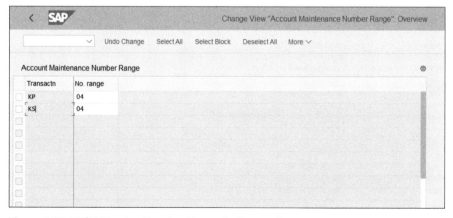

Figure 5.29 GR/IR Clearing Number Range Assignment

The last step is to define the numbers for this number range by following menu path **Materials Management • Logistics Invoice Verification • Clearing Account Maintenance • Maintain Number Assignment for Account Maintenance Documents • Maintain Number Range Interval for Account Maintenance Document**. The name of the number range object is **RE_BELEG**, as shown in Figure 5.30.

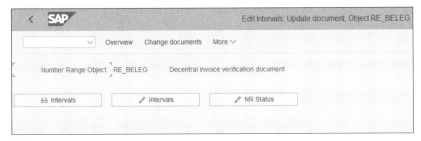

Figure 5.30 Number Ranges for IR Documents

Click the ✏ Intervals button to maintain the ranges as shown in Figure 5.31.

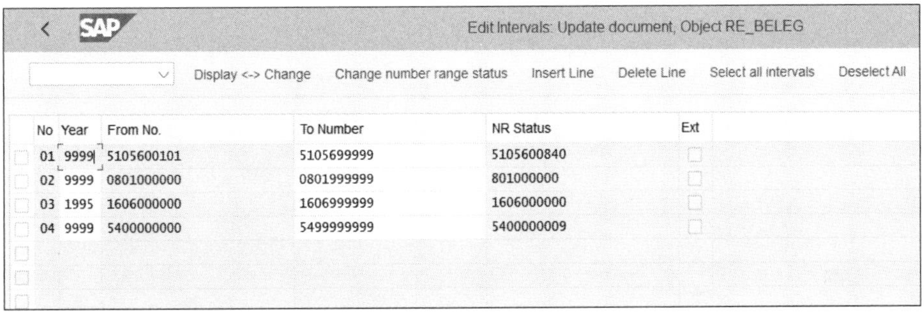

No	Year	From No.	To Number	NR Status	Ext
01	9999	5105600101	5105699999	5105600840	
02	9999	0801000000	0801999999	801000000	
03	1995	1606000000	1606999999	1606000000	
04	9999	5400000000	5499999999	5400000009	

Figure 5.31 Number Ranges Maintenance

Maintain the **From No.** and **To Number** fields of range 04, valid indefinitely (represented by entering a year of 9999). Then save your entries. As with other number ranges, no automatic transport is assigned because number ranges should be maintained in each individual client.

5.2.6 Payment Terms and Outgoing Payments

The next major types of transactions that relate to accounts payable are the payments. The posted invoices, regardless whether they come from financial accounting or MM are posted as open items that await payments.

We will examine the configuration of the payment methods, the automatic payment program, and the electronic bank statement in Chapter 8 in detail; for now, we'll configure the terms of payment and the payment block.

The conditions of the payments depend on the payment terms. The payment term is a configuration object, which is part of the vendor line item. It determines when the payment is due and if some discounts are applicable. For example, payment terms can be payable immediately due net, when there is no grace period and the whole amount should be paid without any discounts. Or it could be 14 days 2%, 30 net, which means that payment is due in 30 days for the full amount, but if made within 14 days, a 2% discount applies.

To configure payment terms, follow menu path **Financial Accounting • Accounts Receivable and Accounts Payable • Business Transactions • Incoming Invoices/Credit Memos • Maintain Terms of Payment** or enter Transaction OBB8.

As shown in Figure 5.32, SAP provides many standard payment terms, which cover many of the standard payment conditions that occur in business. You can of course create new ones, and as with other configuration objects, it's good practice to name them in a custom number range starting with Z or Y.

Figure 5.32 Payment Terms

Let's create a new payment term for payment to a vendor within 14 days for a 4% cash discount, or within 45 days due net. First, find a similar standard payment term, such as 0002. Check the checkbox to the left of 0002, then select **Copy As...** from the top menu, which will result in the screen shown in Figure 5.33.

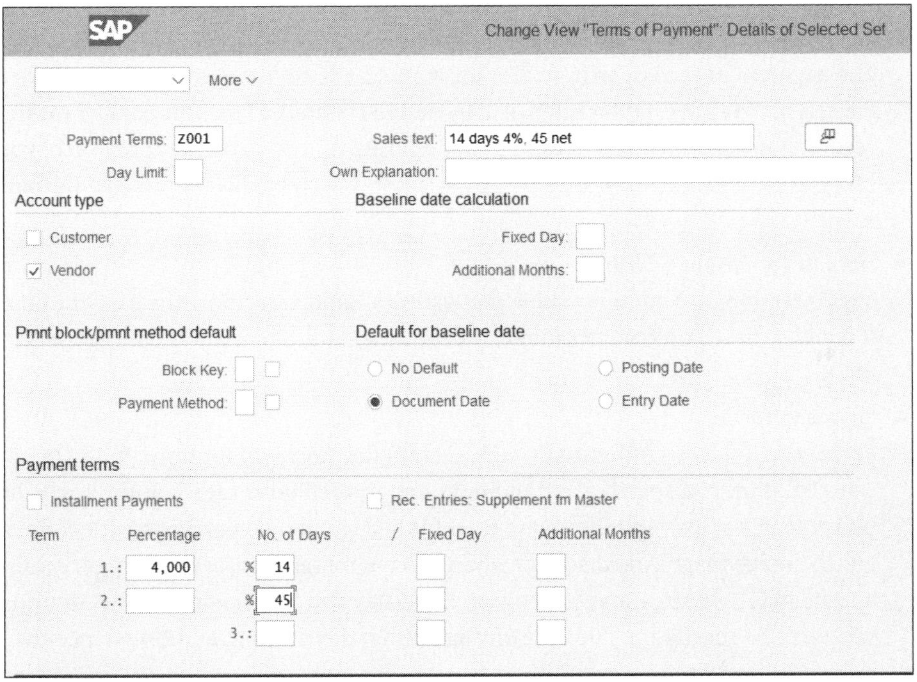

Figure 5.33 New Payment Term Creation

As shown in Figure 5.33, following fields need to be configured:

- **Payment Terms**
  Specify the new payment term code—in this case, Z001.
- **Sales Text**
  Enter a meaningful description of the payment terms.
- **Account Type**
  Determines for which types of accounts the payment term should be valid: customers, vendors, or both. The payment term is an object used both in accounts payable and accounts receivable, but here we can restrict its use if not needed in the other application.

- **Baseline Date Calculation**
  The baseline date is the date used to start calculation of the payment term from. Normally it's determined from the open item itself. However, here you can specify a different determination to start from a fixed date and/or to add additional months to the baseline date.

- **Pmnt Block/Pmnt Method Default**
  The payment of each open item also depends on the payment method, which we'll discuss in detail in Chapter 8. It's usually derived from the business partner master record, but here you can specify a default based on the payment term. Also, you can specify a default payment block indicator, which blocks the invoice for payment.

- **Default for Baseline Date**
  In this section, you configure how the system should determine the baseline date. It could be based on the document date, posting date, or the entry date of the document.

- **Payment Terms**
  In this section, you configure the actual rules for the payment term. In the **No. of Days** column, you specify the valid days and in the **Percentage** column the valid percentage. For example, here we set 4% for 14 days, and no percentage for 45 days, which means there is no discount when 14 days are reached, and the invoice must be paid in full within 45 days. Using the **Fixed Day** and **Additional Months** columns, you can also specify the due date on a particular day and add additional months.

After that, save your entries, and you're ready to start using the new payment term.

Now let's configure the payment blocks. By default, when invoices are posted they are open for payment. However, you may want to block for various reasons them until further review or a further event—for example, when there is a mismatch between the goods receipt and the invoice receipt or when there is reason to believe you might have received duplicate invoices.

Because there can be different reasons for payment blocks, we need to configure the payment block reason codes. To do so, follow menu path **Financial Accounting • Accounts Receivable and Accounts Payable • Business Transactions • Outgoing Payments • Outgoing Payments Global Settings • Payment Block Reasons • Define Payment Block Reasons** or enter Transaction OB27.

The list of payment block reasons is presented as shown in Figure 5.34.

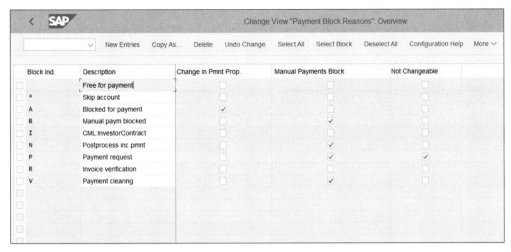

Figure 5.34 Payment Block Reasons

On this screen, you'll configure the following fields:

- **Block Ind.**
  This is the block code, which is entered in the document line item. It could be entered manually or automatically. For example, block A is commonly used manually in financial accounting invoices, whereas block R is the automatic block, set in case of mismatch between GR and IR.

- **Description**
  Here, enter a meaningful description of the blocking reason.

- **Change in Pmnt Prop.**
  A check here allows the block to be manually removed in the payment proposal. The payment proposal is the first step in running the automatic payment program, which will be discussed in detail in Chapter 8. Enabling this option gives you flexibility to make the payment during the payment program process, albeit manually.

- **Manual Payment Block**
  Checking this option prevents clearing items with manual payments.

- **Not Changeable**
  Setting this option prevents manual removal of the blocking reason, meaning it can be removed only as part of a workflow approval process.

In the next step, you'll define default block reasons per payment term by following menu path **Financial Accounting • Accounts Receivable and Accounts Payable • Business Transactions • Outgoing Payments • Outgoing Payments Global Settings • Payment Block Reasons • Define Default Values for Payment Block**.

As shown in Figure 5.35, here you have a list of payment terms and you can assign default payment blocks. Then when posting line items with these payment terms, they will be blocked automatically with the assigned payment block. For example, if you want to block all payables immediately due until further manual review is done, here enter blocking reason "A" in payment term 0001 and save your entry.

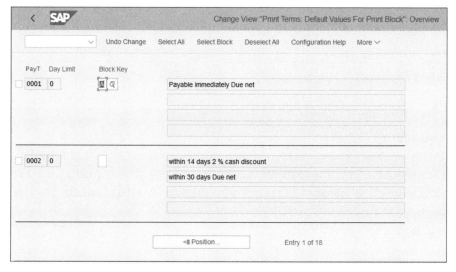

Figure 5.35 Default Payment Block per Payment Term

5.2.7 Integration with Vendor Invoice Management

Vendor invoice management is used to scan paper invoices and automatically post them in SAP S/4HANA. It uses optical character recognition (OCR) to transfer the written information in digital form in the system. Also, a workflow approval process is usually part of the functionality offered in the integration between SAP S/4HANA and such a system, in which invoices undergo an approval process through various chains until they're posted and ready for payment. Thus, VIM applications reduce manual work and costs and improve the control processes within accounts payable. Usually, when a VIM solution is implemented, most vendor invoices are posted through that system.

This is hardly a new idea, and many different applications exist on the market that have a good track record in integration with the SAP ERP system. Commonly used are VIM systems from vendors such as OpenText, ReadSoft, IXOS, and others. SAP Invoice Management by OpenText is the most integrated with SAP solutions.

The configuration of the SAP S/4HANA system with the selected vendor invoice management system lies outside the scope of this book, but when selecting such system and designing the process, keep in mind the following common integration points that should be applied:

1. The invoice is received through email, fax, or paper.
2. The invoice is scanned.
3. It's transferred to an archive server and archived.
4. The document is sent to the SAP S/4HANA VIM application, where it's stored and identified by a designated document processing number.
5. The VIM application extracts data such as vendor number, currency, document date, amount, and so on from the document.
6. The extracted data is validated and allocated to posting fields in SAP S/4HANA.
7. The approval process is performed through a workflow.
8. If there is no exception, the invoice is posted automatically through regular invoice posting transaction.

With that, we've finished our guide to accounts payable business transactions. Let's move on to the various reports available in the information system of accounts payable.

5.3 Information System

Accounts payable in SAP S/4HANA offers a robust and extensive information system, which should meet even the most demanding reporting requirements. We can group the reports into the following categories:

- Master data reports
- Balance reports
- Line item reports

These reports do not require specific configuration and should function out of the box. Let's start by examining the master data reports available for accounts payable.

5.3.1 Master Data Reports

As you already know, vendor master data in SAP S/4HANA is managed using the business partner concept, which is a major improvement compared with previous releases, as now the vendor master record is fully integrated within all modules.

The master data reports for vendors are available at application menu path **Accounting • Financial Accounting • Accounts Payable • Information System • Reports for Accounts Payable • Master Data**.

Report S_ALR_87012086 (Vendor List) provides a simple list of all vendor master data, whereas Report S_ALR_87012087 (Address List) is useful to provide you with snapshot of only the vendor addresses. Report S_ALR_87012089 (Display Changes to Vendors) can show you the creation of and changes to vendors, as shown in Figure 5.36.

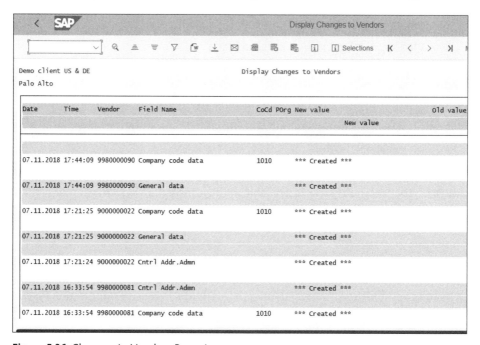

Figure 5.36 Changes to Vendors Report

This is useful for monitoring master data. Especially important are the changes to sensitive fields, such as bank information, which you can monitor using report S_ALR_87012090 (Display/Confirm Critical Vendor Changes). You can configure that certain fields require confirmation from an additional user when changed, and this report can show you what's pending, as shown in Figure 5.37.

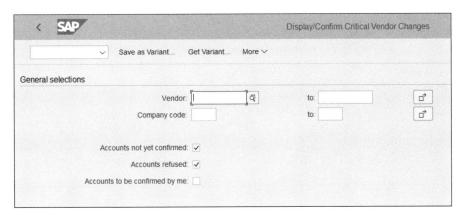

Figure 5.37 Display Vendors to Confirm Report

In addition to the range of vendor accounts and company codes, you can select the following checkboxes for this report:

- **Accounts Not Yet Confirmed**
  By selecting this checkbox, the report will show you vendor accounts that are pending confirmation.
- **Accounts Refused**
  By selecting this checkbox, the report will show you vendor accounts for which confirmations were rejected.
- **Accounts to Be Confirmed by Me**
  By selecting this checkbox, the report will show you vendor accounts that your user has authorization to confirm.

5.3.2 Balance Reports

Balance reports are important because they provide you with summarized information about your payables. They are available at application menu path **Accounting • Financial Accounting • Accounts Payable • Information System • Reports for Accounts Payable • Vendor Balances**.

Report S_ALR_87012079 (Transaction Figures: Account Balance) is quite user-friendly. You can execute it as a drilldown report to navigate through the different characteristics that interest you. For that, you need to select the **Classic Drilldown Report** option, as shown in Figure 5.38.

5 Accounts Payable

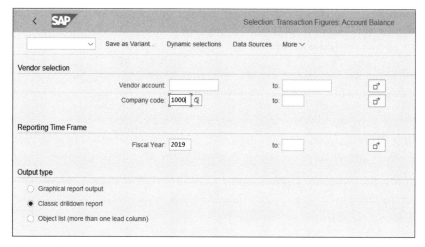

Figure 5.38 Vendors Account Balances Selection Screen

Then the report will show the various characteristics you can drilldown into in the header section, such as vendor, fiscal year, company code, and period. Without selecting any of them, as shown in Figure 5.39, the result section contains the values for all characteristics.

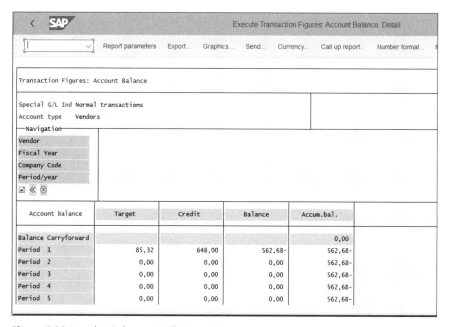

Figure 5.39 Vendor Balances Drill-Down Screen

When you click some of the characteristics, you can display the result only for the selected values. This way you can make a combination of their values, such as interactively displaying the report for period 01 for one specific vendor.

5.3.3 Line Item Reports

Finally, line item reports can provide you with the most detailed analysis on the document level. They are located at application menu path **Accounting • Financial Accounting • Accounts Payable • Information System • Reports for Accounts Payable • Vendors: Items**.

Report S_ALR_87012078 (Due Date Analysis for Open Items) also can run as a drilldown report, which breaks down your payables by due date, as shown in Figure 5.40.

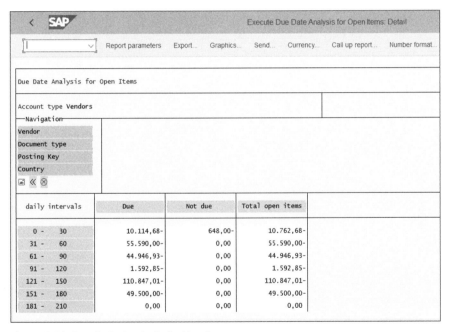

Figure 5.40 Due Date Analysis for Vendors

As you can see, the report breaks down the payables by periods, such as 0–30 days, 31–60 days, 61–90 days, and so on. Again, by selecting characteristics such as vendor, country, and the like, you can see the specific values for which the report is run based on your selections.

5   Accounts Payable

Another very useful report is report S_ALR_87012103 (List of Vendor Line Items), which gives you information about all accounts payable line items.

As shown in Figure 5.41, you can select the following options:

- **Open items**
  Shows only the open items as of the key date you enter here.

- **Cleared items**
  Shows only the cleared items (paid or otherwise cleared, e.g., manually) from within the date range specified.

- **All items**
  Shows all items that are open and cleared within the date range specified.

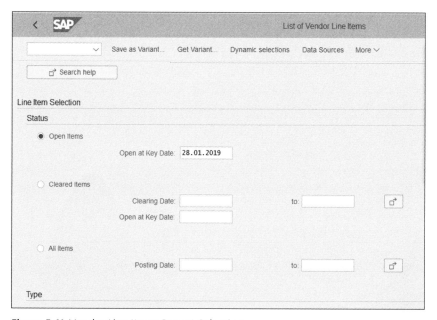

Figure 5.41 Vendor Line Items Report Selections

The report then shows the output list, as shown in Figure 5.42.

Here, as with other standard SAP reports, you have various options to display, summarize, filter, or otherwise change the output. For example, using the **Change layout** button from the top menu, you can select the displayed fields in the report from all the available fields, as shown in Figure 5.43.

5.3 Information System

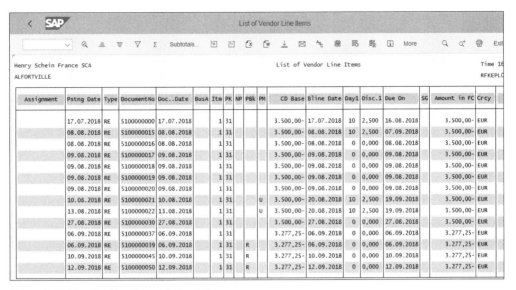

Figure 5.42 Vendor Line Items Output Screen

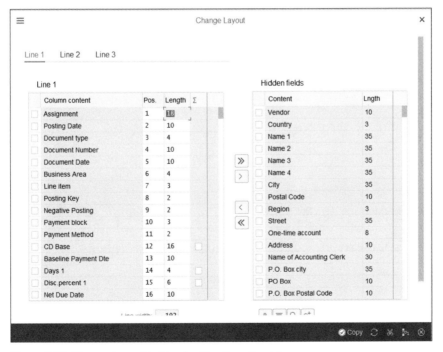

Figure 5.43 Change Layout in Vendor Line Item Report

You can show hidden fields using the **Show Sel. Fields** button ⟨ , and you can hide displayed fields using the **Hide Sel. Fields** button ⟩ . Then you can confirm the selections with the **Copy** button.

Another useful line item report, which is new for SAP S/4HANA, is the line item browser, located at application menu path **Accounting • Financial Accounting • Accounts Payable • Account • FBL1H—Line Item Browser**. As in earlier reports, you select the company code(s), vendor(s), and whether you want to see open items, cleared items, or both, as shown in Figure 5.44.

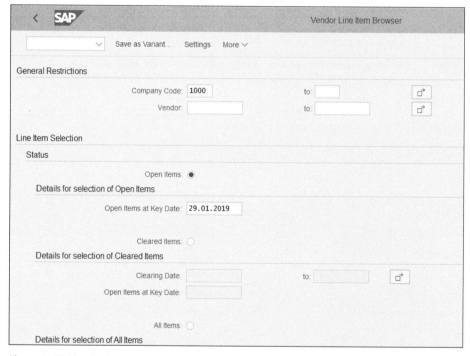

Figure 5.44 Vendor Line Item Browser

After executing the report, you'll see the output, as shown in Figure 5.45.

5.3 Information System

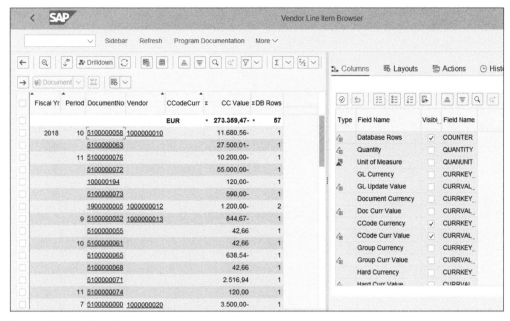

Figure 5.45 Vendor Line Item Browser Output

This is a very user-friendly report. You can change the layout of displayed columns, as shown previously, with the **Change layout** button from the top menu. You can also drill down into further levels of detail. For example, if you double-click the document number, a new window will open with this document displayed. Similarly, if you double-click a vendor account, you'll be taken to the vendor master record in a new window.

Finally, let's look at the classic report FBL1N (Display/Change Line Items), which was widely used in SAP ERP and is also available and useful in SAP S/4HANA. Follow application menu path **Accounting • Financial Accounting • Accounts Payable • Account • FBL1N—Display/Change Line Items** to access it. Again, you select company code(s), vendor(s), and whether you want to see open items, cleared items, or both, as shown in Figure 5.46.

193

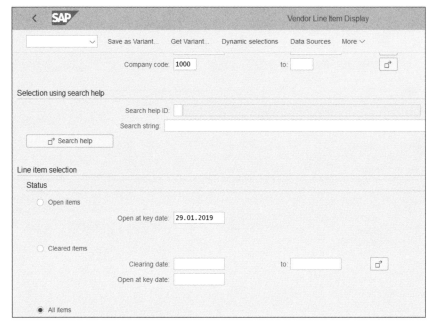

Figure 5.46 Vendor Line Item Display

Then after executing, you'll see the screen shown in Figure 5.47 (familiar if you're coming from an SAP ERP background).

St	Assignment	DocumentNo	Type	Doc..Date	S	DD	Amount in Local Crcy	LCurr	Clrng doc.	Text
		5100000039	RE	06.09.2018			3.277,25-	EUR		
		5100000045	RE	10.09.2018			3.277,25-	EUR		
		5100000050	RE	12.09.2018			3.277,25-	EUR		
*							48.109,00-	EUR		
		2000000003	ZP	23.08.2018			7.000,00	EUR	2000000004	
		2000000004	ZP	23.08.2018			7.000,00-	EUR	2000000004	
		2000000005	ZP	23.08.2018			4.200,00	EUR	2000000005	
		5100000029	RE	23.08.2018			4.200,00-	EUR	2000000005	
		1900000002	KR	23.08.2018			10.000,00-	EUR	2000000006	

Figure 5.47 Vendor Line Item Display Open and Cleared Items

Open items are marked with the symbol in red in the **Cleared/Open Items Symbol** column, whereas cleared items are marked with a green square. In the **Net Due**

Date Symbol column, items that are overdue for payment are marked with the ⚠ symbol in red. Double-clicking a document allows you to drill down to the line item in the respective document, where you can further analyze the details.

That finishes our guide to the information system in accounts payable.

5.4 Summary

In this chapter, we provided an extensive guide on configuring accounts payable in SAP S/4HANA and using its information system. We covered the following main topics:

- Vendor master data, maintained as business partners
- Business transactions
- Information system

One of the key benefits of SAP S/4HANA in the finance area is the full integration of vendors and customers with the business partners concept. We explained how to set up your vendors as business partners and how to fully integrate their financial views and postings.

We covered the main business transactions, posted through accounts payable. We made settings for the processing of invoices both in financial accounting and integrated with MM. With that, we also covered very important integration aspects such as the tax code determination in the purchasing process, the GR/IR clearing, and the integration with a vendor invoice management system. We also covered in detail the configuration required for payment terms and payment blocks. We'll further expand on the payments for outgoing invoices in Chapter 8, in which we'll examine how to set up payment methods and the automatic payment program and create payment file formats.

Finally, we examined in detail the information system of accounts payable, which provides invaluable reports for managing the liabilities of your organization. We looked at how to execute and navigate through the main reports for vendor master data, balances, and line items.

That ends our guide to accounts payable. Next, we'll move to accounts receivable in SAP S/4HANA.

Chapter 6
Accounts Receivable

This chapter gives step-by-step instructions for configuring accounts receivable in SAP S/4HANA and explains the integration with sales and distribution processes. It teaches how to set up business partners as customers, explains the sales flow in SAP S/4HANA, and introduces the reporting capabilities of the receivables subledger.

Now we'll explain in detail how to configure accounts receivable in your SAP S/4HANA system. Accounts receivable is used to manage financial processes related to sales, including posting customer invoices and incoming payments and clearing and reconciling open items. We'll teach you how to set up the required master data, business transactions, and information system for accounts receivable.

Accounts receivable is a mirror of accounts payable in a way. Even in the configuration menu, they share the following common path: **Financial Accounting • Accounts Receivable and Accounts Payable**.

6.1 Business Partner

Accounts receivable uses the business partner as a central object to manage the customer master data, very much like accounts payable. There are separate business partner roles to maintain the business partner master as a customer in financial accounting and as customer in sales. This enables full integration of the customer with its other potential roles, like vendor, credit management, and so on.

The customer is maintained on a general level, which is valid for all company codes and sales areas and provides general data such as name and address, and on a company code level, which provides some of the more specific settings such as the reconciliation account in the general ledger.

Let's start by examining the setup for general data.

6.1.1 General Data

As with vendors, to create, change, or display a customer, enter Transaction BP. As shown in Figure 6.1, select the **Customer Number** option in the **By** field.

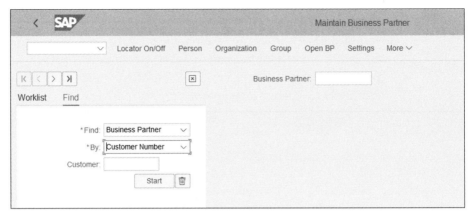

Figure 6.1 Business Partner Search by Customer Number

Then after you click the **Start** button, you'll see a list of business partners created in the customer role. Double-click any customer number to examine its master record.

Figure 6.2 shows the general data of the selected customer in its address tab. You're already familiar with the tab structure of business partner master record from previous chapters. In the **Change in BP Role** field, you can see the role you're maintaining—in this case, general data. Then tabs below group fields with similar functions, such as address fields, control fields, payment transactions, and so on. All these fields are valid for all company codes for that business partner.

You can update fields by selecting **Switch Between Display and Change** in the top menu.

Now select the **Debtor (FI Customer)** role in the **Change in BP role** field, and you'll see more tabs added, which are relevant for the financial accounting customer role.

As shown in Figure 6.3, now you can maintain customer-specific fields such as the tax data, which determines the proper tax treatment when posting integrated transactions for this customer.

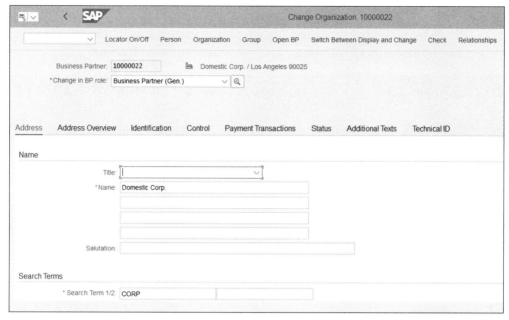

Figure 6.2 Customer Address Data

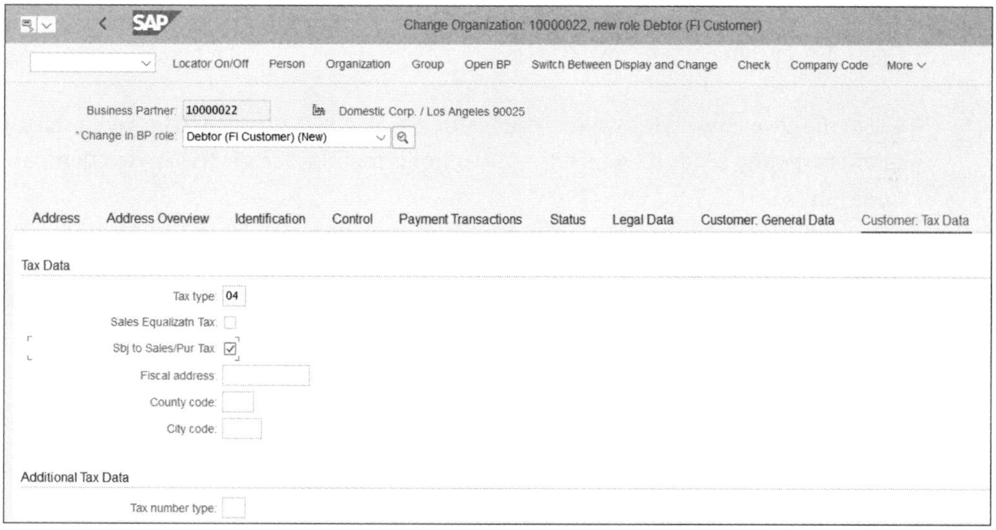

Figure 6.3 Debtor (Financial Accounting Customer) Role General Data

6.1.2 Company Code Data

To view the company code-level data for the customer, select **Company Code** from the top menu, which will bring you to the screen shown in Figure 6.4.

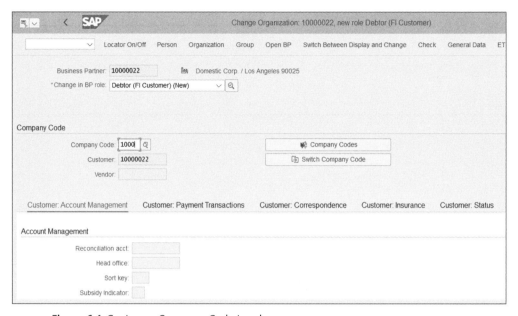

Figure 6.4 Customer Company Code Level

Select the company code as shown on Figure 6.4. You can also switch between the different company codes for which the customer is maintained via the **Switch Company Code** button.

In the **Customer: Account Management** tab, you need to enter a **Reconciliation Acct**, as shown in Figure 6.5.

The reconciliation account is the integral point for the general ledger. Every posting to the customer account also posts to this account in the general ledger.

Click the **Customer: Correspondence** tab and you'll see the screen shown in Figure 6.6.

6.1 Business Partner

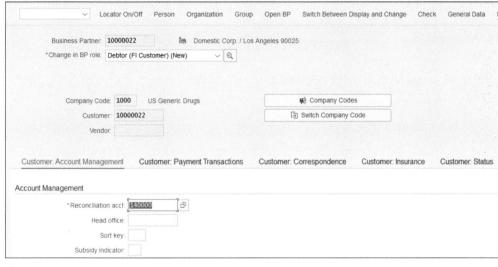

Figure 6.5 Customer Account Management

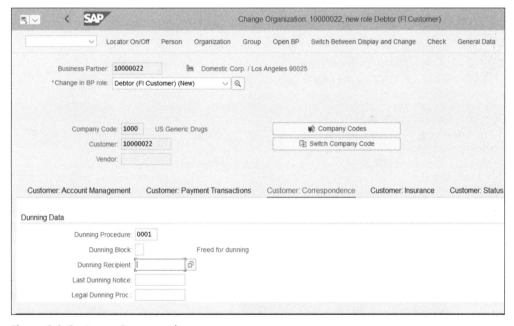

Figure 6.6 Customer Correspondence

In this tab, you maintain important parameters about correspondence with the customer, such as dunning data. *Dunning* is a procedure in SAP for sending reminder letters to customers for overdue items. However, some companies decide to use a more interactive approach to managing their collections from customers by using SAP Collections and Dispute Management.

6.1.3 Sales Data

You also need to maintain the sales data for the customer, which is used by sales and distribution. Most customer invoices are processed through the sales and distribution integrated process, based on the parameters maintained in the sales data for the business partner.

Select the **Customer** role in the **Change in BP role** field, and you see more tabs added that are relevant for the sales processes, as shown in Figure 6.7.

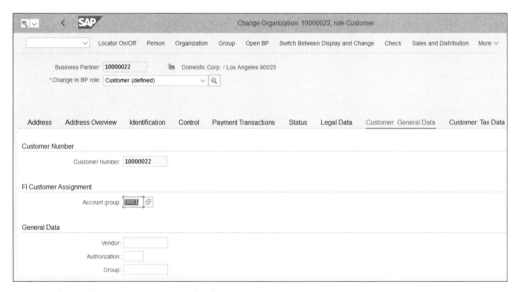

Figure 6.7 Customer General Sales Data

As you can see, the business partner and customer have the same account numbers. This isn't the only option; you can also have different number ranges. But best practice is for them to be the same, and this is controlled with the configuration of the number ranges assigned to the customer account groups and to the business partner account groups, which we'll examine in the next section.

In the **Account group** field, select the financial accounting account group, on which level you maintain the number range for the financial accounting customer accounts. This account group was also used in the old customer creation process in SAP ERP.

To define the sales area level-settings, click the **Sales and Distribution** option in the top menu of Figure 6.7. Similarly to how you can have different settings on the company code level in FI, the different settings on the sales area-level enable different processing for the same customer in various sales areas.

After selecting the sales organization, distribution channel, and division, you're taken to the sales area-level settings of the customer master, as shown in Figure 6.8.

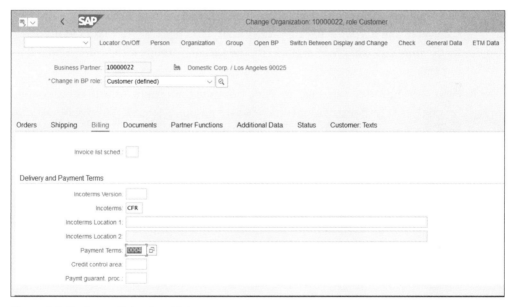

Figure 6.8 Customer Sales Area Data

Here you can enter various settings related to ordering, shipping, billing, and other sales functions. For example, in Figure 6.8 we've selected **Incoterms CFR** for costs and freight and payment terms 0004 for payment immediately due net.

After saving the business partner master record on the general, company code, and sales levels, you're ready to use it in both financial and the sales transactions.

Now let's examine the elements that need to be configured for the business partner as a customer.

6.1.4 Business Partner Configuration

We'll now configure the business partner so it can be used in accounts receivable. We'll configure account groups, number ranges, and the Customer-Vendor Integration (CVI), which we introduced in Chapter 5.

Account Groups

We need to set up the financial accounting customer groups that will be used when creating customers. Figure 6.7 showed how this group is mapped in the business partner master record. It determines the number range for the customer account and the screen layout of the fields in the master record.

To configure account groups, follow menu path **Financial Accounting • Accounts Receivable and Accounts Payable • Customer Accounts • Master Data • Preparations for Creating Customer Master Data • Define Account Groups with Screen Layout (Customers)**.

Figure 6.9 shows the existing customer groups in the system. As with other configuration objects, SAP provides several standard groups, which should meet most business requirements. You can create new groups, and it's good practice to copy them from similar standard groups and name them in a customer number range, starting with Z or Y.

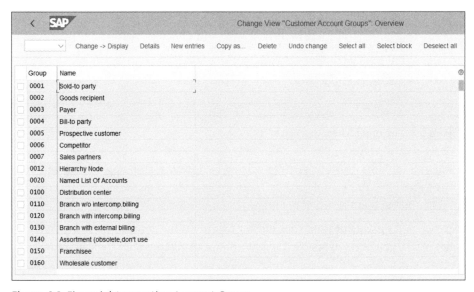

Figure 6.9 Financial Accounting Account Groups

Let's copy the standard group **0001: Sold-to party** to our own group. Select the checkbox on the left of it and select **Copy as...** from the top menu. Then name the new **Account group** "Z001" and give it a meaningful description in the **Meaning** field, as shown in Figure 6.10.

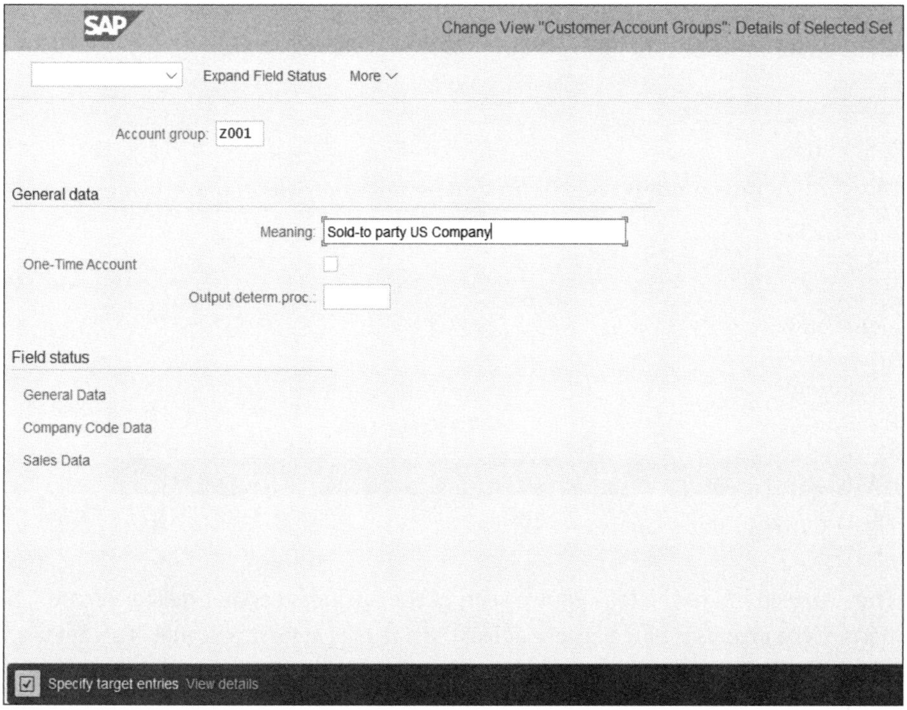

Figure 6.10 Copy Account Group

You can then check and change the related field status by clicking the options in the **Field status** section:

- **General Data**
- **Company Code Data**
- **Sales Data**

First, select **General Data**. You'll see the relevant field status groups for the general data, as shown in Figure 6.11.

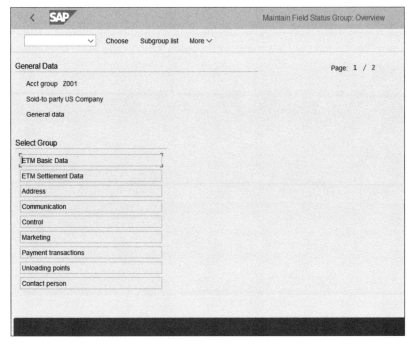

Figure 6.11 Field Status Groups

They correspond to the tabs you've seen in the business partner master record. The text for the groups is blue because at least one field is at least optional. If all fields are hidden within a group, the text of the group shows as black.

Double-click one of the groups—for example **Payment transactions**. This will bring you to the screen shown in Figure 6.12.

As shown in Figure 6.12, here you can set some of the fields as optional, some as required, and some as hidden. In our example, **Bank Details**, **Alternative Payer Account**, and **Alternative Payer in Document** are set as optional, which means that the fields are available for input in the master record but are not required; you can save the record while leaving them empty. **DME Details** is set as suppressed, which means that you would not see this field in the master record.

In this fashion, modify the fields based on your business requirements. Proceed through the company code data and sales data fields for all the customer groups you're going to use.

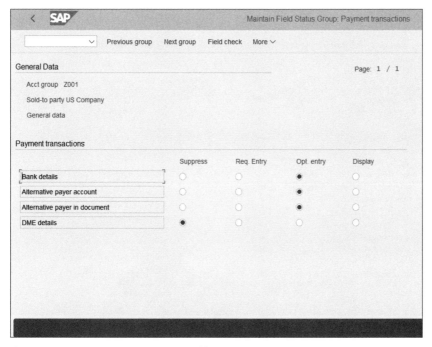

Figure 6.12 Payment Transactions Field Status

In general, it's good practice not to leave too many fields in the master record that aren't to be used, which could cause confusion. It's better to have a clear strategy for which fields are needed and which fields should be required and to hide fields that you know will never be used.

Number Ranges

The next step is to create and assign number ranges for the customer groups by following menu path **Financial Accounting • Accounts Receivable and Accounts Payable • Customer Accounts • Master Data • Preparations for Creating Customer Master Data • Assign Number Ranges to Customer Account Groups**.

In Figure 6.13, the number ranges are assigned to the customer groups. We assign range BP with the idea that this range will match the numbering in the number range assigned to business partner groups.

Then to define the number ranges, follow menu path **Financial Accounting • Accounts Receivable and Accounts Payable • Customer Accounts • Master Data • Preparations for Creating Customer Master Data • Create Number Ranges for Customer Accounts**.

6 Accounts Receivable

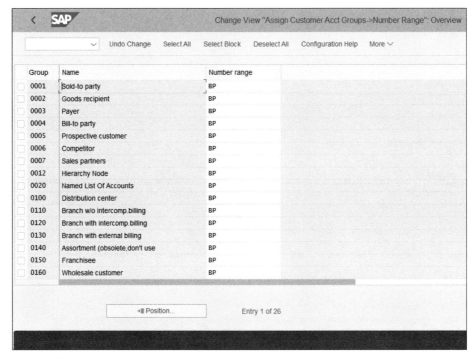

Figure 6.13 Customer Groups with Number Ranges

As shown in Figure 6.14, the **Number Range Object** for customers is DEBITOR. Click the ⌀ Intervals button to change the number ranges.

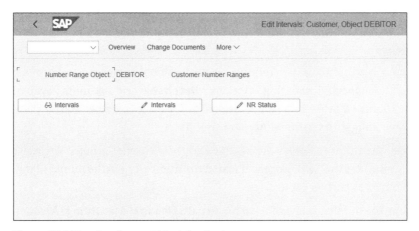

Figure 6.14 Number Range Object for Customers

As shown in Figure 6.15, you maintain the **From** and **To** numbers in the range assigned to customer groups. For range **BP**, which we assigned previously, the numbers start from 0000000001, and this is how the customer accounts will be numbered.

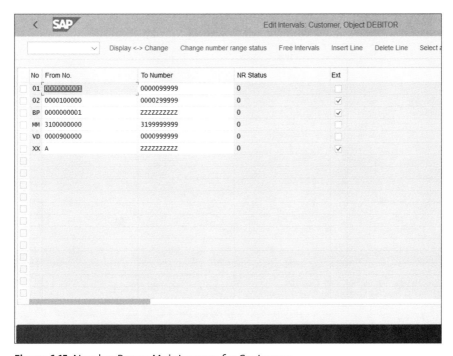

Figure 6.15 Number Range Maintenance for Customers

The next step is to define the business partner groups and their number ranges by following menu path **Cross-Application Components • SAP Business Partner • Business Partner • Basic Settings • Number Ranges and Groupings • Define Groupings and Assign Number Ranges**.

Figure 6.16 shows the business partner groups and their assigned number ranges. These number ranges determine the number assigned when creating a business partner. This number is a different object than the customer number, which is controlled by the customer group. But it's best practice to assign the same number ranges so that both the business partner and the customer number are the same.

To create the number ranges assigned to business partners groups, follow menu path **Cross-Application Components • SAP Business Partner • Business Partner • Basic Settings • Number Ranges and Groupings • Define Number Ranges**. As shown in Figure 6.17, maintain the number ranges, which are assigned in the business partner groups.

209

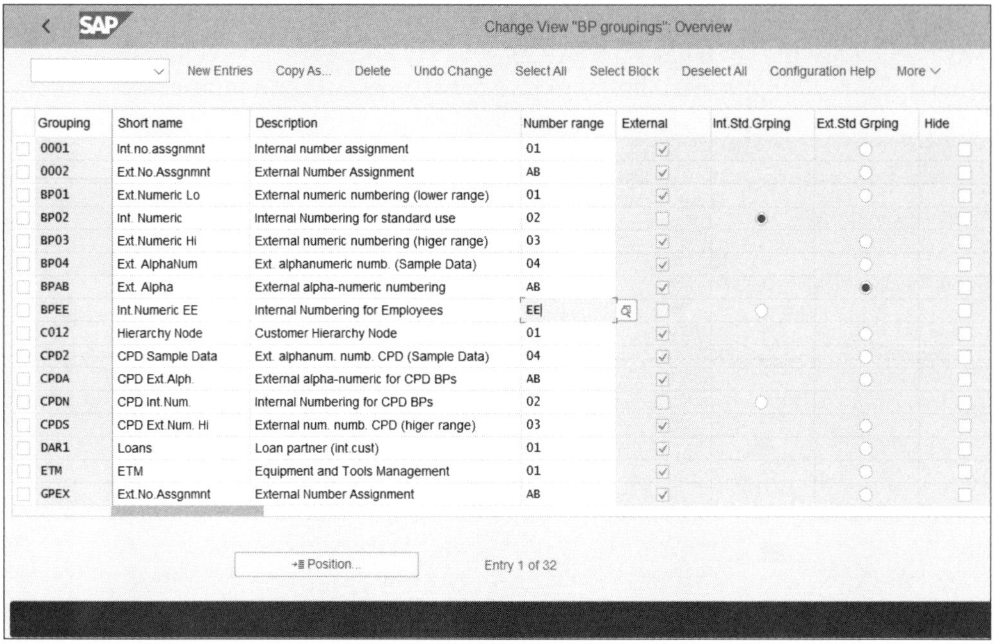

6 Accounts Receivable

Grouping	Short name	Description	Number range	External	Int.Std.Grping	Ext.Std Grping	Hide
0001	Int.no.assgnmnt	Internal number assignment	01	✓	○	○	
0002	Ext.No.Assgnmnt	External Number Assignment	AB	✓	○	○	
BP01	Ext.Numeric Lo	External numeric numbering (lower range)	01	✓	○	○	
BP02	Int. Numeric	Internal Numbering for standard use	02		●	○	
BP03	Ext.Numeric Hi	External numeric numbering (higer range)	03	✓	○	○	
BP04	Ext. AlphaNum	Ext. alphanumeric numb. (Sample Data)	04	✓	○	○	
BPAB	Ext. Alpha	External alpha-numeric numbering	AB	✓	○	●	
BPEE	Int.Numeric EE	Internal Numbering for Employees	EE		○	○	
C012	Hierarchy Node	Customer Hierarchy Node	01	✓	○	○	
CPD2	CPD Sample Data	Ext. alphanum. numb. CPD (Sample Data)	04	✓	○	○	
CPDA	CPD Ext.Alph.	External alpha-numeric for CPD BPs	AB	✓	○	○	
CPDN	CPD Int.Num.	Internal Numbering for CPD BPs	02		○	○	
CPDS	CPD Ext.Num. Hi	External num. numb. CPD (higer range)	03	✓	○	○	
DAR1	Loans	Loan partner (int.cust)	01	✓	○	○	
ETM	ETM	Equipment and Tools Management	01	✓	○	○	
GPEX	Ext.No.Assgnmnt	External Number Assignment	AB	✓	○	○	

Entry 1 of 32

Figure 6.16 Business Partner Groups

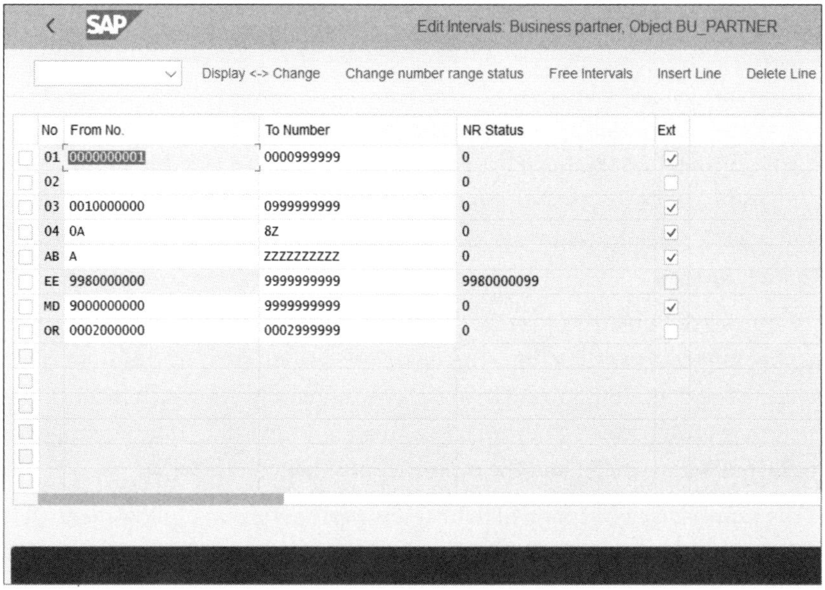

No	From No.	To Number	NR Status	Ext
01	0000000001	0000999999	0	✓
02			0	
03	0010000000	0999999999	0	✓
04	0A	8Z	0	✓
AB	A	ZZZZZZZZZZ	0	✓
EE	9980000000	9999999999	9980000099	
MD	9000000000	9999999999	0	✓
OR	0002000000	0002999999	0	

Figure 6.17 Business Partner Number Ranges

Customer Vendor Integration

Now you need to configure the customer-vendor integration (CVI). As with the vendor accounts, CVI provides the integration between the business partner and the customer.

We configured the synchronization control for business partners in Chapter 5 under menu path **Cross-Application Components • Master Data Synchronization • Synchronization Control • Synchronization Control • Activate PPO Requests for Platform Objects in the Dialog**.

Now let's configure the link between the business partner and customer by following menu path **Cross-Application Components • Master Data Synchronization • Customer/Vendor Integration • Business Partner Settings • Settings for Customer Integration • Set BP Role Category for Direction BP to Customer**.

Figure 6.18 shows a list of the business partner roles related to customers. Double-click **FLCU00**, which is the business partner role for finance customers.

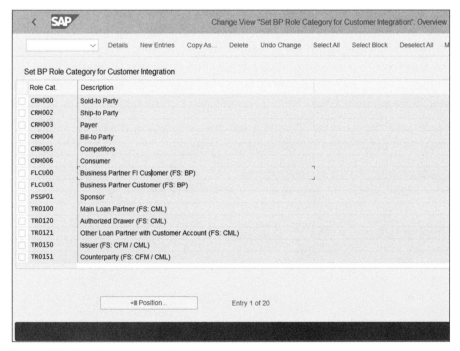

Figure 6.18 Link Business Partner to Customer

As shown in Figure 6.19, you need to check the **Customer-Based** checkbox. This indicates that for this business partner role, it's mandatory to have a customer. Then when you create or change a business partner in this role, the system automatically creates or changes a customer with the relevant data in financial accounting.

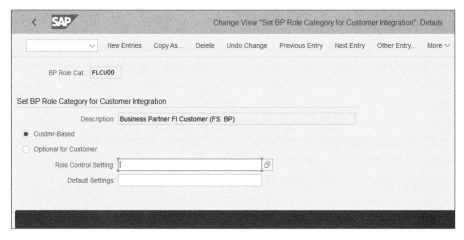

Figure 6.19 Business Partner to Customer Settings for Financial Accounting Customer

Then you have to configure the opposite integration, from customer to business partner, by following menu path **Cross-Application Components • Master Data Synchronization • Customer/Vendor Integration • Business Partner Settings • Settings for Customer Integration • Define BP Role for Direction Customer to BP**.

Select **New Entries** from the top menu and link the relevant customer groups with the business partner roles, as shown in Figure 6.20.

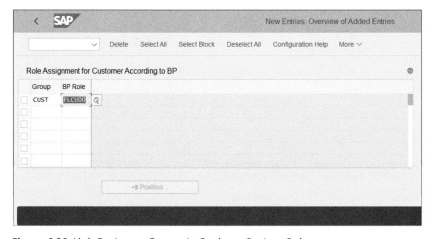

Figure 6.20 Link Customer Groups to Business Partner Roles

Then when you process a customer with the relevant account group, the system also will update the business partner in the business partner roles that are assigned to the customer groups.

6.2 Business Transactions

Accounts receivable is responsible for managing outgoing invoices and credit memos to customers, which are primarily processed and integrated through sales and distribution. It's also possible to process outgoing invoices only in financial accounting, but this option is used rarely.

In this section, you'll learn how to configure the incoming payments and payment terms related to customers.

6.2.1 Outgoing Invoices/Credit Memos in Financial Accounting

Most invoices to customers will be generated out of sales and distribution. They'll reference the sales order, will include materials, and will affect the inventory; we'll discuss those in detail in the next section. However, it's also possible to enter a customer invoice directly in financial accounting. As such, many companies also require posting the customer invoice in financial accounting to generate an actual invoice that can be sent to customers, either printed or sent electronically as a PDF or in some other format.

Similarly to posting vendor invoices, for customer invoices SAP S/4HANA offers Transaction FB70, which provides an easier user interface to post the customer invoice without knowing the specific document types or posting keys to use as all these elements are preconfigured. In standard SAP, this transaction will use the standard document type DR (customer invoice).

You still need to configure the tax codes that can be used when posting outgoing invoices and credit memos through Transaction FB70 because when you create a new tax code it won't be available automatically for this transaction.

To configure tax codes so they can be used in outgoing invoices and credit memos, follow menu path **Financial Accounting • Accounts Receivable and Accounts Payable • Business Transactions • Outgoing Invoices/Credit Memos • Outgoing Invoices/Credit Memos—Enjoy • Define Tax Code per Transaction**.

Enter the **Country Key** in the pop-window, as shown in Figure 6.21.

Figure 6.21 Select Country

Initially, the configuration screen is blank, as shown in Figure 6.22. Select **New Entries** from the top menu.

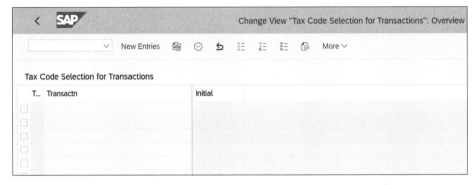

Figure 6.22 Initial Screen Define Tax Codes for Outgoing Invoices and Credit Memos

Next, as shown in Figure 6.23, enter the output sales tax codes, which you will use with customer invoices. In the **Transaction** column, if you select **Relevant for All Transactions**, then the code can be used for all types of transactions. If you check the **Initial** flag, this tax code will be the default for the transaction.

You can also customize the settings to generate printed invoices from within financial accounting customer invoices. You need to configure the correspondence for the customer invoice. SAP provides several predefined forms and print programs to generate various documents from financial accounting, such as customer invoices, balance confirmations, open item lists, and so on.

First you need to configure the correspondence type by following menu path **Financial Accounting • Accounts Receivable and Accounts Payable • Business Transactions • Outgoing Invoices/Credit Memos • Make and Check Settings for Correspondence • Define Correspondence Types**. Then you'll see a warning, as shown in Figure 6.24.

6.2 Business Transactions

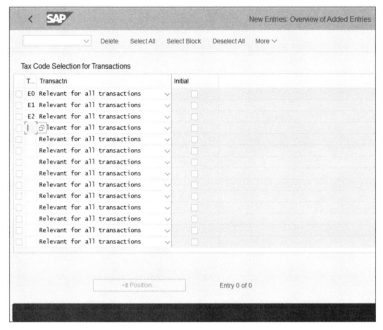

Figure 6.23 Define Tax Codes for Outgoing Invoices

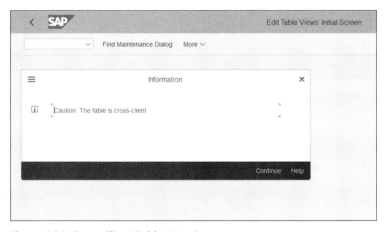

Figure 6.24 Cross-Client Table Warning

SAP issues this warning when you enter a customizing transaction that updates cross-client tables. Most configuration tables are valid only for the current client and need to be transported to other clients of the same system. However, some tables are cross-client, which means that changing them immediately changes all other clients of the same SAP instance. Therefore, SAP warns you to proceed with caution because these usually are the most sensitive configuration transactions.

Proceed by clicking the **Continue** button. Then you'll see a list of the defined correspondence types, as shown in Figure 6.25.

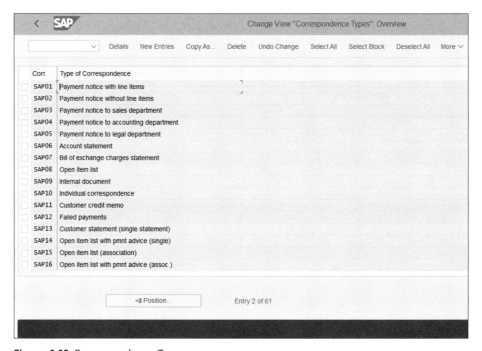

Figure 6.25 Correspondence Types

SAP delivers numerous standard corresponding types that start with *SAP*, followed by two digits. For a customer invoice, double-click **SAP19**: **Customer Invoice**, which will result in the screen shown in Figure 6.26.

The **Doc. Necessary** checkbox makes sure that the correspondence—in this case, a customer invoice—is always generated for the given document number for which it's requested.

You can also copy this standard correspondence type to your own, starting with Z or Y. This may be needed, for example, when you need to develop your own custom program for invoice printing. The programs assigned to correspondence types are assigned in a configuration transaction under menu path **Financial Accounting** • **Accounts Receivable and Accounts Payable** • **Business Transactions** • **Outgoing**

6.2 Business Transactions

Invoices/Credit Memos • **Make and Check Settings for Correspondence** • **Assign Programs for Correspondence Types**. Here select **SAP19**: **Customer Invoice** for the **Correspondence** field.

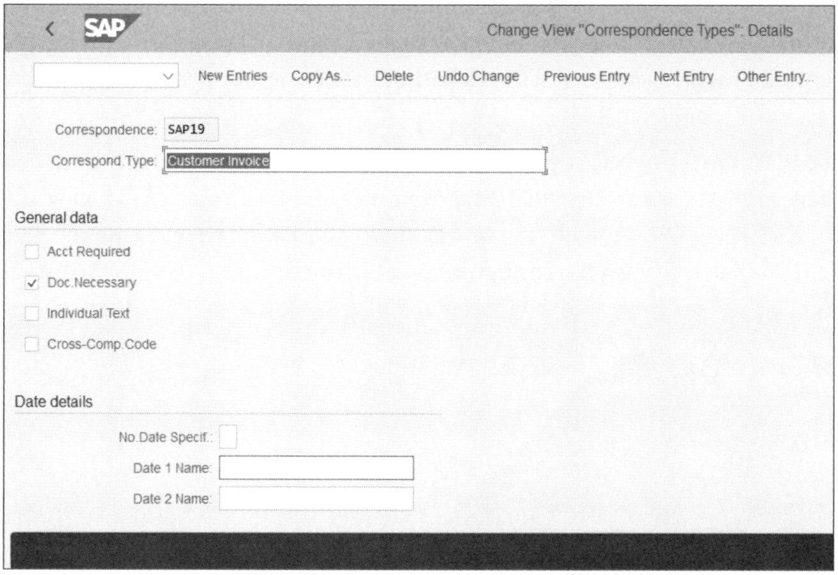

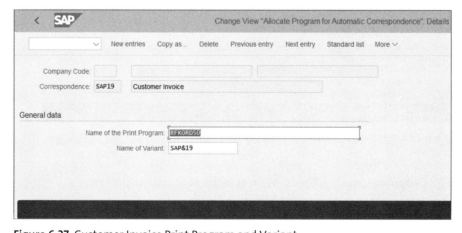

Figure 6.26 Correspondence Type Customer Invoice

As shown in Figure 6.27, standard print program **RFKORD50** and standard variant **SAP&19** are assigned to correspondence type **SAP19**. You can modify those with your customer program and variant.

Figure 6.27 Customer Invoice Print Program and Variant

You can also define your own forms for the invoices and credit memos generated in financial accounting, and this is usually required because the company logo at least should be adapted from the standard provided one.

To assign your form to the print program used, follow menu path **Financial Accounting • Accounts Receivable and Accounts Payable • Business Transactions • Outgoing Invoices/Credit Memos • Make and Check Settings for Correspondence • Define Form Names for Correspondence Print**.

As shown in Figure 6.28, you assign form names to the programs for the various correspondence types, such as RFKORD50 for customer invoices. You also can have different forms per company code if you maintain the company code in the first column. With that, we finish the configuration of invoice printing. Now let's discuss the outgoing invoices generated from sales and distribution.

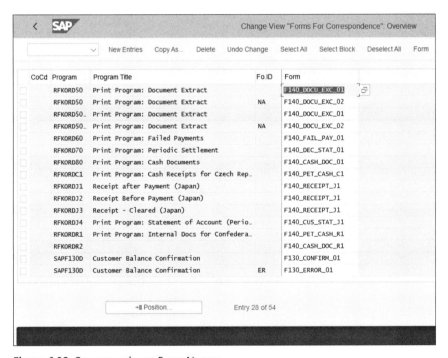

Figure 6.28 Correspondence Form Names

6.2.2 Outgoing Invoices/Credit Memos in Sales and Distribution

As already discussed, the vast majority of the outgoing customer invoices and credit memos will be posted through the integration with sales and distribution (SD). This is because the customer invoices normally should follow sales orders and should reference material(s) in the system. Financial documents for the customer invoices will be posted automatically in financial accounting, as well as the goods issue to customers based on the account determination, set via condition techniques, similar to what we discussed regarding tax code determination in Chapter 5.

Most of the configuration for the SD billing documents, which generate the customer invoice posting in financial accounting and the relevant postings to controlling objects, lies within the sales and distribution configuration. Most important is the pricing procedure, which determines the values posted. The account determination, which is determined based on the relevant condition records, also is part of the SD configuration in the system. However, typically it's a joint responsibility of sales and financial consultants because finance should say which accounts should be posted in which cases.

To assign revenue general ledger accounts for integral sales invoices, follow menu path **Sales and Distribution • Basic Functions • Account Assignment/Costing • Revenue Account Determination • Assign G/L Accounts**.

Figure 6.29 shows a list of tables, which are the condition tables that are used in the account determination process.

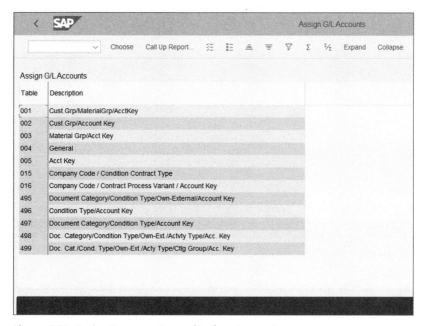

Figure 6.29 Assign Revenue General Ledger Accounts

6   Accounts Receivable

They are a combination of characteristics configured by sales consultants, such as customer group/material group/account key, or just customer group/account key, as shown by the text description of the table. The system first checks the most detailed condition table, and if it finds an assigned account based on all its elements it stops and determines that account. If not, it goes to the next table until it reaches the most generic condition table. If no account can be determined, the billing document in SD is still posted, but not released to accounting. This typically happens when master data such as customer or material data isn't properly maintained or if something is entered incorrectly in the sales order. When the issue is resolved, the document can be released to accounting.

6.2.3   Pricing Procedure in Sales

We mentioned earlier that the values of the customer invoices that are generated by sales and distribution are determined by the sales pricing procedure, a combination of condition types that represent various items such as price, discounts, and taxes. It comes from complex configuration in SD and is part of the sales order. From there, it flows into the billing document and determines the values in the financial document. Figure 6.30 shows the pricing procedure in a sales order.

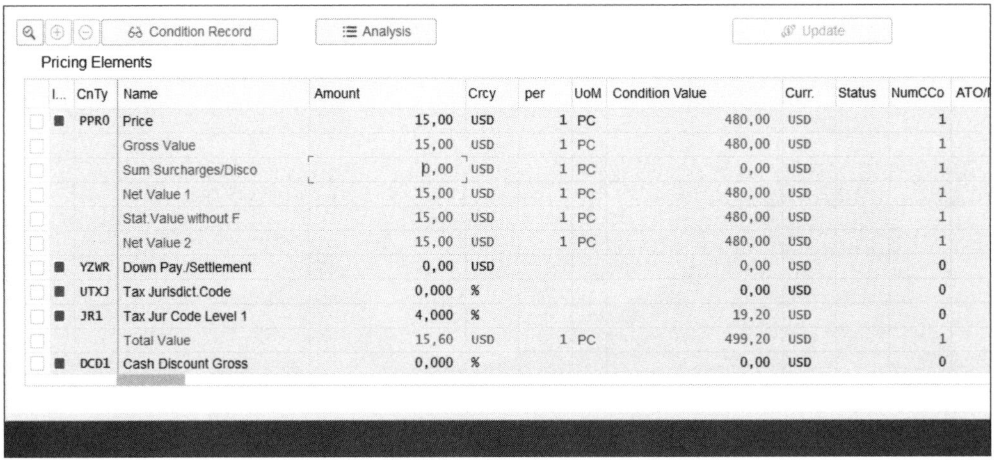

I...	CnTy	Name	Amount		Crcy	per	UoM	Condition Value		Curr.	Status	NumCCo	ATO/
■	PPR0	Price	15,00		USD	1	PC	480,00		USD		1	
		Gross Value	15,00		USD	1	PC	480,00		USD		1	
		Sum Surcharges/Disco	0,00		USD	1	PC	0,00		USD		1	
		Net Value 1	15,00		USD	1	PC	480,00		USD		1	
		Stat.Value without F	15,00		USD	1	PC	480,00		USD		1	
		Net Value 2	15,00		USD	1	PC	480,00		USD		1	
■	YZWR	Down Pay./Settlement	0,00		USD			0,00		USD		0	
■	UTXJ	Tax Jurisdict.Code	0,000		%			0,00		USD		0	
■	JR1	Tax Jur Code Level 1	4,000		%			19,20		USD		0	
		Total Value	15,60		USD	1	PC	499,20		USD		1	
■	DCD1	Cash Discount Gross	0,000		%			0,00		USD		0	

Figure 6.30 Pricing Procedure in Sales Order

The sales price is determined by condition **PR00**: **Price**. The tax is determined by condition **JR1**: **Tax Jur Code Level 1**, which determines the tax based on jurisdiction codes.

To see how these values were determined, click the **Analysis** button above the conditions. The system will show a tree-like structure of the various conditions, as shown in Figure 6.31. You can expand each condition and see how its value was determined. When the system finds a condition value based on the access sequences performed, the log on the right side shows **Condition Record Has Been Found**.

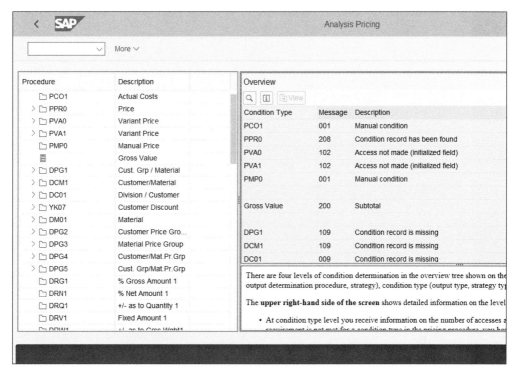

Figure 6.31 Analysis Pricing in Sales Order

6.2.4 Incoming Payments and Payment Terms

The next major types of transactions that relate to accounts receivable are incoming payments from customers. All outgoing invoices, posted from sales and distribution or directly in financial accounting, are posted as open items, which await incoming payments.

The configuration of the payment methods, the automatic payment program, and the electronic bank statement will be covered in detail in Chapter 8 for both incoming and outgoing payments.

Now we'll examine the configuration of the payment terms. The payment terms are used both in accounts receivable and accounts payable. It depends on the configuration of the payment term itself whether it can be used for customer accounts.

To configure payment terms, follow menu path **Financial Accounting • Accounts Receivable and Accounts Payable • Business Transactions • Outgoing Invoices/Credit Memos • Maintain Terms of Payment**.

Figure 6.32 shows the **Change View "Terms of Payment" Overview** screen, familiar from Chapter 5; in fact, the configuration transaction is the same. However, let's create payment term SEPA, which is specifically designed for customer payments, according to SEPA direct debit. SEPA direct debit is a payment system that allows companies to collect payments from customer accounts. This is a procedure in which customers agree that the money for their invoices will be automatically deducted from their bank accounts, for which they sign an agreement (SEPA mandate). We'll examine the SEPA direct debit configuration in detail in Chapter 8.

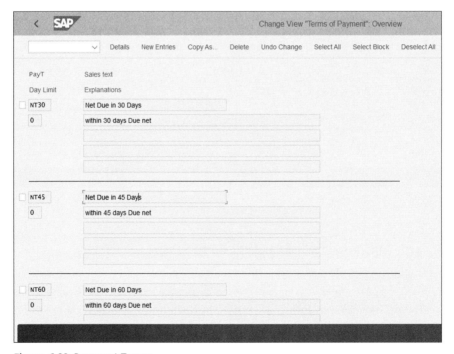

Figure 6.32 Payment Terms

Select **New Entries** from the top menu and configure payment term **SEPA** as shown in Figure 6.33.

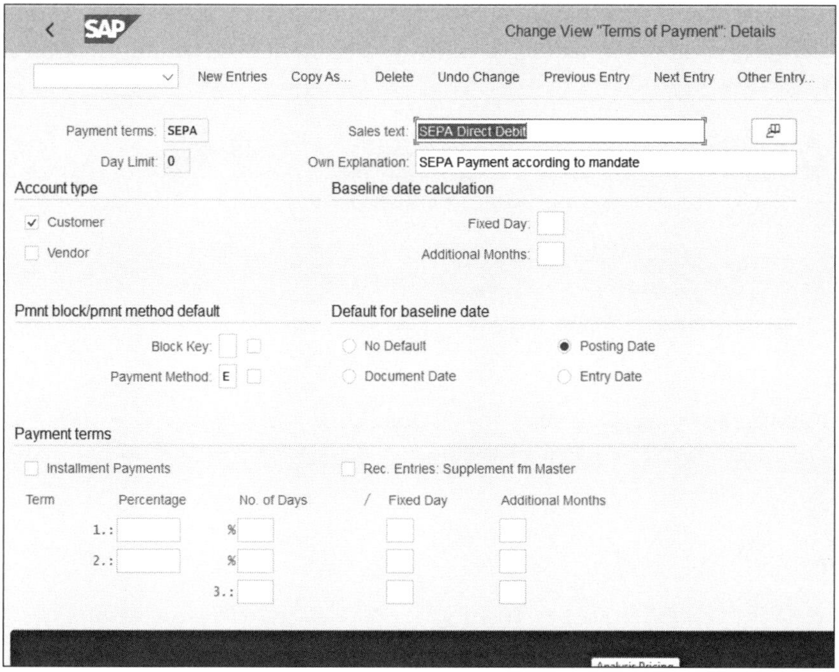

Figure 6.33 SEPA Direct Debit Payment Term

The **Customer** checkbox in the **Account Type** section makes sure this payment term can be used only for customers. A specific payment method is assigned, which is used for SEPA direct debit. The default for baseline date calculation is set to posting date.

6.3 Taxes

Now let's examine the taxes in the accounts receivable processes. Sales (output) tax codes are used, which usually are entered manually in financial accounting invoices and automatically determined in integrated sales and distribution billing documents.

6.3.1 Taxes in Financial Accounting Invoices

When you enter an outgoing invoice directly in financial accounting, normally you enter the tax code manually, unless there's some substitution in place that can determine it automatically.

When posting a customer invoice in financial accounting using Transaction FB70, in the header section there's a **Tax** tab, in which you can enter the tax information as shown in Figure 6.34.

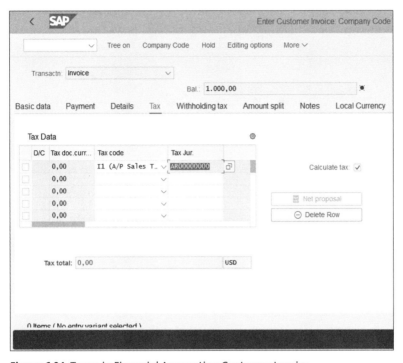

Figure 6.34 Taxes in Financial Accounting Customer Invoice

Select the sales output tax code and if necessary the tax jurisdiction code, which will use a tax rate based on a specific US jurisdiction. If you check the **Calculate Tax** checkbox, the taxes will be calculated automatically by the system.

There is no additional configuration to make this work other than properly setting up your tax codes, as explained in Chapter 3, and enabling the relevant tax codes in the outgoing financial accounting invoice Enjoy screen, as shown in Section 6.2.1.

6.3.2 Tax Determination in the Sales Process

The tax codes are automatically determined in the sales process using the same condition technique used by other elements of the pricing procedure we discussed. In fact, the tax is a separate condition type in the pricing procedure, which is on the sales order level.

Let's revisit Figure 6.30. There, select condition type **JR1: Tax Jur Code Level 1** and click the **Condition Record** button. As shown in Figure 6.35, the system determines tax codes O1 and jurisdiction code GA00000000 based on the condition record found. That determines a sales tax rate of 4%, as per the jurisdiction code setup.

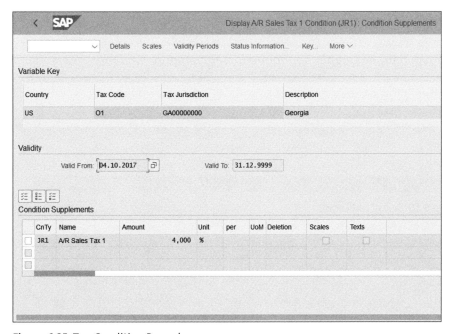

Figure 6.35 Tax Condition Record

The tax determination in sales is using a similar condition technique (condition, access sequences, condition tables) as purchasing to automatically determine the taxes. There are special pricing conditions for taxes, which are used to determine the proper tax code based on fields configured to be used in the condition tables.

To define a condition type, follow menu path **Sales and Distribution • Basic Functions • Pricing • Pricing Control • Define Condition Types**, then select the **Set Pricing Condition Types** activity. The system shows the sales condition defined, as shown in Figure 6.36. The standard SAP condition for output tax is MWST, which you can copy as your own condition starting with Z or Y if needed.

6 Accounts Receivable

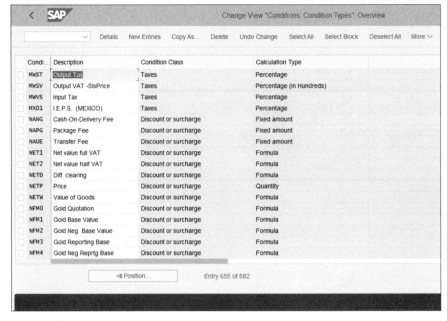

Figure 6.36 Sales Conditions

Double-click **MWST** to see its configuration, as shown in Figure 6.37.

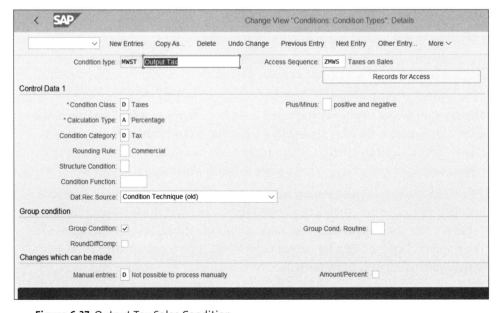

Figure 6.37 Output Tax Sales Condition

The following fields are important:

- **Condition Class**
  This determines the usage of the condition type. Condition class is set to **D Taxes**, which defines this condition as a tax condition.

- **Calculation Type**
  Specifies the type of calculation and is set to **A Percentage** because the output sales tax should be calculated as a percentage.

- **Condition Category**
  Classifies the conditions and for tax conditions should be **D Tax**.

- **Item Condition**
  Should be checked for taxes because it ensures the calculation happens at the item level, which is required for taxes.

Using the **Records for Access** button, you can see what records for access are maintained for the condition types, which determine the tax codes.

Here we've defined our own custom access sequence ZMWS, which is a copy of the standard MWST but accesses different tables. To set this, go to menu path **Sales and Distribution • Basic Functions • Pricing • Pricing Control • Define Access Sequences**, then select the **Maintain Access Sequences** activity.

This is a cross-client table, and after confirming the warning message, you'll see a list of defined access sequences, as shown in Figure 6.38.

Mark **MWST** and select **Copy As...** from the top menu. Then rename the new access sequence to "ZMWS" and change its **Description** to "Taxes on Sales". Now select **Accesses** from the left side of the screen and modify the condition tables to access, as shown in Figure 6.39.

Here, select four condition tables; the sequence of their access is determined by the step number in the first **Access** column. The **Exclusive** checkbox controls whether the system will stop searching for a record after the first successful access finds a value.

Select one of the steps and click **Fields** from the left side of the screen to see the fields that are being used by the condition table access.

6   Accounts Receivable

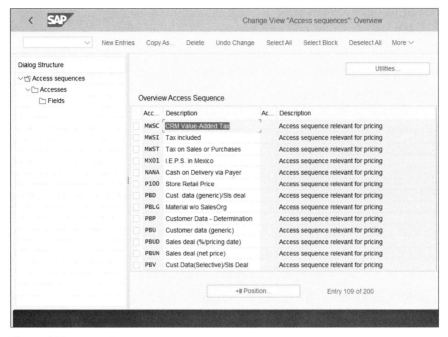

Figure 6.38 Access Sequences

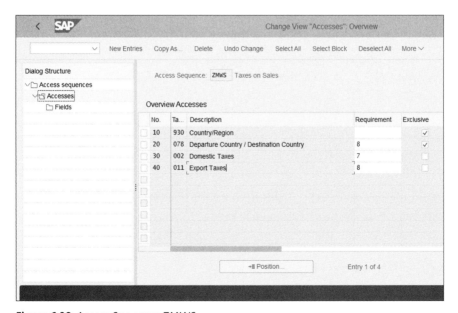

Figure 6.39 Access Sequence ZMWS

6.3 Taxes

As shown in Figure 6.40, the fields **Country (ALAND)** and **Destination Country (LAND1)** are used in step 20 of the access sequence. Therefore when searching using this step, the system will look for matches in both the sending and destination countries of the goods. If there's a match, the tax code maintained for the combination will be determined.

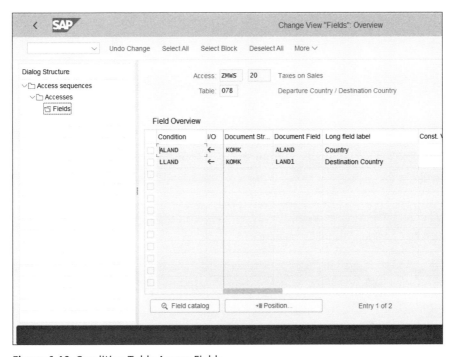

Figure 6.40 Condition Table Access Fields

You can maintain these condition records using Transaction VK11. They're considered application data, and you can also navigate to these settings through the application menu path **Logistics • Sales and Distribution • Master Data • Conditions • Select Using Condition Type • VK11—Create**. Now enter condition type "MWST" to see a popup window asking for the combination of characteristics for which you want to create records, as shown in Figure 6.41. These characteristics are dependent on the configuration of the condition record.

Select the **Departure Country/Destination Country** to see the next screen, in which you enter the condition records. As shown in Figure 6.42, in the header you enter the source country from which goods or services are sent, and then in the lines below you can have multiple records per destination country.

229

6 Accounts Receivable

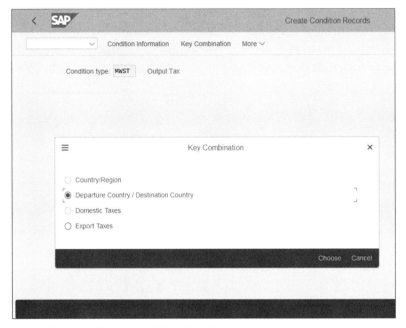

Figure 6.41 Condition Record Combinations

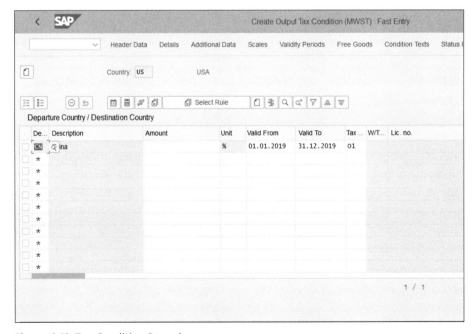

Figure 6.42 Tax Condition Record

The following fields are considered:

- **Destination Country**
  This is the country to which the goods/services are supplied.
- **Tax Classification 1**
  The tax classification indicator, which is maintained in the customer master records, normally specifies whether this business partner is liable for or exempt from taxes.
- **Tax Classification Material**
  The tax classification indicator, which is maintained in the material master records, normally specifies whether this material is taxable at the full rate, taxable at a reduced rate, or not taxable.
- **Valid From and Valid To**
  Here you specify the validity period of the tax code. Tax rates change from time to time, so this way you have the flexibility to add more condition records when the tax rates change.
- **Tax Code**
  This is the tax code, which should be assigned automatically when the criteria in the condition record are met.

Save your entries, and the sales documents that match these condition criteria will automatically receive the assigned tax codes.

6.4 Information System

Accounts receivable has a powerful information system, which provides several standard reports that enable organizations to manage their receivables proactively. We'll examine reports in the following categories:

- Master data reports
- Balance reports
- Line item reports

These reports do not require specific configuration to use.

6.4.1 Master Data Reports

The master data reports provide information that's maintained in the customer business partner master records in a well-structured form.

The master data reports for customers are available at application menu path **Accounting • Financial Accounting • Accounts Receivable • Information System • Reports for Accounts Receivable Accounting • Master Data**.

Report S_ALR_87012179 (Customer List) provides information from the master records, and you can specify on the selection screen which areas of the master record should be displayed.

As shown in Figure 6.43, select the checkboxes for account control, bank data, and payment data, and the relevant areas of the master records will be output.

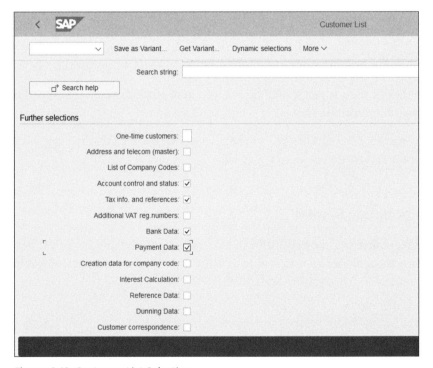

Figure 6.43 Customer List Selections

As you know, customers can be created at the financial accounting level and the sales level. You can use report S_ALR_87012195 (Customer Master Data Comparison) to check which customers have been created in financial accounting but not in sales, and vice versa.

As shown in Figure 6.44, you can check for missing records in financial accounting or in sales and distribution.

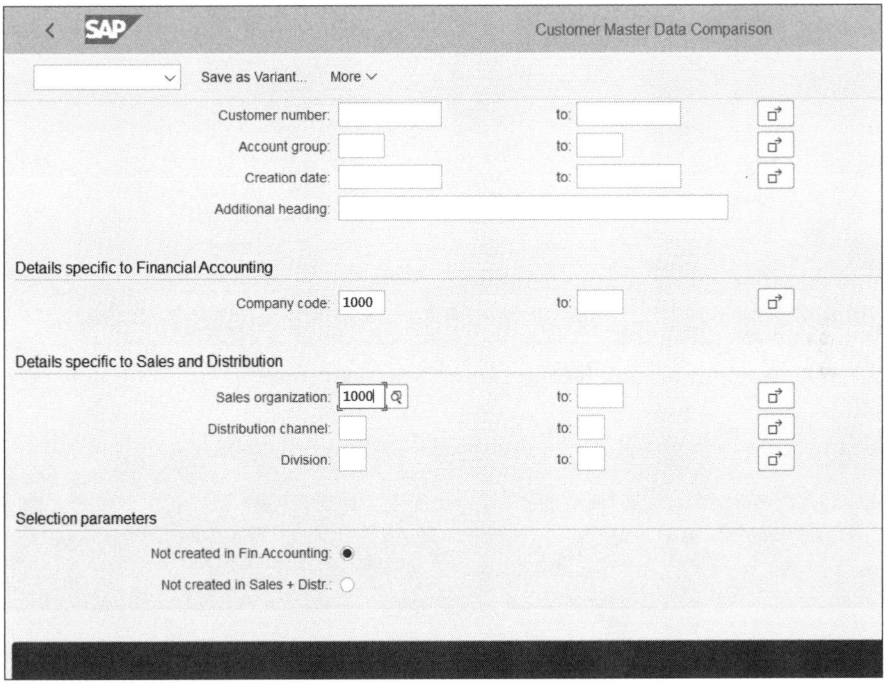

Figure 6.44 Customer Master Data Comparison

6.4.2 Balance Reports

Balance reports provide summarized information about the receivables from your customers. They are available at application menu path **Accounting • Financial Accounting • Accounts Receivable • Information System • Reports for Accounts Receivable Accounting • Customer Balances**.

Report S_ALR_87012169 (Transaction Figures: Account Balance) can be executed as a drilldown report, which is very convenient for online analysis. If you need to download it into Excel, select the **Object List (More than One Lead Column)** option, as shown in Figure 6.45.

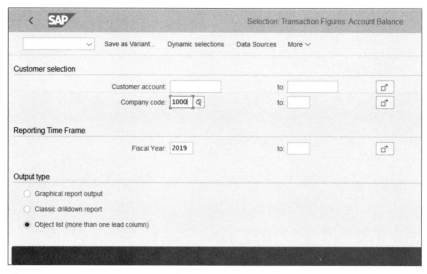

Figure 6.45 Customer Account Balances Report Selection Screen

Then the output screen will be column-based, as shown in Figure 6.46.

Year	Customer	Period/year	Total open items Debit	Total open items Credit	Total open items Balance	Total open items Accumulate...
2019	1000000	001.2019	1.876,00	0,00	1.876,00	1.876,00
2019	1000000	002.2019	0,00	0,00	0,00	1.876,00
2019	1000000	003.2019	0,00	0,00	0,00	1.876,00
2019	1000000	004.2019	0,00	0,00	0,00	1.876,00
2019	1000000	005.2019	0,00	0,00	0,00	1.876,00
2019	1000000	006.2019	0,00	0,00	0,00	1.876,00
2019	1000000	007.2019	0,00	0,00	0,00	1.876,00
2019	1000000	008.2019	0,00	0,00	0,00	1.876,00
2019	1000000	009.2019	0,00	0,00	0,00	1.876,00
2019	1000000	010.2019	0,00	0,00	0,00	1.876,00
2019	1000000	011.2019	0,00	0,00	0,00	1.876,00
2019	1000000	012.2019	0,00	0,00	0,00	1.876,00
2019	1000000	013.2019	0,00	0,00	0,00	1.876,00
2019	1000000	014.2019	0,00	0,00	0,00	1.876,00
2019	1000000	015.2019	0,00	0,00	0,00	1.876,00
2019	1000000	016.2019	0,00	0,00	0,00	1.876,00

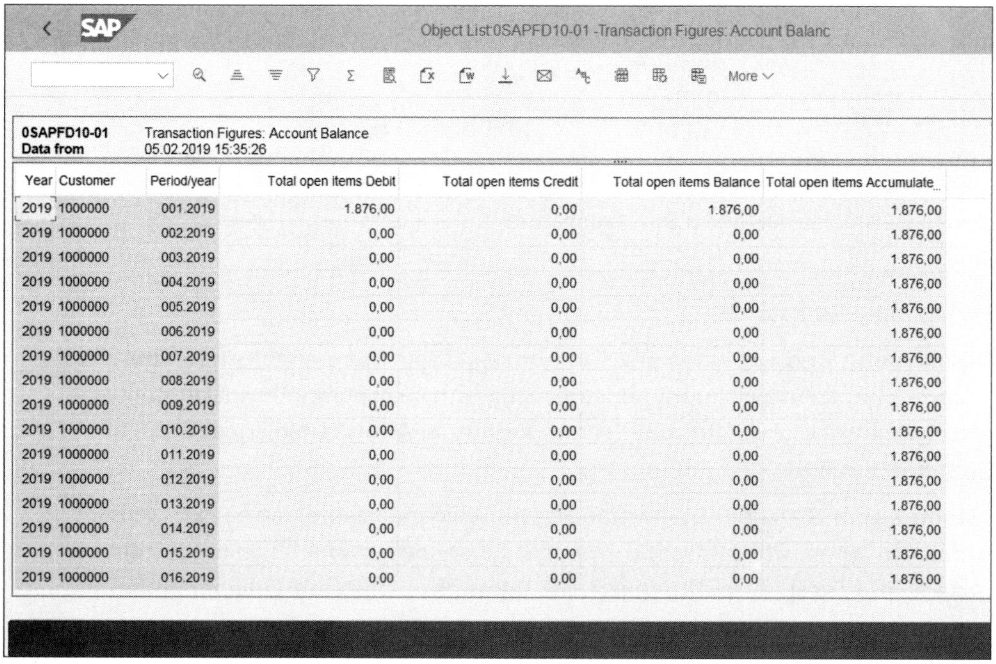

Figure 6.46 Customer Balance Report Output

You can easily export this report to Excel by clicking the ⌧ button on the top toolbar menu. The system will ask you to select a format, as shown in Figure 6.47.

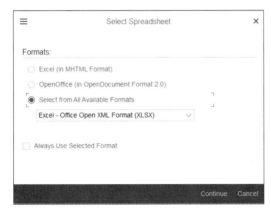

Figure 6.47 Select Spreadsheet Format

Select **Excel—Office Open XML Format (XLSX)**. The system will prompt you for a file name on your local drive to save the Excel file under, and you will see the Excel output, as shown in Figure 6.48.

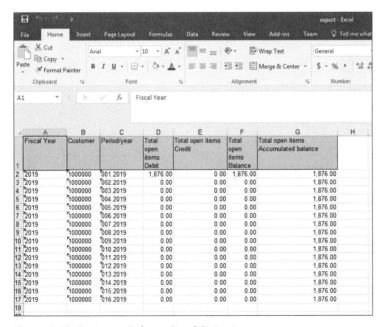

Figure 6.48 Customer Balance Excel Output

With the object list option, you can easily analyze the data further in Excel and send it to other colleagues in Excel format.

6.4.3 Line Item Reports

Accounts receivable line item reports provide very detailed information by which you can analyze every single invoice and payment. They are located at application menu path **Accounting • Financial Accounting • Accounts Receivable • Information System • Reports for Accounts Receivable Accounting • Customers: Items**.

One of the most useful reports is Report S_ALR_87012168 (Due Date Analysis for Open Items), which provides you with aging data for open items. Aging is a common accounts receivable analysis, in which open items are grouped by their maturity.

In this report, select the drilldown option, as shown in Figure 6.49; this is the most useful option for online analysis.

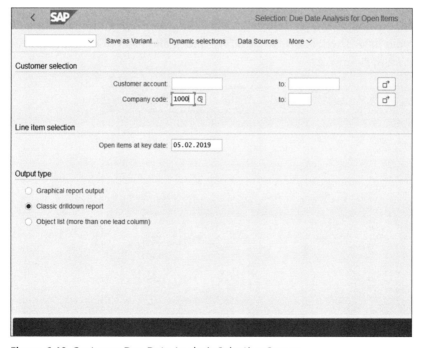

Figure 6.49 Customer Due Date Analysis Selection Screen

If you've run this report already, the system will ask you if you want to run it as a new selection or want to retrieve data from a previous run, as shown in Figure 6.50.

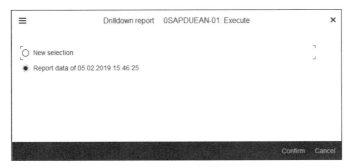

Figure 6.50 Report Selection Option

Select **New Selection** and proceed with the **Confirm** button. As shown in Figure 6.51, you can analyze the amount of receivables due in 0 to 30 days, in 31 to 60 days, and so on. This is very important in forecasting the cash liquidity and to properly manage receivables so that no big items are left overdue for a long period of time.

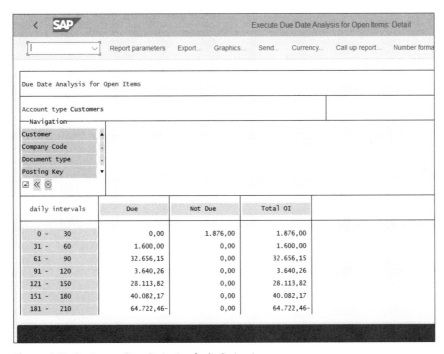

Figure 6.51 Customer Due Date Analysis Output

6.5 Summary

In this chapter, we taught you how to configure accounts receivable in SAP S/4HANA and how to use its information system. Now you should be well prepared in the following main areas:

- Customer master data, maintained as business partners
- Business transactions
- Information system usage

We explained how the customer master data is integrated with the business partner in SAP S/4HANA and what are the best practices in setting up customer groups and number ranges. Now you should be able to configure the Customer-Vendor Integration so that business partners and customer masters are fully integrated.

We covered the main business transactions in accounts receivable, such as outgoing invoices and credit memos, related correspondence, payments, and payment terms. We explained in detail how the account determination and tax determination work in integrated sales documents using the condition technique.

Finally, we discussed the most important reports that the accounts receivable information system offers in the areas of master data reports, balance reports, and line item reports. You've seen how easily you can analyze various aspects of accounts receivable data and how to export your reports in Excel format.

With that, we finish our guide to accounts receivable. Next, we'll continue our SAP S/4HANA journey with the configuration of fixed assets.

Chapter 7
Fixed Assets

This chapter will teach you how to configure fixed assets in SAP S/4HANA. The new asset accounting concept will be discussed in detail. Using the new asset accounting in SAP S/4HANA is required, and you'll learn how it will enable real-time valuations in multiple accounting frameworks.

Fixed assets is a part of financial accounting that records and manages all asset accounting data. It provides functionalities for the whole life cycle of fixed assets, from acquisition to retirement. Sometimes it's referred to as asset accounting, such as in the SAP Reference IMG (Transaction SPRO). These terms can be used interchangeably, but we'll mostly use fixed assets, as it's called in the SAP application menu.

In this chapter, you'll learn about all configuration areas necessary to have fully functional fixed assets. We'll start with the organizational structure section, in which the main foundation blocks of the fixed assets, such as chart of depreciation and depreciation areas, are configured. Then we'll look at how to set up the master data for assets and classify them into well-structured asset classes. Next we'll review the valuation for fixed assets based on various accounting principles, and we'll emphasize how SAP S/4HANA brings tremendous improvements to this process. Then we'll discuss the various business transactions for fixed assets. Finally, we'll conclude with a look at the information system for fixed assets.

Fixed assets empowers companies to achieve flexible reporting based on multiple accounting frameworks. Such reporting is very important for both internal and external purposes for most companies that use SAP, most of which are operating globally and have to comply with various accounting principles. For example, many modern companies need to present their fixed assets based on one or more of the following frameworks:

- International Financial Reporting Standards (IFRS)
- United States Generally Accepted Accounting Principles (US GAAP)

- Local Generally Accepted Accounting Principles (local GAAP), per country implemented
- Local tax rules, per country implemented

The ability to present fixed assets based on various accounting and tax frameworks in SAP is hardly new. Multiple depreciation areas have been used for this requirement since the old R/3 days. With the introduction of the new general ledger in SAP, different subledgers could be updated with accounting principle–specific postings, albeit not in real time. Then with SAP Simple Finance new asset accounting was introduced, which is now required and enhanced in SAP S/4HANA. It's based on the Universal Journal, which means it's completely integrated in real time with the general ledger. There are no redundant data structures to be filled. There are no periodic programs required to post to nonleading ledgers because now the integration works in real time in SAP S/4HANA. Table ACDOCA, the Universal Journal table, has an include called ACDOC_SI_FAA (Universal Journal Entry: Fields for Asset Accounting), which has all the fixed assets fields, such as main asset number (ANLN1), depreciation area (AFABE), and asset value date (BZDAT).

Let's now go over how to configure new asset accounting in SAP S/4HANA in detail. As we do so, we'll emphasize the key differences between implementing a new SAP S/4HANA system (greenfield approach) and migrating an existing SAP system to SAP S/4HANA (brownfield approach).

We'll start with setting up the organization structures for fixed assets.

7.1 Organizational Structures

Organizational structures are the key foundation blocks that need to be set up to represent the company organization from a fixed assets point of view. The top level from a valuation point of view is the chart of depreciation, under which are created one or more depreciation areas. The master data of the fixed assets is stored in asset master records, which are classified based on asset classes. We'll examine each of these objects in detail, starting with the chart of depreciation.

7.1.1 Chart of Depreciation

The chart of depreciation is the highest organizational structure in fixed assets from a valuation point of view. It's used to group the valuation requirements and usually is

7.1 Organizational Structures

country-dependent. Each country typically has its own chart of depreciation, although from a technical point of view multiple countries can be assigned to the same chart of depreciation. This only makes sense if these countries have exactly the same valuation requirements.

The chart of depreciation is independent from other organization objects in SAP, such as company code or controlling area. The relationship with the company code is via the country because the chart of depreciation is assigned to a country and a company code is assigned to a country, which determines the company codes that can use a specific chart of depreciation.

A new chart of depreciation is created by copying existing or SAP reference charts of depreciation. These reference charts of depreciation from SAP already contain predefined depreciation areas based on local specific country requirements.

To configure a chart of depreciation in SAP S/4HANA, follow menu path **Financial Accounting • Asset Accounting • Organizational Structures • Copy Reference Chart of Depreciation/Depreciation Areas**, as shown in Figure 7.1.

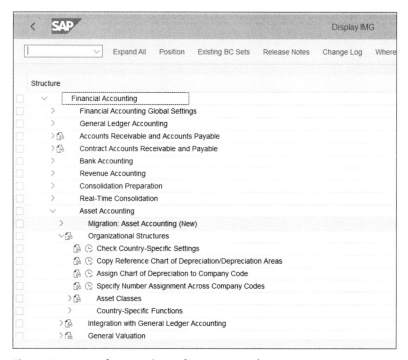

Figure 7.1 Copy Reference Chart of Depreciation/Depreciation Areas

7 Fixed Assets

The next screen presents three options, as shown in Figure 7.2.

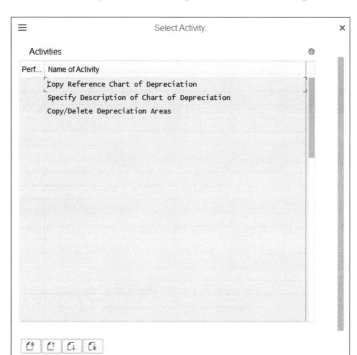

Figure 7.2 Chart of Depreciation Activities

These options are as follows:

- **Copy Reference Chart of Depreciation**
  This is the main activity, which allows you to copy an existing chart of depreciation to a new chart of depreciation.
- **Specify Description of Chart of Depreciation**
  This option provides a convenient transaction to check or change the description of the chart of depreciation.
- **Copy/Delete Depreciation Areas**
  This option enables you to configure the individual depreciation areas within the selected chart of depreciation.

7.1 Organizational Structures

We'll revisit this final option later, but first create a new chart of depreciation by clicking the first option, **Copy Reference Chart of Depreciation**. This leads to the screen shown in Figure 7.3.

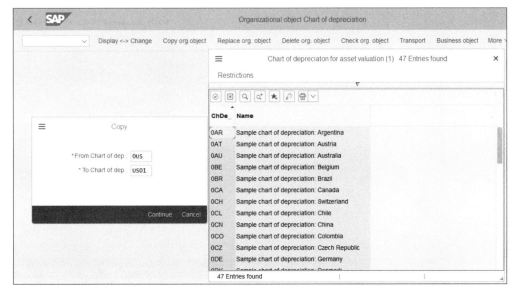

Figure 7.3 Copy Chart of Depreciation

In this screen, select **Copy Org. Object** from the top menu, which lets you select a chart of depreciation to copy from and to enter the code for the new chart of depreciation to be created. As you can see, when displaying the available charts of depreciation, SAP provides several reference charts of depreciation for many countries.

For this example, copy the reference for United States 0US to a new chart of depreciation US01, which will be used for your US company codes. The system will provide an informational message that chart of depreciation 0US is copied to US01.

In the next activity, **Specify Description of Chart of Depreciation**, rename the new chart of depreciation to "USA Chart of depreciation," as shown in Figure 7.4.

Now your new chart of depreciation is created. In the following section, we'll configure the required depreciation areas for it.

243

7 Fixed Assets

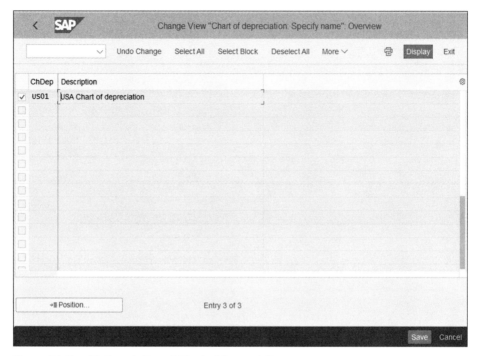

Figure 7.4 Specify Description of Chart of Depreciation

7.1.2 Depreciation Area

Depreciation areas provide valuations for fixed assets based on specific sets of rules. The depreciation rules and useful life for the assets are maintained on the depreciation area level.

The next step is to check the depreciation areas created in the new chart of depreciation for the US. The new chart of depreciation was created via copying and therefore inherits the depreciation areas of the reference chart.

To configure depreciation areas, follow menu path **Financial Accounting • Asset Accounting • General Valuation • Depreciation Areas • Define Depreciation Areas**, as shown in Figure 7.5.

In this transaction, you can delete, modify, or create new depreciation areas. After entering "US01" as the work area, the next screen presents the depreciation areas created for the chart of depreciation. The same configuration transaction is available also via menu path **Financial Accounting • Asset Accounting • Organizational Structures • Copy Reference Chart of Depreciation/Depreciation Areas**, where you choose the **Copy/Delete Depreciation Areas** option.

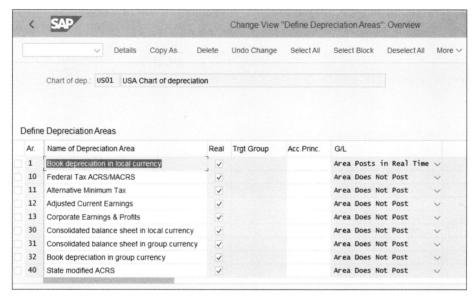

Figure 7.5 Depreciation Areas of Chart of Depreciation US01

Each chart of depreciation requires one depreciation area, which posts to the leading ledger 0L and usually is called 01. It should represent the leading valuation for the group. Normally for a US company it will be US GAAP, whereas for most other countries will be IFRS.

Figure 7.5 shows standard US depreciation areas as required by the US tax rules. For most US companies, these will be needed. However, many global companies may need also to define a depreciation area based on IFRS rules for the US. For other countries, typically an IFRS depreciation area will be needed, as well as depreciation areas for local GAAP and/or local tax rules.

Now let's create an additional depreciation area that represents the IFRS valuation. Start by copying one of the existing areas—for example, 1. Select the area by clicking the checkbox in front of it and select **Copy As...** from the top menu, as shown in Figure 7.6.

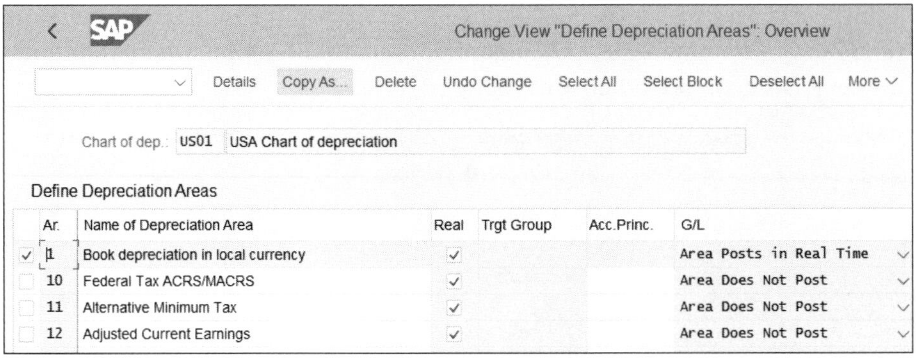

7 Fixed Assets

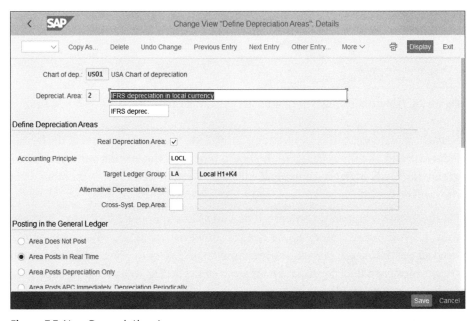

Figure 7.6 Copy Depreciation Area

This opens a new configuration screen, in which the settings are copied from the source area. Change these settings as shown in Figure 7.7.

Figure 7.7 New Depreciation Area

Call the new depreciation area (**Depreciat. Area**) "2" and provide a long and short description. In general, it's good to number the depreciation areas in a logically defined manner. They could be numbered 1, 2, 3, and so on, but also 10, 20, 30, and so on.

Note that you will need a separate depreciation area for each local currency defined for the relevant company codes per accounting principle. The accounting principle is key setting in the depreciation area, which drives the integration with the general ledger. Based on the accounting principle selected, the system automatically assigns the ledger linked to it. This ledger will be posted automatically if you select the **Area Posts in Real Time** option, as shown in Figure 7.7. This is key difference between new asset accounting in SAP S/4HANA and classic asset accounting in SAP ERP. In SAP ERP, there was also an option to post multiple depreciation areas to nonleading ledgers, but the was done periodically by running program RAPERB2000 (Transaction ASKBN). This transaction is now obsolete because the update in the general ledger takes place immediately.

If you're migrating from an existing SAP ERP system to SAP S/4HANA, you may need to set up some new depreciation areas to take advantage of the new features for fixed assets. For example, you may want to introduce a new depreciation area to post in real time based on another accounting principle. The good news is that in SAP S/4HANA 1809, there's a new program that streamlines the process of opening new depreciation areas and inserting them into existing asset master records. This is program RAFAB_COPY_AREA (Subsequent Implementation of a Depreciation Area), which uses BAdI FAA_AA_COPY_AREA (Subsequent Implementation of a Depreciation Area). It's also possible to have your own implementation of the BAdI to influence the depreciation terms and transactional data as the area is being opened.

When migrating from SAP ERP to SAP S/4HANA, also check the readiness of fixed asset settings for the new SAP S/4HANA landscape. Until SAP S/4HANA 1809 this was done using program RASFIN_MIGR_PRECHECK (Check Prerequisites for FI-AA Migration). In 1809, this program is obsolete because these checks now are part of the Software Update Manager (SUM). Before installation of SAP S/4HANA, a simplification check is performed, which checks that the prerequisites for installing SAP S/4HANA are met in fixed assets.

We should also mention that once a chart of depreciation is selected, the next time you enter this or other configuration transactions that are dependent on the chart of depreciation, there will be no prompt from the system. If you need to switch to another chart of depreciation, use configuration path **Financial Accounting • Asset Accounting • General Valuation • Set Chart of Depreciation**.

7.1.3 Asset Class

An *asset class* is the main configuration object that classifies different types of assets. It's used to group them together, assigning numbers from a specific number range. On the asset class level, you configure the account determination, which defines which general ledger accounts are posted from asset transactions. It also defines which fields are available and which fields are required when creating asset master records.

The asset class is not dependent on the chart of depreciation. This means the asset classes defined in the system are available to all countries and company codes in the system. That's why creating asset classes requires a prudent and careful approach. In most companies, twenty to thirty asset classes should be enough to manage the needs of all countries in the system. You shouldn't create new asset classes just for a specific rollout to a new country without very good cause.

To configure asset classes, follow menu path **Financial Accounting** • **Asset Accounting** • **Organizational Structures** • **Asset Classes** • **Define Asset Classes**, as shown in Figure 7.8.

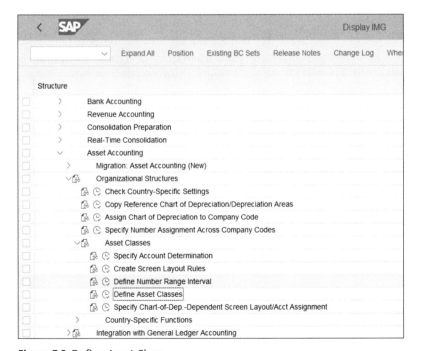

Figure 7.8 Define Asset Class

The asset class contains other important configuration objects also, such as the account determination and the screen layout, which we'll examine in detail in the next section. Once you enter the asset class configuration transaction, you'll see the list of defined asset classes in the system, as shown in Figure 7.9.

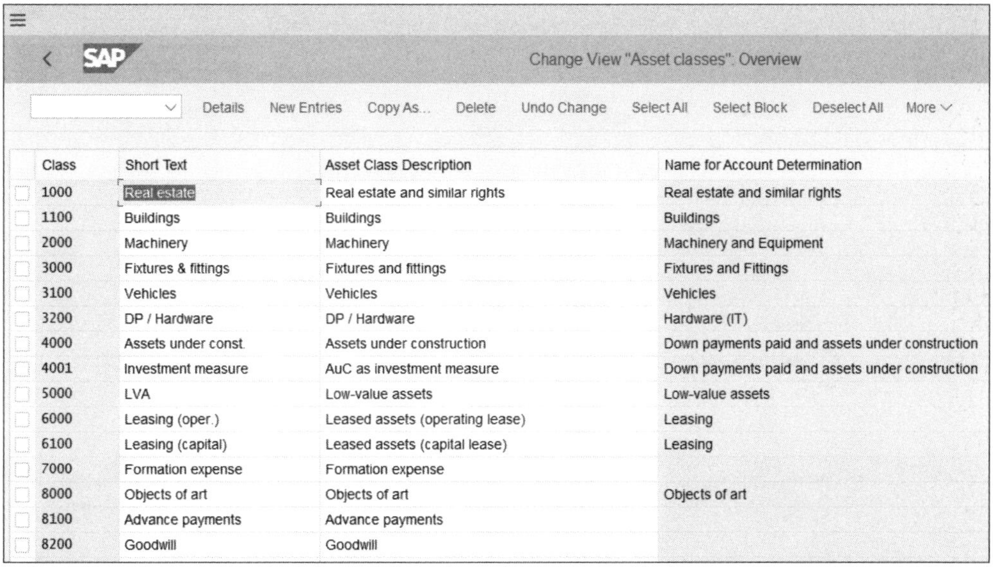
Change View "Asset classes": Overview
Details New Entries Copy As... Delete Undo Change Select All Select Block Deselect All More
Class Short Text Asset Class Description Name for Account Determination
1000 Real estate Real estate and similar rights Real estate and similar rights
1100 Buildings Buildings Buildings
2000 Machinery Machinery Machinery and Equipment
3000 Fixtures & fittings Fixtures and fittings Fixtures and Fittings
3100 Vehicles Vehicles Vehicles
3200 DP / Hardware DP / Hardware Hardware (IT)
4000 Assets under const. Assets under construction Down payments paid and assets under construction
4001 Investment measure AuC as investment measure Down payments paid and assets under construction
5000 LVA Low-value assets Low-value assets
6000 Leasing (oper.) Leased assets (operating lease) Leasing
6100 Leasing (capital) Leased assets (capital lease) Leasing
7000 Formation expense Formation expense
8000 Objects of art Objects of art Objects of art
8100 Advance payments Advance payments
8200 Goodwill Goodwill

Figure 7.9 Asset Classes

These are standard SAP classes delivered by SAP. They could be used as references to create asset classes in your system or used directly. Most companies choose to have similar lists of asset classes, often with four- or six-digit codes. Most companies have classes such as machinery, vehicles, furniture and fixtures, software, goodwill, and so on, in addition specific types of assets typical for each type of company.

To create a new asset class, select one of the existing classes and select **Copy As...** from the menu, which lets you review the settings of the class and change them, as shown in Figure 7.10.

In addition to the **Asset Class** code, the long description, and the **Short Text**, there are some other important settings to be maintained here. The account determination is another configuration object, which is a collection of the general ledger accounts to be posted during various business processes, such as acquisition, retirement, depreciation, and so on. Here you assign the relevant account determination; we'll talk

7 Fixed Assets

about assignment of general ledger accounts in the next section. One account determination can be used for multiple asset classes that share the same accounts.

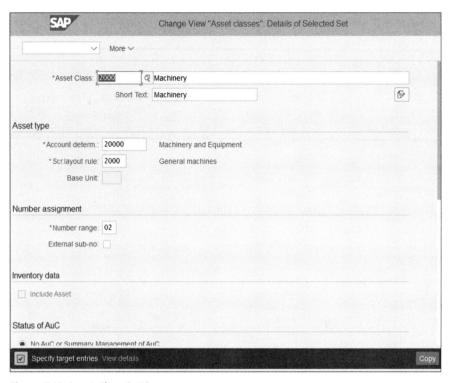

Figure 7.10 Asset Class Settings

The screen layout rule (**Scr. Layout Rule**) determines the required, optional, and suppressed fields when creating asset master records for the class. Another parameter is the number range, which determines the range in which numbers will be assigned sequentially to the created assets in this class. We'll define these objects in the next section on asset master data.

On the asset class level, you also define whether the assets will be managed as assets under construction (AUC) or not. AUCs are special types of assets that aren't purchased but are produced internally over time; we'll discuss them in greater detail in Section 7.4.1.

Also on the asset class level, you can select the **Asset Class Is Blocked** indicator, which would prevent the creation of new assets in this class. This is needed when the company decides that a certain asset class should be deactivated but it isn't possible to delete it because assets already exist in the class.

7.2 Master Data

So far, we've created the main foundation blocks for fixed assets: chart of depreciation, depreciation areas, and asset classes. Now we're ready to start creating the asset master data, which is organized into asset master records and sometime asset subnumbers.

Typically, when implementing SAP S/4HANA greenfield, you'll prepare an Excel file with the assets to be created, which come from a legacy SAP or non-SAP system. Then the migration team will perform the migration using one of the available techniques. In a brownfield implementation, no changes to the asset master data will be required, provided there are no changes in the depreciation areas introduced. But if there are new depreciation areas created to benefit from the increased flexibility of the new asset accounting in SAP S/4HANA, changes may be required. We'll examine these considerations in detail in Chapter 16.

Now let's start with the configuration activities, which are required to set up the asset master records. As discussed in the previous section, the main classifying object for assets is the asset class, but you need to set up the screen layout, account determination, and number ranges assigned to it.

Then we'll cover how to set up user fields in asset master records and how to create asset supernumbers, numbers, and subnumbers, and group assets.

7.2.1 Screen Layout

Each asset master record is organized in tabs, which group together similar fields, such as general data, depreciation, time-dependent data, and so on. SAP provides numerous fields that can appear in these tabs, some required and some optional. Which fields will appear and which will be required is controlled via the screen layout, which is assigned to the asset class and all classes with the same screen layout will share its fields.

To create a screen layout, follow menu path **Financial Accounting • Asset Accounting • Organizational Structures • Asset Classes • Create Screen Layout Rules**, as shown in Figure 7.11.

7 Fixed Assets

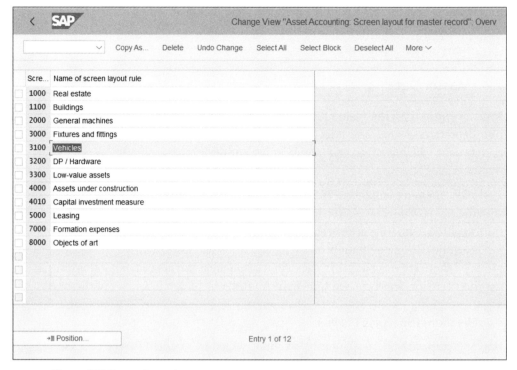

Figure 7.11 Screen Layout

In this transaction, the screen layout is created as a shell. There are no further customizing settings to be done here: just create the code and description. It's good practice to have screen layouts corresponding to the asset classes. Figure 7.11 shows standard screen layouts delivered by SAP. These can be used directly, and you can create additional ones here if there's a need. So let's create a new screen layout rule for the new asset class Machinery and title it the same way. You can select an existing screen layout with similar features, such as **3200 DP/Hardware**, and copy it over to a new one, 3400 Machinery. Select **3200** and choose **Copy As...** from the menu, which allows you to save it as new screen layout 3400.

After you create the new screen layout, you can modify its rules for optional, required, and hidden fields. For this, you need to go to a different configuration transaction. Follow menu path **Financial Accounting • Asset Accounting • Master Data • Screen Layout • Define Screen Layout for Asset Master Data**, then choose

7.2 Master Data

Define Screen Layout for Asset Master Data. This opens a screen in which you can navigate through the already created screen layouts and choose the fields, grouped in logical groups. Select **Layout 3400** and click **Logical Field Groups**, as shown in Figure 7.12.

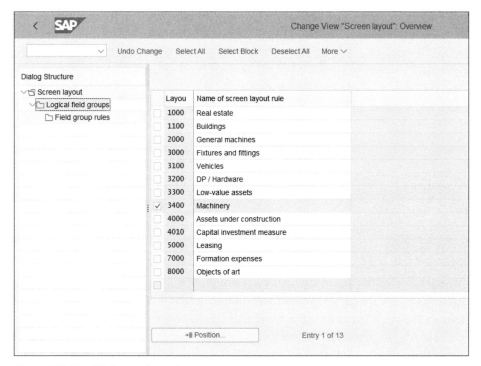

Figure 7.12 Modify Screen Layout

This presents the various logical field groups. They correspond to the tabs in the asset master record. Each logical field group contains multiple fields that logically belong together. For example, in group posting information there are fields such as **Capitalization Date**, **Acquisition Date**, and **Deactivation Date**, which all relate to the postings to the asset.

Let's assume you have a requirement that for machinery the serial number of the machine needs to be entered in the asset master record. You want to make sure that it will never be omitted for these types of assets, so you want to make it a required field in the asset master. First, select the general data group, as shown in Figure 7.13.

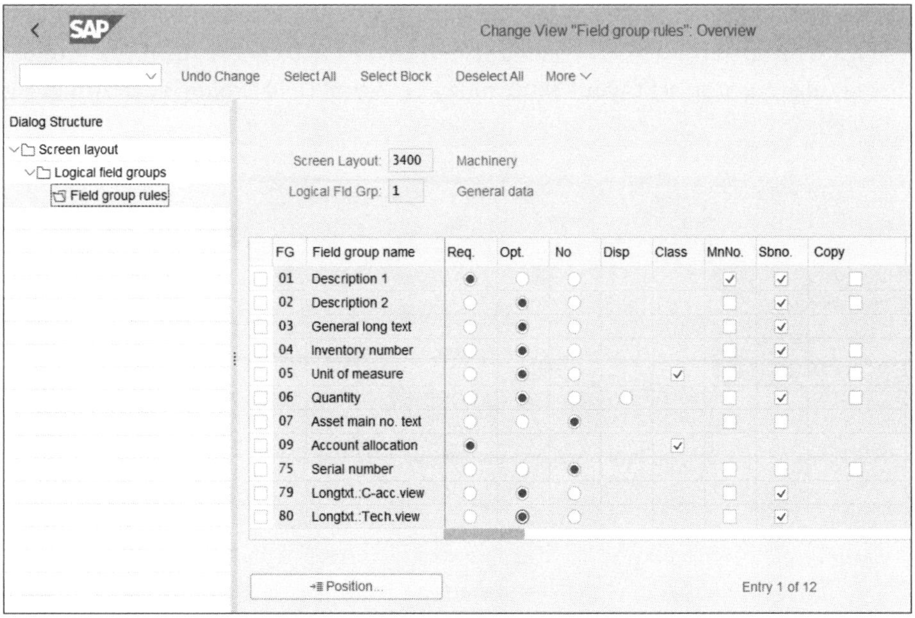

7 Fixed Assets

Figure 7.13 General Data Fields

As you can see, here for every field there are three options: required, optional, and hidden. Change the serial number to required (**Req.**). Now the system won't allow anyone to save a master record under this screen layout without entering a serial number. You also need to specify its maintenance level via the checkboxes on the right: main number (**MnNo.**) and subnumber (**Sbno.**). Select main number, which means that any subnumbers created for this asset will get the value for this field from the main number. The **Copy** checkbox makes sure that when creating an asset with reference to another asset, the value of this field will be transferred. Select that and save. In a similar way, you can modify other fields in the other logical groups based on your requirements.

There is also one more configuration transaction in which fields are selected for the **Depreciation** tab in the asset master record. This is accessed via menu path **Financial Accounting • Asset Accounting • Master Data • Screen Layout • Define Screen Layout for Asset Depreciation Areas**. It's maintained in a similar way, and usually no changes are required because the settings are pretty standard. The available fields are shown in Figure 7.14.

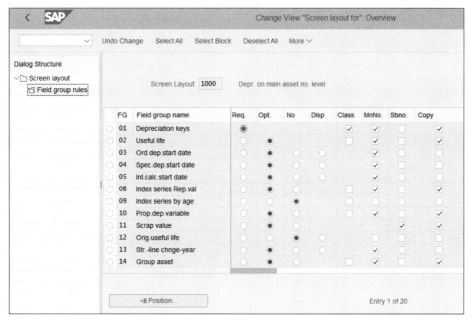

Figure 7.14 Depreciation Screen Layout

With that, we've finished the configuration of the screen layouts. Now let's move to account determination.

7.2.2 Account Determination

Account determination is a very important configuration area, which stipulates which general ledger accounts are posted to from asset transactions. As with the screen layout, it's defined as an object in one configuration transaction where you can create account determinations per asset type, such as vehicles, machines, IT equipment, and so on. Then, in another configuration transaction, you assign accounts per depreciation area and per business process to the defined account determinations.

To define the account determination, follow menu path **Financial Accounting • Asset Accounting • Organizational Structures • Asset Classes • Specify Account Determination**. This brings you to a table with the defined account determinations, as shown in Figure 7.15.

7 Fixed Assets

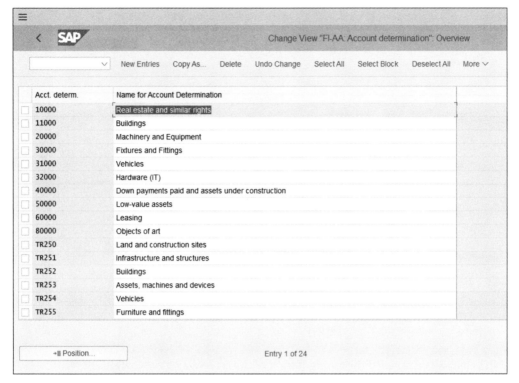

Figure 7.15 Account Determinations

As with screen layouts, these are standard-provided account determinations, which can be used directly or used as references for new objects. They correspond more or less to the asset classes, although it's possible for one account determination to be used by multiple asset classes if you envision that their accounts are and will be exactly the same. We'll use the standard account determination, so we won't create new entries here.

To modify the accounts linked to the account determinations, you need to go to a different transaction via menu path **Financial Accounting • Asset Accounting • Integration with General Ledger Accounting • Assign G/L Accounts**. This transaction is dependent on the chart of accounts. After selecting the chart of accounts, you're presented with the list of account determinations, as shown in Figure 7.16.

7.2 Master Data

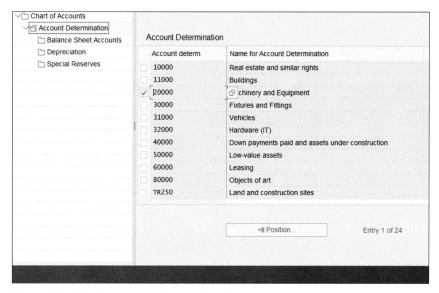

Figure 7.16 Selecting Account Determination

Select **Account Determ. 20000 Machinery and Equipment** by placing a check in the checkbox on the left side. On the left side of the screen, there is tree-like structure in which you can configure different types of accounts, such as balance sheet accounts and depreciation accounts. Click **Balance Sheet Accounts** on the left side of the screen. You'll see a list of the defined depreciation areas, as shown in Figure 7.17.

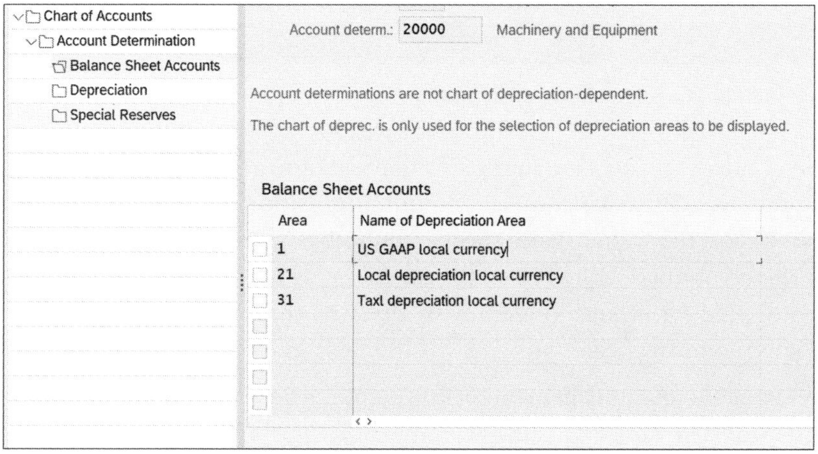

Figure 7.17 Select Depreciation Area

7 Fixed Assets

You need to choose the depreciation area by double-clicking it. The assignment of accounts needs to be done for each depreciation area, which posts to the general ledger. Very often the accounts are the same between depreciation areas, but it's possible to post to different accounts based on the accounting principle used. So double-click **Deprec. Area 1** and to see the configuration screen shown in Figure 7.18.

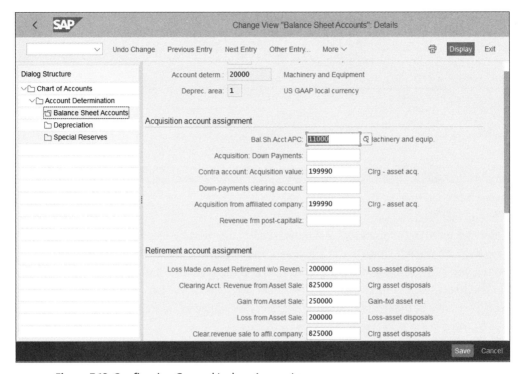

Figure 7.18 Configuring General Ledger Accounts

Here, general ledger accounts are configured to post automatically from the various business processes. The balance sheet account: acquisition and production costs (**Bal.Sh.Acct APC**) is the main balance sheet account that is debited when the asset is acquired. This account is credited when the asset is being retired. A separate account can be customized for acquisitions from an affiliated company in the **Acquisition from Affiliated Company** field.

In the **Retirement Account Assignment** section accounts to be used in the retirement process are provided. Loss and gain accounts are posted with the financial result generated when comparing the sales price with the net book value of the asset. Different accounts can be entered for asset retirement without revenue (scrapping). There are also other sections that are not used often, like accounts for revaluation and investment support.

Then the next important part of the configuration is setting the depreciation accounts. Select the **Depreciation** section on the left side of the screen, select the depreciation area, and you're presented with the configuration screen, as shown in Figure 7.19.

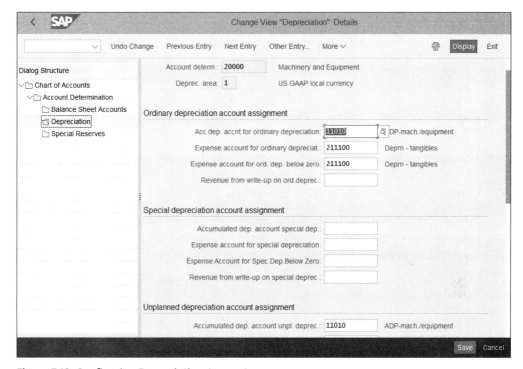

Figure 7.19 Configuring Depreciation Accounts

The depreciation configuration area is also separated into sections based on similar accounts. You need to configure the ordinary depreciation account assignment, which is used by the normal ordinary depreciation run that runs on a monthly basis. The accumulated depreciation account for ordinary depreciation is the balance sheet account, which is posted against expenses during the depreciation run. Then expense accounts need to be configured, which are posted with the depreciation expense. At the same time, the system also posts the cost center from the asset master record. There is also an option to configure a different expense account for depreciation below zero.

In other sections, you can configure accounts for unplanned and special depreciation. *Unplanned depreciation* is depreciation posted in unplanned cases, such as unplanned loss of the value of assets, like obsolescence due to new models, for example. *Special depreciation* is additional depreciation to be posted above the ordinary depreciation in special cases, such as when stipulated by the government. The accounts for these two types of depreciation can be maintained here in a similar fashion as with ordinary depreciation.

After making necessary adjustments to the assignment of general ledger accounts, you can configure number ranges for asset master data.

7.2.3 Number Ranges

Number ranges define the numbers assigned when creating asset master records. From a business point of view, these numbers should be unique per asset class, although technically it's possible for multiple asset classes to share a number range.

To define the asset number ranges, follow menu path **Financial Accounting • Asset Accounting • Organizational Structures • Asset Classes • Define Number Range Interval**. Then enter your company code, as shown in Figure 7.20.

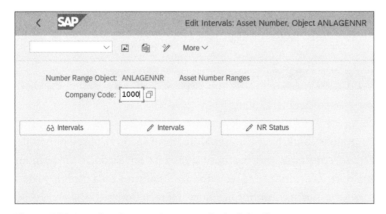

Figure 7.20 Number Ranges Company Code Selection

Click the **Change Intervals** button. This brings you to a transaction in which you can define number ranges, as shown in Figure 7.21.

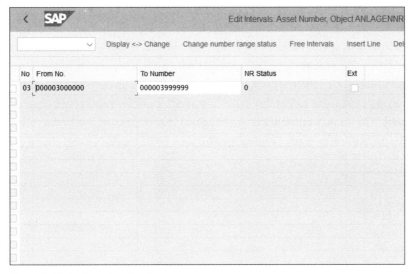

Figure 7.21 Define Number Ranges

The number range is defined with its number (in the example above, 03), which is then assigned in the asset class. You define the **From No.** and **To Number** of the range in which assets are created, deciding how long the asset numbers should be. For example, if you enter 000003000000 to 000003999999, the asset numbers will have seven digits. The system will assign the next number consecutively internally, unless you check the **Ext** checkbox on the left side. This means the number range is external and you will be required to enter a number within that range when creating an asset. In the **NR Status** column, the system displays the current number within the range.

To create a new number range, select **Insert Line** from the top menu. This gives you a blank line in which you can define a new number range. Note that asset master number ranges are not transportable, as is also true of other number ranges in the system, and should be created in each target client individually. This is done to avoid inconsistencies.

Now that we've defined number ranges, let's talk about the user fields available in the asset master records.

7.2.4 User Fields and Asset Supernumbers

As you've seen, an asset comes with multiple standard SAP fields, in which you can record all sorts of information, including general data, time-dependent data, origin information, leasing, and so on. But sometimes this isn't enough. Companies may have very unique and specific reporting requirements that aren't provided for with the standard fields. For that reason, SAP provides a flexible option: four four-character fields and one eight-character field, which can be defined as required in the system and then selected from a dropdown list. These are used for reporting purposes. They're available in the **Assignments** tab in the asset master record, as shown in Figure 7.22.

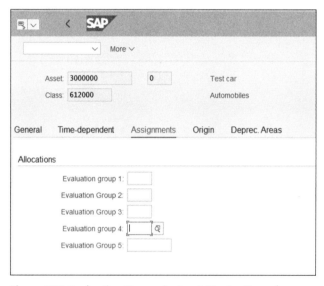

Figure 7.22 Evaluation Groups in Asset Master Record

As you can see, there are four four-character fields available (evaluation groups 1 through 4) and one eight-character field (evaluation group 5). To define the four-character fields, follow menu path **Financial Accounting • Asset Accounting • Master Data • User Fields • Define 4-Character Evaluation Groups**. Click **New Entries** on the screen shown in Figure 7.23.

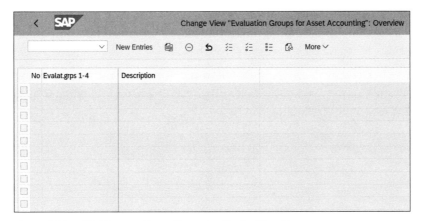

Figure 7.23 Initial Screen: Four-Character Evaluation Groups

You can then define your list, as shown in Figure 7.24.

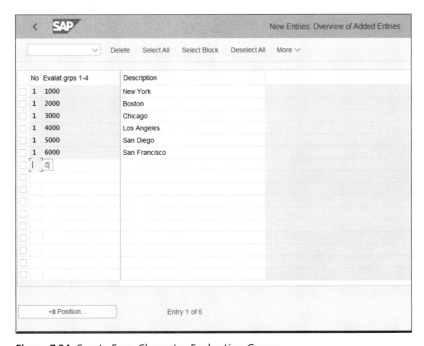

Figure 7.24 Create Four-Character Evaluation Group

In this example, a list of regions is defined in evaluation group 1, which the company may use to report its assets based on geographical location. The number 1 in first column indicates that this is the field group 1 field, then the codes are defined, and the last column holds the descriptions of the values. In similar fashion, you can define other lists for evaluation groups 2 through 4.

For the eight-character evaluation group, follow menu path **Financial Accounting • Asset Accounting • Master Data • User Fields • Define 8-Character Evaluation Groups**. Click **New Entries** on the screen shown in Figure 7.25.

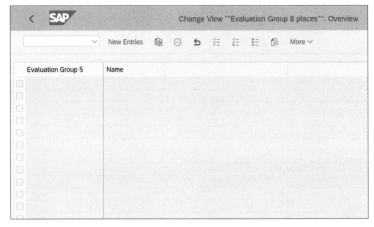

Figure 7.25 Initial Screen: Eight-Character Evaluation Group

In the example shown in Figure 7.26, employees are defined in the eight-character evaluation group 5.

Evaluation Group 5	Name
10000001	Smith
10000002	Johnson
10000003	Watkins
10000004	Peterson

Figure 7.26 Define Eight-Character Evaluation Group

Then in the asset reports you can sort, filter, and summarize by these fields.

Another field that can be used to classify assets by user-defined fields for reporting purposes is the asset supernumber. This is used to assign multiple assets to the same asset supernumber—for example, in the case of different machinery pieces in an assembly line. Then all elements can be reported together using the supernumber. To configure an asset supernumber, follow menu path **Financial Accounting** • **Asset Accounting** • **Master Data** • **User Fields** • **Define Asset Super Number**. Then click **New Entries** to define asset supernumbers, as shown in Figure 7.27.

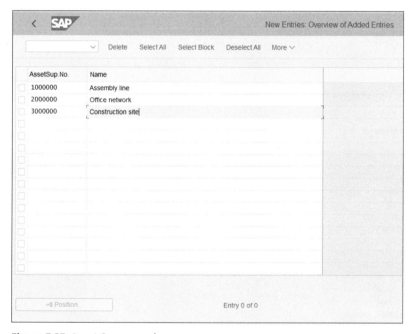

Figure 7.27 Asset Supernumbers

You enter asset supernumbers and their description here. Asset supernumbers are defined for large assets. These large assets, in turn, have a number of other assets assigned to them. An example is an assembly line, to which could belong numerous machinery and equipment assets.

7.2.5 Asset Numbers, Subnumbers, and Group Numbers

Assets are usually created using asset numbers, which are either internally assigned by the system from the number range of the asset class, or are externally entered by the user in the case of external number assignment. However, there are also two other options: to create subnumbers for asset numbers or to create group numbers, which can serve as valuation levels for the assets assigned to them.

Subnumbers are used for assets that belong to higher-level assets, which should be depreciated together at the main asset number level. To create asset subnumbers, no additional configuration is required. From the application menu, follow menu path **Accounting • Financial Accounting • Fixed Assets • Asset • Create • Subnumber • AS11- Asset**, as shown in Figure 7.28.

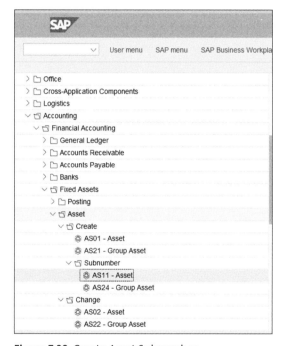

Figure 7.28 Create Asset Subnumber

Then the information in the asset subnumber will be defaulted from the main number, with the opportunity to change some of the fields (not the depreciation terms, however, because the valuation level will be the main asset number). After saving, the system will assign a number in the format main asset number-1, -2, and so on, starting with 1 for the first subnumber. So for example, if the main asset number is 200000, then the first subnumber will be 200000-1, the second 200000-2, and so on.

To use group asset numbers, they first need to be enabled at the depreciation area level. To configure this, follow menu path **Financial Accounting • Asset Accounting •**

General Valuation • Group Assets • Specify Depreciation Areas for Group Assets. In this transaction, you select the company code and depreciation area for which group assets are allowed. You also can configure for specific asset classes that they consist entirely of group assets in menu path **Financial Accounting • Asset Accounting • General Valuation • Group Assets • Specify Asset Classes for Group Assets**.

In the screen shown in Figure 7.29, you can select asset classes, which consist of group assets, by placing a checkmark in the **Class Consists Entirely of Group Assets** column.

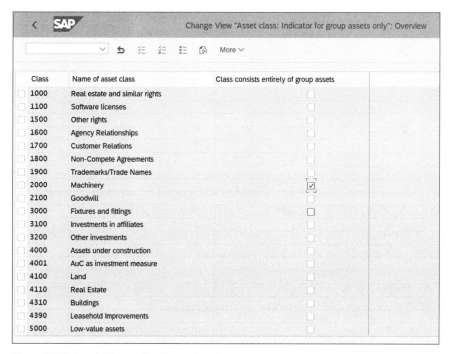

Figure 7.29 Asset Classes for Group Assets

After enabling the group assets, their creation is straightforward using application menu path **Accounting • Financial Accounting • Fixed Assets • Asset • Create • AS21-Group Asset**. It's also possible to create subnumbers for group assets.

So now you know that there are flexible options to group your assets using different valuation levels such as main number, subnumber, and group number.

Now that we've covered the master data configuration, we'll cover the various business transactions that are performed with fixed assets and the required configuration for them.

7.3 Business Transactions

In terms of business transactions posted to assets in SAP S/4HANA, there are no changes compared to SAP ERP: the main transactions are still acquisitions, retirements, transfers, and manual value corrections. However, there is a key underlying difference compared to SAP ERP: now each transaction posts in real time in all ledgers, which are mapped to the depreciation areas, which post to the general ledger (Section 7.1.2). In the old SAP ERP world, they were posted in real time in the leading ledger (0L) only. Then any differences with the nonleading ledgers were stored as delta values and posted as periodic jobs. This real-time integration is one of the key benefits of the new asset accounting. It requires some additional settings, which need to be performed as part of the migration from SAP ERP to SAP S/4HANA, which we'll examine in the following section.

7.3.1 Acquisitions

Acquisition is the process of acquiring assets, either procuring them from external sources or creating them internally. Initially when an asset master record is created, it's just a shell without values. Then when the acquisition is posted the asset receives its values and depreciation can start based on its depreciation rules.

The most common process of acquisition of assets is an integrated process through purchasing. Usually there will be a purchase order (and sometimes a purchase requisition before that), in which an asset number is assigned. In this case, the asset is capitalized either at the time of the goods receipt (in the case of valuated GR) or at the time of the invoice receipt (in the case of nonvaluated GR).

It's also possible to enter asset acquisition directly in financial accounting. In this case, a financial vendor invoice is posted, which debits the asset, typically using asset transaction type 100 for external asset acquisition.

The third main method of acquiring assets is the assets under construction process. This is used when the company doesn't procure the asset from external sources but creates it internally.

We already configured the general ledger accounts to be posted by the various transactions when we discussed the account determination in Section 7.2.2. However, there are some other SAP S/4HANA-specific configuration transactions that need to be performed. If you're migrating from SAP ERP to SAP S/4HANA, you need to make sure that these configuration activities are completed because they're new, required customizing that didn't exist before in SAP ERP.

To enable the acquisition processes in SAP S/4HANA, you need to configure a technical clearing account for integrated asset acquisition. To do so, follow menu path **Financial Accounting • Asset Accounting • Integration with General Ledger Accounting • Technical Clearing Account for Integrated Asset Acquisition • Define Technical Clearing Account for Integrated Asset Acquisition**. Then select **New Entries** from the menu and assign the account for your chart of accounts, as shown in Figure 7.30. Enter a chart of accounts and the general ledger account that should be used as the technical clearing account for integrated asset acquisitions.

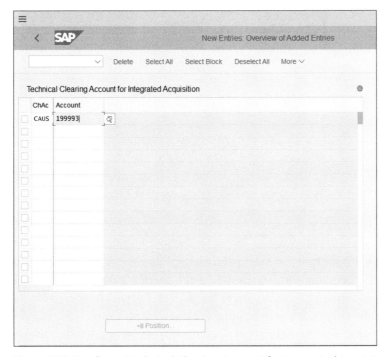

Figure 7.30 Configure Technical Clearing Account for Integrated Asset Acquisition

When posting the acquisition, the system will offset the vendor line item with this technical account. The invoice posting will be the same for all accounting principles. However, the valuation part may be different in the different ledgers because the different accounting principles may require different valuation approaches. Therefore the system will post separate documents per ledger, in which the asset line items will be offset by the same technical clearing account. At the end, the balance of this account will always be zero. It won't be included in the financial statements but only in the notes for the financial statements with zero balance as an auxiliary account.

You also need to define alternative document types for accounting principle-specific documents. In this activity, you specify which document types should be used when posting the ledger-specific part of the acquisition documents. To configure them, follow menu path **Financial Accounting • Asset Accounting • Integration with General Ledger Accounting • Integrated Transactions: Alternative Doc. Type for Acctg-Princpl-Spec. Docs • Specify Alternative Document Type for Acctg-Principle-Specific Documents**, as shown in Figure 7.31.

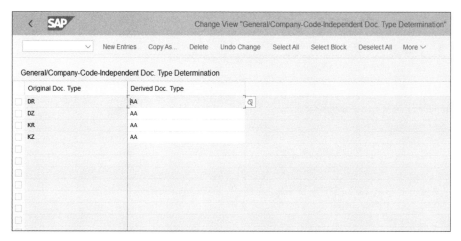

Figure 7.31 Alternative Document Types

Here in the **Original Doc. Type** column is the original document type used in the entry view of the document—for example, KR for vendor invoice. Then in the **Derived Doc. Type** column, this is mapped to the document type to be used for the accounting principle-specific entry to be done per nonleading ledger. Some companies may require that these are different document types with separate number ranges; if not, then it's possible to use the standard asset document type AA.

You have also the option to configure these alternative document types per company code in menu path **Financial Accounting • Asset Accounting • Integration with General Ledger Accounting • Integrated Transactions: Alternative Doc. Type for Acctg-Princpl-Spec. Docs • Define Separate Document Types by Company Code**.

7.3.2 Transfers

Asset transfer is the process of transferring an asset to another asset number, which may be needed either if there is a change in its cost assignment or if the asset is being transferred to another company code.

If you need to change the cost assignment of an asset, use asset transfer within the company code, which is done via **Accounting • Financial Accounting • Fixed Assets • Posting • Transfer • ABUMN—Transfer within Company Code**.

This transaction retires the asset that is being transferred from and capitalizes the new asset, which you could either create before using the transaction or create automatically via the transaction. The system uses predefined transaction types for the acquisition and the retirement, which you can configure by following menu path **Financial Accounting • Asset Accounting • Transactions • Transfer Postings • Define Transaction Types for Transfers**. Then you can check the settings of the transaction types for retirement and acquisition. Typically, transaction types 300 (for prior year acquisitions) and 320 (for current year acquisitions) are used for retirement, and types 310 (prior year) and 330 (current year) for acquisitions.

Normally, standard asset transaction types should fulfill your needs, but if you need to change some settings, best practice is to make a copy of a standard asset transaction type to a new one starting with Z, so it doesn't get overwritten in future updates of the system, and make the changes there. Figure 7.32 shows the settings behind a transaction type.

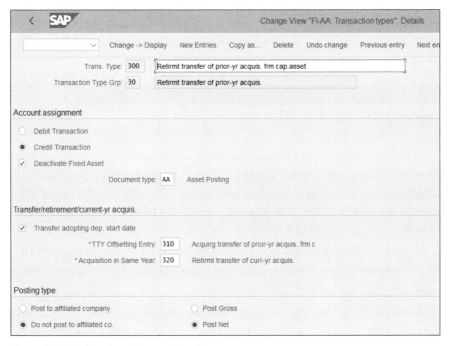

Figure 7.32 Define Asset Transaction Type

The asset transaction type always belongs to a transaction type group (in Figure 7.32, for type 300 it belongs to group 30). These groups are standard and not to be changed. When creating a new transaction type, you should select from the existing standard groups, depending on the posting requirements of the transaction type.

Then there a is setting to determine whether this transaction type creates a credit or debit (credit in the case of retirement), and the **Deactivate Fixed Asset** checkbox is checked because the asset should be deactivated after retirement. Other settings provided are the document type used for the transaction, whether it's an intercompany transaction or not (in this case not, because **Do Not Post to Affiliate Co** is selected), and the **Post Net** indicator. This means that the net value will be transferred—that is, acquisition value minus posted depreciation.

Further settings are shown in Figure 7.33.

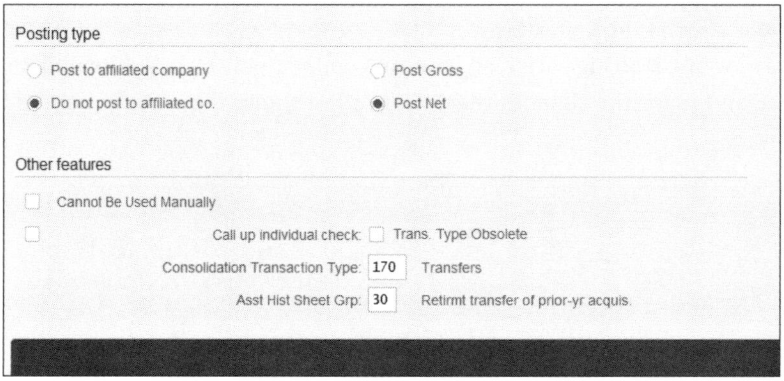

Figure 7.33 Other Transaction Type Settings

Here you have the option to set a transaction type as not to be used manually, meaning it can only be selected internally by the system. Also you have to select the related consolidation transaction type, which is used by the consolidation functions of the system, and the asset history sheet group, which is used for the history sheet report, which we'll examine in Section 7.5.3.

Intercompany asset transfers are made to transfer an asset from one legal entity to another within the group. The process usually also involves issuing intercompany invoices.

You also need to define an asset transfer variant by following menu path **Financial Accounting • Asset Accounting • Transactions • Intercompany Asset Transfers • Automatic Intercompany Asset Transfers • Define Transfer Variants**. Then click the **Define**

Transfer Variant option, which will bring you to the list of transfer variants, as shown in Figure 7.34.

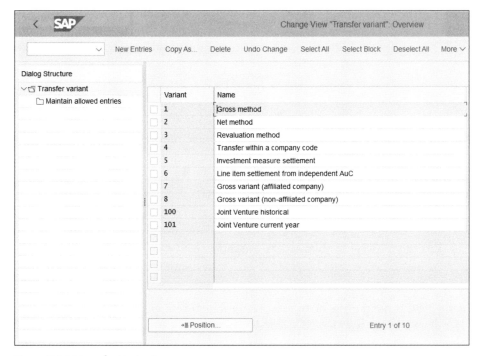

Figure 7.34 Transfer Variants

Figure 7.34 shows the SAP standard-provided variants, which can be copied to custom specific variants in which necessary adjustments can be made. Variants can be gross or net, depending whether the gross values are being transferred (including the posted depreciation) or the net values (without the posted depreciation). For example, select the intercompany transfer variant **7 Gross Variant (Affiliated Company)**, then select **Maintain Allowed Entries** from the left side of the screen, which will bring you to the screen shown in Figure 7.35.

Here you configure the retirement and acquisition transaction types to be used in the case of legally independent units and in case of transfer of one legally-independent unit. In the **Rel. Type** column, 1 is used for legally independent and 2 for legally one unit. You can specify a cross-system depreciation area or enter *, which is valid for all cross-system depreciation areas. Then select a transfer method in the **Trans. Meth. Field** and enter retirement and acquisition transaction types.

7 Fixed Assets

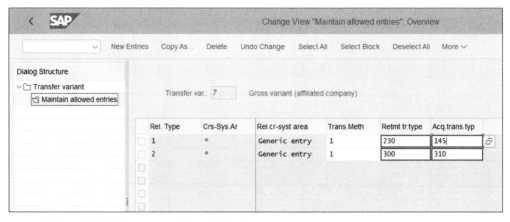

Figure 7.35 Intercompany Gross Variant

Similarly, you can define your own transfer variant that uses your own transaction types.

7.3.3 Retirement

Retirement is the process of an asset being sold or scrapped and deactivated in the system. When the asset is sold, usually this involves posting and issuing a customer invoice, which records the revenue posted. At this moment, the system compares the net book value of the asset to the selling price and posts gain or loss to the configured accounts. In the case of scrapping, there is no revenue, and if there is outstanding net book value, it will be posted to the loss account assigned.

To check and configure the general ledger accounts for retirement postings, follow menu path **Financial Accounting** • **Asset Accounting** • **Transactions** • **Retirements** • **Assign Accounts**. Then select the relevant chart of accounts, account determination, and depreciation area, and you will see the retirement section only of the account determination, as shown in Figure 7.36.

The first account, **Loss Made on Asset Retirement w/o Reven.**, is used to post the loss in case of scrapping. **Clearing Acct. Revenue from Asset Sale** is used to clear the net revenue from asset sales. The **Gain from Asset Sale** account is posted to when the revenue from the asset sale is higher than the net book value of the asset. The **Loss from Asset Sale** account is posted to when the revenue from the asset sale is less than the net book value of the asset. The last account, **Clear.revenue sale to affil.company**, is used to clear the revenue from asset sales to affiliated companies.

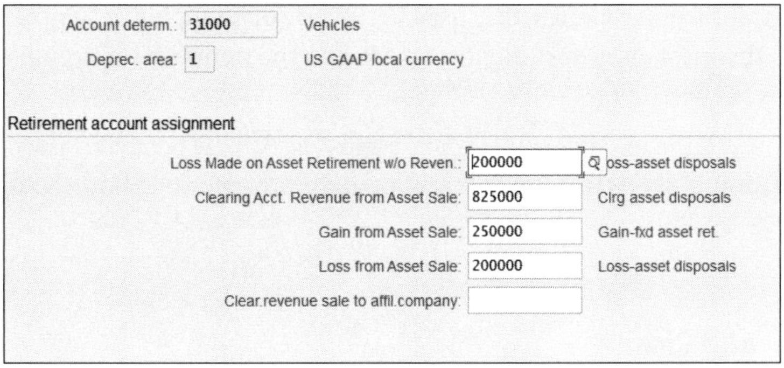

7.3 Business Transactions

Account determ.: 31000	Vehicles
Deprec. area: 1	US GAAP local currency

Retirement account assignment

Loss Made on Asset Retirement w/o Reven.:	200000	Loss-asset disposals
Clearing Acct. Revenue from Asset Sale:	825000	Clrg asset disposals
Gain from Asset Sale:	250000	Gain-fxd asset ret.
Loss from Asset Sale:	200000	Loss-asset disposals
Clear.revenue sale to affil.company:		

Figure 7.36 Configure Retirement Accounts

We covered the transaction type configurations when we covered the asset transfers transaction, but if you want to check the retirement transaction types you can do so by following menu path **Financial Accounting • Asset Accounting • Transactions • Retirements • Define Transaction Types for Retirements**. This gives you list of only the retirement transaction types, which you can check or define your own. For example, double-click **Trans. Type 200 Retirement without Revenue**, as shown in Figure 7.37.

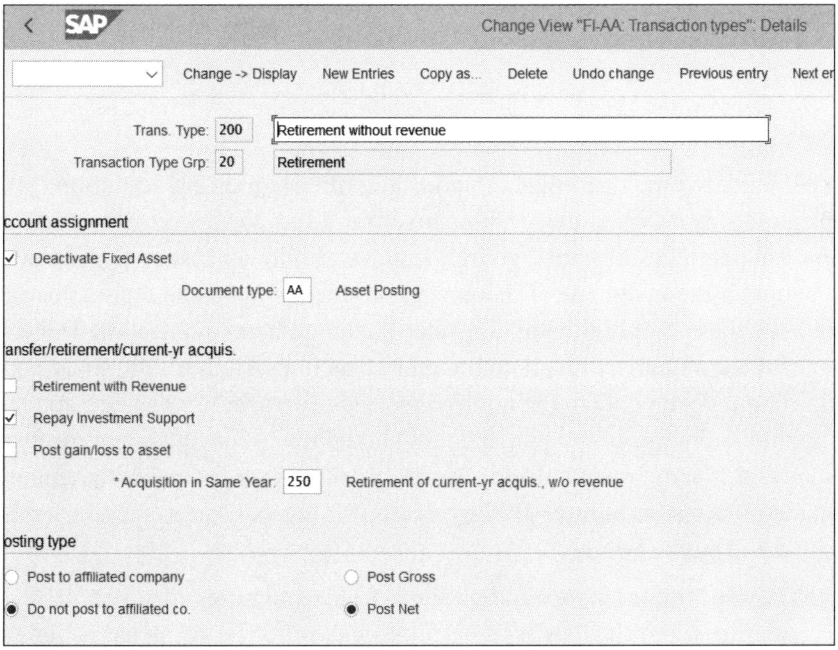

Change View "FI-AA: Transaction types": Details

Change -> Display   New Entries   Copy as...   Delete   Undo change   Previous entry   Next er

Trans. Type: 200   Retirement without revenue
Transaction Type Grp: 20   Retirement

ccount assignment

☑ Deactivate Fixed Asset
    Document type: AA   Asset Posting

ansfer/retirement/current-yr acquis.

☐ Retirement with Revenue
☑ Repay Investment Support
☐ Post gain/loss to asset
    *Acquisition in Same Year: 250   Retirement of current-yr acquis., w/o revenue

osting type

○ Post to affiliated company   ○ Post Gross
● Do not post to affiliated co.   ● Post Net

Figure 7.37 Transaction Type Retirement without Revenue

This is a standard transaction type, which is used in the case of scrapping. As you can see, it deactivates the asset, uses document type AA, and the option for retirement without revenue is not selected.

Now that we've covered the main business transactions that can be posted to assets, let's look the valuation of assets and the different periodic procedures that are being run in fixed assets.

7.4 Valuation and Closing

Asset valuation is the process to determine the fair market value of assets. Usually it's determined by running a periodic depreciation program that gradually reduces the value of assets based on predefined rules (depreciation terms). These rules are different per asset class (type of assets), but also can be different per various sets of accounting rules that a company may need to use to prepare its reports and financial statements. This is exactly where we see the big benefit of the new asset accounting provided by SAP S/4HANA, which we'll examine in detail in this section. We'll start with a detailed discussion of new asset accounting, then we'll discuss multiple valuation principles. After that, we'll configure the depreciation, revaluation, and other closing activities in fixed assets.

7.4.1 New Asset Accounting Concept

SAP has been able to provide parallel valuation according to multiple accounting and tax principles using different depreciation areas for a very long time. However, the solution wasn't perfect. The classic general ledger was only for asset reporting purposes. Then with the invention of the new general ledger, you could post the differences from these different valuation rules in the different nonleading ledgers. However, this wasn't perfect real-time integration as the system would would post delta adjustment documents to the nonleading ledgers to provide the value according to the alternative valuation approach they should represent. This was a cumbersome solution. It did provide the opportunity to prepare financial statements according to different accounting frameworks, but it involved a certain amount of workaround and manual work.

First with the SAP Simple Finance add-on for SAP ERP, and now with SAP S/4HANA Finance, SAP introduced the new asset accounting option. The key benefits of new asset accounting included the real-time integration of asset accounting with the general ledger and the posting of separate documents from the different depreciation areas in the nonleading ledgers. This means that when there are different values resulting from an acquisition or depreciation document, the system automatically will post separate documents in separate number ranges to leading ledger 0L and to the nonleading ledgers in real time.

With SAP S/4HANA, new asset accounting is mandatory, whereas it was optional with the SAP S/4HANA Finance add-on. This means that if you're upgrading from SAP ERP to SAP S/4HANA, you need to plan the time and project activities to convert from classic asset accounting to new asset accounting. Program RASFIN_MIGR_PRECHECK will help you check that all migration prerequisites are met, as shown in Figure 7.38.

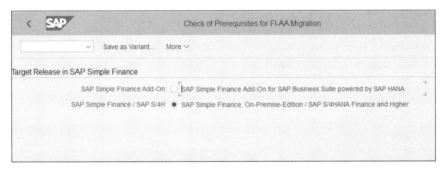

Figure 7.38 Migration Check Program

This program checks the readiness of the customizing and transaction data for new asset accounting. This is important because there are underlying technical changes in the landscape of fixed assets. For example, with new asset accounting the actual line items will be posted to the Universal Journal, table ACDOCA. Tables from classic asset accounting like tables ANEK, ANEP, ANLC, and ANLP become redundant but still exist as compatibility views. Custom programs that rely on these tables should continue to work if they only read from these tables (but nevertheless should be checked and tested), whereas custom programs that write to these tables will need to be updated.

7.4.2 Multiple Valuation Principles

It's very common that the same fixed assets should be treated differently based on different valuation principles. For example, the useful life or the method of depreciation for certain categories of assets should be different. In the US, the half-year convention of depreciation is common, which means that depreciation for half a year should be taken in the year of acquisition of the asset and in the year of its retirement, regardless of when it's required. This of course will result in different depreciation amounts compared to IFRS or local GAAP, for which no such method is used.

The management of different valuation principles in SAP is done using the different depreciation areas. Depreciation area 01, which is linked with the leading ledger, should represent the leading valuation from the point of view of the group. Normally for US companies this should be US GAAP or IFRS. For European companies, it will almost always be IFRS. Other commonly used depreciation areas are for local GAAP and local tax. Thus the local specific accounting principles and the local tax rules can be managed for each country, if needed.

The link between the depreciation area and the ledger is the accounting principle. Let's revisit the configuration of the depreciation area, which you can access via menu path **Financial Accounting • Asset Accounting • General Valuation • Depreciation Area • Define Depreciation Areas**. When you click a depreciation area, you see its definition, as shown in Figure 7.39.

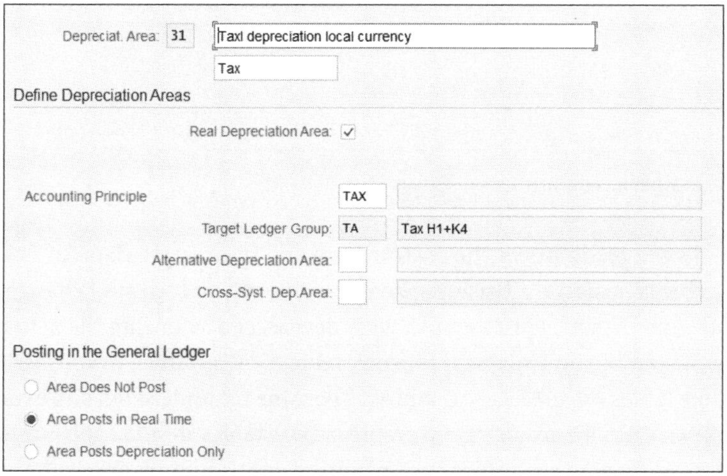

Figure 7.39 Accounting Principle in Depreciation Area

As shown in Figure 7.39, you assign the **Accounting Principle** in the depreciation area—in this case, "TAX", which represents local tax valuation. Those accounting principles are assigned to ledgers in **Financial Accounting • Financial Accounting Global Settings • Ledgers • Parallel Accounting • Assign Accounting Principle to Ledger Groups**.

Now that we've ensured that we have depreciation areas that cover multiple accounting principles, let's look at how the system calculates depreciation differently using depreciation keys.

7.4.3 Depreciation Key

The depreciation key contains various settings required to calculate the depreciation amounts, such as the method of depreciation and the time control. This configuration object bundles together multiple calculation methods of various types: base methods (such as linear depreciation), declining-balance methods (when depreciation should be faster in the beginning of the asset life), multilevel methods (which involve change during the useful life), and so on. SAP provides myriad standard calculation methods, which you can check or copy and create your own by following menu path **Financial Accounting • Asset Accounting • Depreciation • Valuation Methods • Depreciation Key • Calculation Methods**. Usually the standard methods should suffice to meet any business requirements in your organization.

Another important setting at the depreciation key level is the period control. This setting determines when the depreciation should start and end. Its configuration is available at menu path **Financial Accounting • Asset Accounting • Depreciation • Valuation Methods • Period Control**. Here you can define period control methods and define their calendar assignments. Open the **Define Calendar Assignments** transaction, as shown in Figure 7.40.

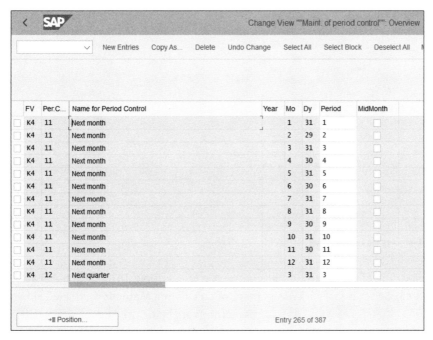

Figure 7.40 Period Control Calendar Assignments

Here per fiscal year variant and period control method, you define the period after which the depreciation will start in case of acquisition and after which it will end in case of retirement. In Figure 7.40, for example, for method **11 Next Month**, the depreciation should start at the beginning of the month following the acquisition and it should finish in the next month, following the retirement. So for example for acquisitions, that happen between dates 1–31 of period 01 it will start in the beginning of the following period 02.

Now let's look at the configuration of the depreciation key itself via menu path **Financial Accounting • Asset Accounting • Depreciation • Valuation Methods • Depreciation Key • Maintain Depreciation Key**. This shows a list of the existing depreciation keys on the right side of the screen and navigation to their underlying calculation methods on the left side, as shown in Figure 7.41.

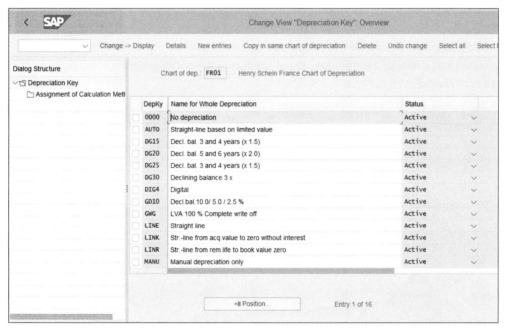

Figure 7.41 Depreciation Keys

Now select one of the more complicated keys—for example, **DG15 Decl. bal. 3 and 4 years (x 1.5)**. Select it and click **Assignment of Calculation Methods** on the left side of the screen. This shows what calculation methods are being used by the depreciation key, as shown in Figure 7.42.

7.4 Valuation and Closing

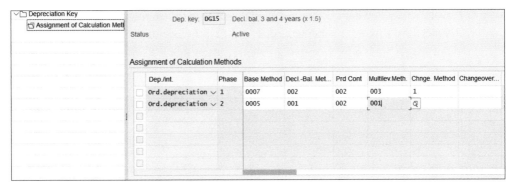

Figure 7.42 Assignment of Calculation Methods

Here you see two lines because this depreciation key has two phases. Initially, the first phase's methods will be valid, until the changeover happens, as defined by the changeover method defined previously. The changeover will switch to the methods of the next phase. There are various changeover methods, such as changeover when a specific net book value percentage is reached, changeover after the end of useful life, and so on. It's also possible to do a manual changeover. Different types of calculation methods are assigned in the method columns.

As with other configuration objects, it's wise to copy existing SAP provided keys and rename them in the Z name range if you need to develop your own depreciation keys so that new settings won't be overwritten in future upgrades.

Now that we've covered the settings needed for depreciation, let's look at the depreciation run and how it differs in SAP S/4HANA compared to SAP ERP.

7.4.4 Depreciation Run

The depreciation run posts the calculated depreciation values to the assets on a periodic basis (usually monthly). In SAP S/4HANA new asset accounting (and the SAP S/4HANA Finance add-on) there's a new depreciation program, program FAA_DEPRECIATION_POST, which replaces the old program RAPOST2000 from SAP ERP classic asset accounting. For the users it's the same Transaction AFAB, but behind the scenes it's the new program, which significantly increases performance because it posts precalculated plan values rather than calculating them at run time. Planned depreciation values are updated in table FAAT_PLAN_VALUES.

281

7 Fixed Assets

Now let's look at what options are available during the depreciation run. To start the depreciation run, follow application menu path **Accounting • Financial Accounting • Fixed Assets • Periodic Processing • Depreciation Run • AFAB-Execute**, as shown in Figure 7.43.

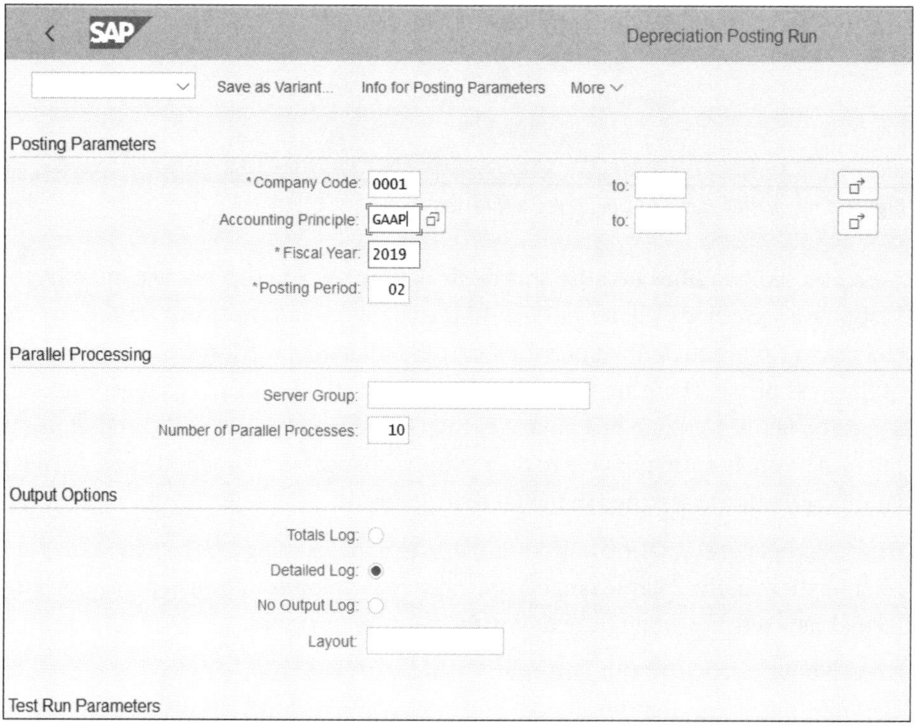

Figure 7.43 Depreciation Run

There are some key differences in the new depreciation program in SAP S/4HANA. Now the depreciation run posts financial documents, which are updated at the asset level in table ACDOCA. Also, now the depreciation program can be run for multiple company codes at the same time. There is also an added field in the selection screen, **Accounting Principle**, which allows you to run the depreciation either for a specific principle or for multiple principles at the same time. This might be required if the accounting principles have different periodic requirements. Another improvement is that now the depreciation run can be run multiple times for the same period, which replaces the repeat and restart run options in SAP ERP. As previously in SAP ERP, a test run can be run, and as before it's limited to 1,000 assets at the same time. Also as before, the real run has to be executed in background mode.

After executing the depreciation run, you can view the log by executing Transaction AFBP (Display Log). You can see that the depreciation posts separate documents per accounting principle and ledger. Thus all postings are done in real time and are fully integrated with the general ledger.

7.4.5 Revaluation

Revaluation is a process required in some countries where companies operate in a high-inflation environment and must periodically revalue assets to represent them at the current fair value. You need to configure general ledger accounts in which revaluation will be posted using menu path **Financial Accounting • Asset Accounting • Special Valuation • Revaluation of Fixed Assets • Maintain Accounts for Revaluation**. After selecting the proper chart of accounts and depreciation areas, you can assign revaluation accounts for acquisition and production costs (APC) and depreciation, as shown in Figure 7.44.

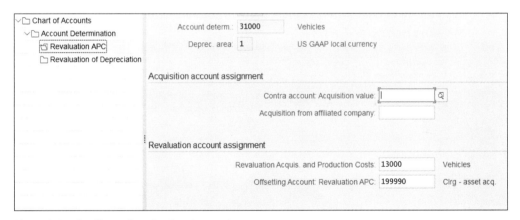

Figure 7.44 Configure Revaluation Accounts

You also need to configure which depreciation areas are relevant for revaluation by following menu path **Financial Accounting • Asset Accounting • Special Valuation • Revaluation of Fixed Assets • Revaluation for the Balance Sheet • Determine Depreciation Areas**. There you can select whether APC, depreciation, or both are relevant for depreciation, as shown in Figure 7.45.

7 Fixed Assets

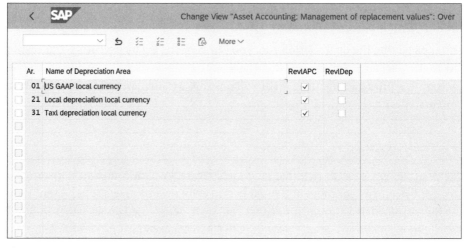

Figure 7.45 Determine Depreciation Areas

You also have to configure revaluation measures, which contain the rules for revaluation, using menu path **Financial Accounting • Asset Accounting • Special Valuation • Revaluation of Fixed Assets • Revaluation for the Balance Sheet • Define Revaluation Measures** (see Figure 7.46).

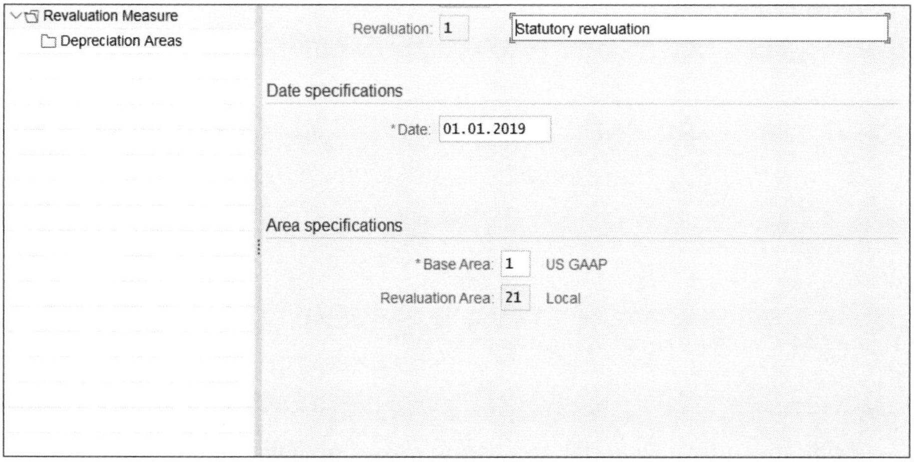

Figure 7.46 Revaluation Measure

Here you configure the date on which the system will post the revaluation to fixed assets, the base depreciation area that will provide the values, and the revaluation area in which the revaluation will be posted.

The actual calculation rules usually are too complicated to be configured, so they need to be programmed using a custom implementation of BAdI `FIAA_REAVLUATE_AS`, which you can set by following menu path **Financial Accounting • Asset Accounting • Special Valuation • Revaluation of Fixed Assets • Revaluation for the Balance Sheet • Implementations for Add-In for Revaluation and New Valuation**.

7.4.6 Manual Value Correction

We've covered various automatic valuation processes, such as depreciation and revaluation. It's also possible to perform manual value correction, in which you can increase or decrease the value of an asset manually on an asset-by-asset basis.

The transactions to perform manual value correction are located at application menu path **Accounting • Financial Accounting • Fixed Assets • Posting • Manual Value Correction**. There you have the following four transactions at your disposal:

- **ABZU—Write-Up**
  Used to manually increase the value of an asset
- **ABMA—Manual Depreciation**
  Used to post manual depreciation to a specific asset
- **ABAA—Unplanned Depreciation**
  Used to post additional unplanned depreciation
- **ABMR—Transfer of Reserves**
  Used to create reserves for some asset transactions, such as sale of undervalued assets

To configure these processes, you need to assign general ledger accounts. We assigned general ledger accounts for other processes previously, and we use the same menu path here: **SPRO • Financial Accounting • Asset Accounting • Integration with General Ledger Accounting • Assign G/L Accounts**.

In the depreciation section of the account determination, shown in Figure 7.47, you optionally can assign the general ledger accounts for the manual value correction.

7 Fixed Assets

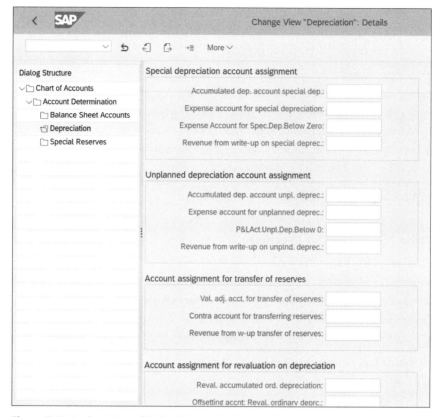

Figure 7.47 Assign Manual Value Correction Accounts

7.4.7 Year-End Closing Activities

In SAP S/4HANA, the year-end closing process for fixed assets has been greatly simplified compared to SAP ERP.

Balance carryforward is a year-end closing procedure that transfers the year-end account balances as beginning balances for the next year. In SAP S/4HANA, the balance carryforward for fixed assets is part of the balance carryforward transaction for the general ledger, which is available under application menu path **Accounting • Financial Accounting • General Ledger • Periodic Processing • Closing • Carrying Forward • FAGLGVTR—Balance Carryforward**, as shown in Figure 7.48.

7.4 Valuation and Closing

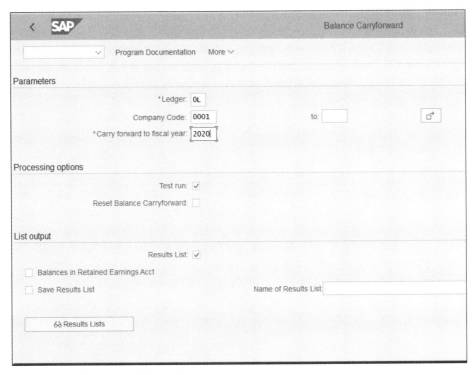

Figure 7.48 Balance Carryforward

In SAP S/4HANA 1809, there are some further improvements of the carryforward program for fixed assets. In previous releases, the new fiscal year was opened at the start of processing before the processing of the individual assets. In 1809, the new fiscal year is opened after the program has processed all assets successfully. Also, now the program not only processes the active assets but also processes deactivated assets.

Because fixed assets now is fully integrated with the general ledger, there's no need for additional asset accounting year-end closing programs, and the fiscal year change program RAJAWE00 (Transaction AJRW) from SAP ERP is obsolete and is no longer available. As APC postings in asset accounting are now posted to the general ledger in real time, in all depreciation areas periodical APC postings are obsolete and SAP ERP Transaction ASKB is no longer supported. Also, you no longer need to reconcile the general ledger with fixed assets due to the real-time integration, so programs RAABST01 and RAABST02 are obsolete.

Now that we've covered the various valuation functions for fixed assets, it's time to look at the information system and the powerful asset reports in SAP S/4HANA.

7.5 Information System

SAP S/4HANA comes with myriad standard asset reports, which greatly benefit from the streamlined real-time integration with the general ledger. We'll look at the main reports that the information system provides for assets and point out the key benefits of SAP S/4HANA's new asset accounting.

7.5.1 Asset Explorer

Asset explorer is the main report when you want to analyze a single asset. It provides great integrated visibility as you can drill down from the asset values to the asset master record to the related documents to the related account assignment objects, such as general ledger account, cost center, and so on. To access the asset explorer, follow application menu path **Accounting** • **Financial Accounting** • **Fixed Assets** • **Information System** • **Reports on Asset Accounting** • **Individual Asset** • **AW01N—Asset Explorer**. Then enter a **Company Code** and **Asset**, as shown in Figure 7.49.

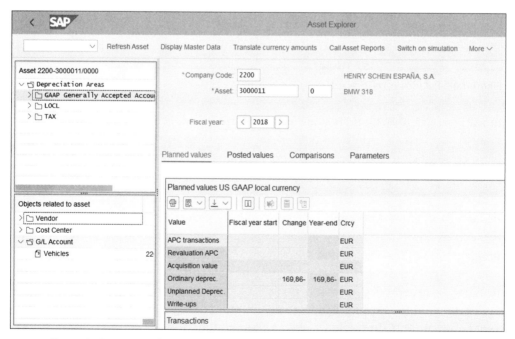

Figure 7.49 Asset Explorer

On the top-left side of the screen, you see the accounting principles and the related depreciation areas. You can navigate through them to switch between the different valuations and analyze their values. In the lower-left section of the screen, you see the objects related to the asset, such as the vendor from which the asset was acquired and the assigned cost center and general ledger account. On the right side, you see the values of the asset, separated by tabs for planned values, posted values, comparison, and depreciation parameters. In the planned section, you see the depreciation planned for the selected year, whereas in the posted section you see only the depreciation already posted through the performed depreciation runs.

Next, in the **Comparison** tab you can compare the values between the depreciation areas. Below that you can see the posted documents for the asset, and by double-clicking you can enter the document display overview. Once in a financial document related to the asset, you can view the asset accounting overview of the document by following menu path **More—Asset Accounting Document Display**. This shows you the real-time integration of the asset documents with the general ledger, as shown in Figure 7.50.

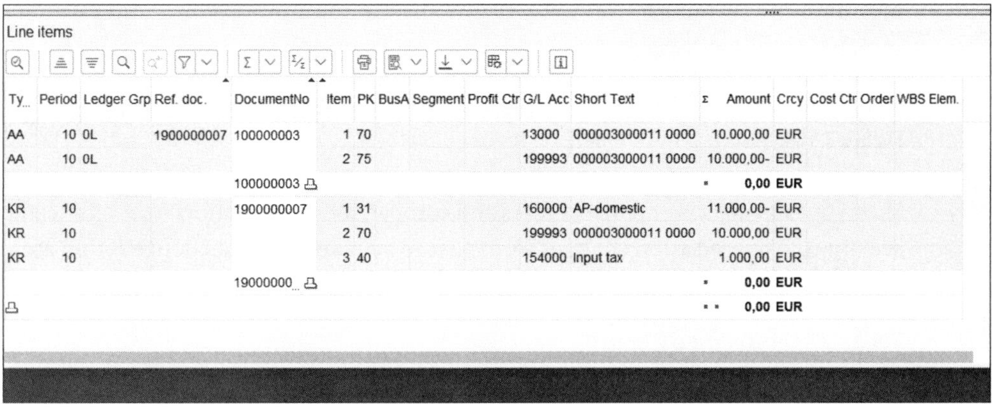

Ty	Period	Ledger Grp	Ref. doc.	DocumentNo	Item	PK	BusA	Segment	Profit Ctr	G/L Acc	Short Text	Σ	Amount	Crcy	Cost Ctr	Order	WBS Elem.
AA	10	0L	1900000007	100000003	1	70				13000	000003000011 0000		10.000,00	EUR			
AA	10	0L			2	75				199993	000003000011 0000		10.000,00-	EUR			
				100000003									0,00	EUR			
KR	10			1900000007	1	31				160000	AP-domestic		11.000,00-	EUR			
KR	10				2	70				199993	000003000011 0000		10.000,00	EUR			
KR	10				3	40				154000	Input tax		1.000,00	EUR			
				19000000									0,00	EUR			
													0,00	EUR			

Figure 7.50 Asset Overview of Document

In the old SAP ERP, the system would create separate asset document numbers. Now everything is integrated with the general ledger and all asset fields are part of the Universal Journal, table ACDOCA. As explained in Section 7.3.1, this acquisition document has two parts: invoice and valuation. In our example, this is accomplished using the clearing account 199993, which is posted against the vendor in the invoice part and offset against the asset in the valuation part. You see the document posted in the leading ledger OL, but if you drill down from another depreciation area, you'll see the document posted in the linked nonleading ledger and therefore see the valuation according to its accounting principle.

Quite useful also is the depreciation comparison, which you can access in the **Comparison** tab in the asset explorer report. It compares the depreciation values in multiple depreciation areas, as shown in Figure 7.51.

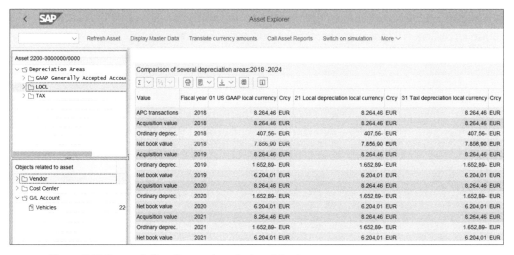

Figure 7.51 Depreciation Comparison in Asset Explorer

Here you can select comparison years and comparison depreciation areas and readily see how the depreciation changes from year to year and between depreciation areas.

7.5.2 Asset Balance Reports

There are many useful asset balance reports that can help you analyze many or all assets simultaneously per depreciation area. These reports are available in the application menu under **Accounting • Financial Accounting • Fixed Assets • Information System • Reports on Asset Accounting • Asset Balances • Asset Lists • Asset Balances**. Let's start one of them—for example, report S_ALR_87011966 (... by Cost Center), which can provide a valuable overview of the assets per cost center. After selecting a company code, cost centers, and a key date to run the report (it should be the last date of a period), you'll see the output, as shown in Figure 7.52.

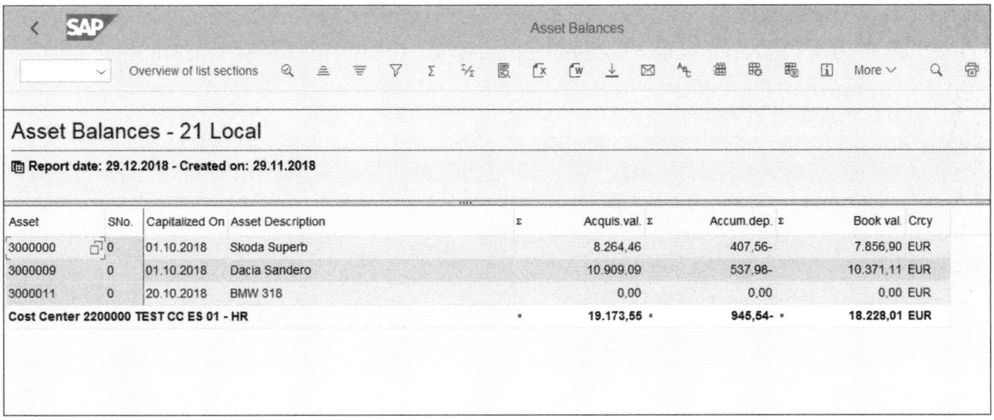

7.5 Information System

![Asset Balances Report]

Figure 7.52 Asset Balances Report

Here you can see the acquisition value, accumulated depreciation, and net book value per asset. The report provides totals per asset class and cost center. You can double-click any line, which allows you to drill down to the asset explorer for this asset.

The report is linked with other asset reports. You can navigate from the **More • Standard Asset Report** menu and select from a list of other standard asset reports to be automatically run for the same selection, as shown in Figure 7.53.

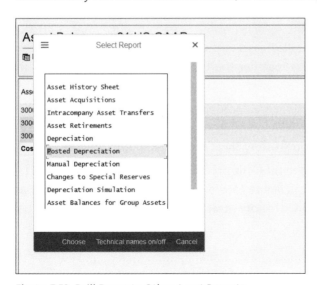

Figure 7.53 Drill Down to Other Asset Reports

You can double-click any of the linked reports and it will be executed for the selected asset. For example, select **Posted Depreciation** to see this report for the selected asset, as shown in Figure 7.54.

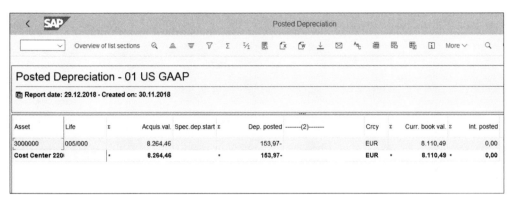

Figure 7.54 Posted Depreciation Report

7.5.3 Asset History Sheet

Asset history sheets are important reports that show the full history of assets from their acquisition to their retirement. It shows all transactions and therefore is suitable when there is a need for very detailed reporting on assets. It's a powerful and flexible report because unlike other asset reports it provides the opportunity to configure your own layout with the so-called history sheet version. In SAP S/4HANA, there is also an SAP Fiori app called Asset History Sheet that works with key figure groups, which similarly groups together the transactions in specific cells of the report.

To configure history sheet versions, you need to configure history sheet groups. To do so, follow menu path **Financial Accounting • Asset Accounting • Information System • Asset History Sheet • Define History Sheet Groups**, as shown in Figure 7.55.

Transaction types are assigned to these history sheet groups in the definition of the transaction type. For example, in Figure 7.56 you see the definition of asset transaction type 100. It has been assigned history sheet group **10 Acquisition**.

7.5 Information System

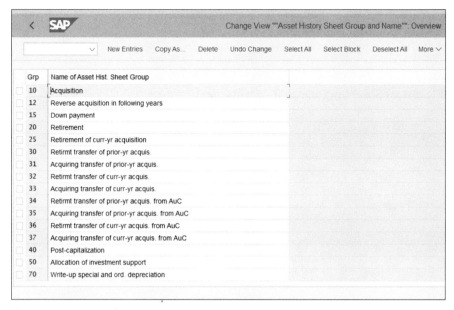

Figure 7.55 History Sheet Groups

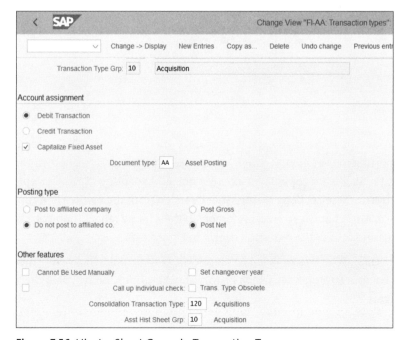

Figure 7.56 Hisotry Sheet Group in Transaction Type

Transaction types are logically assigned to history sheet groups such as acquisition, retirement, postcapitalization, and so on.

The next step is to configure the history sheet versions by following menu path **Financial Accounting • Asset Accounting • Information System • Asset History Sheet • Define History Sheet Versions**, as shown in Figure 7.57.

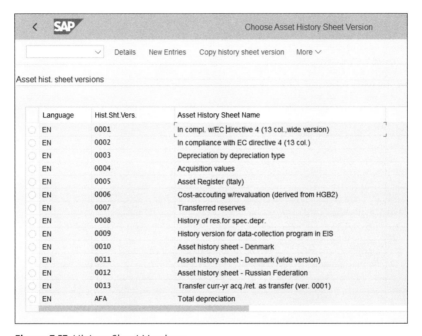

Figure 7.57 History Sheet Versions

As you can see, SAP provides many standard history sheet versions, which already cover asset reporting requirements in many countries. You can create your own version, however, preferably by copying an existing one and adapting it. For example, select version 0001 and select **Copy History Sheet Version** from the menu. Then copy and set your own four-character name, preferably in the Z name range. Then you can see the columns and cells of the report, as shown in Figure 7.58.

You can double-click any cell to be taken its definition. For example, click **Acquisition** under line 02, column 10. You'll see which history sheet groups should appear there, as shown in Figure 7.59.

7.5 Information System

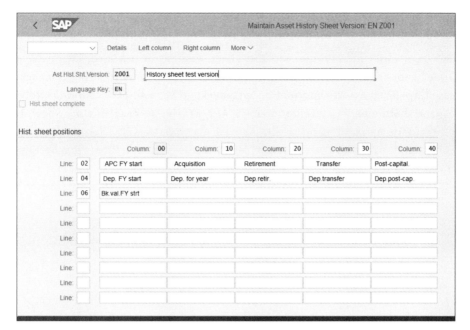

Figure 7.58 History Sheet Structure

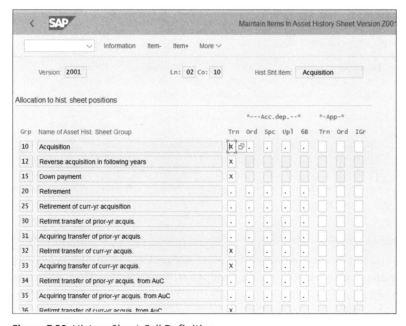

Figure 7.59 History Sheet Cell Definition

History sheet groups indicated with **X** appear in this cell, whereas those indicated with a period (.) appear elsewhere in the report.

Once you are done modifying the history sheet version, you can save. Then you can run the history sheet report from the application menu path **Accounting • Financial Accounting • Fixed Assets • Information System • Reports on Asset Accounting • Notes to Financial Statements • International • S_ALR_87011990—Asset History Sheet**. In addition to the standard report selections, such as company code, asset classes, asset numbers, and so on, you also can select history sheet versions in the **Further Settings** section of the selection screen, where you can select your own version or one of the standard ones. After executing you'll see the output based on the history sheet version selected, as shown in Figure 7.60.

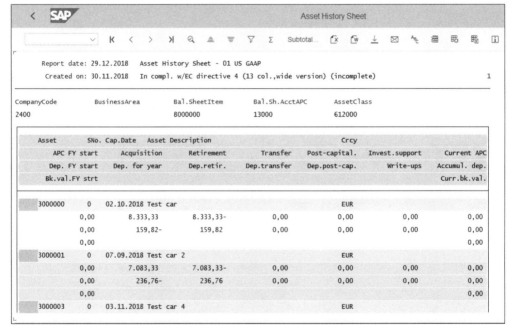

Figure 7.60 History Sheet Report

If you're going to use the Asset History Sheet app, you can configure key figure groups in menu path **Financial Accounting • Asset Accounting • Information System • Asset History Sheet • Define Key Figure Groups for Asset History Sheet (Fiori)**, as shown in Figure 7.61.

7.6 Summary

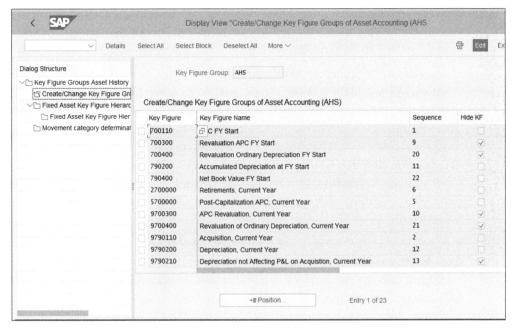

Figure 7.61 Define Key Figure Group

Similarly to history sheet versions, key figure groups group together the transactions for report purposes. Each key figure is defined with its sequence in the report. You can use standard key figure groups or define your own. You can use standard groups **AHS: Asset History Sheet (Posted Values, No Hierarchy)** and **AHS_HRY: Asset History Sheet (Posted Values, Wth Hierarchy)** for posted values. For planned values, you can use **AHS_PLAN: Asset History Sheet (Plan Values, No Hierarchy)** and **AHS_HRY_PL: Asset History Sheet (Plan Values, With Hierarchy)**.

With that, we finish our discussion of the asset history sheet report and the information system for assets. Of course, there are many other standard reports that you can explore in the information system. They work similarly to the reports we discussed and don't require additional configuration.

7.6 Summary

We covered a lot of material, and now you should be able to configure and use fixed assets in SAP S/4HANA. You've learned the most important customizing transactions and recommended settings for them, so you can either configure fixed assets in new greenfield implementation or adapt your current settings in a brownfield implementation from an existing SAP ERP system.

We explained the fundamental differences between the new asset accounting provided by SAP S/4HANA and the classic asset accounting from SAP ERP. Now you understand how immensely beneficial and powerful SAP S/4HANA is with its real-time integration between fixed assets and the general ledger. After properly configuring fixed assets, you'll be able to have powerful real-time reports on your assets, based on multiple accounting and tax frameworks. This is invaluable for most of today's global businesses that want to have multiple countries and regions in the same ERP system but need to have transparent reporting based on different accounting standards.

Now you should have a very good understanding of the organizational structure in fixed assets and what options you have to represent the various countries and valuations using charts of depreciation and depreciation areas. You know how to set up the asset master data efficiently in well-defined asset classes. You know how to configure and process various asset transactions. You understand how to properly set up multiple valuation principles and execute your depreciation runs. Last but not least, you are familiar with the various reports in fixed assets, including the highly customizable asset history sheet report.

With that, we finish our guide to configuring fixed assets, and we'll move next to bank accounting.

Chapter 8
Bank Accounting

This chapter gives step-by-step instructions for configuring bank processes in SAP S/4HANA, including generating SEPA-compliant outgoing payment files and reconciling open items automatically using the electronic bank statement.

Bank accounting refers to the outgoing and incoming payment processes. It's not a separate area in SAP S/4HANA—it's part of accounts payable and accounts receivable—but its processes and settings are quire specific and deserved to be explained is chapter of its own rather than discussing the outgoing payments as part of accounts payable and the incoming payments as part of accounts receivable.

In this chapter, we'll help you configure the following:

- Bank master data—house banks and bank accounts
- Automatic payment program
- Payment files
- Electronic bank statement

Let's start with setting up the master data required for the bank processes.

8.1 Master Data

Banks need to be defined as master data objects in the system, which are called *bank keys*. To assign a bank in the customer or vendor master record, it first needs to be created as bank key. In addition, the banks at which your organization has accounts need to be defined as house banks, which require a lot more configuration. We'll also discuss the creation of bank accounts and International Bank Account Numbers (IBANs).

8.1.1 Bank Keys

The first step is to create bank keys for all the needed banks. It's good practice to automatically load all the banks in the country where you implement the system to avoid having missing banks later on when creating vendor and customer master data. In most countries, a list of banks in the country can be obtained from the central bank or from another financial institution. SAP provides the necessary programs to upload such a file automatically, which will load the banks as bank keys in the system. They are located at menu path **Cross-Application Components • Bank Directory • Bank Directory Data Transfer**. Here you can find two programs: **Transfer Bank Directory Data—International**, which uses international SWIFT format, and **Transfer Bank Directory Data—Country-Specific**, which can work with country-specific formats.

For the US, select **Transfer Bank Directory Data—Country-Specific**. Figure 8.1 shows the selection screen of the program to transfer bank data.

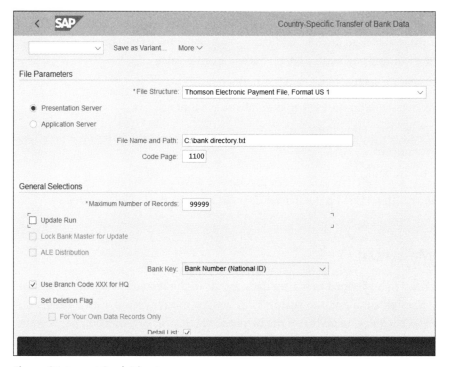

Figure 8.1 Import Bank Directory

In the **File Structure** field, you can select from multiple formats that SAP provides. For example, for the US the Thomson Electronic Payment File, Format US1, is provided.

You can select a file either from the application or the presentation server and its code page. When you check the **Update Run** checkbox, the system creates the bank keys; otherwise, it runs in test mode.

You can also define additional file formats using menu path **Cross-Application Components • Bank Directory • Bank Directory Data Transfer • Define File Formats for Country-Specific Bank Directories**. The system issues a warning that this is a cross-client table, and after confirming that you'll see a list of available bank import formats, as shown in Figure 8.2.

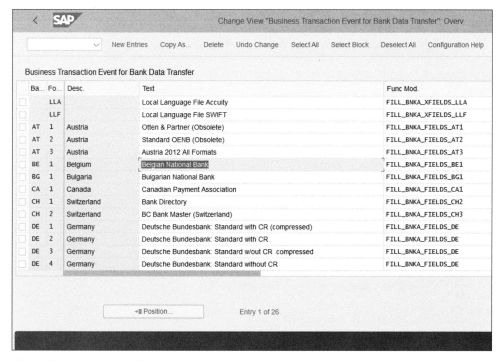

Figure 8.2 Bank Import File Formats

The formats are defined per country. You can create a new format using the **New Entries** option from the top menu. If you need additional nonstandard formats, you'll need to work with your development team to develop a new functional module, which you would assign in the **Func. Mod.** column.

You can also create bank keys manually using Transaction FI01. This is needed when a bank needs to be used in the business partner master record, and the bank key does not exist. You need to enter the bank country and bank key as shown in Figure 8.3.

8   Bank Accounting

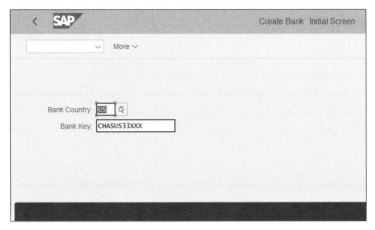

Figure 8.3 Create Bank Manually

Then press ⌜Enter⌝ and you are transferred to the bank key screen, where you can enter the details of the bank, as shown in Figure 8.4.

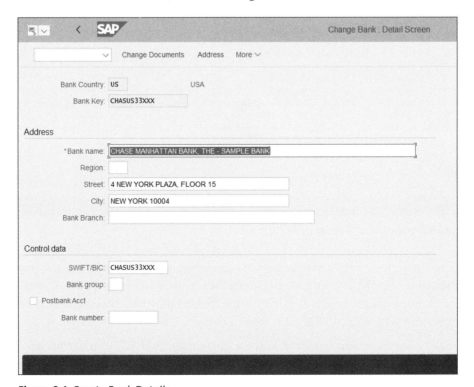

Figure 8.4 Create Bank Details

The following fields need to be configured:

- **Bank name**
  The legal name of the bank.
- **Street**
  The legal address of the bank.
- **City**
  City of the bank.
- **SWIFT/BIC**
  This is the bank code—SWIFT, BIC, or US routing number.

Save your entry and you can use the bank key in your business partner master data.

8.1.2 House Banks

A *house bank* is a bank your company is doing business with. You need to create the bank key as a house bank on the company code level by following application menu path **Accounting • Financial Accounting • Banks • Master Data • Master Data • FI12_HBANK—Manage House Banks or Transaction FI12_HBANK**. As shown in Figure 8.5, first you need to select the company code.

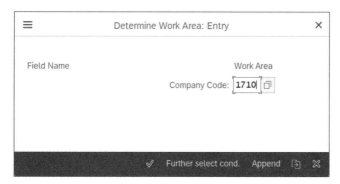

Figure 8.5 Select Company Code

Proceed with the ✓ button. On the next screen, shown in Figure 8.6, you define house banks per company code.

8 Bank Accounting

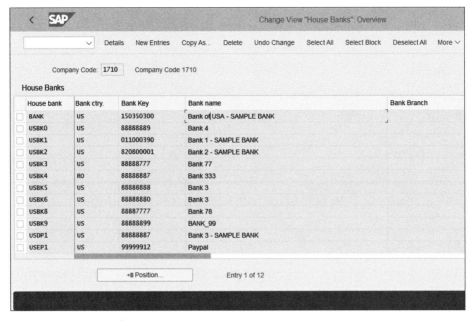

Figure 8.6 House Banks

The following fields should be configured:

- **House Bank**
  This is a freely definable code that defines the house banks. It's good practice to follow a naming convention that includes some letters from the bank name or some other logical sequence within the company code.

- **Bank Ctry**
  The country key of the bank.

- **Bank Key**
  This is the bank key you defined in the previous step.

- **Bank Name**
  The bank name associated with the key, which is populated automatically.

You can create new house banks with the **New Entries** option from the top menu or copy existing house banks with the **Copy As...** option. Then you can fill in or adapt the details of the house bank, as shown in Figure 8.7.

Click the **Address** button to expand the **Address** and **Control Data** sections, as shown in Figure 8.8.

8.1 Master Data

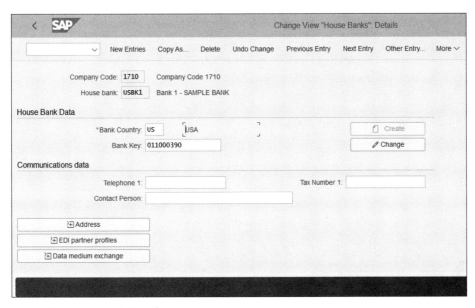

Figure 8.7 House Bank Details

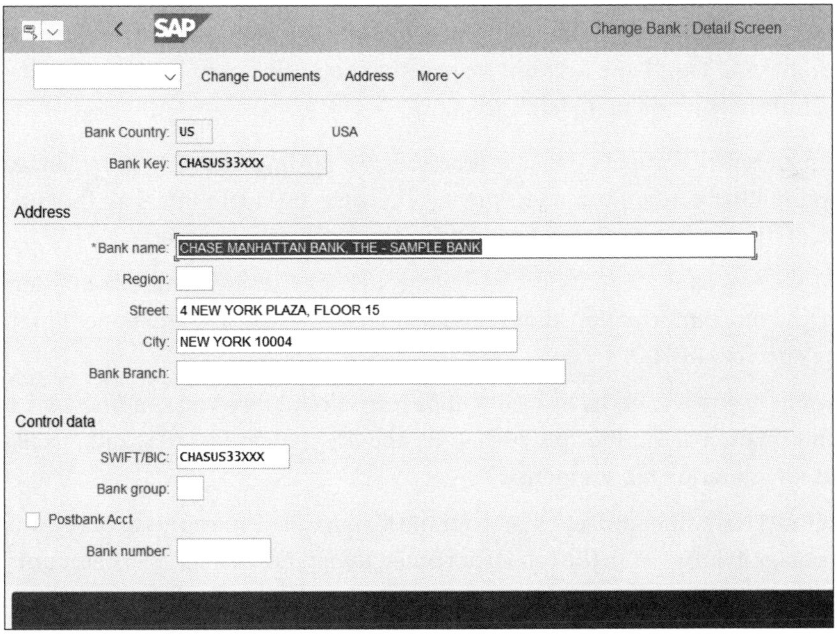

Figure 8.8 Address and Control Fields of Bank Key

Here you can provide additional information about the address and the bank number.

There are also other settings related to house banks in the payment program configuration, but we'll examine those in Section 8.2.

8.1.3 Bank Accounts and IBANs

Bank account maintenance has been changed dramatically in SAP S/4HANA. In the older SAP ERP releases, bank accounts were transported between the clients, whereas in SAP S/4HANA they're considered master data and their maintenance is considered the responsibility of treasury departments. As such, SAP now offers the SAP Fiori Manage Bank Accounts app to maintain the bank accounts, but this isn't all. In fact, there are two options:

- **Bank Account Management Lite (BAM Lite)**
  This option is included in the overall SAP S/4HANA license. With it, the maintenance of bank accounts happens exclusively through the SAP Fiori Bank Account Management Lite app.

- **Bank Account Management (BAM)**
  This option requires an additional license for SAP Cash Management, but in addition to the SAP Fiori Bank Account Management app, you can also use workflow-based processes to maintain bank accounts.

For companies with complex treasury processes, we recommend SAP Cash Management and full Bank Account Management; otherwise, BAM Lite offers all the functionality of the previous GUI transactions plus the SAP Fiori user interface.

Now let's see how bank accounts are created through the Manage Bank Accounts app. You can find more information about this app in the SAP Fiori apps reference library at *http://s-prs.co/v485700*.

As shown in Figure 8.9, in the SAP Fiori apps reference library you can find a lot of useful information about the app, such as its app ID, product features, and installation and configuration requirements.

Start the SAP Fiori launchpad and search for the Manage Bank Accounts app. To do so, click the magnifying glass in the top-right corner, then type "Manage bank accounts" in the text box. Click the **Manage Bank Accounts** tile and the app will open.

8.1 Master Data

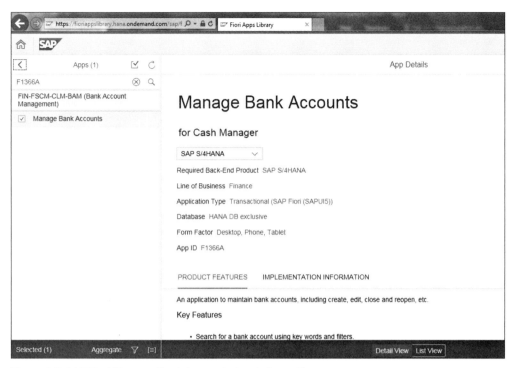

Figure 8.9 SAP Fiori Manage Bank Accounts App Information

You can create a new bank account using the **New Bank Account** button, shown in Figure 8.10.

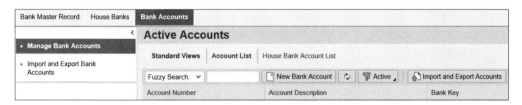

Figure 8.10 Manage Bank Accounts

On the next screen, enter the bank account details, as shown in Figure 8.11. Fields marked with * are required. You have to enter an account opening date, company code, account holder, bank country, bank key (this is the key we created in the previous section), currency, IBAN, account number, account description, and account type.

8   Bank Accounting

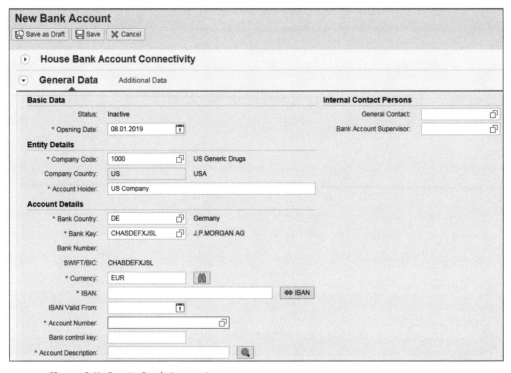

Figure 8.11 Create Bank Account

Using the **IBAN** button, the system can propose an IBAN for you based on the account number and bank key. The IBAN is the international standard for naming bank accounts.

Once you're ready, you can save with the **Save** button or save as a draft with the **Save as Draft** button.

As mentioned, bank accounts are master data. But you can find more information about transferring bank accounts between clients in SAP Note 2437574.

8.2 Automatic Payment Program

The automatic payment program is a central point of all payment functions in SAP S/4HANA, as it was in previous SAP ERP releases. All automatic payment processes—both outgoing and incoming—are managed through the payment program. It serves to close open vendor and customer items that match its selection criteria, to post payments, and to generate payment media such as payment files or checks, as well as supporting documentation such as payment advices and payment lists.

In this section, we'll discuss the parameter and global settings of the automatic payment program. We'll then move on to discuss payment methods and bank determination, before closing out our discussion with a look at some of the common issues with such payment programs.

8.2.1 Automatic Payment Program Parameters

The automatic payment program is available at application menu path **Accounting • Financial Accounting • Accounts Payable • Periodic Processing • F110—Payments**. Figure 8.12 shows the parameter selections of the payment program.

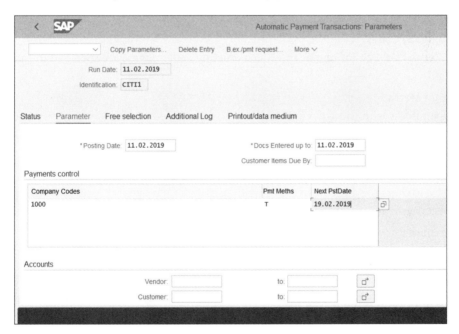

Figure 8.12 Automatic Payment Program

The following fields on this screen are important:

- **Run Date**

  When you run the payment program, you create a payment run, which includes certain selections of open items to be paid. This payment run is uniquely identified with two parameters: run date and identification. The run date is not the posting or payment date; it's just used to identify the payment run and normally is the creation date of the payment run.

- **Identification**
  This is the other field that in combination with the run date uniquely identifies the payment run. Normally in one day there would be multiple payment runs, so the identification name is important. It makes sense to use a good, logical naming convention for the runs so that they can be easily identified—for example, starting with an abbreviation of the bank name or the country for which the payments are relevant.

- **Posting Date**
  This is the posting date of the payment documents that clear the open items.

- **Docs Entered up to**
  Open items with posting dates up to this date will be selected.

- **Company Codes**
  The run could be for one or multiple company codes, which are specified in this field.

- **Pmnt Meths**
  Open items with these payment methods will be selected. The payment method determines how an open item should be paid (e.g., bank transfer or check). We'll discuss the configuration of the payment methods shortly.

- **Next PstDate**
  Used to check the due date of the open items. If an item is overdue on the date of the next payment run or will lose its cash discount, it will be paid in this payment run.

- **Accounts**
  Here you enter a range of vendor or/and customer accounts for which open items are to be processed.

The automatic payment program consists of two steps: payment proposal and payment run. The payment proposal is like a test run; it provides you with a list of proposed payments, which you can check and analyze before actually making payments and closing open items. It also provides error information for any items that couldn't be paid due to various reasons such as missing payment methods or a missing bank account in the master record. The payment run makes postings to bank accounts and clears open items, then generates payment files or checks (although you also have the option to generate the payment media during the payment proposal).

Let's examine the configuration needed for the payment program.

8.2.2 Automatic Payment Program Global Settings

The payment program configuration comprises multiple steps on the country and company code levels. The settings are located at menu path **Financial Accounting • Accounts Receivable and Accounts Payable • Business Transactions • Outgoing Payments • Automatic Outgoing Payments • Payment Method/Bank Selection for Payment Program**. However, there are numerous configuration transactions, and it's easy to get lost within them. That's why it's good to use Transaction FBZP, with which SAP provides a convenient interface in one place for the configuration activities for the payment program.

As shown in Figure 8.13, here you can navigate through the various settings required for the payment program. First click the **All Company Codes** button.

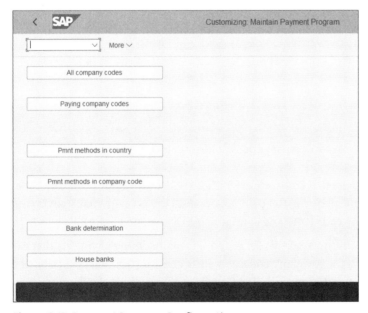

Figure 8.13 Payment Program Configuration

In this configuration activity, you specify the paying company code per company code. It's possible that payments for multiple company codes are managed through one company code, and you establish this relationship here. Double-click company code 1000 and configure it as shown in Figure 8.14.

In the **Paying Company Code Field**, enter "1000" for a 1:1 relationship. On this screen, also select the special general ledger transactions that can be paid (see Chapter 5, Section 5.2.2).

8 Bank Accounting

Figure 8.14 Payment Program Company Codes Configuration

Go back to the main screen of Transaction FBZP and select the **Paying Company Codes** button. Here, make settings for the company codes that actually make payments. Double-click company code 1000 and configure it as shown on Figure 8.15.

Here, specify the minimum amounts for outgoing and incoming payments that can be processed through the company code, as well as control parameters such as permitting exchange rate differences, bills of exchange, and direct debit prenotifications. Another important parameter is the SEPA creditor identification number, which is a required field for making SEPA payments, which are relevant for Europe and which will be examined in detail in Section 8.3.1.

Click the **Forms** button to expand the form settings, and click the **Sender Details** button to expand the settings for the sender of the forms, as shown in Figure 8.16.

Here you specify the forms for the payment advices and accompanying sheet forms, which could be in the older SAPscript form or in the newer PDF form. Then in the sender details for SAPscript forms, you specify the text elements, which are written to be input in the form's header, footer, signature, and sender areas.

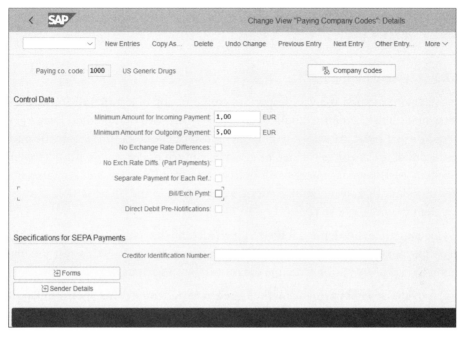

Figure 8.15 Paying Company Codes Configuration

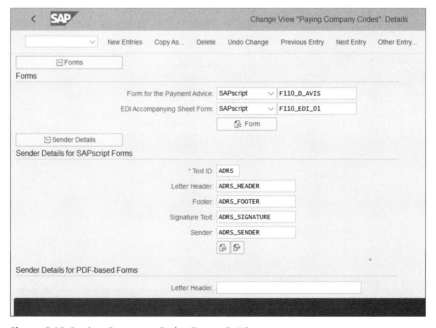

Figure 8.16 Paying Company Codes Forms Settings

8   Bank Accounting

Now we'll configure the payment method and continue with payment program configuration on the payment method level.

8.2.3 Payment Method

The payment method is a very important configuration object in SAP S/4HANA. It determines how payments should be made for items with that method. Each open item to be paid should have a payment method. It can be determined from the master record of the business partner, or it can be entered or derived on the document line item level. Payment methods differ from country to country, and typical examples include wire transfer; bank transfer, such as ACH in the US or SEPA in Europe; payment by check; and so on.

Payment methods are defined at the country level and the company code level. In the main payment program configuration screen shown in Figure 8.13, select the **Pmt methods in country** button to define the country-level settings.

SAP provides several payment methods, relevant per country. As shown in Figure 8.17, for the US payment methods are defined for payments with checks and various types of bank transfers.

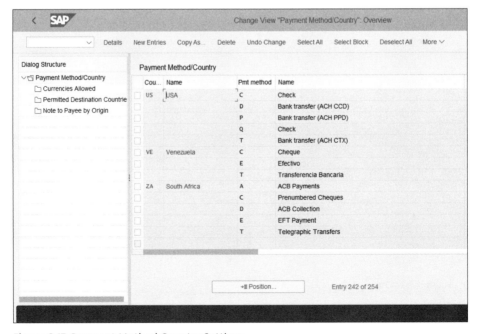

Figure 8.17 Payment Method Country Settings

8.2 Automatic Payment Program

Double-click payment method **C** for the US to check it and modify its settings if necessary. This brings you to the screen shown in Figure 8.18.

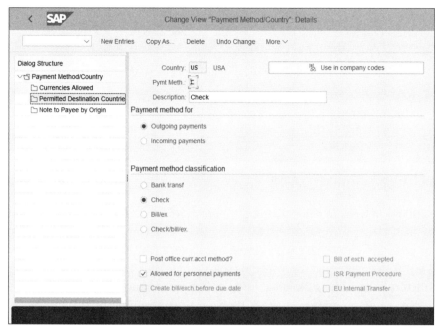

Figure 8.18 Payment Method for Checks

The following fields are important:

- **Payment method for**
  Here you select whether the payment method is for outgoing or incoming payments. Separate payment methods are created for outgoing and incoming payments; it isn't possible to have the same method for both.

- **Payment method classification**
  This determines the type of payment, such as check, bank transfer, or bill of exchange.

- **Required Master Data Classification**
  The group of fields checked here will be required to be maintained in the master record of the business partner; otherwise there will be an error with the payment in the automatic payment program.

- **Posting Details**
  Here you specify the document types used for the payments and clearing with this payment method.

8 Bank Accounting

Scroll down to maintain the payment medium settings for the payment method, as shown in Figure 8.19.

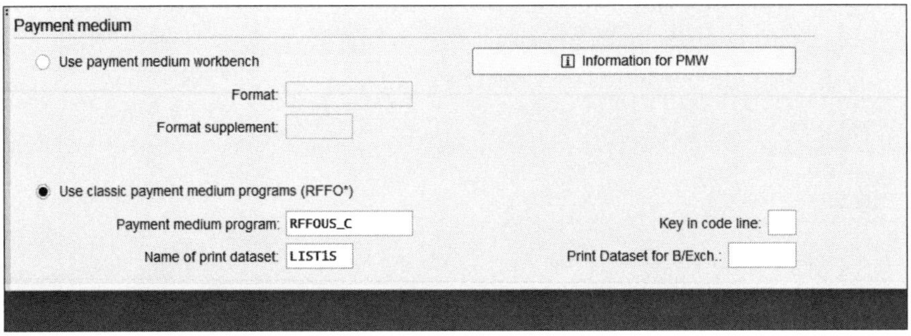

Figure 8.19 Payment Method Payment Medium Settings

There are two options to generate the payment medium (bank files or checks): use the Payment Medium Workbench (PMW) or use classic payment medium programs such as program RFFOUS_C for checks and program RFFOUS_T for payment bank files. PMW is the newer and more flexible solution for bank files from SAP, in which the structure of the file is designed by selecting and arranging specific elements, rather than being programmed as in the older program RFFOUS_T. However, for checks the classic program RFFOUS_C still is the most common solution. We'll talk more about the payment files in Section 8.3.

You can also restrict the use of the payment method for specific currencies by selecting the **Currencies Allowed** option from the left side of the screen and restrict the use of destination countries by selecting the **Permitted Destination Countries** option from the left side of the screen.

Select **Currencies Allowed** and select **New Entries** from the top menu. Enter USD and CAD, as shown in Figure 8.20. As explained in the configuration screen itself, those currencies are then permitted for this payment method. Leaving this table empty means that all currencies are permitted.

Save your entries with the **Save** button, then go back to the main configuration screen of the payment program.

The next step is to define the payment methods per company code. Click the **Pmnt Methods in Company Code** button, which leads you to the payment methods defined on the company code level. Scroll down to find company code 1000, as shown in Figure 8.21.

316

8.2 Automatic Payment Program

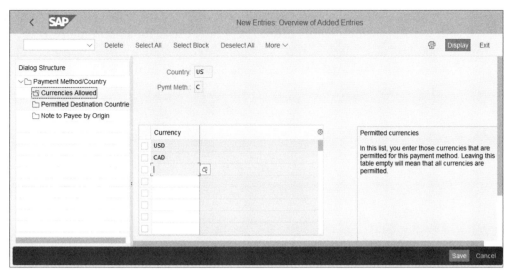

Figure 8.20 Payment Method Permitted Currencies

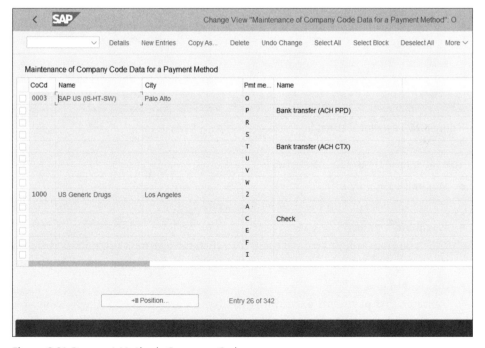

Figure 8.21 Payment Methods Company Code

Payment methods defined per company code are listed in the **Pmt method** column. Only those payment methods defined here can be used for a given company code. Double-click the **C Check** method for company code 1000, which brings you to the configuration screen on the company code level, shown in Figure 8.22.

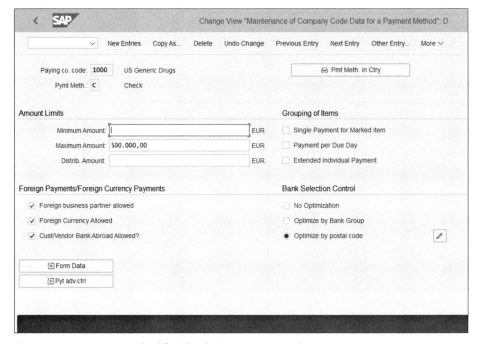

Figure 8.22 Payment Method for Checks in Company Code

You can configure the following fields:

- **Minimum Amount**
  The minimum amount per payment for which the payment method can be used.

- **Maximum Amount**
  The maximum amount per payment for which the payment method can be used. Here we configure that no more than 500,000 EUR can be paid with a check.

- **Distrib. Amount**
  Indicates that payments exceeding this amount are checked if they can be split into several payments totaling a maximum of this amount.

- **Single Payment for Marked Item**
  If you select this checkbox, there will be a separate payment for each line item and the system will not try to group line items.

- **Payment per Due Day**
  Specifies that only items that are due on the same day will be paid with a single payment.
- **Foreign Payments**
  In these fields, you select if foreign currency and/or partners are allowed to be paid with this method.
- **Bank Selection Control**
  Specifies if the system should optimize the payments by selecting the optimal pair of banks in case you divide the banks into groups.

Extend the form data for the method by clicking the **Form Data** button.

As shown in Figure 8.23, you specify the form name for printing checks—in this case, the standard SAP form F110_PRENUM_CHCK for prenumbered checks—but you can develop your own form and assign it here. You also maintain the details for the drawer of the check to be printed on the checks.

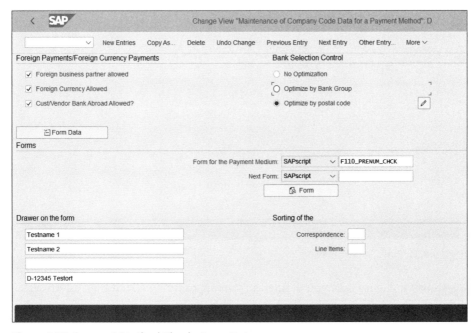

Figure 8.23 Payment Method Checks Form Data

Save your entries with the **Save** button, then go back to the main configuration screen of the payment program.

8.2.4 Bank Determination

The next step in the automatic payment program configuration is to define the bank determination. This determines the sequence of selection of house banks per payment method from which payments are to be made. Also, you specify the bank accounts here used per house bank and the general ledger accounts to be posted during the payment process.

From the main payment program configuration screen shown in Figure 8.13, select the **Bank Determination** button.

As shown in Figure 8.24, you'll see a list of paying company codes. Select company code 1000 and click **Ranking Order** on the left side of the screen. Then, using the **New Entries** option from the top menu, you can create the ranking order of house banks for the company code.

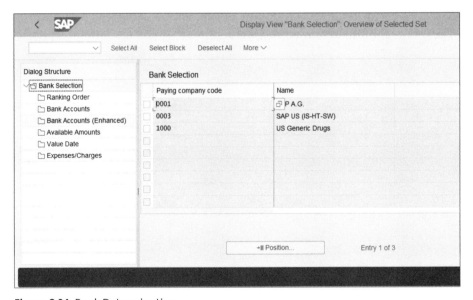

Figure 8.24 Bank Determination

As shown in Figure 8.25, you define the ranking order by which the system selects house banks per payment method and, optionally, currency. If there are more than one house bank specified per combination of payment method and currency, they are marked in the **Rank. Order** column sequentially: 1, 2, and so on. Then the house bank with ranking order 1 will be selected first, and only if payment can't be made from this house bank due to an amount limitation will the system select the next house bank.

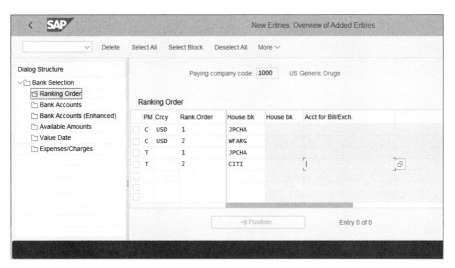

Figure 8.25 House Banks Ranking Order

In the next step, select **Bank Accounts** from the left side of the screen. On the screen shown in Figure 8.26, you determine the sequence of bank accounts per house bank from which payments are made. Click the **New Entries** option from the top menu.

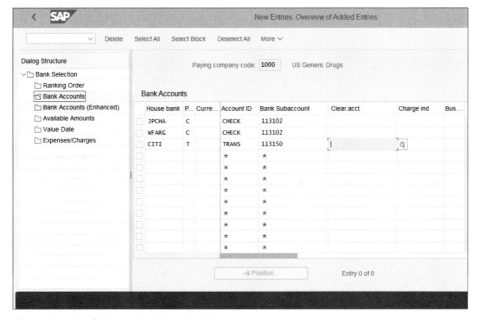

Figure 8.26 Bank Accounts Determination

8 Bank Accounting

In this configuration screen, you select the house banks and payment methods and assign to them bank account IDs and general ledger accounts to be posted to. The bank account ID represents an actual bank account in your bank from which outgoing payments are made and which can receive incoming payments.

You can also check the bank account IDs when you select the **House Banks** button from the main payment program configuration screen, shown in Figure 8.13.

Select company code 1000 and click **House Banks** on the left side of the screen, as shown in Figure 8.27. Then select one of the house banks and click **Bank Accounts** on the left side of the screen to see a list of the defined bank accounts per house bank. If you click **Create Bank Account** from the top menu you'll be taken to the Bank Account Management SAP Fiori app to create bank accounts, which we discussed in Section 8.1.3.

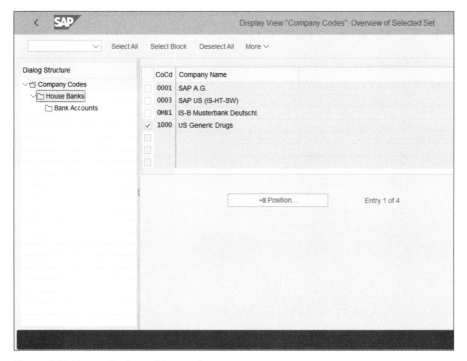

Figure 8.27 House Banks and Accounts

With that, we complete the configuration of the payment program. Now we'll review some common issues when running it because sometimes the error log isn't very useful and you may benefit from some practical advice on how to tackle some of the issues you may encounter.

8.2.5 Common Issues with the Payment Program

When you run the payment program, as already discussed, you first run a payment proposal, which is like a test run. Then you can check the proposed payments for any possible errors. If you find any errors, you can delete the payment proposal and, after resolving the issue(s), run it again.

As discussed, the automatic payment program is available at application menu path **Accounting • Financial Accounting • Accounts Payable • Periodic Processing • F110—Payments**. From the main screen of the payment program shown in Figure 8.28, select **Edit Proposal** from the top menu.

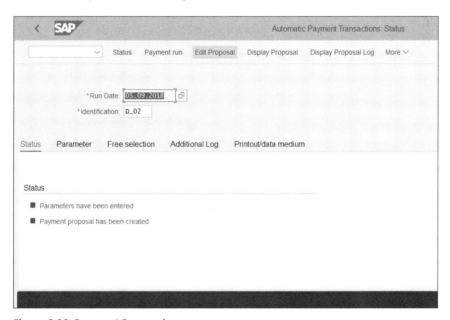

Figure 8.28 Payment Proposal

Then the system asks you whether you want to see the items per specific accounting clerks, as maintained in the business partner master data, or all accounting clerks, as shown in Figure 8.29.

8 Bank Accounting

Figure 8.29 Accounting Clerks Selection

Proceed with the **Continue** button, and you'll see the list of proposed payments and exceptions, as shown in Figure 8.30.

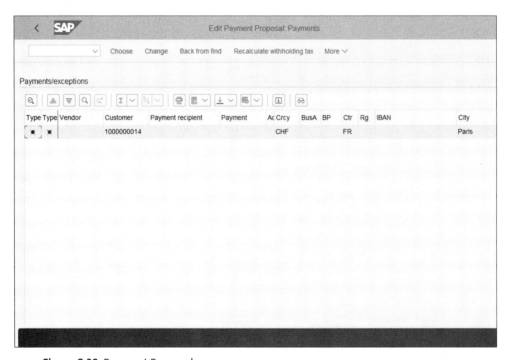

Figure 8.30 Payment Proposal

Payment proposals which are OK are marked in green, whereas exceptions are marked in red. Double-click a red line to see its details. If there are more than one item per vendor/customer, they will be listed here. In the case shown in Figure 8.31, there is one item.

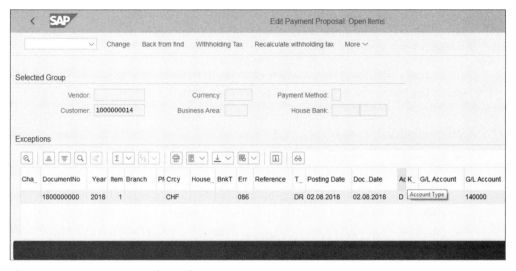

Figure 8.31 Payment Proposal Details

Double-click this item and you will see the error, as shown in Figure 8.32.

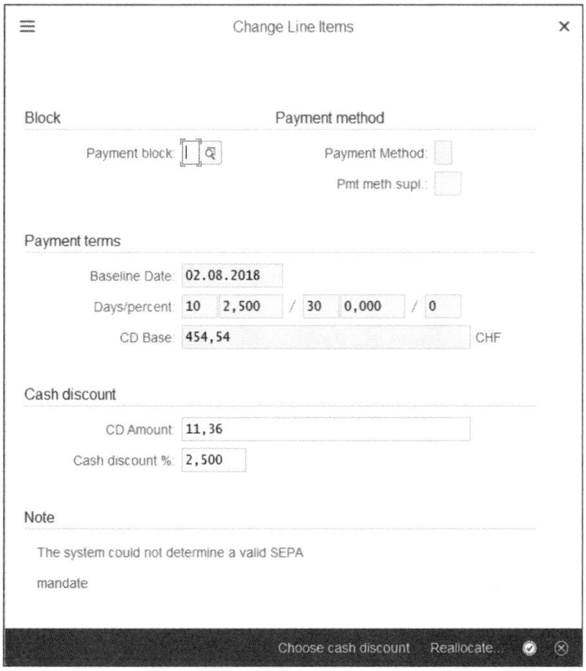

Figure 8.32 Payment Proposal Error

So you see that in this case the error is that the system could not determine a valid SEPA mandate. This means you have to check the customer master record and create a SEPA mandate for this customer.

You can also get an overview of the encountered issues by clicking the **Display Proposal Log** button from the main screen of the payment program.

Here are some other common error messages from the payment program and potential resolutions:

- **No Valid Payment Method Found**
  This means that the system can't determine the payment method to use. Check the document and the business partner master. The payment method needs to be either derived from the master record or entered in the document.

- **Company Code [xxx] Does Not Appear in Proposal [xxx] (Run Date and ID)**
  This isn't a very explanatory message. Check if there are open items based on the parameters of the run. But this message also could appear for other reasons; check the proposal log for more clues.

- **Account or Item Is Blocked for Payment**
  This is an easy message. It means that either the business partner master record or the open item itself contains a payment block.

- **Account [xxx] Blocked by Payment Proposal [xxx]**
  This means that the selected account is also participating in another payment run, so the system locks it in other payment proposals. You have to finish the previous payment run first.

- **No Payment Possible because Items with a Debit Bal. Still Exist**
  This is caused by a vendor having a debit balance. For payment, only vendors with credit balances are considered.

- **Required Details Are Not Maintained in Business Partner Master**
  As per the configuration of the payment program, some fields are required in the business partner master data. If they're not maintained, you'll get this error.

8.3 Payment Files

Payment files are an important part of the configuration of the bank accounting in SAP S/4HANA. They're used to initiate payments from and to your house banks automatically using data from the SAP S/4HANA system.

The system can generate both incoming and outgoing payment files, based on various common standards, such as SEPA, which is now the standard for payments in the eurozone; ACH in the United States; and many other country-specific formats.

8.3.1 SEPA Payment Files

Single Euro Payments Area (SEPA) is a project of the European Union (EU), which aims to simplify payments in euros between member states of the EU. It's becoming the de facto standard for electronic payments in Europe, so it's very important for SAP S/4HANA systems to manage.

There are two types of payments that should be managed by SAP S/4HANA, as follows:

- **SEPA credit transfers**
  SEPA credit transfers are used for outgoing payments. SAP S/4HANA generates outgoing payment files, which you can send to your bank to pay your vendors.
- **SEPA direct debits**
  SEPA direct debits are used to receive payments from customers automatically from their accounts. This is done based on a specific written agreement, called a *SEPA mandate*, which needs to be stored in the customer master record. Then the payment program can generate the SEPA direct debit files.

You need to have separate payment methods defined for SEPA credit transfer and direct debit. Then from the main payment program configuration screen, shown in Figure 8.13, select the **Pmnt Methods in Country** button and create payment methods for SEPA credit transfer and direct debit.

Select **Use Payment Medium Workbench** and enter "CGI_XML_CT" in the **Format** field for SEPA credit transfer, as shown in Figure 8.33.

For direct debit, enter "CGI_XML_DD" in **Format**. These are standard formats provided by SAP, which replace older standard formats SEPA_CT and SEPA_DT. These formats are used to generate XML payment files. The format structure is based on ISO 20022. They correspond to the implementation template defined by banks and corporate customers through the Common Global Implementation (CGI) initiative and are fully SEPA compliant.

Using the **Format settings** button, you can check the settings of the payment format. In Figure 8.34, you can see the settings of payment format CGI_XML_CT. It's a tree-like structure in which you can define XML nodes and the data that should go into them, which could be derived from various fields from the payment structure of the system.

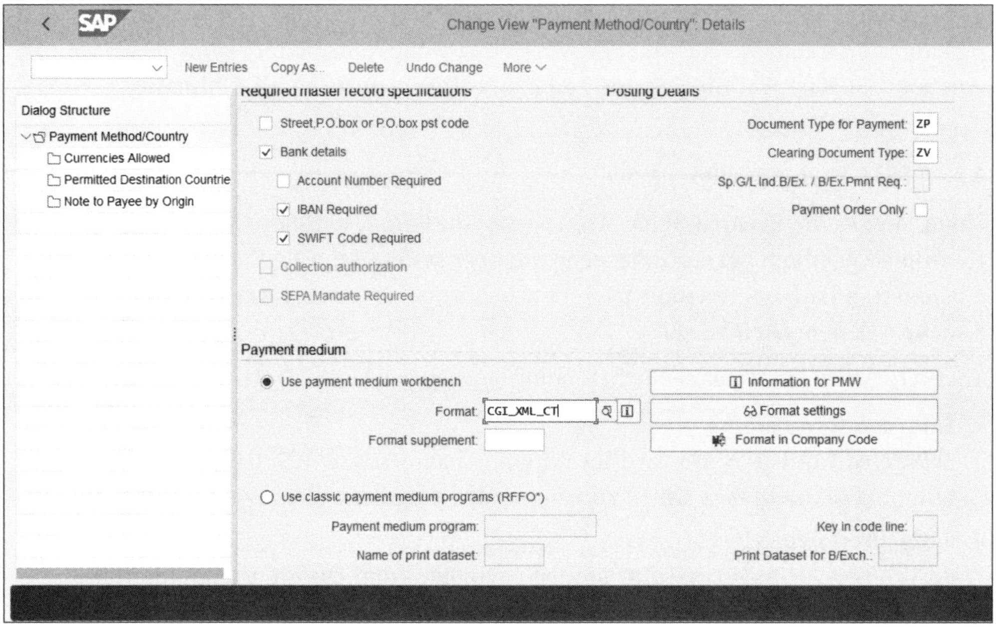

8 Bank Accounting

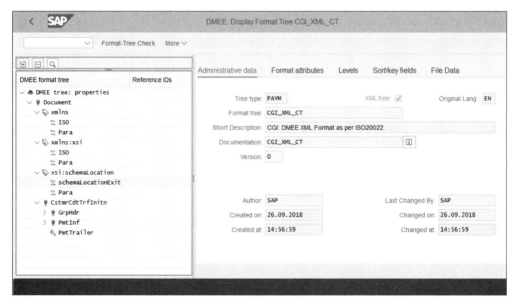

Figure 8.33 Payment Method Medium Configuration

Figure 8.34 Payment Format CGI_XML_CT

If you need to modify the format, you should make a copy, starting with Z, and modify your custom format. Transaction DMEE allows you to maintain the payment format.

As shown in Figure 8.35, you have to select **PAYM** for the **Tree Type**, which is used for payment files, and enter the **Format Tree** name. Then with the **Change** button, you can make changes as appropriate.

Figure 8.35 Payment Format Maintenance

Then you need to assign your payment format to the payment method, as shown in Figure 8.33.

8.3.2 Other Common Formats

SAP provides myriad bank formats for various countries, which can be used either out of the box or with minor modifications, as required by specific banks. The newer method of providing such formats is using the Payment Medium Workbench, rather than older classic programs, in which one program provides one payment file. PMW offers a flexible, easy-to-use interface in which most requirements can be accomplished with configuration of the XML nodes and no additional ABAP programming is needed.

To view the existing standard formats provided, check Transaction DMEE with tree type PAYM, as was shown in Figure 8.35. Using the [🔍] button to the right of the **Format Tree** field or with the [F4] button, you can view the available formats.

Figure 8.36 shows standard formats defined for the United States. As you can see, formats for direct debit and XML formats conforming to ISO 20022 are provided. Also, there's a payment format for positive pay, which is used to clear checks.

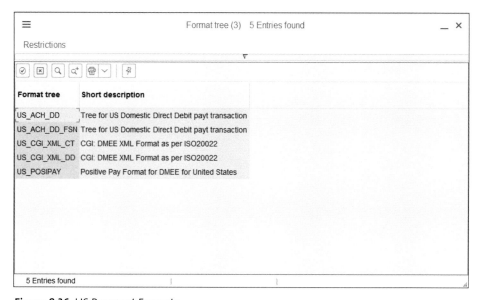

Figure 8.36 US Payment Formats

SAP also provides formats for many countries based on the CGI: DMEE XML format as per ISO 20022, as shown in Figure 8.37.

The format for each country starts with the country code and is a variation of the common CGI_XML format for debit and credit transfers.

The best practice when setting up payment file formats as part of your SAP S/4HANA project is to generate sample files using the standard formats provided and get in touch with house banks to make sure the formats are accepted. If any changes are needed, you should use the standard formats as a basis to create new ones in the Z name range and modify them appropriately.

8.4 Electronic Bank Statements

Figure 8.37 CGI XML Formats

If you're doing a brownfield SAP S/4HANA implementation, you need to make sure that the formats used are still supported. For example, now the older SEPA formats SEPA_CT and SEPA_DT are obsolete and need to be replaced in the settings of the payment methods with the CGI-based formats CGI_XML_CT and CGI_XML_DD.

8.4 Electronic Bank Statements

The electronic bank statement is a file that you receive from the house bank, which contains the various bank transactions posted for a given period, such as deposits, withdrawals, bank charges, and so on. SAP S/4HANA provides functionality to read

8 Bank Accounting

this file and automatically make postings to bank accounts, business partner accounts, and other general ledger accounts based on this information. Thus a lot of manual work, time, and effort is saved, and transparency into and security of the bank postings in the system is provided.

We'll start by providing an overview of the electronic bank statement process. Then we'll delve into its configuration by defining the settings for account symbols, posting rules, and transaction types.

8.4.1 Overview

Importing a bank statement is done with Transaction FF_5, available at application menu path **Accounting** • **Financial Accounting** • **Banks** • **Input** • **Bank Statement** • **FF_5 – Import**.

As shown in Figure 8.38, SAP again provides various standard formats for electronic bank statements. Some of them are international and some are country-specific. SWIFT MT940 is the most common international format. You need to check what format your bank is using and make changes if necessary.

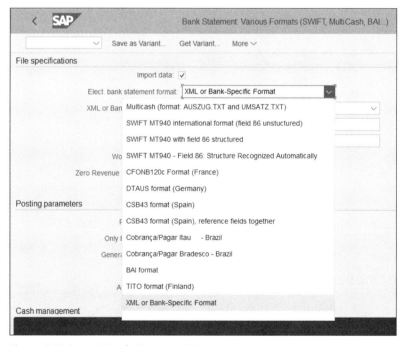

Figure 8.38 Import Bank Statement Formats

332

Let's see now what configuration settings are required for the electronic bank statement by following menu path **Financial Accounting • Bank Accounting • Business Transactions • Payment Transactions • Electronic Bank Statement • Make Global Settings for Electronic Bank Statement**.

The system asks you for a chart of accounts, as shown in Figure 8.39. All global settings for electronic bank statements are dependent on the chart of accounts.

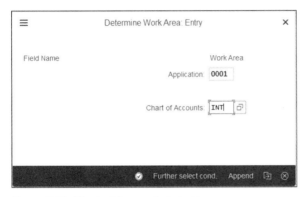

Figure 8.39 Chart of Accounts Selection

Enter your **Chart of Accounts**, "INT", and confirm with the ✓ button. On the next screen, you define the account symbols.

8.4.2 Account Symbols

Account symbols are groupings under which you classify similar bank accounts. Figure 8.40 shows the screen in which you define the various settings for the electronic bank statement by selecting them from the left side. Double-click **Create Account Symbols** and the defined account symbols will be displayed on the right side of the screen. Typical account symbols include deposits, interest, checks received, checks out, and so on.

Click **Assign Accounts to Account Symbol** on the left side of the screen, which will result in the screen shown in Figure 8.41.

8 Bank Accounting

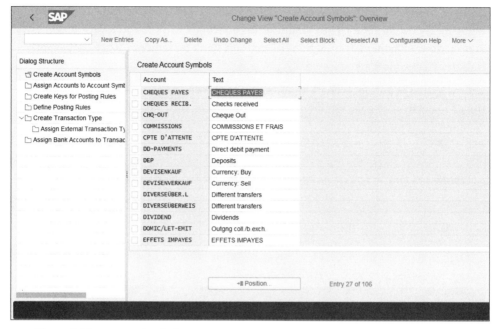

Figure 8.40 Account Symbols

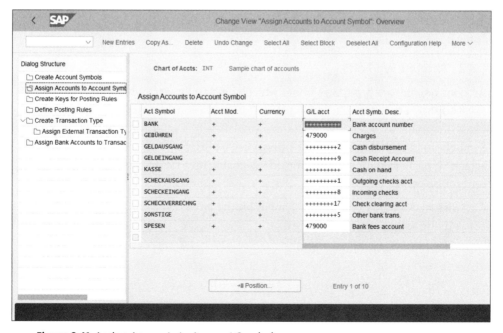

Figure 8.41 Assign Accounts to Account Symbols

In this configuration activity, you assign general ledger accounts to be posted per account symbols. They are assigned using masking, where the **+** sign plays the role of a wildcard. So +++++++++2, for example, means accounts ending in 2.

8.4.3 Posting Rules

In the next step, click **Create Keys for Posting Rules** from the left side of the screen. This will result in the screen shown in Figure 8.42.

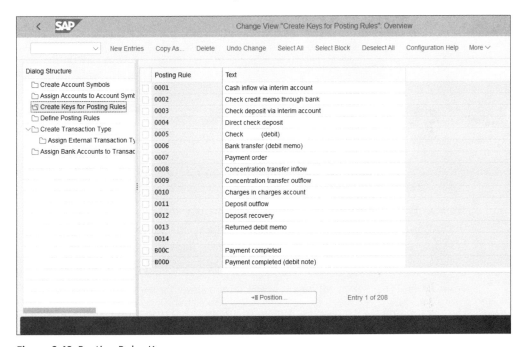

Figure 8.42 Posting Rules Keys

Posting rules define how bank transactions should be posted. They are identified with four-character codes, which we define here by selecting **New Entries** from the top menu.

Then click **Define Posting Rules** from the left side of the screen. As shown in Figure 8.43, you define per posting rule which account symbols should be posted on the debit and credit side and using which posting keys. You also specify the document type to be used for the posting and the type of account, such as general ledger account or subledger account.

8   Bank Accounting

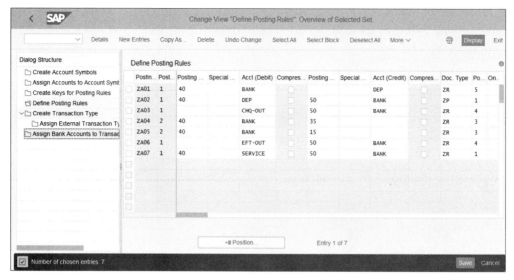

Figure 8.43 Define Posting Rules

8.4.4 Transaction Types

Transaction types are used to classify similar bank transactions. You define them in the SAP S/4HANA system and then assign to them external transaction types, which are bank-defined and are present in the electronic bank statement file.

From the main configuration screen of the electronic bank statement, click **Create Transaction Type**, which will result in the screen shown in Figure 8.44.

Here you name the transaction types. To create the transaction types, you need to have researched the external transaction types provided by your bank and have a clear strategy for how you will assign them.

Then select a transaction type and click **Assign External Transaction Types to Posting Rules** from the left side of the screen.

As shown in Figure 8.45, assign the external transaction types (in this case, 009 and 051) to your transaction types. Then assign the posting rule defined before, which will be used for posting. You can also set a + or – sign to be applied to the amount from the bank statement and interpretation algorithm, which is used to identify open items in SAP S/4HANA. The processing type defines what type of amount is this: opening balance, inflow, outflow, closing balance, and so on.

336

8.4 Electronic Bank Statements

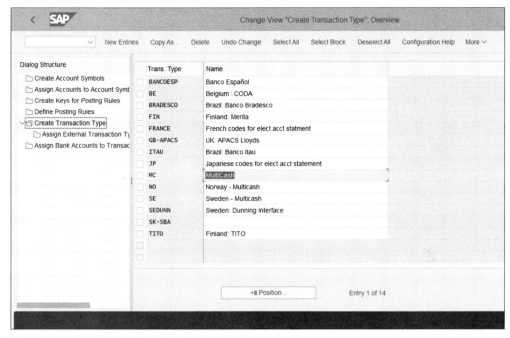

Figure 8.44 Create Transaction Types

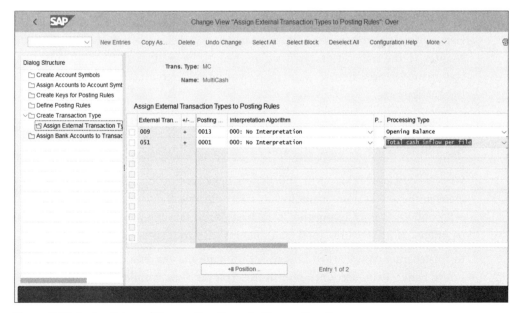

Figure 8.45 Assign External Transaction Types to Transaction Types

8   Bank Accounting

You then need to assign bank accounts to transaction types, as shown in Figure 8.46.

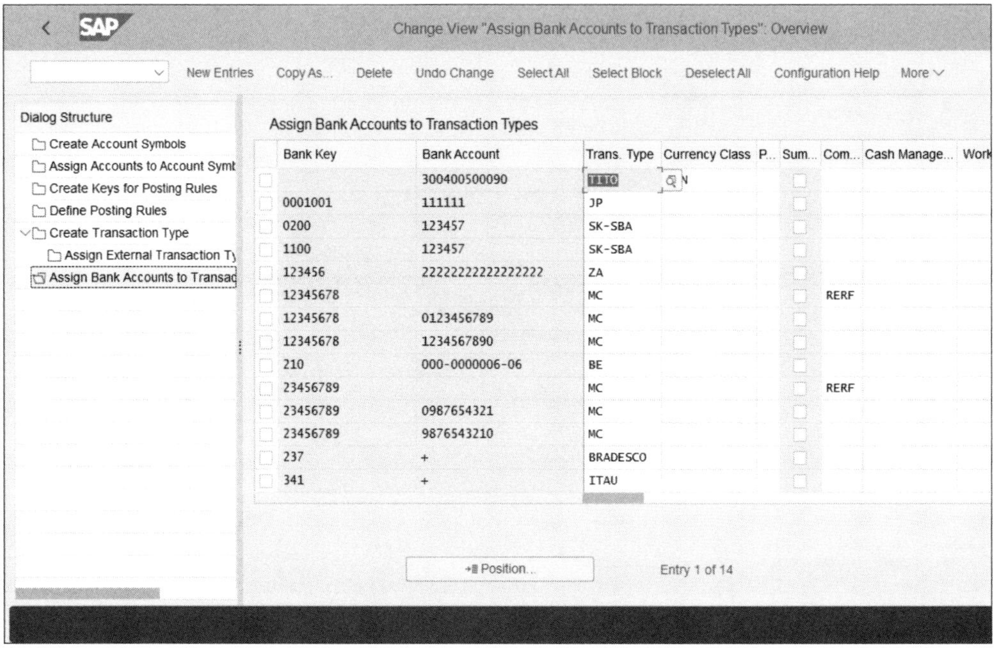

Bank Key	Bank Account	Trans. Type	Currency Class	P...	Sum...	Com...	Cash Manage...	Work
	300400500090	TTT0	N					
0001001	111111	JP						
0200	123457	SK-SBA						
1100	123457	SK-SBA						
123456	2222222222222222	ZA						
12345678		MC					RERF	
12345678	0123456789	MC						
12345678	1234567890	MC						
210	000-0000006-06	BE						
23456789		MC					RERF	
23456789	0987654321	MC						
23456789	9876543210	MC						
237	+	BRADESCO						
341	+	ITAU						

Figure 8.46   Assign Bank Accounts to Transaction Types

Here you assign the bank keys and bank accounts, which we defined previously in the chapter, to the transaction types we defined in the settings of the electronic bank statement. So the system now can know which bank accounts to update based on the transaction type.

8.5   Summary

We covered a lot of material in this chapter. Bank accounting is important to multiple areas in the SAP system, including accounts payable and accounts receivable. Properly configured bank accounting is the basis for streamlined outgoing and incoming payments, which ensures good relationships with your customers and vendors.

In this chapter, you learned how to set up bank master data, including bank keys, house banks, bank accounts, and IBANs. You are now familiar with the new Bank Account Management SAP Fiori app to manage bank master data, which is required in SAP S/4HANA.

We configured the automatic payment program both for outgoing and incoming payments. The automatic payment program is the backbone of the bank processes in SAP S/4HANA. You learned how to set up payment methods on both country and company code levels. Now you know how to link the house banks and accounts with the payment methods and ensure smooth payment operations.

You also learned how to set up payment files for incoming and outgoing automatic payments. We paid special attention to the SEPA payments, which are now standard for euro payments.

Finally, we configured the electronic bank statement, which enables you to import bank statements from your bank and thus automatically close vendor and customer open items and update your bank accounts.

With that, we finish our guide on financial accounting processes and move on with the controlling configuration.

Chapter 9
General Controlling and Cost Element Accounting

This chapter gives step-by-step instructions for configuring general controlling settings and settings for cost element accounting, which is now fully integrated with financial accounting in SAP S/4HANA. You'll learn how to create cost elements as general ledger accounts and cost element groups and learn about the actual postings in cost element accounting.

We finished the configuration of financial accounting in SAP S/4HANA, so now we'll move to the other big financial area: controlling, commonly referred simply as CO. Controlling provides invaluable information and enables flexible analysis for internal managerial accounting. However, in today's business world the requirements of external legal accounting and internal managerial accounting often interrelate and are becoming increasingly more demanding. Therefore SAP S/4HANA provides fully integrated financial accounting and controlling, sharing the same tables and data structures, yet enabling separate, state-of-the-art analysis and reports that meet the requirements both of external users of accounting information such as tax authorities and auditors and of internal managers and controllers.

We'll start the configuration of controlling with the general controlling settings, which are the foundation of all other controlling areas, such as product costing and profitability analysis. Then we'll set up the required master data for cost element accounting and examine the actual postings on the cost element accounting level.

9.1 General Controlling Settings

By *general controlling settings*, we mean settings that apply to all controlling components, such as maintaining the controlling area, number ranges, versions, and so on.

The general controlling settings are located in the customizing menu under **Controlling • General Controlling**, as shown in Figure 9.1.

9 General Controlling and Cost Element Accounting

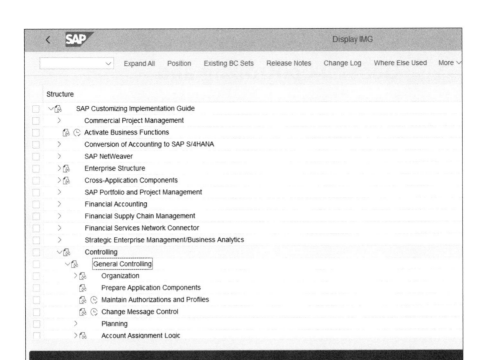

Figure 9.1 General Controlling Settings

We'll start with the definition of the controlling area.

9.1.1 Maintain Controlling Area

The controlling area is the main organizational object in controlling. The main business decision is whether to use one controlling area across multiple company codes or to use separate controlling areas per company code. In most cases, cross-company code controlling makes sense because the controlling function for most companies is executed across legal entities, countries, and regions.

You've seen how a new controlling area can be created or copied from an existing controlling area in Chapter 3, Section 3.2.3. We also already assigned our company code 1000 to controlling area 0001. Now, to modify the controlling area general settings, follow menu path **Controlling • General Controlling • Organization • Maintain Controlling Area** and select the **Maintain Controlling Area** activity, as shown in Figure 9.2.

9.1 General Controlling Settings

Figure 9.2 Controlling Area Activities

Next, select the checkbox next to controlling area 0001, as shown in Figure 9.3, and click **Activate Components/Control Indicators** on the left side of the screen.

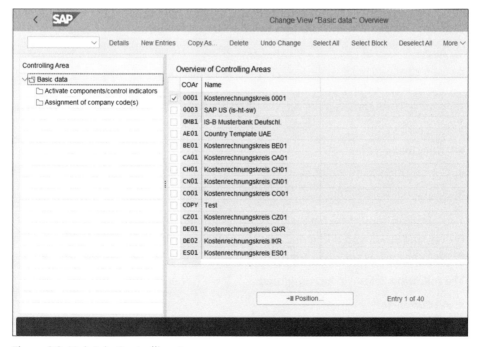

Figure 9.3 Maintain Controlling Area

343

9 General Controlling and Cost Element Accounting

In this configuration transaction, you activate and deactivate the various controlling components and set control indicators, as shown in Figure 9.4.

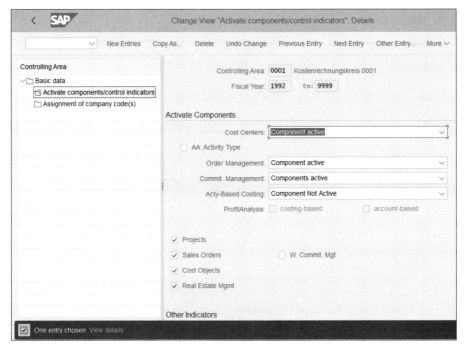

Figure 9.4 Controlling Area Components and Indicators

You can activate components such as the following:

- Cost centers
- Order management
- Commitment management
- Activity-based costing

You also can indicate whether account-assignment fields such as projects, sales orders, and so on should be updated in controlling. Scrolling down, you configure further settings, as shown in Figure 9.5.

The following fields can be configured:

- **All Currencies**

  Controls whether values are updated only in the controlling area currency, or also in transaction and object currency (the currency of the controlling objects such as cost center or internal order). Normally you should check this indicator.

- **Variances**
  This indicator activates the calculation of variances for primary cost postings.
- **CoCd Validation**
  This indicator applies to cross-company code postings. If you set it, postings to an account assignment object such as a cost center or an order can be made only from the company code defined in the master data for the account assignment object.

We covered the assignment of company codes in Chapter 3, Section 3.2.3.

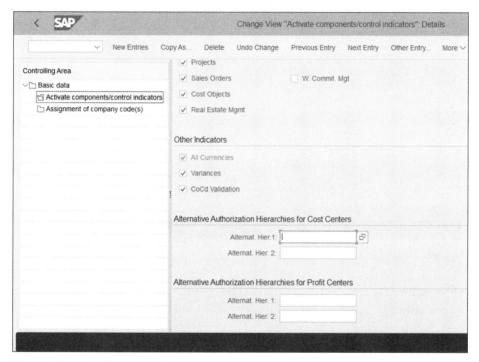

Figure 9.5 Controlling Area Other Settings

9.1.2 Number Ranges

Number ranges provide document numbers for the controlling transactions. In general, in controlling there are no legal requirements that affect the numbering of documents. Still, it makes sense to use continuous, internal number assignment.

To set up the number ranges for controlling, follow menu path **Controlling • General Controlling • Organization • Maintain Number Ranges for Controlling Documents**. In the initial screen, shown in Figure 9.6, you can see that the number range object for controlling documents is RK_BELEG.

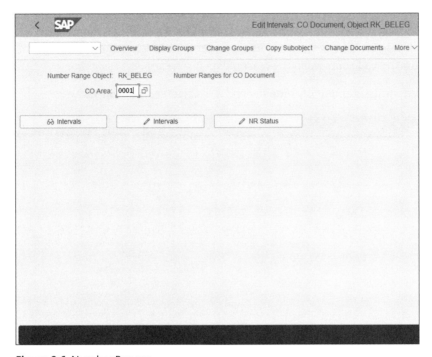

Figure 9.6 Number Ranges

Enter controlling area 0001 and click the **Change Intervals** button ✎ Intervals. In the screen shown in Figure 9.7, set up number ranges and define their from and to numbers. In this example, documents posted in number range 01 will start with number 100000000 and will be assigned internally consecutively because the **Ext** checkbox is not checked.

Now go back, and from the screen shown in Figure 9.6, select **Change Groups** from the top menu.

Figure 9.8 shows the screen in which you can assign the various controlling transactions to groups, which represent the previously created number ranges. So for example, you can see that transactions **COIN: CO Through-Postings from FI** and **KAZO: Down Payment** are assigned to number range 01. Below you have other groups and other assigned transactions. And on top of the list, you have the nonassigned transactions in the section **Nonassigned Elements**.

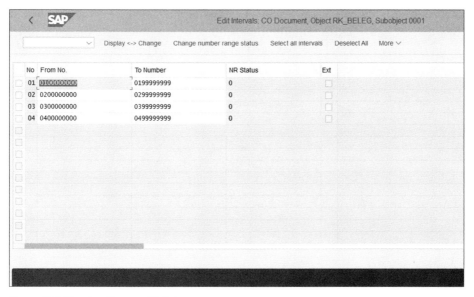

Figure 9.7 Number Ranges Maintenance

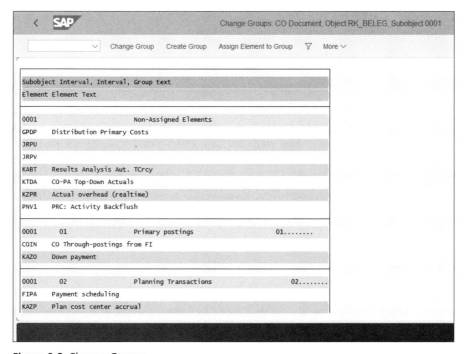

Figure 9.8 Change Groups

9 General Controlling and Cost Element Accounting

Let's assign **GPDP: Distribution Primary Costs** to number range 03 in the **Assessment, Distribution, Order Settlement Internal** section. Position the cursor on **GPDP Distribution Primary Costs** and select **Assign Element to Group** from the top menu. Then a pop-up screen appears, as shown in Figure 9.9.

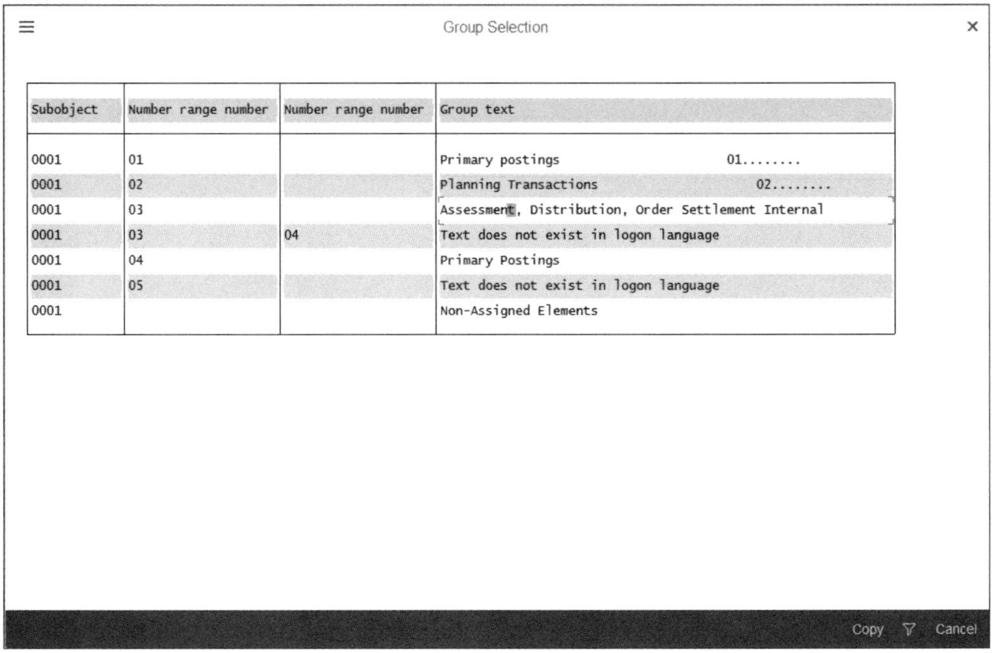

Subobject	Number range number	Number range number	Group text
0001	01		Primary postings 01........
0001	02		Planning Transactions 02........
0001	03		Assessment, Distribution, Order Settlement Internal
0001	03	04	Text does not exist in logon language
0001	04		Primary Postings
0001	05		Text does not exist in logon language
0001			Non-Assigned Elements

Figure 9.9 Assign Transactions to Groups

Here, double-click the group you want to assign the transaction to—in this case, **Assessment, Distribution, Order Settlement Internal**. Then **GPDP Distribution Primary Costs** disappears from the nonassigned transactions list and appears in the **Assessment, Distribution, Order Settlement Internal** list.

After assigning the nonassigned transactions, save your entries with the **Save** button.

9.1.3 Versions

A version is a customizing object that contains fiscal year–dependent indicators for plan and actual data for one controlling area. They are used to distinguish between actual and plan postings in controlling. In your controlling area, you should have a few versions to manage the actual and/or plan data.

9.1 General Controlling Settings

To maintain versions, follow menu path **Controlling • General Controlling • Organization • Maintain Versions**. In the configuration screen, shown in Figure 9.10, you'll see a list of the controlling versions defined, which are standard versions, and you can create or copy additional versions.

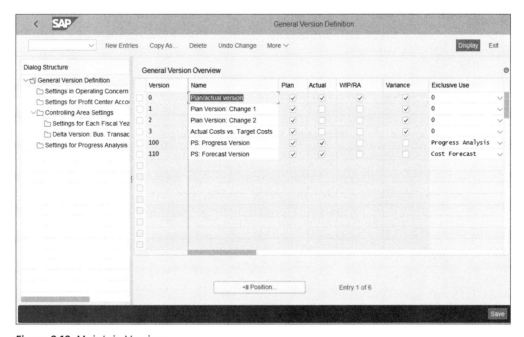

Figure 9.10 Maintain Versions

The following fields are important:

- **Version**
  Version key to identify the version. Normally version 0 is used to handle both actual and plan data and is the main version for controlling.

- **Name**
  Meaningful description of the version.

- **Plan**
  A check in this column indicates that the version contains plan data.

- **Actual**
  A check in this column indicates that the version contains actual data.

- **WIP/RA**
  Controls whether data from results analysis or work in progress (WIP) calculation can be written to the version.

9 General Controlling and Cost Element Accounting

- **Variance**

  Controls whether data from the variance calculation can be written to the version.

- **Exclusive Use**

  Select an option from the dropdown list if the version is to be used exclusively by a specific application.

> **Note**
>
> We'll customize the settings under the **Settings in Operating Concern** option on the left side of the screen in Chapter 13.

Now select version 0 and click the **Settings in Operating Concern** option from the left side of the screen.

Figure 9.11 shows the year-dependent settings for version 0. You can lock the version per year. You also can enable integrated planning per year, and you can allow copying, which means that the version can be used as reference for another version in a copy procedure.

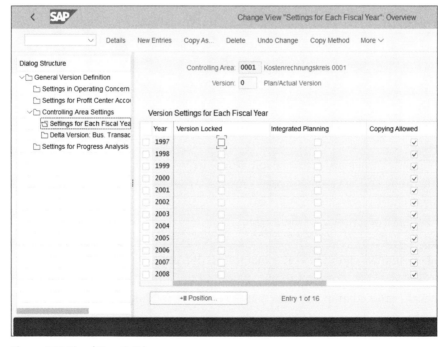

Figure 9.11 Fiscal Year Settings

9.1 General Controlling Settings

Select the last available year and click the **Copy As...** option from the top menu. Then you can maintain the settings for the latest years, as shown in Figure 9.12.

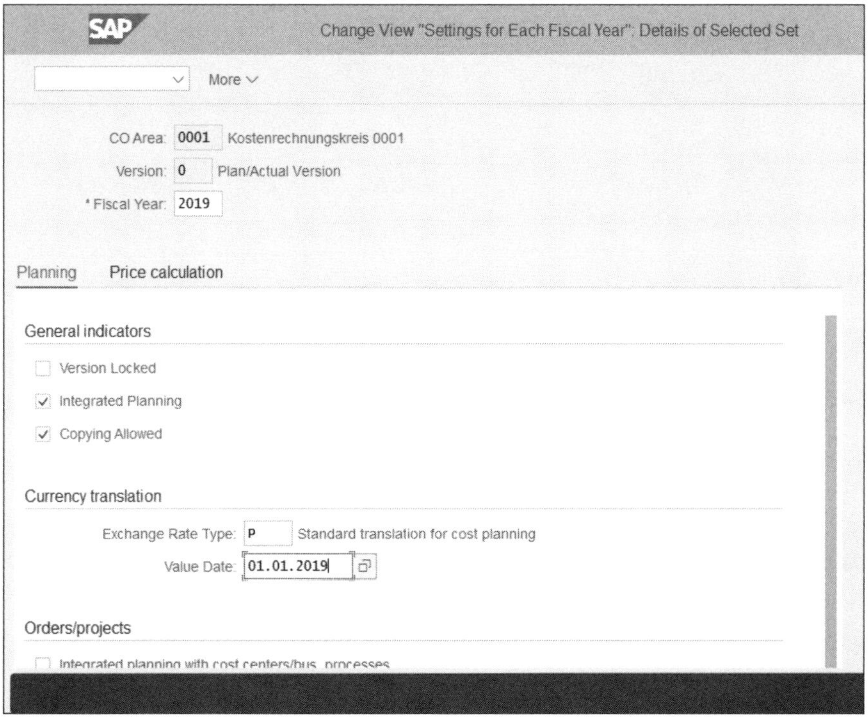

Figure 9.12 Copy Fiscal Year Settings

Then maintain the next few years in a similar fashion.

In the next step, you need to specify which general ledger will provide the data to the versions that handle actual data by following menu path **Controlling • General Controlling • Organization • Define Ledger for CO Version**.

In the screen shown in Figure 9.13, you assign a ledger to a controlling version.

Version 0 has to be assigned to the 0L leading ledger. If you have other versions that manage actual data, you need to create a new entry here to assign the relevant general ledger.

351

9 General Controlling and Cost Element Accounting

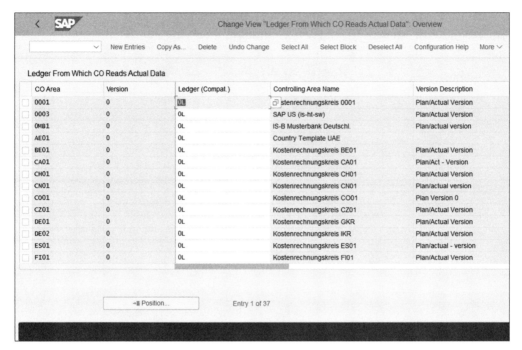

Figure 9.13 Ledger for Controlling Postings

9.2 Master Data

Controlling uses various master data objects such as cost elements, cost centers, internal orders, and so on. In this section, we'll focus just on the cost element master data. In the following chapters, we'll set up cost centers, internal orders, and other controlling master data objects as we cover the various controlling areas.

9.2.1 Cost Elements

Cost elements are the accounts of the controlling modules. All controlling postings flow into cost elements. Normally all profit and loss accounts are set as cost elements, and there are also secondary cost elements, which are used not purely in financial accounting postings but also in internal controlling postings.

The big difference in SAP S/4HANA compared with previous SAP releases is that the cost elements are fully integrated with the general ledger. There is now in SAP

S/4HANA no separate cost element master data object. Instead the cost elements are types of general ledger accounts with special settings in the master record.

Let's look at the master record of a cost element. The application menu path **Accounting • Controlling • Cost Element Accounting • Master Data • Cost Element • Individual Processing • FS00 - Edit Cost Element** now points to Transaction FS00, which is used to maintain general ledger accounts. If you enter the old Transaction KA01/KA02/KA03 to create/change/display a cost element, you will be redirected to Transaction FS00 to maintain general ledger accounts.

Enter "430000" in the **G/L Account** field for **Salaries** in company code 1000. As you can see in Figure 9.14, this account is defined as a profit and loss general ledger account.

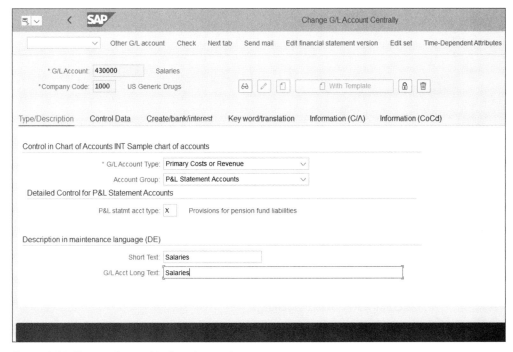

Figure 9.14 Change General Ledger Account

Select the **Control Data** tab, as shown in Figure 9.15, and scroll down to see the controlling settings for the account.

9 General Controlling and Cost Element Accounting

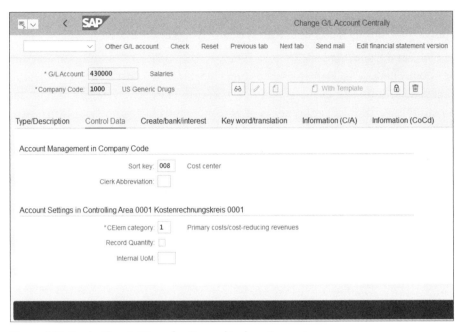

Figure 9.15 Controlling Settings for General Ledger Account

Now in SAP S/4HANA in the general ledger account master record, there's a section for account settings in the controlling area. Here we have three fields, which were previously available in the cost element master record:

- **CElem Category**
  The cost element category determines for which business transactions the cost element can be used. There are separate categories for primary and secondary cost elements, as follows:
  - 01: Primary costs/cost-reducing revenues
  - 03: Accrual/deferral per surcharge
  - 04: Accrual/deferral per debit = actual
  - 11: Revenues
  - 12: Sales deduction
  - 22: External settlement
- **Record Quantity**
  Defines whether the system issues a message if no quantity or quantity unit is specified when posting to the cost element.

354

- **Internal UoM**
  Controls whether an internal unit of measure (UoM) can be used for this cost element.

After maintaining these settings, the account is integrated between the general ledger and controlling.

9.2.2 Cost Element Groups

Cost element groups are used to group together similar cost elements for reporting and data-processing needs. Here there is no difference in SAP S/4HANA compared with previous releases: Transactions KAH1, KAH2, and KAH3 to create, change, and display cost element groups are still valid.

To create a cost element group, follow application menu path **Accounting • Controlling • Cost Element Accounting • Master Data • Cost Element Group • KAH1—Create**.

Enter a meaningful name for the cost element group, as shown in Figure 9.16. You can create a cost element group with reference to an existing group, which you specify together with the chart of accounts in the **Reference** section. Proceed with the **Enter** button.

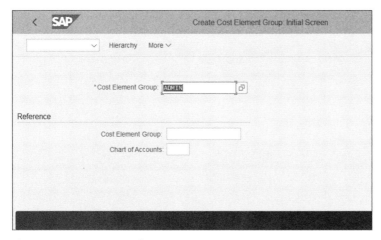

Figure 9.16 Create Cost Element Group

Initially you see only the top level of the cost element group. Enter a meaningful description, as shown in Figure 9.17.

9 General Controlling and Cost Element Accounting

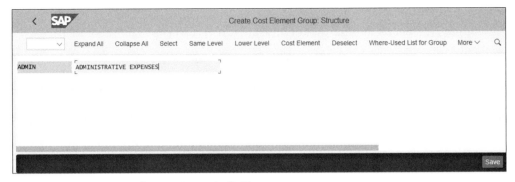

Figure 9.17 Cost Element Group Definition

Then you have three options from the top menu:

- **Same Level**
  Inserts cost element group at the same level.

- **Lower Level**
  Inserts cost element group at a lower level.

- **Cost Element**
  Inserts cost element.

Using these options, you can build your cost element groups as shown in Figure 9.18.

Figure 9.18 Cost Element Group Structure

In this case, we created a few cost element groups under the main cost element group and then assigned a range of cost elements to the cost element group. In this fashion, you can build very complex cost element groups.

9.3 Actual Postings

As part of the cost element accounting application menu, SAP provides some actual postings, which are used to reclassify costs within the controlling. In this section, we'll look at manual reposting and activity allocation.

9.3.1 Manual Reposting

Manual reposting of costs is used to correct posting errors in controlling. For example, if a posting is entered to the wrong cost element, you can report the cost to another cost element with this function.

There are three transactions provided, which are available at the application menu path **Accounting • Controlling • Cost Element Accounting • Actual Postings • Manual Reposting of Costs**, as shown in Figure 9.19.

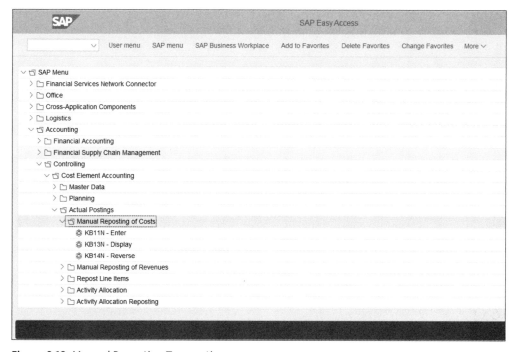

Figure 9.19 Manual Reposting Transactions

9　General Controlling and Cost Element Accounting

You can use three transactions:

- **KB11N—Enter**
  With this transaction you enter manual repostings of costs.

- **KB13N—Display**
  With this transaction you can display already entered reportings of costs.

- **KB14N—Reverse**
  This transaction is used to reverse already entered reportings of costs.

Let's see how to report costs. Enter Transaction KB11N. In the initial screen, shown in Figure 9.20, in the **Scrn Var.** field you can select a screen variant depending on what type of sending and receiving controlling object you will repost from and to. The default is the last screen variant used.

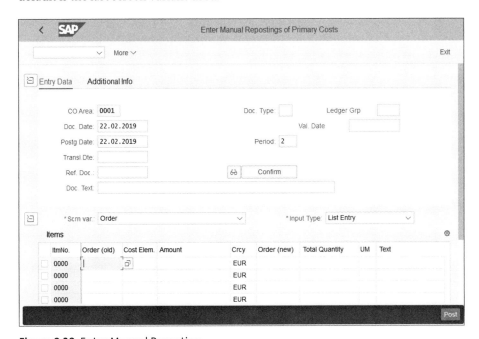

Figure 9.20 Enter Manual Reposting

You can choose from the following screen variants, as shown in Figure 9.21:

- Cost center
- Cost center/order/pers. no.
- Customer Project
- Order
- Real estate objects
- Sales order/cost object
- WBS element/network
- WBS element/order

These screen variants determine which of these fields will be available for reposting.

Figure 9.21 Screen Variants

Select **Cost center** and the screen changes to include the old cost center (original cost center to repost from) and new cost center. Enter the necessary data, as shown in Figure 9.22.

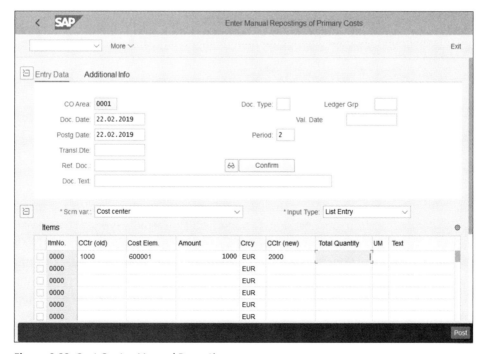

Figure 9.22 Cost Center Manual Reposting

The following fields are important:

- **CO Area**
  The controlling area where the reporting is taking place.
- **Doc. Date**
  Document date of the reposting. For CO internal documents, normally it's the same as the posting date.
- **Postg Date**
  Posting date of the reposting, which is used to determine the period.
- **CCtr (Old)**
  The cost center that was originally incorrectly posted. With other screen variants, you can repost from internal orders, WBS elements, and other controlling objects.
- **Cost Elem.**
  The cost element originally posted. In the reposting the same cost element is used.
- **Amount**
  Amount to be reposted.
- **Crcy**
  Currency of the reposting.
- **CCtr (New)**
  This cost center will receive the reposting.

After that, save your entry with the **Post** button. The result is that the original cost center that was debited in the original document is now credited, and the new cost center gets the debit and is charged with the cost. The same cost element is used.

9.3.2 Activity Allocation

Activity allocation is another function in cost element accounting, which allows you to reallocate costs from one cost center to another using activity types. Activity types are objects used to record activities in SAP S/4HANA, such as labor hours. We'll talk about them in detail in the next chapter; for now, note that activity types are tracing factors, which are used as allocation bases during activity allocation. Let's see now activity allocation, which is also sometimes referred to as direct activity allocation, works.

There are three transactions provided, which are available at the application menu path **Accounting • Controlling • Cost Element Accounting • Actual Postings • Activity Allocation**, as shown in Figure 9.23.

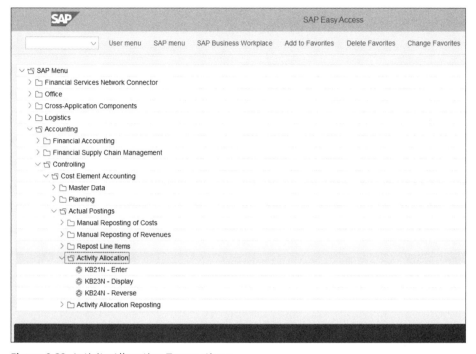

Figure 9.23 Activity Allocation Transactions

You can use the following three transactions:

- **KB21N—Enter**
  With this transaction you enter an activity allocation.
- **KB23N—Display**
  With this transaction you can display already entered activity allocations.
- **KB24N—Reverse**
  This transaction is used to reverse already entered activity allocations.

Let's see how to enter an activity allocation. Enter Transaction KB21N and enter the necessary data, as shown in Figure 9.24.

9   General Controlling and Cost Element Accounting

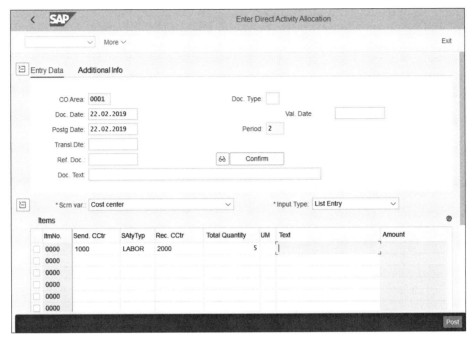

Figure 9.24 Enter Activity Allocation

The following fields should be filled in:

- **CO Area**
  The controlling area where the reporting is taking place. By default, it's the last controlling area used. If you need to change it, you can set a different controlling area with Transaction OKKS.

- **Doc. Date**
  Document date of the activity allocation.

- **Postg Date**
  Posting date of the activity allocation, which is used to determine the period.

- **Send. CCtr**
  The cost center that currently bears the cost.

- **SAtyTyp**
  Sender activity type, which is used to allocate the cost.

- **Rec. CCtr**
  This cost center will receive the cost after the activity allocation.

- **Total Quantity**
  Quantity associated with the activity.

So in our example, we allocate five hours of labor using the LABOR activity type from cost center 1000 to cost center 2000.

To use the activity allocation, you need to create activity types, which are controlling master data objects. Also, both the sender and receiver cost center should be set up with the same sender activity type. We'll cover this setup in Chapter 10.

9.4 Summary

In this chapter, we configured SAP S/4HANA's general controlling settings:

- Controlling area
- Number ranges
- Versions

Now you know how to configure your controlling area so that the controlling components you need are activated and versions are set up and able to receive actual and/or plan data for the fiscal years needed.

We then covered the master data in cost element accounting, which consists of cost elements and cost element groups. As you learned, with SAP S/4HANA cost elements are fully integrated with general ledger accounting, and this is one of the most fundamental imporvments that SAP S/4HANA Finance offers. As such, there is no separate cost element master record object, and you define the cost elements by their cost element category, which is part of the general ledger account master record.

You learned about various types of actual postings in cost element accounting, such as manual reposting of costs and activity allocation. Although these transactions do not require special customizing to be used, it's helpful to learn how you can repost and allocate costs and how you can define screen variants to select the desired sender and receiver objects. We also got a glimpse of some functionalities offered in cost center accounting, such the activity types.

Now that we've defined the global settings and the cost elements, we're ready to delve deep into the secrets of controlling in SAP S/4HANA. We'll start with cost center accounting.

Chapter 10
Cost Center Accounting

This chapter gives step-by-step instructions for configuring cost center accounting, including the settings for mater data, actual postings, and planning. In this chapter, you'll learn how to configure and perform various periodic allocations.

Cost center accounting, together with internal order accounting, are part of overhead costing, which serves to track and analyze costs that aren't directly attributable to the production of goods and services. These costs, which are commonly referred to as *overhead costs*, are posted to cost objects such as cost centers or internal orders, and then some of them are reallocated to other cost objects such as profitability segments or to other cost centers to provide better analysis of the costs and profitability of your company.

In this chapter, we'll guide you through how to configure cost center accounting. We'll start by setting up the required master data, such as cost centers, cost center groups, activity types, and statistical key figures.

Next, we'll examine the various actual postings that can be performed in cost center accounting. After that, you'll learn how to perform various allocation functions, such as distribution and assessment. Then we'll delve into the planning functions for cost centers. We'll finish with a guide to the information system in cost center accounting.

10.1 Master Data

As with other financial modules, we'll start with setting up the required master data for cost center accounting. The following master data objects are used in cost center accounting:

- Cost centers
- Cost center groups
- Activity types
- Statistical key figures

We'll start by setting up the required configuration for cost centers.

10.1.1 Cost Centers

A *cost center* depicts an organizational unit within a company from a cost point of view. For example, cost centers can represent various departments, such as human resources, finance, or information technology (IT). The cost center is a master data object, which is created in the application menu.

Before you can create cost centers, you need to define cost center categories. The category of a cost center defines certain characteristics during cost center creation.

To define cost center types, follow menu path **Controlling • Cost Center Accounting • Master Data • Cost Centers • Define Cost Center Categories**.

Figure 10.1 shows the various cost center categories defined in the system. They correspond to the different functions that various departments in the organization can perform.

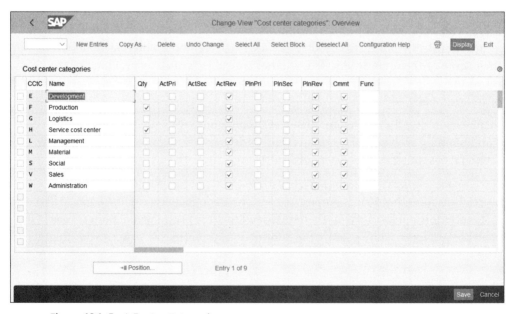

Figure 10.1 Cost Center Categories

The following fields are to be configured:

- **Qty**
  The quantity indicator enables managing quantities in a cost center. If it's set, a message is issued when posting.

- **ActPri**

  This indicator locks the cost center of this category for posting of actual primary costs. If the indicator is active, no actual primary costs can be posted to the cost center.

- **ActSec**

  This indicator locks the cost center of this category for posting of actual secondary costs. If the indicator is active, no actual secondary costs can be posted to the cost center.

- **ActRev**

  This indicator locks the cost center of this category for posting of actual revenues. If the indicator is active, no actual revenues can be posted to the cost center. The revenues posted on cost centers are for statistical purposes only—for reporting analysis.

- **PlnPri**

  This indicator locks the cost center of this category for posting of plan primary costs. If the indicator is active, no plan primary costs can be posted to the cost center.

- **PlnSec**

  This indicator locks the cost center of this category for posting of plan secondary costs. If the indicator is active, no plan secondary costs can be posted to the cost center.

- **PlnRev**

  This indicator locks the cost center of this category for posting of plan revenues. If the indicator is active, no plan revenues can be posted to the cost center.

- **Cmmt**

  This indicator locks the cost center of this category for commitments. A commitment is a contractual obligation that isn't yet recorded in financial accounting but that will lead to expenses in the future.

- **Func**

  Here you can specify a default functional area. The functional area is used to record the cost of sales.

Make changes as required and save your entries. Now you're ready to create cost centers by following application menu path **Accounting • Controlling • Cost Center Accounting • Master Data • Cost Center • Individual Processing • KS01—Create**.

10 Cost Center Accounting

Figure 10.2 shows the initial screen of the cost center creation.

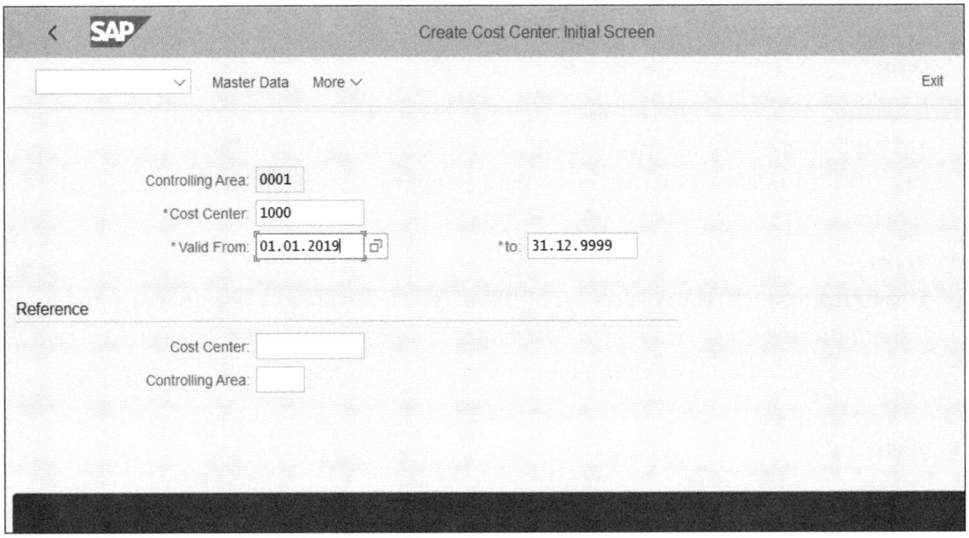

Figure 10.2 Create Cost Center

Cost centers are always created for a controlling area and time period. They are valid only within that time period. You can also add additional periods and change some time-sensitive parameters of the cost center.

Enter **Controlling Area** "0001", **Cost Center** number "1000", and a validity period from the beginning of the current year to 31.12.9999. This means that the validity is infinite.

In terms of cost center numbering, you need to come up with a logical concept that encompasses all the cost centers of the organization. You need to decide how many digits the cost centers will have; four- or six-digit numbers are common. Then they should be well structured based on similar functions; for example, all production-related cost centers should be in the same range, starting with the same number.

You can also create the cost center with reference to an existing cost center, which you would enter together with its controlling area in the **Reference** section of the initial screen.

Proceed by pressing ⌈Enter⌋. On the next screen, you maintain the various cost center fields, organized in tabs. As shown in Figure 10.3, the first tab is the **Basic data** tab. Some fields are required, marked with red asterisks. The others are optional.

10.1 Master Data

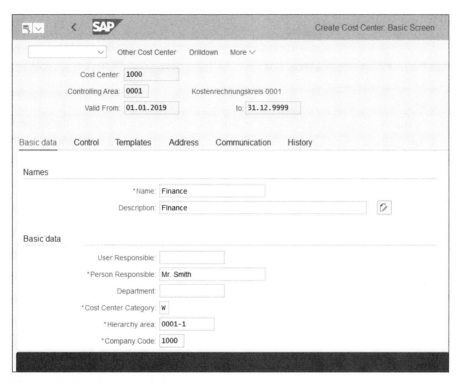

Figure 10.3 Cost Center Basic Data

Maintain the fields as follows:

- **Name**
  Meaningful short description of the cost center.
- **Description**
  Meaningful long description of the cost center.
- **Person Responsible**
  The name of the responsible manager of the department.
- **Cost Center Category**
  In this field, select one of the cost center categories configured previously.
- **Hierarchy area**
  Cost centers need to be assigned to a tree-like structure that organizes them in a hierarchy. Here, assign the cost center to the appropriate node. We'll define the cost center hierarchy in the next section.

369

- **Company Code**
  A cost center is always assigned to a company code.

You need to scroll down to see the rest of the fields. They can be maintained as follows:

- **Business Area**
  Here you can assign the cost center to a business area, which is an organizational unit of financial accounting that represents a separate area of operations or responsibilities.
- **Functional Area**
  Here you can assign the cost center to a functional area, which is used to create a profit and loss report using the cost-of-sales accounting method.
- **Currency**
  In this field, specify the currency of the cost center.
- **Profit Center**
  Here you need to assign the cost center to a profit center, which is an organizational unit in accounting based on the management-oriented internal structure of the organization. Then the profit center is derived from the cost center, not only in the respective expense line items but also in offsetting balance sheet line items, using the document splitting technique you learned in Chapter 4.

Figure 10.4 shows the control fields of the cost center master record. Here you can lock certain types of postings to the cost center. We covered these types of postings in Figure 10.1. In fact, they are defaulted based on the settings of the selected cost center category in the **Basic data** tab. Usually for most centers you need to lock actual and plan revenues because cost centers are cost objects, and sometimes you also may need to lock some or all of the other categories if you need to prevent cost postings to the cost center.

On the **Templates** tab, shown in Figure 10.5, you can assign templates to the cost center. These are used in planning and for process allocations. Templates are created independently of cost centers and enable multiple cost centers to be assigned simultaneously. A template contains columns and rows. Costs or quantities could be entered as variables, and the costs are calculated automatically based on these variables. They are mostly used in process cost accounting.

10.1 Master Data

Figure 10.4 Cost Center Control Data

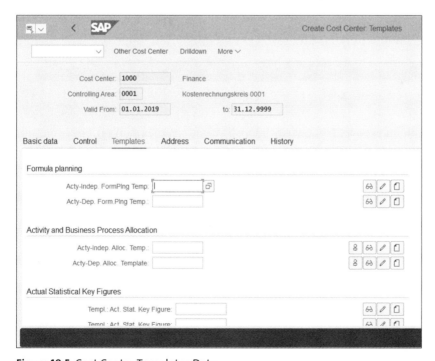

Figure 10.5 Cost Center Templates Data

In the **Address** tab, shown in Figure 10.6, you can enter various address information relevant for the cost center, such as street address, region, country, and so on.

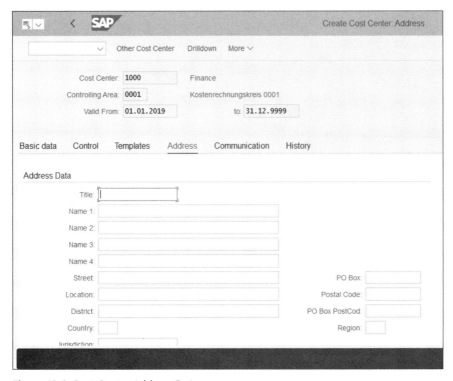

Figure 10.6 Cost Center Address Data

In the **Communication** tab, shown in Figure 10.7, you can maintain various communication information relevant for the cost center, such as language key, telephone, fax, and so on.

In the **History** tab, shown in Figure 10.8, the system stores information about the user who created the cost center, the date of creation, and also any changes to the cost center master, which you can view with the **Change document** button later on when you display the cost center.

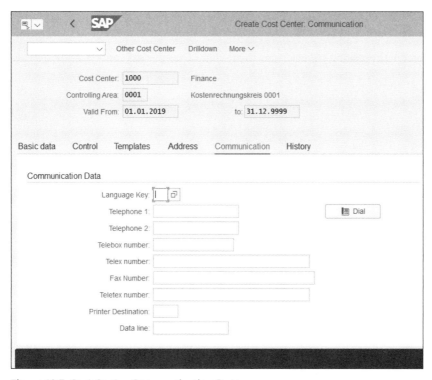

Figure 10.7 Cost Center Communication Data

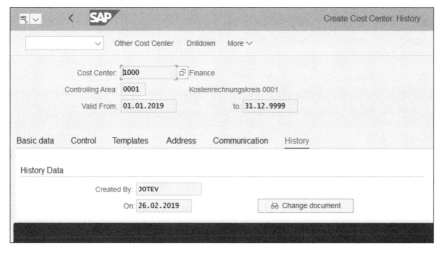

Figure 10.8 Cost Center History Tab

10 Cost Center Accounting

Once you're done, save the cost center with the **Save** button.

In this fashion, you can create the rest of the cost centers also. It's convenient here to use collective processing transactions to create or change cost centers, shown in Figure 10.9.

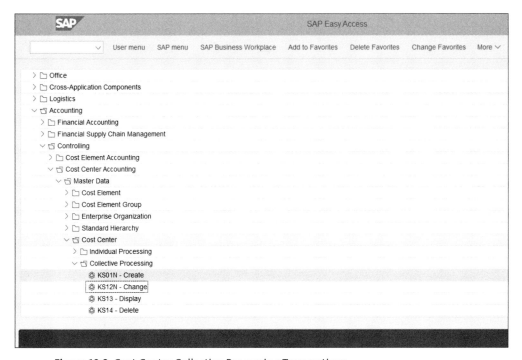

Figure 10.9 Cost Center Collective Processing Transactions

You can create, change, or display cost centers in mass mode. For example, to change the validity end date of multiple cost centers collectively, enter Transaction KS12N (Change). The system initially shows a warning message, shown in Figure 10.10.

You're advised to proceed with caution when using collective maintenance transactions. You can hide this message in the future by ticking the **Do not display this warning again** checkbox.

Proceed with the ✓ button. On the next screen, shown in Figure 10.11, you can limit the selected cost centers for mass maintenance by their validity period, by controlling area, cost center number, and language.

10.1 Master Data

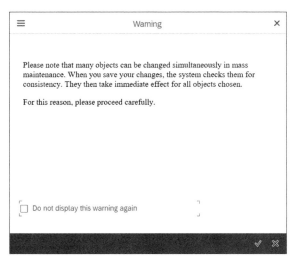

Figure 10.10 Mass Maintenance Warning

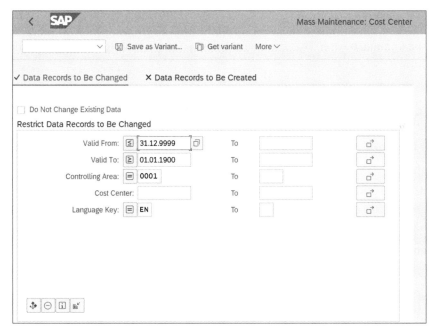

Figure 10.11 Mass Maintenance Selection Screen

Proceed with the **Execute** button. On the next screen, shown in Figure 10.12, you can enter a new value that will be changed in all selected cost centers.

375

10 Cost Center Accounting

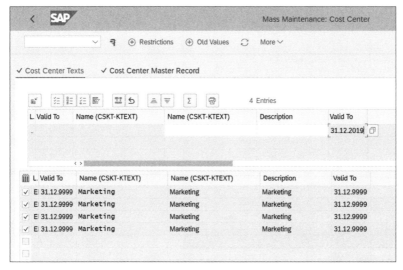

Figure 10.12 Mass Maintenance Execution Screen

In this example, we entered a new validity period and executed it with the 🔲 (**Perform Mass Change**) button. That changes the date en mass, and we can save with the **Save** button.

10.1.2 Cost Center Groups

The cost center group is used to classify cost centers with similar functions. It is a node in the cost center standard hierarchy that groups together all cost centers of a controlling area. The cost center group, which is in fact a hierarchy node, is a required field in the cost center master record, as we showed in Figure 10.3.

Let's see how to maintain the standard hierarchy and the associated cost center groups by following application menu path **Accounting • Controlling • Cost Center Accounting • Master Data • Standard Hierarchy • OKEON—Change**.

As shown in Figure 10.13, the screen is separated into multiple windows for easier navigation. On the left side you have a search function, with which you can search by cost centers or cost center groups. On the right side on the top, you see the actual cost center standard hierarchy; at the bottom, you see details of the selected node.

The standard hierarchy is a tree-like structure, which you can expand using the arrows > on the left of each node. You can also expand the whole hierarchy from the node where you're positioned with the 🔲 (**Expand Subtree**) button, and collapse the whole hierarchy from the node where you're positioned with the ⊟ (**Collapse Subtree**) button. You can close the details for the selected node with the ⊞ (**Close Detail Area**) button and get a better view of the hierarchy, as shown in Figure 10.14.

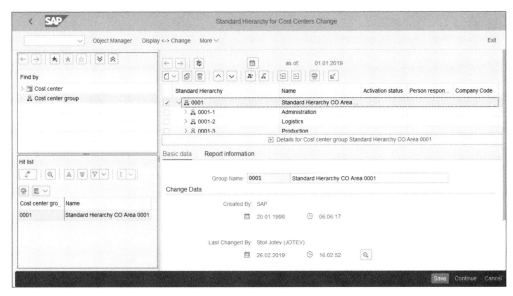

Figure 10.13 Cost Center Standard Hierarchy

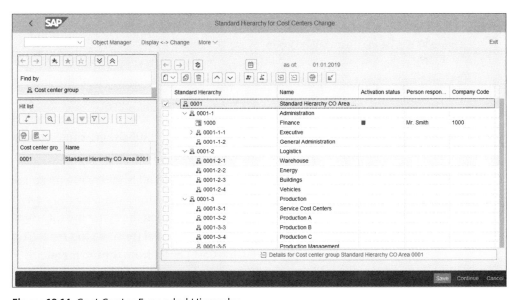

Figure 10.14 Cost Center Expanded Hierarchy

10 Cost Center Accounting

Here, cost center groups have the 🗄 symbol on the left side and cost centers have the 📋 symbol. Double-clicking on a cost center group or cost center shows its details in the lower section of the screen.

You can also maintain cost center groups individually and not through the standard hierarchy using the transactions available at menu path **Accounting • Controlling • Cost Center Accounting • Master Data • Cost Center Group** or by entering Transaction KSH2 (Change).

As shown in Figure 10.15, in the first screen you select the **Cost Center Group**. The controlling area is defaulted from the last used controlling area and can't be changed here. If you need to switch the controlling area here, and in general, you can use Transaction OKKS to set the controlling area.

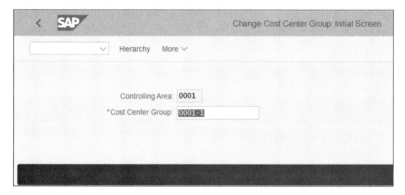

Figure 10.15 Change Cost Center Group

Proceed by pressing [Enter]. As shown in Figure 10.16, the interface of this transaction is different from the standard hierarchy Transaction OKEON shown in Figure 10.13. But it's still a tree-like structure, where you see on top the cost center group entered on the previous screen. Below you see all assigned subgroups, which you can expand with the ⊞ button, shown on the left side of each expandable node.

You can expand the whole strucutre be selecting **Expand All** from the top menu and collapse the whole structure with **Collapse All** from the top menu. You can assign additional groups on the same level as you're positioned in the hierarchy with **Same Level** from top menu and on lower levels with **Lower Level** from the top menu. You can assign cost centers to groups by selecting **Cost Center** from the top menu.

10.1 Master Data

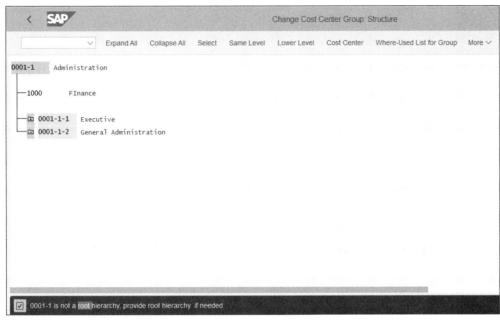

Figure 10.16 Cost Center Group Structure

Once you're done, save the cost center group with the **Save** button.

10.1.3 Activity Types

Activity types are master data objects used to record activities such as labor hours. They are tracing factors, which are used as allocation bases during activity allocation. Activity types are associated with rates that you maintain per activity type/cost center combination.

To create an activity type, follow application menu path **Accounting • Controlling • Cost Center Accounting • Master Data • Activity Type • Individual Processing • KL01—Create**.

On the initial screen, shown in Figure 10.17, you give the name of the activity type, which could be up to six characters long, and a validity period, as we did for cost centers. You have also the option to copy from an existing activity type, specified in the **Copy from** section.

379

10   Cost Center Accounting

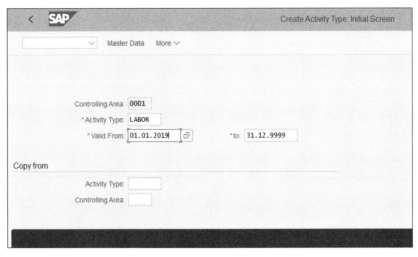

Figure 10.17 Create Activity Type Initial Screen

After you press ⌈Enter⌉, fill in the required fields, as shown in Figure 10.18.

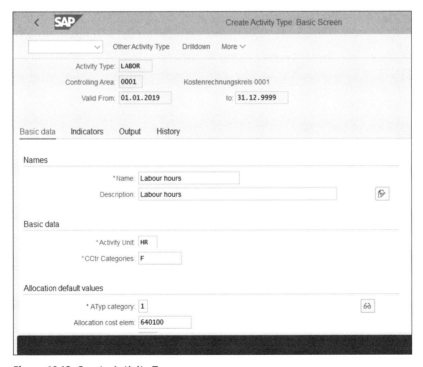

Figure 10.18 Create Activity Type

The following fields should be maintained:

- **Name**
  Meaningful short description of the activity type.
- **Description**
  Meaningful long description of the activity type.
- **Activity Unit**
  Here select the type of unit used to measure the activity such as hours, kilograms, and so on.
- **CCtr Categories**
  In this field, select one of the cost center categories.
- **ATyp category**
  In this field, select the method of activity quantity planning and activity allocation.
- **Allocation cost elem.**
  Here you select a secondary cost element, under which the activity type will be allocated.
- **Price Indicator**
  Determines how the price is being determined.

Once you're done, save the activity type with the **Save** button.

In this step, you created the activity type as a master data object. You also need to enter activity type rates per cost center. To do so, follow application menu path **Accounting • Controlling • Cost Center Accounting • Planning • Activity Output/Prices • KP26—Change**.

In the initial screen, shown in Figure 10.19, you need to select the controlling version, from and to period, and fiscal year for which to maintain the activity price. You also enter the cost center, range of cost centers, or cost center group. You have two options: you can maintain the same price for the whole period using the **Overview Screen** button or per period using the **Period Screen** button.

10 Cost Center Accounting

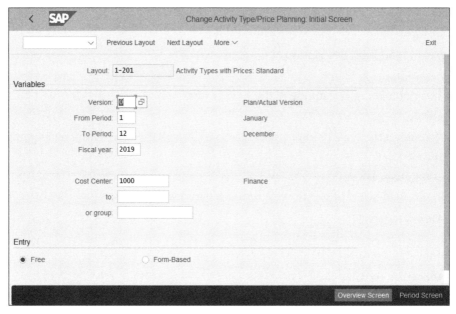

Figure 10.19 Change Activity Prices

Let's maintain the overview screen, shown in Figure 10.20.

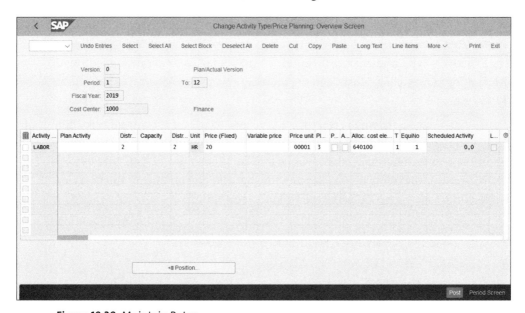

Figure 10.20 Maintain Rates

The parameters maintained in the activity type master record are transferred here. In this example, we maintain a fixed price of 20, which means the labor costs 20 dollars per hour regardless of the quantity. You can also maintain a variable price, which depends on the activity output.

After maintaining the rates, save with the **Post** button.

10.1.4 Statistical Key Figures

Statistical key figures represent values that are used to measure cost centers, internal orders, and other controlling objects. For example, one statistical key figure would be the number of employees per cost center which perform various activities. Statistical key figures can be used as allocation bases for periodic allocations such distributions and assessments, which we'll configure later in this chapter.

To create a statistical key figure, follow application menu path **Accounting • Controlling • Cost Center Accounting • Master Data • Statistical Key Figures • Individual Processing • KK01—Create**.

Figure 10.21 shows the initial screen to create a statistical key figure. Here, you need to give name of the key figure, which could be up to six characters long. You have also the option to copy from an existing statistical key figure, specified in the **Copy from** section.

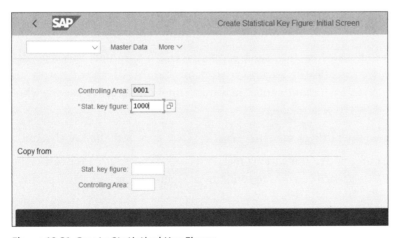

Figure 10.21 Create Statistical Key Figure

After you proceed with the **Enter** button, fill in the required fields, as shown in Figure 10.22.

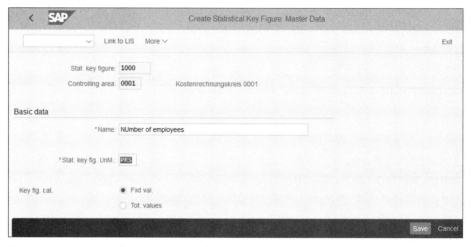

Figure 10.22 Statistical Key Figure Detailed Screen

In this screen, you enter a meaningful description of the statistical key figure and relevant unit of measure. You also have to select from two types of key figure categories:

- **Fxd val.**
  In this case, the values of the statistical key figure are not totaled. Every period will be maintained a fixed value. The statistical key figure for employees, for example, should be marked as a fixed value because each month you need to record the number of employees; you shouldn't sum up the values per month.

- **Tot. values**
  If you select this checkbox, the values per period will be summed. This is needed, for example, for statistical key figures for consumption, in which you have a specific consumption each month and you want to total the values to see the whole consumption for the year to date.

After maintaining the settings, save the statistical key figure with the **Save** button.

This concludes our guide to the various master data objects required for cost center accounting. Now we'll teach you how to configure and perform various actual postings.

10.2 Actual Postings

Most postings into cost centers come from integrated documents that automatically determine the cost center. For example, the cost center is derived from the asset master record during asset postings.

When entering manual postings in financial accounting, usually you need to enter the cost center manually in the expense line item. However, there's also a configuration transaction for default cost object determination, which often is used to determine the correct cost center.

We'll start by configuring default cost objects, provided by the automatic account assignment. Then we'll look into the cost center substitution, which enables you to substitute cost centers based on specific predefined criteria.

10.2.1 Automatic Account Assignment

The automatic account assignment in controlling provides default cost objects based on certain criteria. To configure default cost objects, enter Transaction OKB9 or follow menu path **Controlling • Cost Center Accounting • Actual Postings • Manual Actual Postings • Edit Automatic Account Assignment**.

Figure 10.23 shows how you can maintain default cost object assignments per cost element.

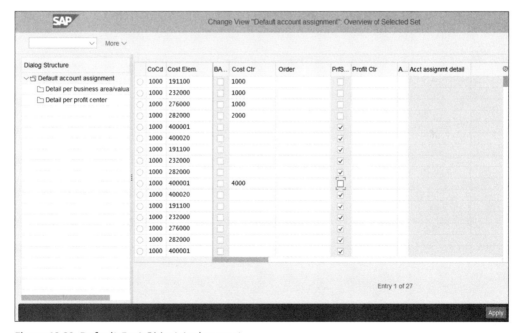

Figure 10.23 Default Cost Object Assignment

10 Cost Center Accounting

You maintain this configuration table per company code and cost element, and in the following columns you can assign default cost objects:

- Cost Center
- Order
- Profitability Segment
- Profit Center

Then when posting to this cost element, if there is no other cost object determination, such as from material master record or asset, for example, the system will assign the default cost object maintained here.

You can also maintain multiple cost objects per profit center or business area if you select indicator **2 Business Area Is Mandatory** or **3 Profit Center Is Mandatory** in the **Acct assignment detail** column. Then click **Detail per business area/valuation area** on the left side of the screen to maintain per multiple business areas, or click **Detail per profit center** to maintain per multiple profit centers' cost objects.

As shown in Figure 10.24, here for company code 1000 and cost element 416300, you can maintain different default cost centers depending on the profit center.

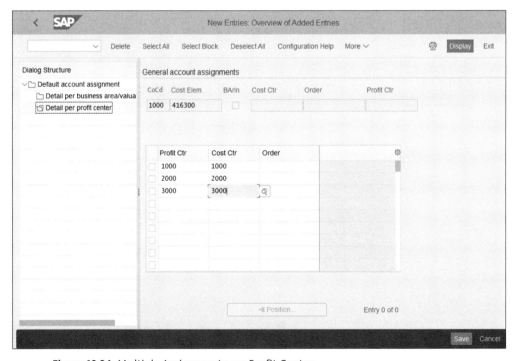

Figure 10.24 Multiple Assignments per Profit Center

In such fashion, maintain the default cost object assignments for all needed cost elements, and save your entries with the **Save** button. It's good practice to have an extensive list of cost elements maintained in this configuration transaction to avoid posting errors during the producutive use of the system in case no other automatic account assignments are in play.

10.2.2 Substitutions for Account Assignment

There is also another technique that could be used to determine cost centers during actual postings. It is called *substitution* and is a tool provided by SAP to populate certain fields based on certain rules. It's commonly used throughout the different areas of financial accounting and controlling. Here we'll just briefly describe a case in which the cost center is derived using substitution.

To define substitution, enter Transaction GGB1. Figure 10.25 shows the initial screen.

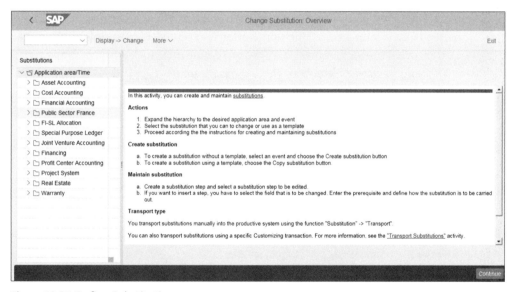

Figure 10.25 Define Substitutions

There's a tree-like structure, and on the left side of the screen you can select the application for which to define substitution. Navigate through **Cost Accounting • Line Item**, where you define substitutions working on the line item of the controlling document. SAP provides sample substitutions such as **0_CO_1: Example Substitution for CO**. Copy this with the **Copy Substitution** function from the top menu, then give a name and meaningful description for the new substitution, as shown in Figure 10.26.

10 Cost Center Accounting

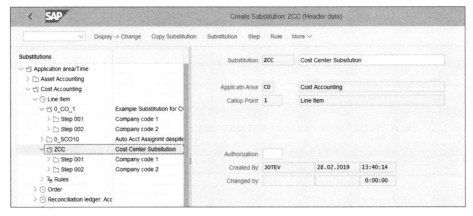

Figure 10.26 Define Cost Center Substitution

Each substitution can have one or multiple steps. Expand the steps and you'll see that they contain prerequisite and substitution sections. When the criteria defined in the prerequisite section are met, the substitution is executed. For example, in the prerequisite section, shown in Figure 10.27, you can define rules such as *when company code is 1000* and *when cost element is within a specific range*.

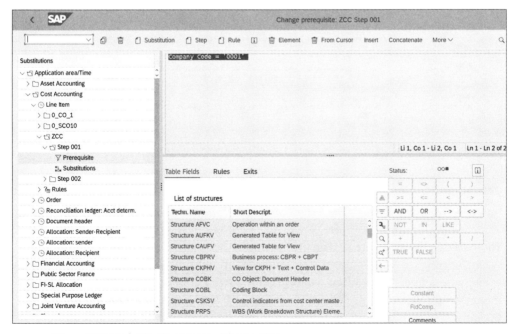

Figure 10.27 Substitution Prerequisite Section

Then in the substitution section, you can define the cost center as a field to be substituted using the ⊕ (**Insert Subst. Etnry**) button. Then a new window opens with the list of the substitutable fields, as shown in Figure 10.28.

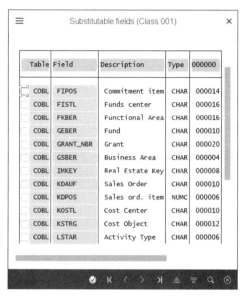

Figure 10.28 Substitutable Fields

Select the cost center field (technical name KOSTL), then the system asks you for the substitution method, as shown in Figure 10.29.

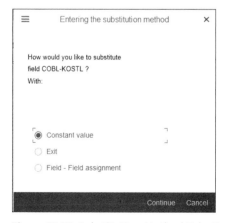

Figure 10.29 Substitution Method Selection

10　Cost Center Accounting

You have three options:

- **Constant value**

  With this option, the cost center will be substituted with the fixed value specified in the next screen.

- **Exit**

  This option uses a user exit, which is custom-programmed code you can apply if the requirement is more complex.

- **Field-Field assignment**

  With this option, the value of the substitutable field will be substituted with a value from another field.

Select **Constant value** for this example. As shown in Figure 10.30, provide a constant value to be applied when the conditions are met. Then save your entry with the **Save** button.

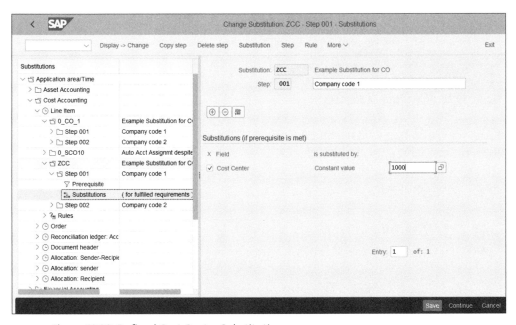

Figure 10.30 Defined Cost Center Substitution

10.3 Periodic Allocations

In controlling, costs often need to be reallocated from some cost objects to others. For example, administrative costs are posted to administrative cost centers such as finance, IT, human resources, and so on. However, to provide an accurate profitability picture of the company based on lines of business and operations, management may want to assign these costs to the various departments that sell various products and thus assess properly the real operating margins and profitability per product. This is a task of controlling, which provides various types of periodic allocation procedures that can be used to automatically reallocate costs based on predefined criteria.

In the following sections, we'll explain the following types of periodic allocations:

- Accrual calculation
- Distribution
- Assessment
- Activity allocation

10.3.1 Accrual Calculation

Accrual calculation is a process of evenly spreading irregularly occurring costs over periods and distributing them by cause. It isn't a commonly used procedure because most companies prefer to have the accrual posted in specific periods only.

Accrual costs are posted to accrual cost centers or internal orders and then distributed. You need to create a special accrual cost center using menu path **Controlling • Cost Center Accounting • Actual Postings • Period-End Closing • Accrual Calculation • Create Accrual Cost Centers** or an accrual internal order using menu path **Controlling • Cost Center Accounting • Actual Postings • Period-End Closing • Accrual Calculation • Create Accrual Orders**.

Follow menu path **Controlling • Cost Center Accounting • Actual Postings • Period-End Closing • Accrual Calculation • Determine Order Types for Accrual Orders** to define the accrual order type. SAP provides standard accrual order type 9A00, as shown in Figure 10.31.

10   Cost Center Accounting

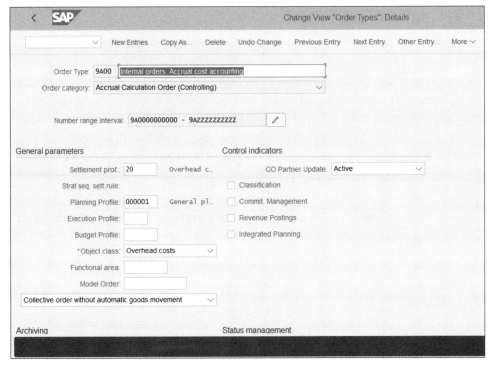

Figure 10.31 Accrual Order Type

If needed, you can create your own order type for that purpose. We'll examine order type creation in detail in Chapter 11.

There are two methods for accrual calculation: the target = actual method and the percentage method.

In the *target = actual method*, the planning of the primary costs is made with an accrual cost element. The system calculates target costs and enters these costs in the actual value fields. This method is used when accrual can be planned periodically. In the *percentage method*, you use specific percentages to calculate accruals. SAP provides enhancement **COOM 0002 User Exits for cost center accrual calculation**, with which you can program user exits to perform the accrual calculation.

10.3.2 Distribution

Distribution and assessment are the two main methods to allocate costs out of cost centers. Distribution uses the original cost elements posted to allocate cost to the

10.3 Periodic Allocations

sender cost centers. This way, you can see the original cost element posted from the sender cost center on the receiving cost center. Distribution is used to allocate only primary costs.

Assessment uses secondary assessment cost elements with cost element 43 to allocate costs. In the receiver cost center, the original cost element from the sender isn't available. Assessment allocates both primary and secondary costs.

Both distribution and assessment use similar techniques, in which you define cycles and segments. The cycle is a periodic allocation run, which can contain one or more segments. Each segment contains relationships between sender and receiver cost centers, as well as various control parameters.

To create a distribution cycle, follow menu path **Controlling • Cost Center Accounting • Actual Postings • Period-End Closing • Distribution • Define Distribution**, then select **Create Actual Distribution**. This transaction is also available in the application menu path via **Accounting • Controlling • Cost Center Accounting • Period-End Closing • Current Settings • S_ALR_87005757—Define Distribution**.

On the initial screen, shown in Figure 10.32, you enter a meaningful cycle name that will enable you easily to identify the purpose of the cycle and a start date, which normally is the beginning of the fiscal year. Each year, you usually define your cycles again, copying the cycles from the previous year. You can create a cycle with reference to an existing cycle, specified in the **Copy from** section. After that, proceed with the **Execute** button.

Figure 10.32 Create Distribution Cycle

393

10 Cost Center Accounting

The next screen, shown in Figure 10.33, is the header of the cycle, which applies to all segments.

Figure 10.33 Distribution Cycle Header

The following fields should be maintained:

- **Start Date and To**
  You already specified the start date on the previous screen, but here you also need to define an end date after which the cycle is no longer valid. If you create new cycles every year, this should be the end of the fiscal year.

- **Text**
  Meaningful long description of the cycle.

- **Iterative**
  This indicator enables iterative sender/receiver relationships. In this case, the iteration is repeated until each sender is fully credited.

- **Cumulative**
  If you check this indicator, the sender amounts posted are allocated based on tracing factors, which are cumulated for all periods.

- **Derive Func. Area**
  By checking this indicator, the functional area proposed in the cycle definition for the receiver is ignored and is derived again.

- **Field Groups**
  In general, only amounts in the controlling area currency are used in periodic allocations. However, you can choose other types of currencies in this section.

After maintaining the cycle definition, you need to attach at least one segment. To do so, select **Attach segment** from the top menu. The segment, as shown in Figure 10.34, is organized into tabs.

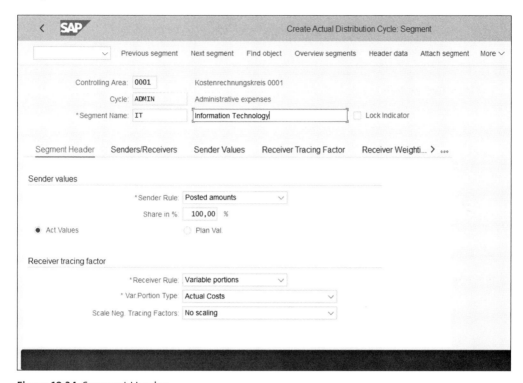

Figure 10.34 Segment Header

In the first tab, **Segment Header**, you specify rules for the distribution calculation as follows:

- **Sender Rule**
  This field determines how the sender values are calculated. The following options are available:
  - **Posted amounts**
    Amounts posted on the sender are the sender values.
  - **Fixed amounts**
    For this option, you need to define fixed amounts for senders in the selection criteria on the **Sender Values** screen.

10 Cost Center Accounting

– Fixed rates
On the **Sender Values** screen, you need to enter fixed prices for the senders, which are multiplied by the receiver tracing factors.

- Share in %
Determines the percentage of the sender values to be distributed.

- Act. Values/Plan Val.
Defines whether the cycle distributes actual or plan values.

- Receiver Rule
Determines how the receiver tracing factors are determined, with the following options:

    – Variable portions
    The receiver tracing factors automatically calculate the amounts to be distributed.

    – Fixed amounts
    Fixed amounts are distributed, which are specified in the **Tracing Values** tab.

    – Fixed percentages
    With this option, you define fixed percentages for the receivers in the **Receiver Tracing Factor** screen. The value from the sender is distributed to the receivers according to these percentages.

    – Fixed portions
    Here you define fixed portions for the receivers in the **Receiver Tracing Factor** screen. This is similar to the fixed percentage option, but the amount is not limited to 100.

- Var.Portion Type
Determines the basis of the distribution.

- Scale Neg. Tracing Factors
Controls how negative tracing factors are treated. Negative tracing factors are possible when the tracing factors are not entered as fixed portions or percentages but are derived.

In the next tab, **Senders/Receivers**, you define the sender and receiver cost objects, as shown in Figure 10.35.

The costs posted originally on the cost objects defined in the **Sender** section will be distributed to the objects defined in the **Receiver** section, according to the tracing factors defined in the **Receiver Tracing Factor** tab. You also need to specify the sender cost elements.

Click the next tab, **Sender Values**, as shown in Figure 10.36.

10.3 Periodic Allocations

Figure 10.35 Sender/Receiver Settings

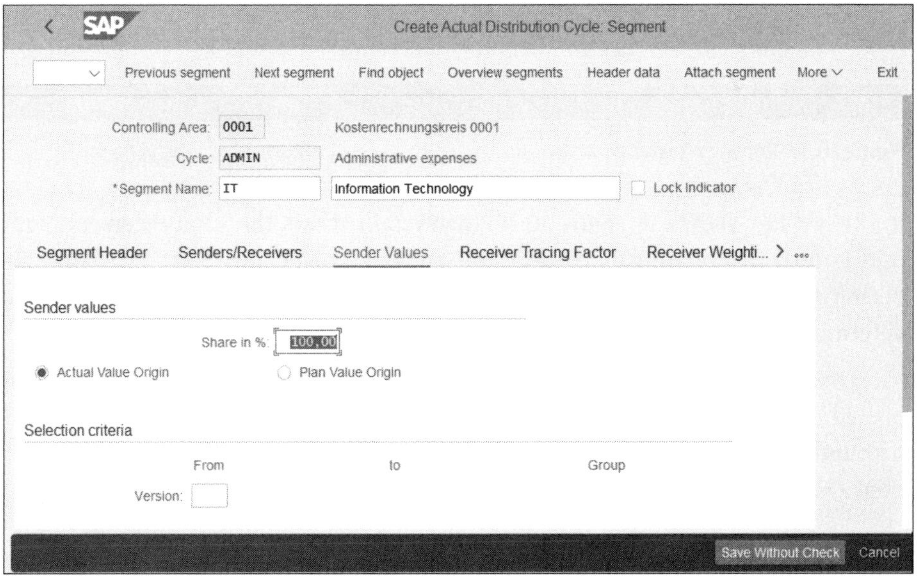

Figure 10.36 Sender Values

10 Cost Center Accounting

Here you can specify the percentage of sender values to be distributed (usually 100%), as well as whether the origin of the values is actual or plan data. You can also specify a version.

In the next tab, shown in Figure 10.37, specify the settings for the **Receiver Tracing Factor**, which are transferred from the **Segment Header** tab.

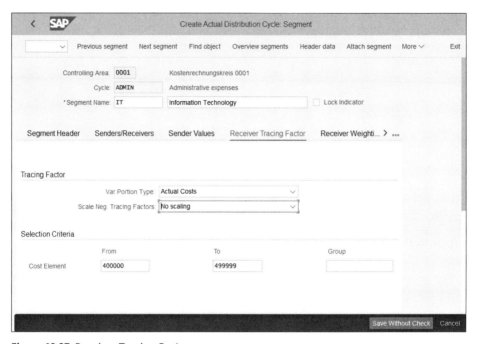

Figure 10.37 Receiver Tracing Factor

In the last tab, shown in Figure 10.38, the system shows the valid receivers found based on the parameters entered in the **Sender/Receivers** tab. You can change the default factor for each receiver, which is 100%. The factor is used to multiply the determined amount.

After that, save your cycle, and it's ready to be used as part of the period-end closing procedures. It's executed with Transaction KSV5 in the application menu path **Accounting • Controlling • Cost Center Accounting • Period-End Closing • Single Functions • Allocations. • KSV5—Distribution**, shown in Figure 10.39.

10.3 Periodic Allocations

Figure 10.38 Receiver Weighting Factors

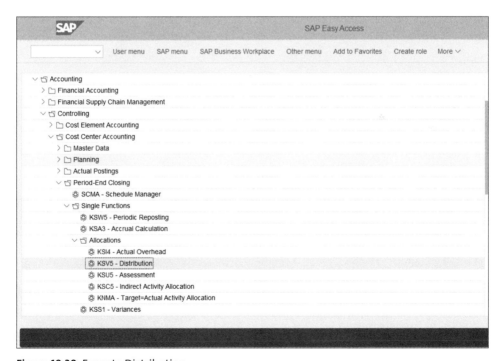

Figure 10.39 Execute Distribution

10.3.3 Assessment

Assessment is another allocation technique, which uses secondary assessment cost elements instead of the original cost elements to allocate costs. It can be used to allocate not only primary but also secondary costs.

Assessment is also set up using cycles and segments. To create an assessment cycle, follow menu path **Controlling • Cost Center Accounting • Actual Postings • Period-End Closing • Assessment • Maintain Assessment**, then select **Create Actual Assessment**. This transaction is also available in the application menu path via **Accounting • Controlling • Cost Center Accounting • Period-End Closing • Current Settings • S_ALR_87005742—Define Assessment**.

On the initial screen, shown in Figure 10.40, you enter a cycle name and start date.

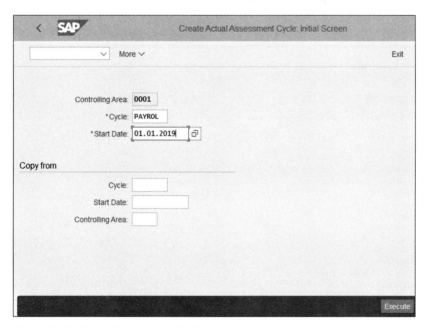

Figure 10.40 Create Assessment Cycle

The same guidance applies as for distribution cycles. You can create a cycle with reference to an existing cycle, specified in the **Copy from** section. After that, proceed with the **Execute** button.

The header screen of the cycle is shown in Figure 10.41.

10.3   Periodic Allocations

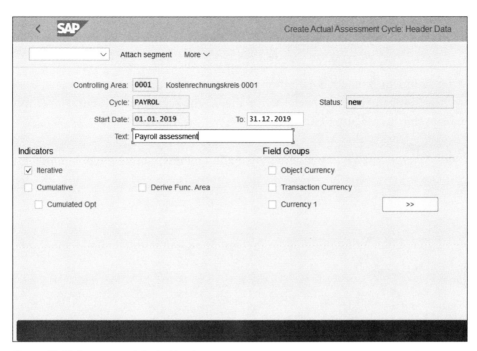

Figure 10.41 Assessment Cycle Header

The following fields can be configured:

- **Start Date and To**
  Start and end date of validity of the cycle. As with distribution cycles, if you create new cycles each year, the end date here should be the end date of the fiscal year.
- **Text**
  Meaningful long description of the cycle.
- **Iterative**
  This indicator enables iterative sender/receiver relationships. In this case, the iteration is repeated until each sender is fully credited.
- **Cumulative**
  If you check this indicator, the sender amounts posted are allocated based on tracing factors, which are cumulated for all periods.
- **Derive Func. Area**
  By checking this indicator, the functional area proposed in the cycle definition for the receiver is ignored and is derived again.

10 Cost Center Accounting

- **Field Groups**
  In general, only amounts in the controlling area currency are used in periodic allocations. However, you can choose other types of currencies in this section.

After maintaining the cycle definition, you need to attach at least one segment. To do so, select **Attach Segment** from the top menu. The segment is also organized into tabs, as shown in Figure 10.42.

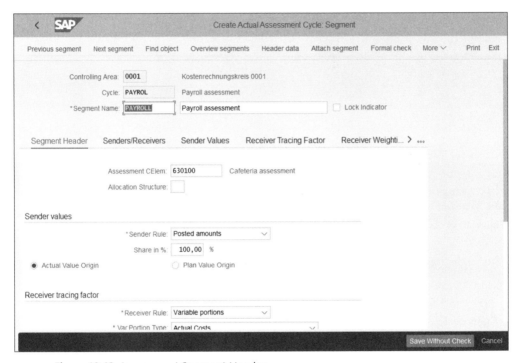

Figure 10.42 Assessment Segment Header

In the first tab, **Segment Header**, shown in Figure 10.42, you specify rules for the assessment calculation as follows:

- **Assessment CElem**
  This is a key difference of this distribution. Here you need to specify either an assessment cost element in this field or an allocation structure in the next field. Specified assessment cost elements are used to post the assessment instead of the original cost elements.

- **Allocation Structure**
  The allocation structure defines cost elements based on certain criteria and is used

in various areas of controlling. We'll configure allocation structures in the next chapter. In this field, you can specify an allocation structure to determine the assessment cost elements based on the originally posted cost elements.

- **Sender Rule**
  This field determines how the sender values are calculated. The following options are available:
  - **Posted amounts**
    Amounts posted on the sender are the sender values.
  - **Fixed amounts**
    For this option, you need to define fixed amounts for senders in the selection criteria on the **Sender Values** screen.
  - **Fixed rates**
    On the **Sender Values** screen, you need to enter fixed prices for the senders, which are multiplied by the receiver tracing factors.
- **Share in %**
  Determines the percentage of the sender values to be distributed.
- **Actual Value Origin/Plan Value Origin**
  Defines whether the cycle allocates actual or plan values.
- **Receiver Rule**
  Determines how the receiver tracing factors are determined, with the following options:
  - **Variable portions**
    The receiver tracing factors automatically calculate the amounts to be assessed.
  - **Fixed amounts**
    Fixed amounts are assessed, which are specified in the **Tracing Values** tab.
  - **Fixed percentages**
    With this option, you define fixed percentages for the receivers in the **Receiver Tracing Factor** screen. The value from the sender is assessed to the receivers according to these percentages.
  - **Fixed portions**
    Here you define fixed portions for the receivers in the **Receiver Tracing Factor** screen. This is similar to the fixed percentage option, but the amount is not limited to 100.
- **Var. Portion Type**
  Determines the basis of the assessment.

10 Cost Center Accounting

- **Scale Neg. Tracing Factors**
  Controls how negative tracing factors are treated. Negative tracing factors are possible when the tracing factors are not entered as fixed portions or percentages but are derived.

In the next tab, **Senders/Receivers**, define the sender and receiver cost objects, as shown in Figure 10.43.

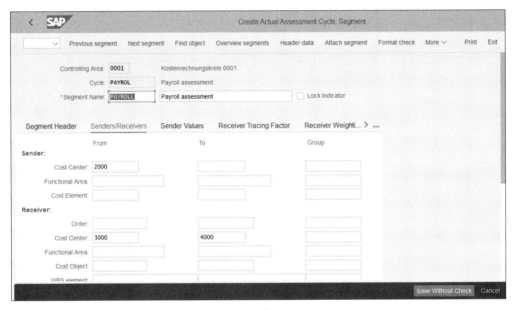

Figure 10.43 Assessment Segment Sender/Receiver Settings

The costs posted originally on the cost objects defined in the **Sender** section will be distributed to the objects defined in the **Receiver** section, according to the tracing factors defined in the **Receiver Tracing Factor** tab. You also need to specify the sender cost elements.

Click the next tab, **Sender Values**, as shown in Figure 10.44.

Here you can specify the percentage of sender values to be distributed (usually 100%), as well as whether the origin of the values is actual or plan data. You can also specify a version.

In the next tab, shown in Figure 10.45, specify the settings for the **Receiver Tracing Factor**, which are transferred from the **Segment Header** tab.

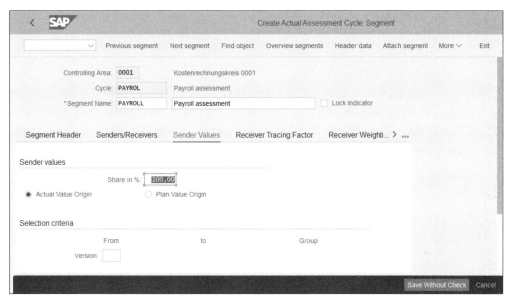

Figure 10.44 Assessment Sender Values

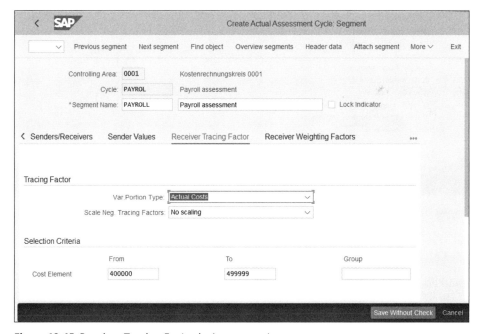

Figure 10.45 Receiver Tracing Factor in Assessment

10 Cost Center Accounting

In the last tab, **Receiver Weighting Factors**, the system shows the valid receivers found based on the parameters entered in the **Sender/Receivers** tab. You can change the default factor for each receiver, which is 100%. The factor is used to multiply the determined amount.

After that, save your cycle, and it's ready to be used as part of the period-end closing procedures. It's executed with Transaction KSU5 in the application menu path **Accounting • Controlling • Cost Center Accounting • Period-End Closing • Single Functions • Allocations • KSU5—Assessment**, shown in Figure 10.46.

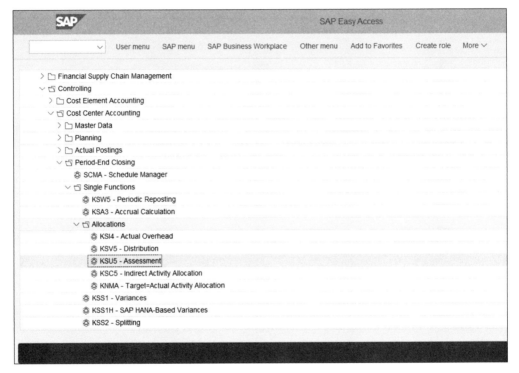

Figure 10.46 Execute Assessment

10.3.4 Activity Allocation

Activity allocation is used to allocate activity costs performed on cost centers. There is direct activity allocation and indirect activity allocation. We looked at the direct activity allocation (Transaction KB21N) in Chapter 9, Section 9.3.2. The indirect activity allocation also allocates activity costs posted to a cost center and activity type—

not manually, but again using cycles and segments, similarly to the distribution and assessment.

To create a distribution cycle, follow menu path **Controlling • Cost Center Accounting • Actual Postings • Period-End Closing • Activity Allocation • Indirect Activity Allocation • Define Indirect Activity Allocation**, then select **Create actual indirect activity allocation**.

The setup of the activity allocation cycle is very similar to the distribution and assessment cycle, so we won't go through every single screen. Again, you define a name, description, and start and end date for the cycle. Then you attach segments, which have very similar control fields as the cycles we already looked at. However, you need to specify a sender **Activity Type** in the **Sender** section, as well as a **Cost Center** in both the **Sender** and the **Receiver** sections, as shown in Figure 10.47.

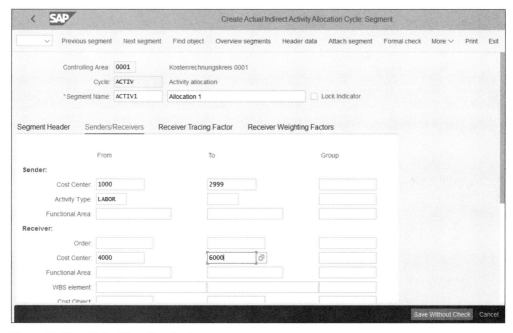

Figure 10.47 Activity Allocation Cycle

After that, you can execute the activity allocation with Transaction KSC5 in the application menu path **Accounting • Controlling • Cost Center Accounting • Period-End Closing • Single Functions • KSC5—Indirect Activity Allocation**, as shown in Figure 10.48.

10 Cost Center Accounting

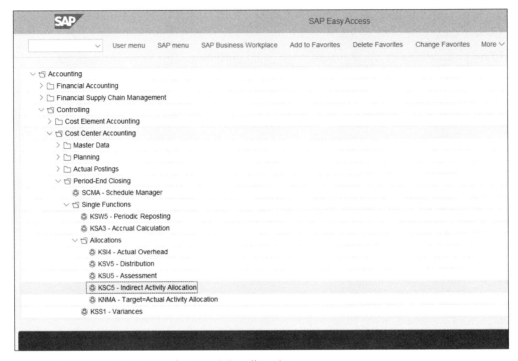

Figure 10.48 Execute Indirect Activity Allocation

With that, we finish our guide on periodic allocations and move to cost center planning.

10.4 Planning

Cost center accounting is part of management accounting, which aims to provide management with a detailed analysis of how the business is performing, how profitable it is, and where there's room for improvement. As such, planning is very important because it enables management to compare actual costs with plan costs. Because cost centers are main cost objects in most companies, cost center planning has tremendous significance in the overall controlling planning process.

We'll start by configuring the basic settings for planning. Then, you'll learn how to perform manual planning.

10.4.1 Basic Settings for Planning

The basic settings for planning include defining the settings for the controlling versions in which you'll perform planning. To do so, follow menu path **Controlling • Cost Center Accounting • Planning • Basic Settings for Planning • Define versions**, then select **Maintain Version Settings in Controlling Area**.

In the screen shown in Figure 10.49, you define the versions relevant for planning.

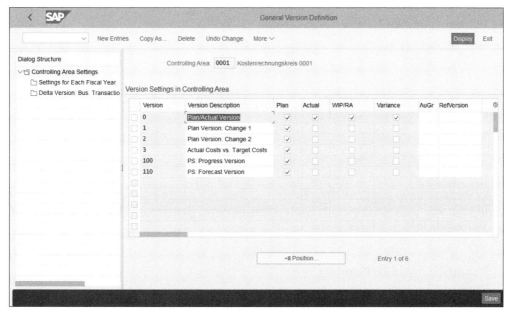

Figure 10.49 Maintain Versions for Planning

The versions that you will be planning in need to be marked with a check in the **Plan** column. Select one of the plan versions and click **Settings for Each Fiscal Year** from the left side of the screen.

Then you need to maintain the settings per each fiscal year, as shown in Figure 10.50. You need to check the **Integrated Planning** checkbox if you're going to use the plan data in other components also. You also need to specify a dedicated exchange rate type for cost planning (as standard exchange rate type P).

After maintaining all relevant settings for planning versions, save your entries with the **Save** button.

10 Cost Center Accounting

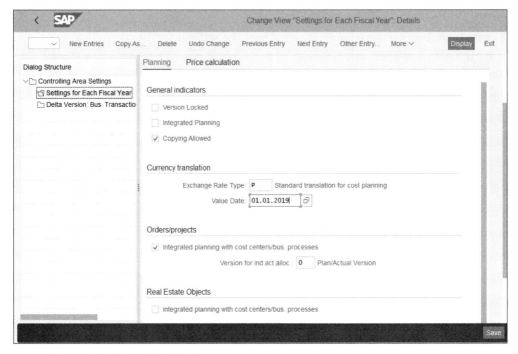

Figure 10.50 Fiscal Year Settings

10.4.2 Manual Planning

Manual planning is the process of manually entering plan values for the relevant cost centers. There are three main processes:

- Planning on the cost center/cost element level
- Planning on the cost center/activity type level
- Planning on the statistical key figure level

To plan on the cost center/cost element level, use Transaction KP06, as shown in Figure 10.51.

To manually plan on both cost center and activity type levels, use Transaction KP26, as shown in Figure 10.52.

10.4 Planning

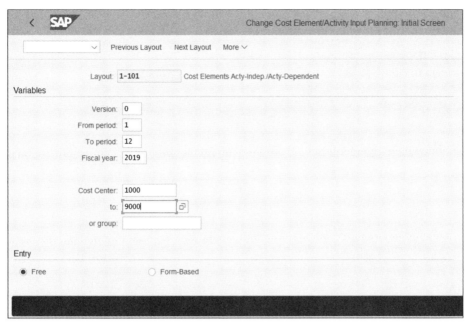

Figure 10.51 Manual Planning Cost Centers

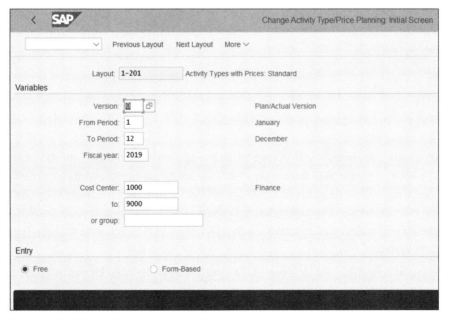

Figure 10.52 Manual Planning Activity Types

You enter fiscal year and periods for which the planning values are valid, and select the cost centers to enter planning values for.

Similarly, you can use Transaction KP46 to plan on the statistical key figure level, as shown in Figure 10.53.

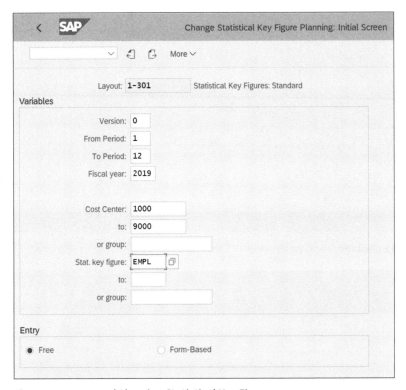

Figure 10.53 Manual Planning Statistical Key Figures

As you can see, these transactions use different planning layouts. Figure 10.51 uses layout **1-101**, and Figure 10.52 uses layout **1-201**. To define planning layouts for activity type planning, follow menu path **Controlling • Cost Center Accounting • Planning • Manual Planning • User-Defined Planning Layouts • Create Planning Layouts for Activity Type Planning**. Here you can create a new layout by selecting **Create Activity Type Planning Layout**.

In the initial screen, shown in Figure 10.54, create a new planning layout by copying an existing standard layout, such as 1-201.

10.4 Planning

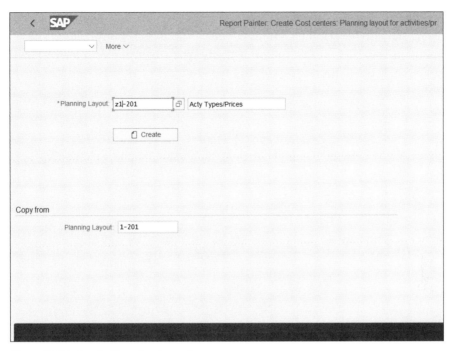

Figure 10.54 Create Planning Layout

Click the **Create** button. What you see next is the layout definition screen, shown in Figure 10.55.

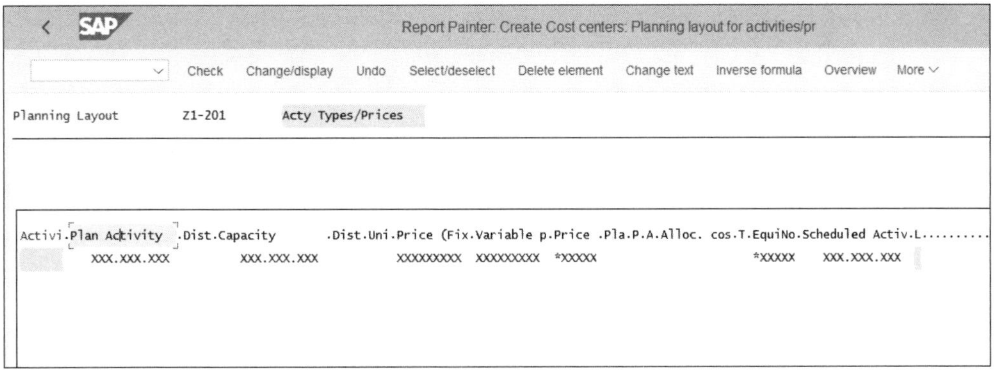

Figure 10.55 Layout Definition

10 Cost Center Accounting

This is created using Report Painter, which is an SAP tool used to define a layout for reporting, but also for planning purposes. You can double-click any column and see its definition. Let's click the first column, **Activity Type**. On the screen shown in Figure 10.56, you see the definition of the **Activity Type** column.

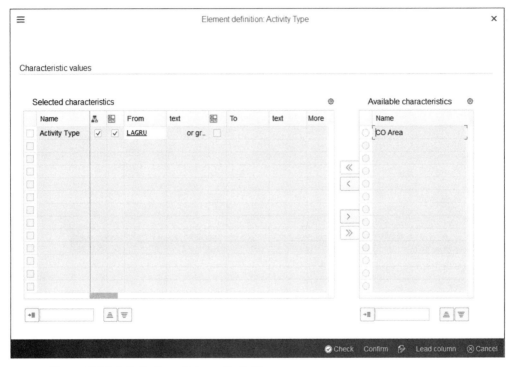

Figure 10.56 Activity Type Column Definition

This column contains the **Activity Type** characteristic. A characteristic means that the column contains field values, such as different activity types. You can also add the **CO Area** characteristic by moving it from the **Available characteristics** section to the **Selected characteristics** section using the left arrow ⟨ button. Next to the name of the characteristic, there are two columns ♣ (**Set**) and ▦ (**Variable**), which are both checked. **Set** means that the field is interpreted as a group value. By checking the **Variable** option, you make sure the value of the field is derived from the selection screen. In that case, you need to enter a variable name in the next column.

Go back and click another column, **Plan Activity**. As shown in Figure 10.57, in this column we don't have characteristics but key figures. The key figure here represents value fields such as actual costs, plan costs, or activity output. In this case, the key figure represents the plan activity.

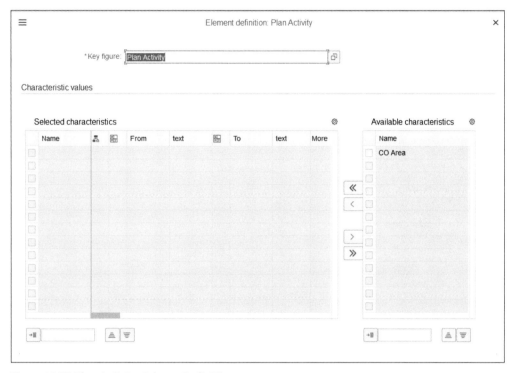

Figure 10.57 Plan Activity Column Definition

In such fashion, you can define the whole structure of the layout. After that, save it with the **Save** button. Now you can use it in your manual planning.

10.5 Information System

The information system for cost center accounting is very rich and provides myriad standard reports that should meet even the most demanding user requirements. In addition, SAP provides a user-friendly tool to generate user-defined reports for cost centers: the Report Painter, which we briefly mentioned in the previous section.

Let's first look at the standard reports available before moving on to user-defined reports.

10.5.1 Standard Reports

You can access the information system for cost center accounting under application menu path **Accounting • Controlling • Cost Center Accounting • Information System**, as shown in Figure 10.58.

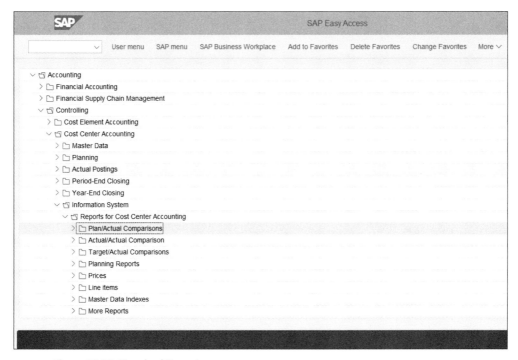

Figure 10.58 Standard Reports

As you can see, the reports are grouped into various categories such as plan/actual comparisons, actual/actual comparisons, line items, and so on. We won't go into detail on all reports, but we'll show you one of the most commonly used ones: report S_ALR_87013611 (Cost Centers: Actual/Plan/Variance) in the **Plan/Actual Comparisons** section.

Figure 10.59 shows the output screen of the report. For the selected cost centers, you see the costs per cost element with separate columns for actual costs, plan costs, and variance between the two.

You can double-click any line and the system will offer you the ability to drill down using another of the linked reports, as shown in Figure 10.60.

10.5 Information System

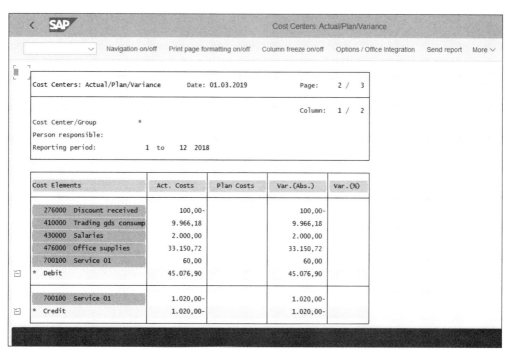

Figure 10.59 Actual Plan Comparison Report

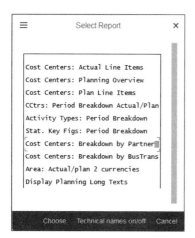

Figure 10.60 Drilldown Reports

If you select, for example, **Cost Centers: Actual Line Items**, the system automatically executes a line item report for the selected line from the previous report.

10.5.2 User-Defined Reports

SAP S/4HANA provides a convenient tool to create new user-defined reports for cost centers. The reports in cost center accounting are created using Report Painter, which you saw in play when defining the planning layout.

To create a user-defined report, follow menu path **Controlling • Cost Center Accounting • Information System • User-Defined Reports • Create Reports**, then select **Create Report**.

In the initial screen, shown in Figure 10.61, you need to select a report library, which provides you with collection of characteristics and key figures you can use in your report. Standard library **1AB Cost Centers: Variance Analysis** can be used for cost center reports to provide comparison between actual and plan costs. It's always a good idea to copy from an existing standard report and adapt it to your needs. Select standard report **1ABW-001 Cost Centers: Variances** in the **Copy from Report** field and click the **Create** button.

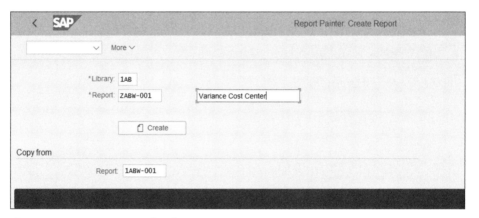

Figure 10.61 Create User-Defined Report

Figure 10.62 shows the report layout, which you can change. Click the first column, **Input Side: Cost Elements**.

Figure 10.63 shows that two characteristics are selected: cost element and activity type. The cost element is set and variable, which means that the range of cost elements should be entered on the selection screen of the report. For the activity type, enter "*", which means that all activity types will be selected. The list of available characteristics you see on the right side of the screen is defined in the library.

10.5 Information System

Figure 10.62 Report Layout Definition

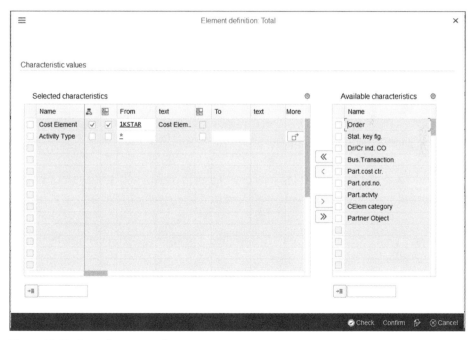

Figure 10.63 Cost Elements Column

10 Cost Center Accounting

Click the second column, **Actual Costs**. The screen shown in Figure 10.64 shows the definition of the **Actual Costs** column.

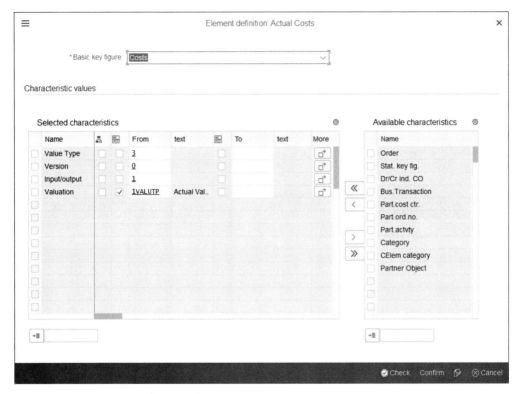

Figure 10.64 Actual Costs Column

Here, the basic key figure **Costs** is selected, and the column will display the amounts for it based on the characteristics selected and their values (e.g., above version 0 only).

The **Target Costs** column is similarly designed, but uses the plan version because it relates to plan costs.

Go back to the previous screen, shown in Figure 10.65. Click the **Var. (Abs.)** column and you'll see the screen shown in Figure 10.66.

10.5 Information System

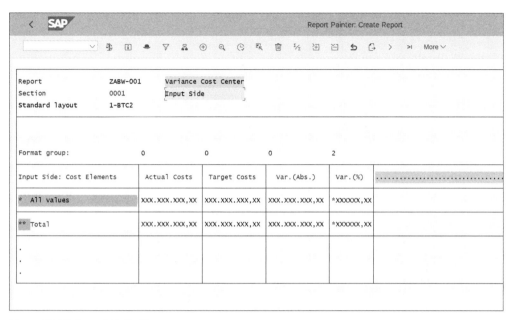

Figure 10.65 Report Definition

Figure 10.66 Variance Column

10 Cost Center Accounting

The variance column is defined as a formula. In this formula editor screen, you define that it is equal to column 1 – column 2. You can also make more complex calculations using the formula editor with various mathematic operations, both with columns and rows.

Select **General data selection** from the top menu (refer back to Figure 10.62) to see the screen shown in Figure 10.67.

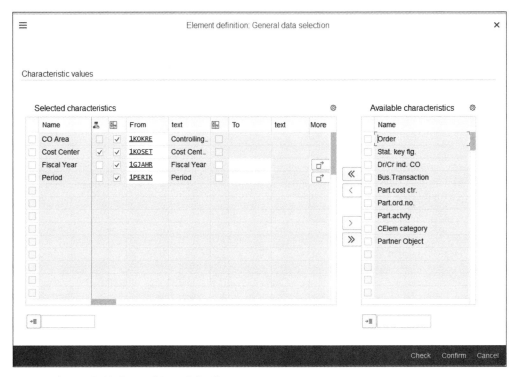

Figure 10.67 General Data Selection

The general data selections refer to the report as a whole, not just to specific columns or rows. In our example, the controlling area, cost center, fiscal year, and period are defined as variables, which means that a user has to enter them in the selection screen of the report, and then the report output will contain values restricted by those selections.

After modifying your report, save it with the **Save** button. In this way, you can easily enhance the standard reports provided by SAP to cover business-specific requirements with your user-defined reports.

10.6 Summary

This was an extensive chapter in which we covered a lot of ground. We delved deep into the secrets of cost center accounting, which is one of the core controlling and overhead accounting areas. After reading this chapter, you should have a good understanding of the master data objects for cost center accounting, such as cost centers, cost center groups, activity types, and statistical key figures.

You now understand the flow of values to cost centers, both in manual postings and in integrated documents. We explained how to maintain the automatic determination of cost centers and substitutions that also can be used to derive cost centers.

You learned how to set up various allocation procedures, such as accrual calculation distribution, assessment, and activity allocation. These are major parts of cost center accounting. Costs often don't stay in the original cost center because for management analysis needs, they need to be allocated to portray the purpose of the costs more accurately.

You also learned how to set up planning for cost centers, including setting up its basic settings and creating planning layouts. Finally, we guided you through the information system for cost centers, including how to use standard reports and drill down from them, as well as how to create user-defined reports with the powerful Report Painter tool.

With that, we finish our guide to cost center accounting. Next we'll move to the other core overhead accounting area: internal orders.

Chapter 11
Internal Orders

This chapter gives step-by-step instructions for configuring internal orders in SAP S/4HANA and using them for planning and budgeting in an organization.

Internal orders are controlling objects used to represent internal projects, such as marketing campaigns, information technology implementation projects, or research and development projects. They are fundamentally different than cost centers, which usually represent departments within the organization, whereas internal orders are created for specific projects with a specific duration.

Therefore, internal orders are created with a limited time span. The cost centers represent a more or less stable structure, whereas internal orders are created, budgeted for, accumulate costs, and, at some point in time, closed.

There is also another option to represent projects: the project system in SAP S/4HANA. This system is used to depict very complex projects with a lot of steps, milestones, and complex cost structures. The project system lies outside of the scope of this book because it's a cross-point between finance and logistics and deserves a whole book of its own. Just keep in mind that you should use internal orders for smaller projects, whereas for very complex projects you should explore the project system.

We'll start by setting up the required master data for internal orders. Then you'll learn how to configure and enter budgets for internal orders. Budgeting is one of the important functions that helps to manage costs tracked on an internal order level. In the next section, we'll examine the actual postings and periodic allocations that are performed on internal orders. Then we'll continue with internal orders planning, and we'll finish with a guide to the internal orders information system.

11.1 Master Data

The internal order is a master data object created from the application menu in a way similar to how we created cost centers. Before we get to creating the internal order, we must first activate internal orders in SAP S/4HANA and then configure order types, which classify the internal orders and control their settings.

The first step in setting up internal orders is to ensure that order management is activated for your controlling area. To do that, follow menu path **Controlling • Internal Orders • Activate Order Management in Controlling Area**.

Check the checkbox to the left of your controlling area (in our case, 0001), and click **Activate components/control indicators** on the left side of the screen, as shown in Figure 11.1.

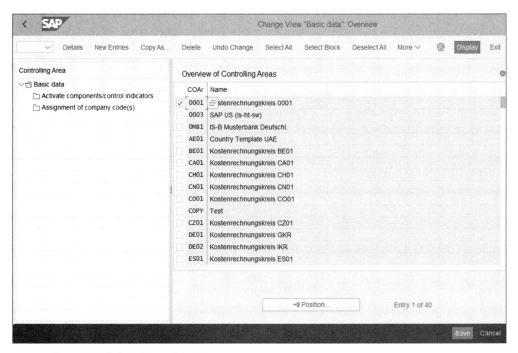

Figure 11.1 Select Controlling Area

On the screen shown in Figure 11.2, make sure that **Order Management** is set to **Component active**. Then save your entry with the **Save** button.

11.1 Master Data

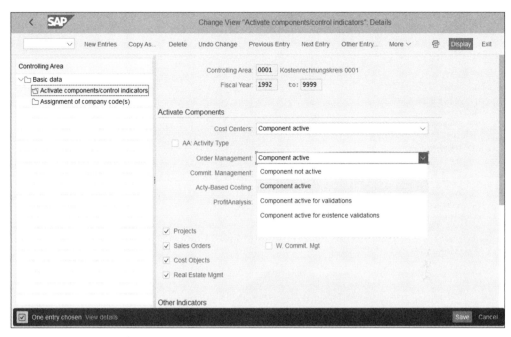

Figure 11.2 Activate Order Management

The next step is to create the relevant master data.

11.1.1 Order Types

Order types are used to classify internal orders by grouping those with similar functions and purposes. They control various parameters, such as number ranges and the settlement profile.

To configure order types, follow menu path **Controlling • Internal Orders • Order Master Data • Define Order Types**.

In the first configuration screen, you see a list of defined order types, as shown in Figure 11.3. As with other configuration objects, there are several predefined standard order types delivered by SAP to cover the main business requirements for internal orders, such as marketing, development, construction, and so on. It's good practice to use the standard order types as reference and copy them into your own order types in custom name ranges starting with Z or Y.

427

11 Internal Orders

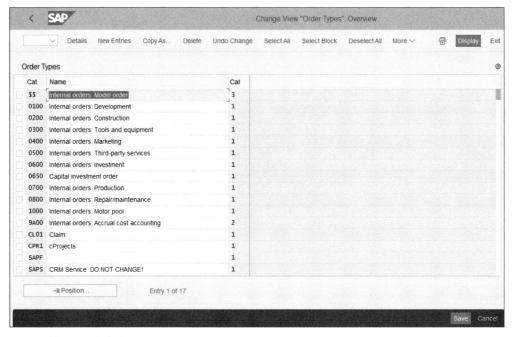

Figure 11.3 Order Types

Check the checkbox to the left of standard order type **0400 Internal orders: Marketing**, then select **Copy As...** from the top menu. In the resulting screen, shown in Figure 11.4, change the order type name to start with Z and enter a meaningful description for the order type.

Below that, you'll see the number range interval, which we'll cover in detail in the next section. Then there are several configuration options, as follows:

- **Settlement prof.**
  The settlement profile controls how the internal order costs will be settled to other cost receivers. The internal order is a transitory cost object; it accumulates costs, but at the end usually the costs are settled to another object, such as a fixed asset, cost center, and so on. We'll configure the settlement profile later in this section.

- **Planning Profile**
  The planning profile controls various parameters in the planning process for internal orders. We'll configure it in Section 11.4.

- **Execution Profile**
  The execution profile is used to trigger the creation of orders, reservations, and goods issues.

11.1 Master Data

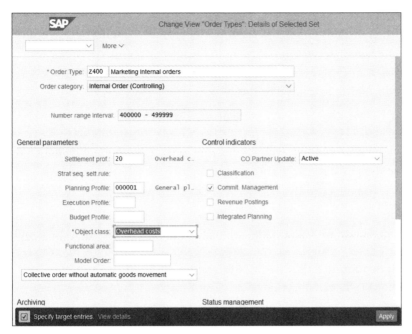

Figure 11.4 Define Order Type

- **Budget Profile**
  You need to specify a budget profile when you're going to use budgeting for internal orders of this order type. It contains customizing settings that affect the budgeting process.
- **Object class**
  The object class classifies the order based on its purpose. You can choose from overhead costs, production, investment, or profitability analysis.
- **Functional area**
  The functional area is used to prepare a profit and loss statement based on the cost-of-sales method. You can assign a functional area to an internal order, and here you can specify a default functional order for the order type.
- **Model Order**
  A model order can be used to predetermine certain parameters of the internal order, and here you can specify a default model order for the order type.
- **Collective order with/without automatic goods movement**
  This indicator controls whether an automatic goods movement is allowed, if production orders are assigned to collective orders.

429

- **CO Partner Update**

  The CO partner update defines how the totals records for the sender-receiver combinations are updated. You can choose from the following options:

  - **Not Active**

    There is no update of totals records.

  - **Partially Active**

    Totals records are generated only for settlements between internal orders.

  - **Active**

    For all settlements, a totals record is generated.

  In general, you should choose an active or partially active update.

- **Classification**

  The classification in SAP S/4HANA is no longer supported, so this indicator should not be set. It's been replaced by the general object summarization. You can read further information in SAP Notes 339863 and 188231. You can also use enhancement COOPA003 for classification.

- **Commit. Management**

  Check this indicator when you want to record commitments on the internal orders of this order type.

- **Revenue Postings**

  Check this indicator when you want to record revenue postings on the internal orders of this order type (with general ledger accounts with cost element category 11 or 12).

- **Integrated Planning**

  This indicator should be checked if orders from this order type should participate in integrated planning. *Integrated planning* means that activity inputs planned are updated directly on the sending cost center. Make sure you check this indicator from the beginning if needed; it's very difficult to set it retrospectively.

Scroll down and you can configure further fields, as shown in Figure 11.5.

In the **Archiving** area, you can configure the following fields:

- **Residence Time 1**

  In this field, you specify the period that needs to pass between setting the deletion flag (which can be reset) and setting the deletion indicator (which cannot be reset) for an internal order.

- **Residence Time 2**

  In this field, you specify the period that needs to pass between setting the deletion indicator and reorganizing the internal orders.

Figure 11.5 Order Type Further Fields

In the **Status management** area, you can configure the following fields:

- **Status Profile**
  The status profile controls the various statuses you can set for internal orders (such as **Created**, **Released**, **Closed**, and so on).

- **Release Immediately**
  If you set this indicator, orders from this order type are immediately released when created.

- **Status dependent field select.**
  You can use a status-dependent field status, meaning that various fields are required, optional, or suppressed based on the status of the order (created, released, and so on).

In the **Master data display** area, you can configure the following fields:

- **Order Layout**
  The order layout determines which fields are required, which fields can be changed, and which fields are display only.

- **Print form**
  If you want to print the internal order, you can assign a print form here. As standard, you can use form CO_ORDER.

- **Field selection**
  With this button, you can define a user-defined field selection. For example, you can set that a specific field is required for this order type.

After configuring the fields, save your order type with the **Save** button.

11.1.2 Screen Layouts

Internal order master data is organized into tabs, which follow a standard order. You can control which sections of data appear on which tabs using screen layouts. To configure a screen layout, follow menu path **Controlling • Internal Orders • Order Master Data • Screen Layout • Define Order Layouts**.

Figure 11.6 shows the standard screen layout **SAP0**.

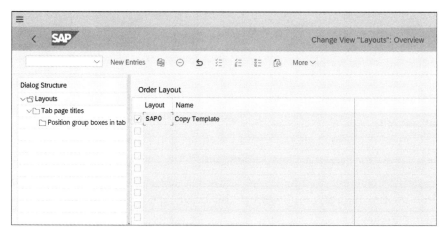

Figure 11.6 Define Screen Layout

Select this layout and copy it using the (**Copy As...**) button from the top menu. Then give a custom name for the copied layout, as shown in Figure 11.7.

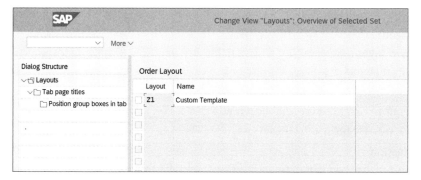

Figure 11.7 Copy Custom Layout

Continue with the **Apply** button. The system issues the message shown in Figure 11.8, which notifies you of all the dependent entries copied.

11.1 Master Data

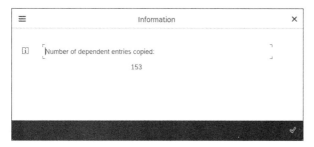

Figure 11.8 Dependent Entries Message

Proceed with the ✓ button. Then select the **Tab page titles** option on the left side of the screen, which brings you to the tabs definition, shown in Figure 11.9.

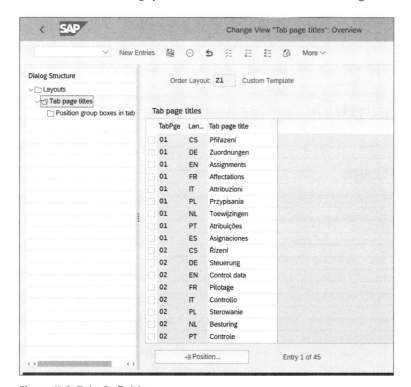

Figure 11.9 Tabs Definition

In the **TabPge** column, you see the sequence number of the tab. Tabs are maintained separately per language, which is shown in the **Language** column. In the **Tab page title** column, the description of the tab is defined.

433

11 Internal Orders

To maintain the group boxes within a tab, select it and click **Position group boxes in tab pages** on the left side of the screen. This brings you to the screen shown in Figure 11.10.

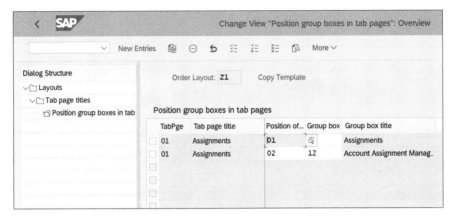

Figure 11.10 Group Boxes in Tabs

Here you can assign different group boxes and change their order in the tab.

Modify the layout in whatever way will be the most convenient for your users, then save your entries with the **Save** button.

11.1.3 Number Ranges

As with other master data objects, we need to specify number ranges for internal orders to determine the numbers assigned when creating these orders.

To configure order number ranges, follow menu path **Controlling • Internal Orders • Order Master Data • Maintain Number Ranges for Orders**. Figure 11.11 shows the initial screen of the number ranges configuration transaction. The number range object for internal orders is AUFTRAG.

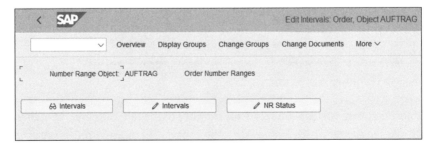

Figure 11.11 Internal Order Number Ranges

11.1 Master Data

Here, you define the various number ranges to be used with the internal order types. The first column, **No**, is the number range number, which is then assigned to an order type. The **From No.** and **To Number** columns define the lower and upper limit of numbers within that number range. In the **NR Status** column you'll see the current number assigned in the system. The checkbox in the **Ext** column defines that the number range is external; that is, a user has to enter the internal order number manually, within the defined limits.

So, for example, as defined in Figure 11.12, number range 02 uses external number assignment within the A and ZZZZZZZZZZZZ limits. That means an internal order number using this number range can be up to 12 letters long, and all letters are allowed. Number range 04 uses internal number assignment between 000002000000 and 000002999999. In this case, the first internal order will have the number 2000000 assigned by the system, the second 2000001, and so on.

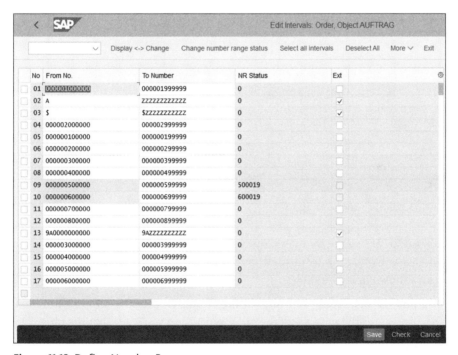

Figure 11.12 Define Number Ranges

After configuring your number ranges, save your entries with the **Save** button. Then go back to the initial number ranges screen, shown in Figure 11.11. Select **Change Groups** from the top menu.

435

11 Internal Orders

On the screen shown in Figure 11.13, you can see the various number ranges and the order types assigned to them. On the top, the unassigned order types are listed. After scrolling down, you'll see the new order type Z400 assigned to group (number range) 08. This is because you copied it from order type 0400, which is also assigned to group 08.

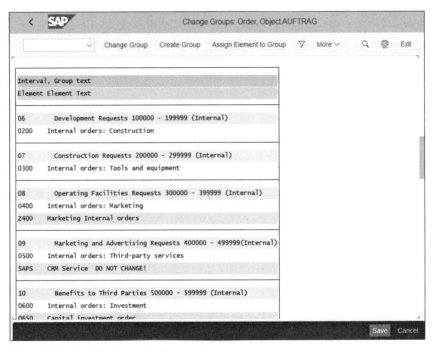

Figure 11.13 Number Range Assignment

If you want to change the number range assignment, select the order type and select **Assign Element to Group** from the top menu. Then you see a list of the available groups, as shown in Figure 11.14.

Select group **09 Marketing and Advertising Requests 400000–499999 (Internal)**, which better suites the purpose of the order type. Then you'll see that the order type is reallocated, as shown in Figure 11.15.

After that, save your entries with the **Save** button.

Click the *Intervals* (**Change Intervals**) button and you'll see the screen shown in Figure 11.12.

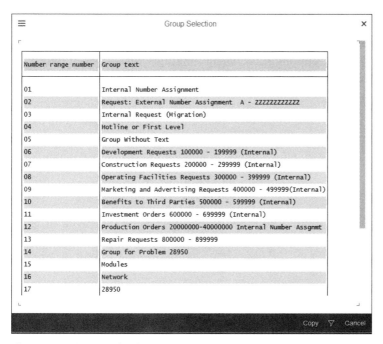

Figure 11.14 Group Selection

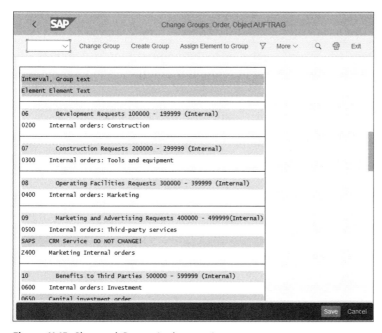

Figure 11.15 Changed Group Assignment

11.1.4 Create Internal Order

Now that we've set up the order types, we can create internal orders by following application menu path **Accounting • Controlling • Internal Orders • Master Data • Special Functions • Order • KO01—Create**.

On the initial screen, shown on Figure 11.16, you need to enter your controlling area and order type. Enter the order type you created previously for marketing orders, Z400. Also, you can create the order with reference to an existing order, specified in the **Reference** section.

Figure 11.16 Create Internal Order

As with other master data objects created up to this point, the internal order master record is organized into tabs, as shown in Figure 11.17. On the header level, enter the internal order description.

Below that, in the first tab, **Assignments**, configure the following fields:

- **Company Code**
  Each internal order needs to be created for a company code, which you specify here.
- **Business Area**
  If you use business areas in your project, it's a good idea to assign a default business area for the internal order here.
- **Plant**
  It's possible to assign a plant to an internal order. This is mandatory for production orders.

- **Functional Area**

  The functional area is used to prepare profit and loss statements, based on the cost-of-sales method. You should assign a functional area to the internal order if you use cost of sales accounting.

- **Object Class**

  The object class classifies the order based on its purpose. Here the object class configured in the order type is defaulted.

- **Profit Center**

  When profit center accounting is activated, you have to assign a profit center here. We'll activate and configure profit center accounting in Chapter 12.

- **Responsible CCtr**

  Here you can specify a cost center that's responsible for the internal order. For example, for marketing internal orders, typically this will be the marketing department cost center.

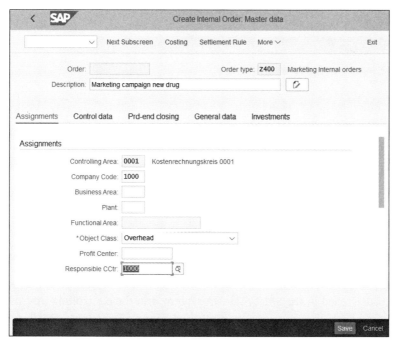

Figure 11.17 Internal Order Assignments

Next, let's move on to the **Control data** tab, shown in Figure 11.18.

11 Internal Orders

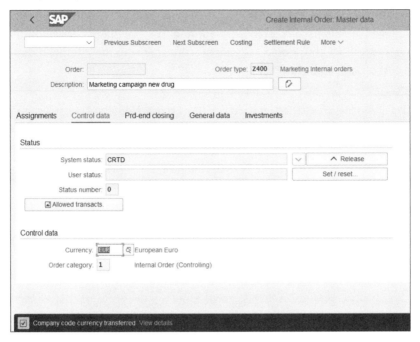

Figure 11.18 Internal Order Control Data

In the **Control data** tab, the following fields are available:

- **System status**
  Here the system shows the current status of the order. When you create the order, the status is *created*, indicated by **CRTD** in this field. This status doesn't allow you to post to the order. After creating it, you need to release it with the ^ **Release** button.

- **Currency**
  In this field, you specify the currency of the internal order. By default, this is the currency of the controlling area.

- **Order category**
  Orders are widely used in the SAP S/4HANA system, and the controlling internal orders are just one of the categories, as indicated here by category **1**, which you can't change. In logistics, other types of orders are widely used, such as production, process, or quality management orders.

In the **Prd-end closing** tab, you can make settings for the period-end closing, such as assigning results analysis keys and costing sheets, as shown in Figure 11.19.

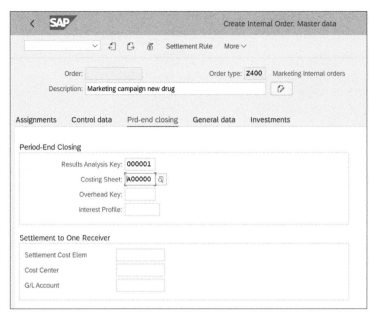

Figure 11.19 Internal Order Period-End Closing Data

In the **General data** tab, you can enter data for information purposes only for the applicant of the order, such as telephone, application date, department, and so on, as shown in Figure 11.20.

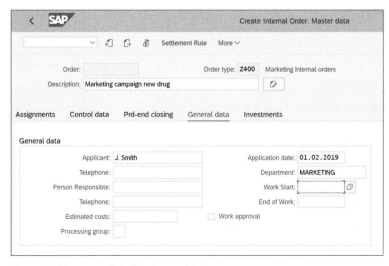

Figure 11.20 Internal Order General Data

The **Investments** tab is relevant if you use investment management functionality. If so, here you make settings for the internal order such as assignment of an investment profile, investment program, and so on, as shown in Figure 11.21.

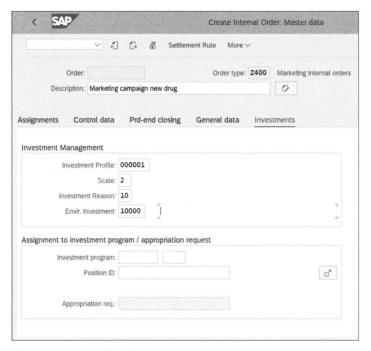

Figure 11.21 Internal Order Investments Data

After completing all the data, save the internal order with the **Save** button. Then you get a message that the order was created with a number from the number range assigned to the order type.

If you go into change mode and view the newly created order, you can release it with the ^ **Release** button in the **Control data** tab. Then the status becomes **REL** (released), as shown in Figure 11.22.

You can go back to the previous status with the ⌵ button. You can go to the next status, which now shows as ^ **Tech. Comple** (technically complete). This way, you can move forward and backward between the statuses.

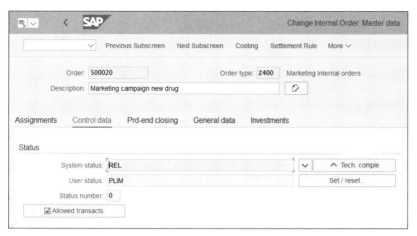

Figure 11.22 Released Internal Order

11.2 Budgeting

A major part of the internal orders process addresses budgeting for internal orders. Internal orders are used for specific projects such as marketing campaigns, research and development projects, or IT implementation projects. As such, usually they're associated with particular budgets that they need to adhere to. If the budget is reached, no more postings should be allowed for the internal order, although there are various settings that can control how stringent that is.

You've seen the budget profile in the order type configuration, shown back in Figure 11.4. Now let's see how to configure a budget profile. To do so, follow menu path **Controlling • Internal Orders • Budgeting and Availability Control • Maintain Budget Profile**, then select activity **Maintain Budget Profile**. You'll see a list of the available budget profiles; the standard one is the **000001 General** budget profile. Select it, and make a copy of it by selecting **Copy As...** from the top menu. Then rename the new copy to Z00001, as shown in Figure 11.23.

On this screen, configure the following fields:

- **Budget profile**
  The name of the budget profile.
- **Text**
  Meaningful description of the budget profile.

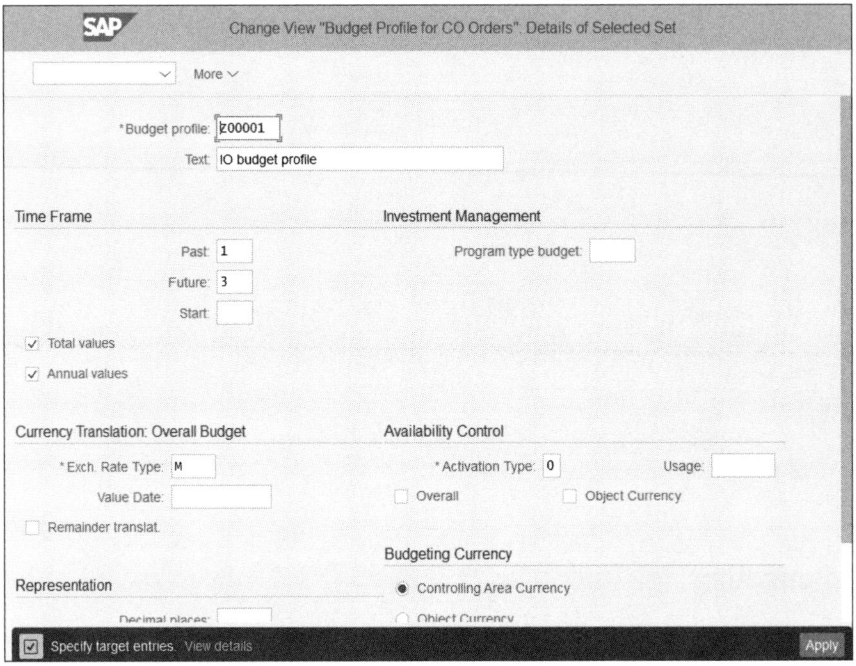

11 Internal Orders

Figure 11.23 Create Budget Profile

- **Past/Future/Start**
  Here you define how long in to the past and far into the future you can budget for. For example, a value of 3 in the **Past** field means you can budget three years before the start year; a value of 5 in the **Future** field means you can budget five years into the future. In the **Start** field, you can specify a different start year from the default, which is the current fiscal year.

- **Total values**
  Specifies that it's possible to budget overall values.

- **Annual values**
  Specifies that it's possible to budget annual values.

- **Exch. Rate Type**
  Here you need to specify the exchange rate type to be used when converting from another currency to the controlling currency.

- **Value Date**
  In this field, you can enter a default value date for exchange rate conversion. If you don't enter a value date, currencies will be translated by period.

- **Activation Type**
  This indicator controls how the availability control should be activated, with three options:
  - **0 = Cannot Be Activated**
    This means that active availability control isn't possible for orders with this budget profile.
  - **1 = Automatic Activation during Budget Allocation**
    Availability control is automatically activated during budgeting, which means that it's activated when you enter a budget for the order.
  - **2 = Background Activation when Usage Exceeded**
    Availability control is activated in the background automatically for all orders in which the commitments exceed the usage number entered in the budget profile.
- **Usage**
  This field is used in conjunction with activation type 2 controls when the availability control should be activated. For example, if you enter 70 here, it means that when 70% of the budget is reached, the availability control will be activated.
- **Overall**
  Indicates that availability checks are conducted against the overall value.
- **Object Currency**
  If you select this checkbox, the availability control will be performed in the object currency.
- **Representation**
  Using the **Scaling Factor** and **Decimal Places** parameters, you can control the layout of values representation.
- **Budgeting Currency**
  In this section, you select in which currency the budgeting should be done from following options:
  - **Controlling Area Currency**
  - **Object Currency**
  - **Transactional Currency**
- **Default Object Currency**
  If budgeting is done in a user-defined transactional currency, and you set this indicator, then the relevant object is budgeted by default in the object currency. If the indicator is not set, then the relevant object is budgeted by default in controlling area currency.

11 Internal Orders

After maintaining the settings, save the budget profile with the **Save** button.

In the next step, you need to maintain number ranges for budgeting by following menu path **Controlling • Internal Orders • Budgeting and Availability Control • Maintain Number Ranges for Budgeting**.

As shown in Figure 11.24, the number range object for budgeting is BP_BELEG. Click the *Intervals* (**Change Intervals**) button to maintain the number ranges.

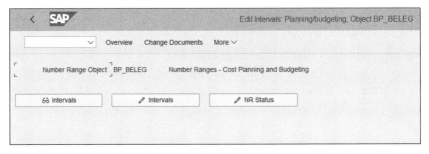

Figure 11.24 Budgeting Number Ranges

As with other number range objects, you provide a number range number (as shown in Figure 11.25; 01, 02, 03, and so on) and the starting and ending numbers of the range. In the **NR Status** column, the system displays the current last assigned number. In the **Ext** column, you can specify that the number range should use an external document number, which is a rarely used option for budgeting.

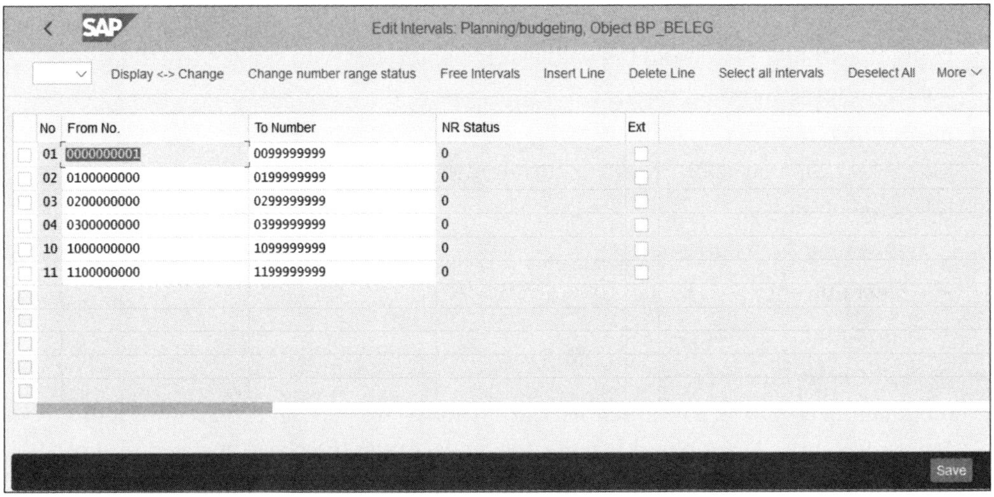

No	From No.	To Number	NR Status	Ext
01	0000000001	0099999999	0	
02	0100000000	0199999999	0	
03	0200000000	0299999999	0	
04	0300000000	0399999999	0	
10	1000000000	1099999999	0	
11	1100000000	1199999999	0	

Figure 11.25 Maintain Number Ranges

After maintaining the number ranges, save your entries with the **Save** button.

In the next step, configure the tolerance limits for availaibility controls by following menu path **Controlling • Internal Orders • Budgeting and Availability Control • Define Tolerance Limits for Availability Control**.

In the screen shown in Figure 11.26, you can configure per controlling area the percentage of the assigned budget that will trigger the relevant action when reached. For this example, specify 90% of the budget for controlling area 0001.

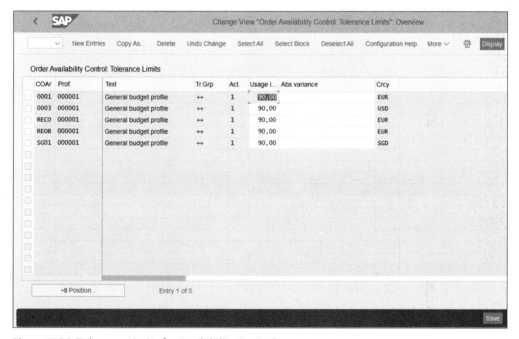

Figure 11.26 Tolerance Limits for Availability Control

Change the percentage as appropriate, then save your entry with the **Save** button.

In the next step, you specify cost elements exempt from availability control. This functionality is used when you need to exclude postings to some cost elements from the availability control calculation. To specify exempt cost elements, follow menu path **Controlling • Internal Orders • Budgeting and Availability Control • Specify Exempt Cost Elements from Availability Control**. Then select **New Entries** from the top menu to enter the exempt cost elements, as shown in Figure 11.27.

11 Internal Orders

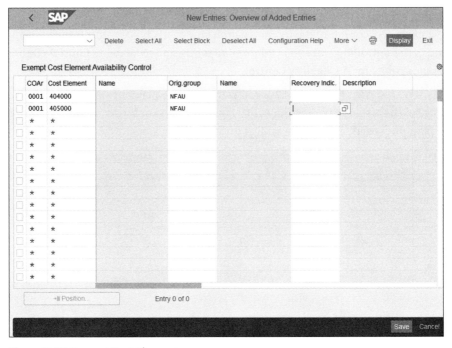

Figure 11.27 Exempt Cost Elements

Enter the cost elements to be exempted. The following fields are to be configured:

- **COAr**
  The controlling area for which you configure cost elements to be exempted.
- **Cost Element**
  The cost elements to be exempted.
- **Orig. group**
  The origin group is used for subdividing material and overhead costs. You can enter "*" to select all origin groups or specify a origin group.
- **Recovery Indic.**
  In this field, you can enter a default value date for the exchange rate. If your company belongs to a joint venture, incurred costs can be shared among partners using a recovery indicator, which can alter the budgeting if specified here.

After entering the cost elements to be exempted, save your entries with the **Save** button.

11.3 Actual Postings and Periodic Allocations

After you've set the budget for your internal orders, you can start accumulating costs on them. The actual postings to internal orders occur very much like postings to cost centers. They can come from integrated postings from logistics or from postings in financial accounting, where the cost object is an internal order.

A big difference in internal order accounting compared to cost centers is that in essence the internal order is a transitionary cost object, which at the end is settled to another object, such as a cost center, fixed asset, profitability segment, and so on. Let's first configure the system for settlement of internal orders and then move on to the topic of periodic reposting.

11.3.1 Settlement

Settlement is a very important function of internal orders accounting that allows you to allocate the costs accumulated on internal orders to other predefined receivers.

To be able to settle an internal order, you need to maintain a settlement rule in the order. There you specify the receivers of the costs and their share. Let's change the internal order created previously by adding a settlement profile. To do so, follow application menu path **Accounting • Controlling • Internal Orders • Master Data • Special Functions • Order • KO02—Change**. In the initial screen, enter the internal order number. Then in the order master data screen, as was shown in Figure 11.17, select **Settlement Rule** from the top menu.

Figure 11.28 shows the settlement rule definition screen. In the lines here, you can enter one or multiple settlement receivers.

The following fields are available for configuration on this screen:

- **Cat**
  The category of the settlement receiver, such as cost center, fixed asset, or profitability segment. The available categories depend on the settlement profile, which we'll configure shortly.

- **Settlement Receiver**
  The receiver cost object. In this example, we provided two different cost centers.

- **%**
  The share of costs the receiver should receive during settlement. If only one receiver is specified, it's 100%.

449

- **Equivalence No.**
  With this option, the costs are distributed to the settlement receivers in proportion to the equivalence numbers defined here.

- **Settlement Type**
  You can use full or periodic settlement. The difference is that, in full settlement, you settle each time all the unsettled costs that have occurred for a sender object for all the periods prior to the settlement.

- **Source Assignment**
  Another option is to use a source assignment structure to allocate the costs. Source assignments define the assignment of debit cost element(s) to a receiver.

- **No.**
  Sequential number of the receivers.

- **From Period/Fiscal Year and To Period/Fiscal Year**
  You can use these fields to limit the validity duration of a settlement rule.

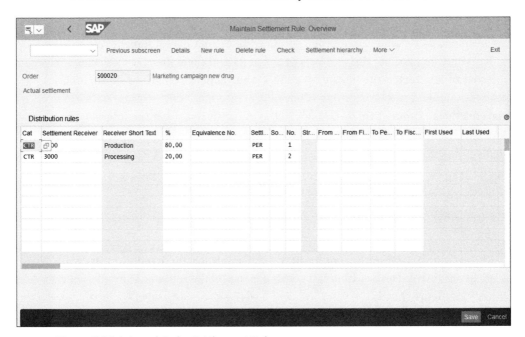

Figure 11.28 Internal Order Settlement Rule

After defining the settlement rule, save your entries with the **Save** button.

Let's now examine the configuration settings required for settlement. First, you need to define a settlement profile by following menu path **Controlling • Internal Orders •**

11.3 Actual Postings and Periodic Allocations

Actual Postings • Settlement • Maintain Settlement Profiles, then choose the **Maintain Settlement Profiles** activity.

You'll see a list of the settlement profiles defined in the system, as shown in Figure 11.29. You may remember that in new order type Z400, you used settlement profile 20, as was shown in Figure 11.4. Double-click settlement profile **20 Overhead costs** now to review its settings.

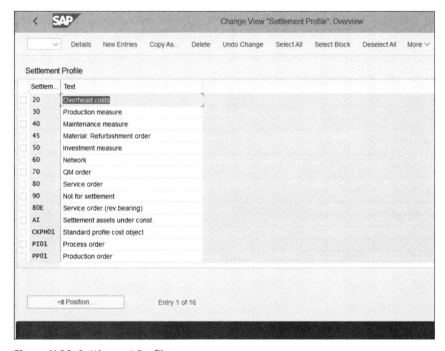

Figure 11.29 Settlement Profiles

Figure 11.30 shows the configuration of the settlement profile. The following fields can be configured on this screen:

- **Settlement Profile**
  Code and description of the settlement profile.
- **Actual Costs/Cost of Sales**
  In this section, you select whether settlement is mandatory, optional, or not possible.
- **Default Values**
  In this section, you enter a default allocation structure, source structure, profitability

11 Internal Orders

analysis transfer structure, and object type. These structures are controlling tools for defining allocation cost elements in various controlling components.

- **Valid Receivers**
  In this section, you define for the various cost objects whether they can receive settlement or if it's mandatory to receive settlement for this settlement profile. This is what defines the possible receivers in the internal order settlement rule screen (see Figure 11.28). For example, here the general ledger account is marked as **Settlement Not Allowed**, so you don't see the general ledger account as a valid receiver in the order from order type Z400, which has settlement profile 20.

- **100%—Validation**
  If you set this indicator, the system will issue a warning if the total to be settled is different from 100%. If it isn't set, you'll get a warning only if the total is more than 100%.

- **%—Settlement**
  If you set this indicator, you can use the settlement rule to determine the distribution rules governing the percentage costs to be settled.

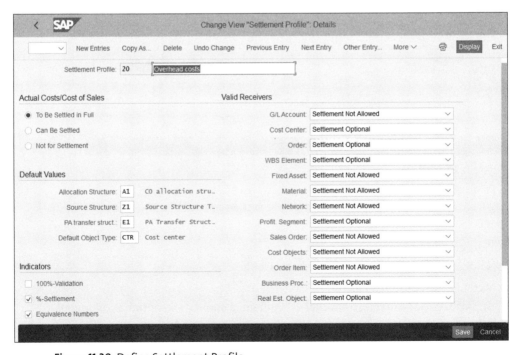

Figure 11.30 Define Settlement Profile

11.3 Actual Postings and Periodic Allocations

- **Equivalence Numbers**
  With this indicator, you can define distribution rules in the settlement rule, according to which costs are settled proportionally.

- **Amount Settlement**
  With this indicator, you can define distribution rules in the settlement rule, according to which costs are settled by amount.

- **Variances to Costing-Based PA**
  If you set this indicator, variances are settled to costing-based profitability analysis.

To view the **Other Parameters** section, scroll down as shown in Figure 11.31.

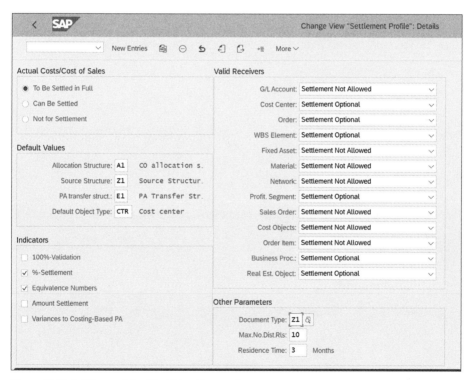

Figure 11.31 Settlement Profile Other Parameters

The following fields are available:

- **Document Type**
  Default document type for settlement runs relevant for accounting.

453

- **Max.No.Dist.Rls.**
  The maximum number of allowed distribution rules.

- **Residence Time**
  Here you define the residence time for the settlement documents in calendar months.

You also need to maintain number ranges for settlement by following menu path **Controlling • Internal Orders • Actual Postings • Settlement • Maintain Number Ranges for Settlement Documents**.

As shown in Figure 11.32, the number range object for settlement documents is CO_ABRECHN. Click the *Intervals* (**Change Intervals**) button to maintain the number ranges.

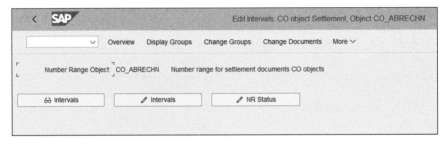

Figure 11.32 Number Ranges for Settlement

For settlement, normally only one number range is required, as shown in Figure 11.33.

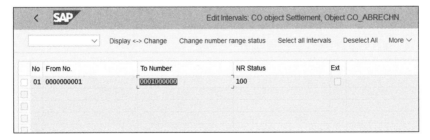

Figure 11.33 Define Number Range for Settlement

Go back and select **Change Groups** from the top menu to assign your controlling area to the number range. If your controlling area shows in the **Nonassigned Elements** section on the top of the list, position the cursor on it and select **Assign Element to Group** from the top menu. Then select group **01**. As shown in Figure 11.34, controlling area **0001** is assigned to group **01**.

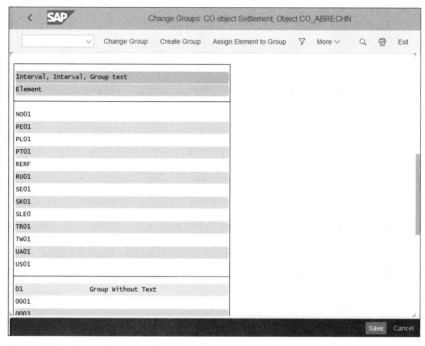

Figure 11.34 Assign Number Range

The settlement of individual internal orders is executed via application menu path **Accounting • Controlling • Internal Orders • Period-End Closing • Single Functions • Settlement • KO88—Individual Processing**. In practice, more often the collective settlement of multiple orders is used via application menu path **Accounting • Controlling • Internal Orders • Period-End Closing • Single Functions • Settlement • KO8G—Collective Processing**, as shown in Figure 11.35.

You also need to set up a variant in which you specify the selections of orders. You can create/change/display variants from this screen with the [ ][ ][ ] buttons to the right of the **Selection Variant** field. You can also set it up from the configuration menu via **Controlling • Internal Orders • Actual Postings • Settlement • Define Selection Variants for Settlement**.

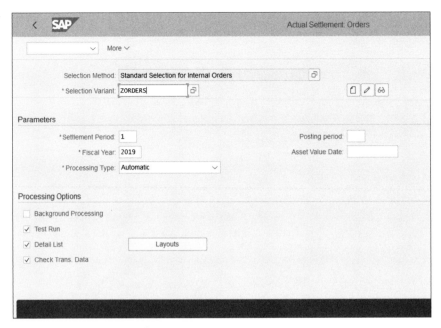

Figure 11.35 Collective Settlement

On the initial screen shown in Figure 11.36, enter a name for your variant and click the **Create** button.

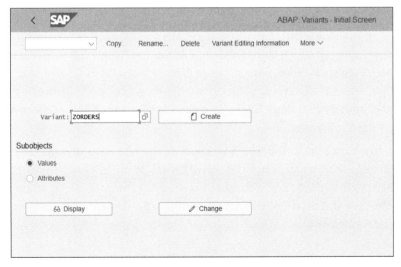

Figure 11.36 Create Selection Variant

11.3 Actual Postings and Periodic Allocations

In the next screen, shown in Figure 11.37, you define the criteria to select internal orders, such as order group, from and to order numbers, controlling area, company code, and so on.

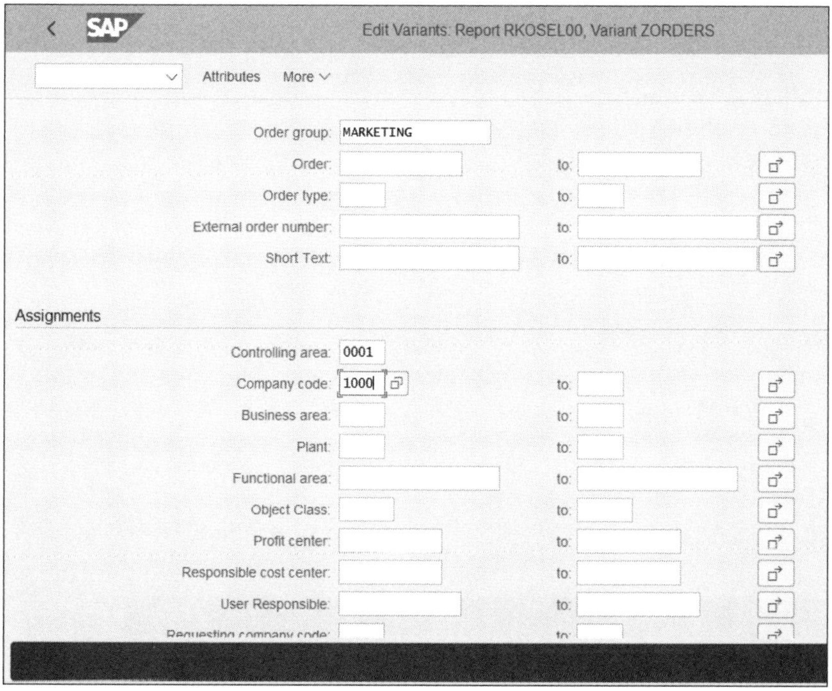

Figure 11.37 Definition of Selection Variant

Now you know how to configure the SAP S/4HANA system to settle your internal orders.

11.3.2 Periodic Reposting

Another method of allocating costs from internal orders is periodic reposting (Transaction KSW5). This allocation method enables you to transfer costs posted on an internal order (as well as on cost centers, WBS elements, and other cost objects) based on rules defined in allocation cycles, rather than fixed settlement receivers, as you've seen in the settlement process.

Periodic reposting uses the cycle-segment technique you know from the cost center distributions and assessments. To create a periodic reposting, follow menu path

Controlling • Internal Orders • Actual Postings • Define Periodic Repostings, then select **Create actual periodic reposting**.

On the initial screen, shown in Figure 11.38, you specify a cycle name and start date, and you can provide an existing cycle to be used as a reference.

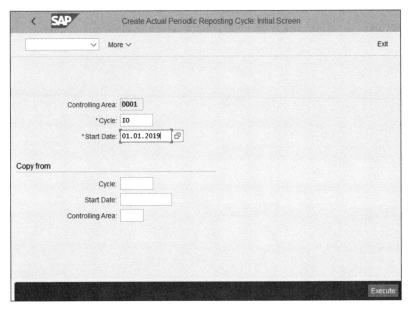

Figure 11.38 Create Periodic Reposting

Click the **Execute** button and you're taken to the cycle header definition screen, shown in Figure 11.39.

On this screen, you can maintain the following fields:

- **Start Date and To**
  The start date and end date, which define the validity of the cycle.

- **Text**
  Meaningful long description of the cycle.

- **Iterative**
  This indicator enables iterative sender/receiver relationships. In this case, the iteration is repeated until each sender is fully credited.

- **Cumulative**
  If you check this indicator, the sender amounts posted are allocated based on tracing factors, which are cumulated for all periods.

- **Derive Func. Area**
  By checking this indicator, the functional area proposed in the cycle definition for the receiver is ignored and is derived again.
- **Field Groups**
  In general, only amounts in the controlling area currency are used in periodic allocations. However, you can choose other types of currencies in this section.

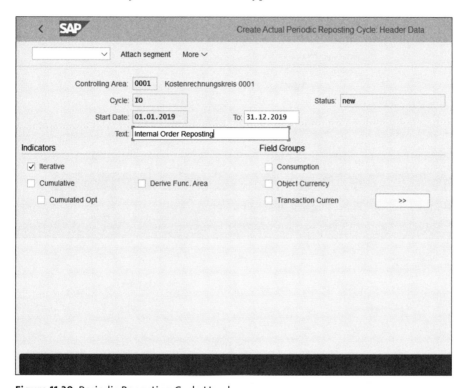

Figure 11.39 Periodic Reposting Cycle Header

After maintaining the cycle definition, you need to attach at least one segment. To do so, select **Attach segment** from the top menu.

Next you define the segment, as shown in Figure 11.40. The fields are the same as for the cost center distribution cycle, as explained in Chapter 10, Section 10.3.2, so we won't explain them again in detail here.

As sender objects, specify the internal orders, as shown in Figure 11.41.

11 Internal Orders

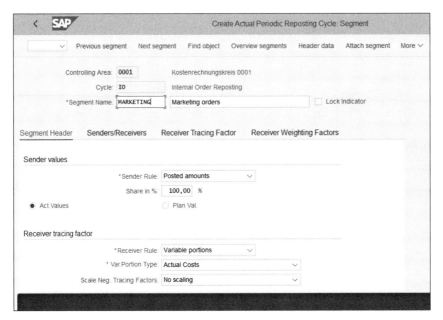

Figure 11.40 Periodic Reposting Segment Definition

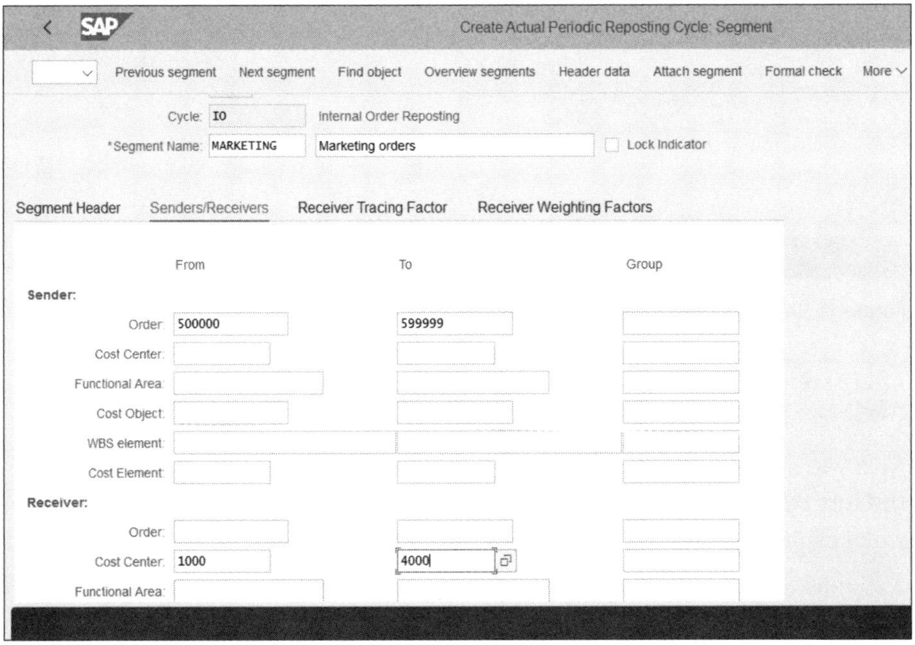

Figure 11.41 Segment Sender/Receiver Rules

After filling in all the tabs, in a similar fashion as you did for cost center allocation cycles, you can save your cycle. To execute it, follow application menu path **Accounting • Controlling • Internal Orders • Period-End Closing • Single Functions • KSW5—Periodic Reposting**.

As shown in Figure 11.42, you need to specify the period and year of execution (normally the reposting cycles are to be run on a monthly basis as part of the period-end closing), select the cycle, and execute it first in test and then in real mode.

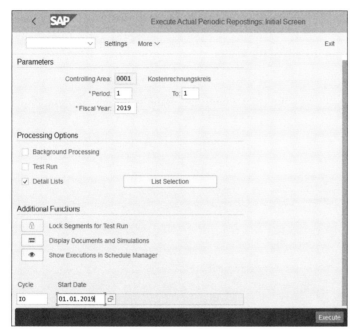

Figure 11.42 Execute Periodic Reposting

11.4 Planning

Another major functionality that internal orders offer is planning. You can enter plan values on the internal order level, then compare them with the actuals for useful management analysis.

In this section, we'll explain how to configure the basic settings required for internal orders planning. Then, we'll teach you how to use statistical key figures for planning. We'll finish with a guide to using periodic allocations in the internal orders planning process.

11.4.1 Basic Settings

The settings for internal order planning are very similar to the settings for cost center planning in Chapter 10, Section 10.4.1, so we'll just recap them briefly.

The controlling versions for which you'll do internal order planning need to be enabled for planning. For that, follow menu path **Controlling • Internal Orders • Planning • Basic Settings for Planning • Define Versions**, then select **Maintain Version Settings in Controlling Area**.

Ensure that in the **Plan** column you've checked the versions you'll be planning in, as shown in Figure 11.43. Also enable planning in each fiscal year using the **Settings for Each Fiscal Year** option on the left of the screen.

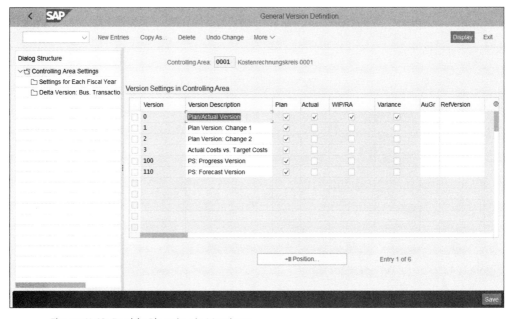

Figure 11.43 Enable Planning in Versions

You also need to assign planning transactions to number ranges in **Controlling • Internal Orders • Planning • Basic Settings for Planning • Assign Planning Transactions to Number Ranges**. Enter your controlling area, as shown in Figure 11.44, and click the *Intervals* (**Change Intervals**) button.

On the screen shown in Figure 11.45, you can set the beginning and ending range of the planning number ranges.

11.4 Planning

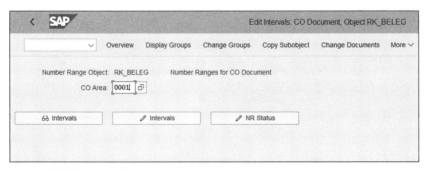

Figure 11.44 Planning Number Ranges

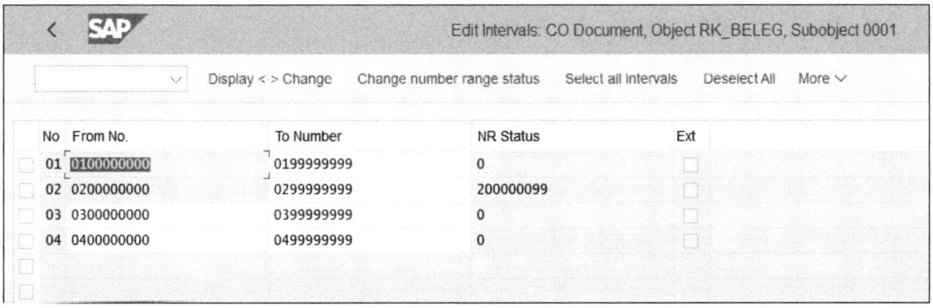

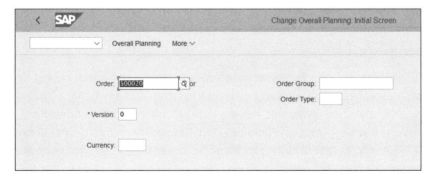

Figure 11.45 Maintain Planning Number Ranges

To record planning values to internal orders, follow application menu path **Accounting • Controlling • Internal Orders • Planning • Overall Values • KO12—Change**. On the initial screen, shown in Figure 11.46, enter your order number, order group, or order type, and your version.

Figure 11.46 Internal Order Planning

463

On the next screen, shown in Figure 11.47, enter the plan costs per year for the order.

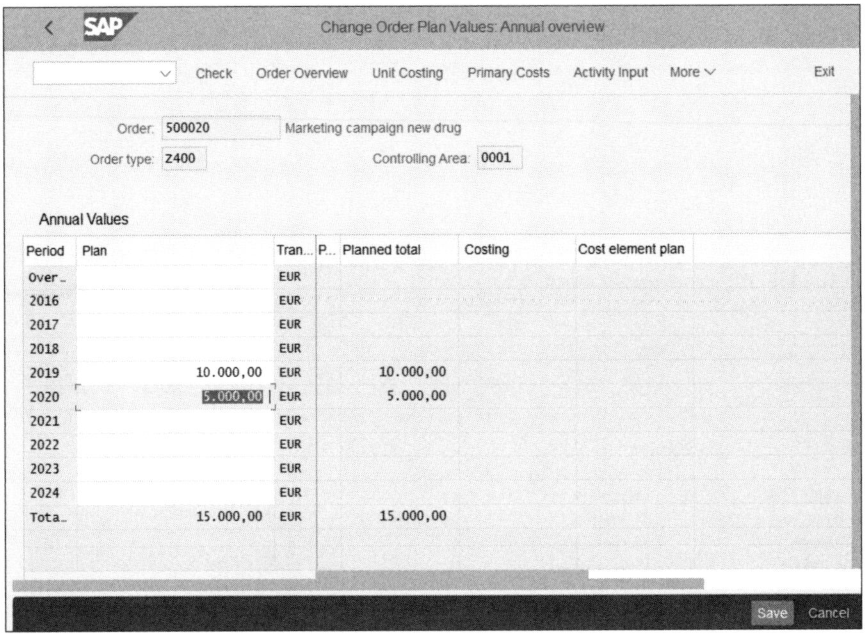
Change Order Plan Values: Annual overview
Check
Order Overview
Unit Costing
Primary Costs
Activity Input
More
Exit
Order: 500020
Marketing campaign new drug
Order type: Z400
Controlling Area: 0001
Annual Values
Period
Plan
Tran...
P...
Planned total
Costing
Cost element plan
Over...
2016
2017
2018
2019 10.000,00 EUR 10.000,00
2020 5.000,00 EUR 5.000,00
2021
2022
2023
2024
Tota... 15.000,00 EUR 15.000,00
Save
Cancel

Figure 11.47 Internal Order Plan Values

Save your entry with the **Save** button. The system generates a planning document, as shown in Figure 11.48.

Document 0200000000 posted

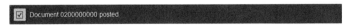

Figure 11.48 Planning Document Number

11.4.2 Statistical Key Figures

You can also plan based on a combination of internal orders and statistical key figures. To do so, you need to use a planning layout for statistical key figure planning. You can use the standard layouts or create your own via menu path **Controlling • Internal Orders • Planning • Manual Planning • User-Defined Planning Layouts • Create Planning Layouts for Statistical Key Figure Planning**; once there, select the **Create Statistical Key Figure Planning Layout** activity.

On the initial screen, shown in Figure 11.49, enter a name and description for the new layout and choose an existing layout to use as a reference. Then click the **Create** button.

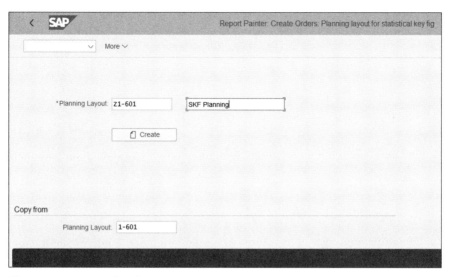

Figure 11.49 Create Planning Layout

Figure 11.50 shows the layout definition screen. It's designed in Report Painter, in a similar fashion as the planning layout we developed for cost centers in Chapter 10. You can double-click any column and change its definition. Click the first column, **Statis**, which represents the statistical key figures.

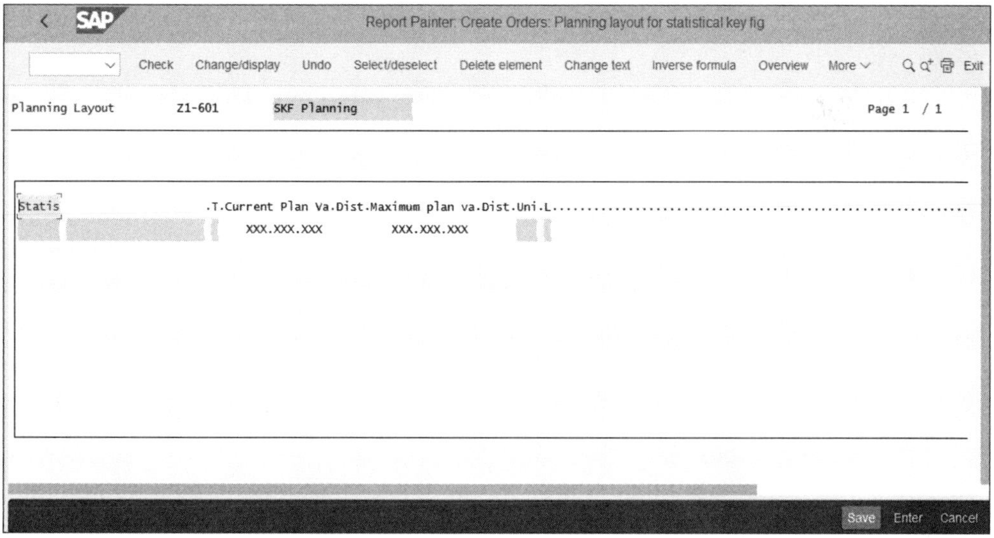

Figure 11.50 Planning Layout Definition

Figure 11.51 shows the definition of the statistical key figure column.

Figure 11.51 Statistical Key Figure Column Definition

This column contains the statistical key figure characteristic. The **Set** and **Variable** columns are checked, which means that a range of statistical key figures need to be entered on the selection screen when entering plan values. You can also add the controlling area.

The next columns hold value fields, which represent the current plan values and maximum plan values. You can add or remove columns based on your planning requirements. After that, save your entries with the **Save** button, and then you can use the new layout in your planning.

11.4.3 Allocations

In internal orders planning, you also can use allocations to allocate plan costs. The settings are very similar to the actual costs allocations we configured, but different transactions are used. To execute a planned settlement, follow application menu path

11.4 Planning

Accounting • Controlling • Internal Orders • Planning • Allocations • Settlement, and then **KO9E—Individual Processing** for individual internal orders or **KO9G—Collective Processing** for multiple orders.

To define plan periodic reposting, follow menu path **Controlling • Internal Orders • Planning • Define Periodic Repostings**, then select **Create Plan Periodic Reposting**.

As shown in Figure 11.52, this looks the same as the actual reporting cycle you created, but it uses Transaction KSW7 (as opposed to Transaction KSW1, used for actual reposting).

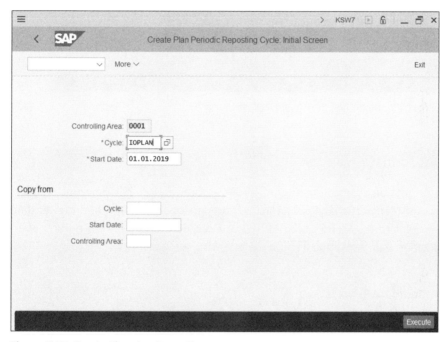

Figure 11.52 Create Planning Reposting

The cycle definition screen shown in Figure 11.53 is somewhat different than the one for the actual cycle because you need to specify a controlling plan version. Then, you need to attach one or more segments, in which you specify plan values as a source. Other than that, the settings are like those for the actual reposting cycle you created.

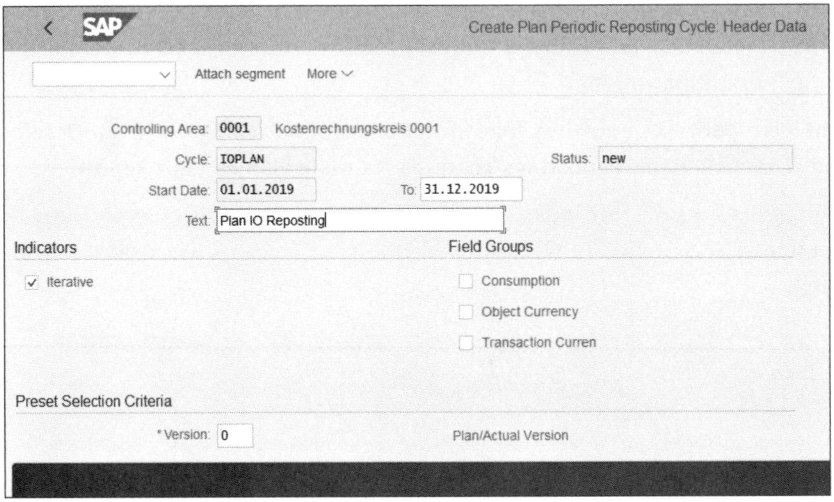

11 Internal Orders

Figure 11.53 Planning Reposting Cycle Definition

11.5 Information System

The information system for internal order accounting offers numerous standard reports for plan/actual and actual/actual comparison, line items, and master data. Also, similar to cost center accounting, you can develop your own reports easily using Report Painter, as you'll see in the following sections.

11.5.1 Standard Reports

The standard reports for internal orders are available at application menu path **Accounting • Controlling • Internal Orders • Information System • Reports for Internal Orders**. We'll look at some of the most commonly used and useful reports in this section.

The plan/actual comparisons are some of the most commonly used reports. They're quite useful to compare actual and plan costs, which enables you to track and analyze how incurred costs on internal orders relate to plan costs. Enter Transaction S_ALR_87012993 (Orders: Actual/Plan/Variance), which compares the actual and planned costs and provides the variance in a separate column.

On the selection screen shown in Figure 11.54, you enter the controlling area, fiscal year, from and to periods, and plan version, as well as internal order and cost element selections. Then execute the report with the **Execute** button.

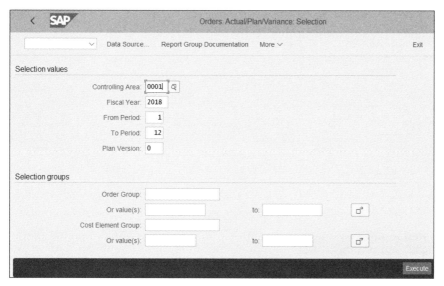

Figure 11.54 Orders Actual/Comparison Selection Screen

On the report output screen, shown Figure 11.55, on the left side you see the internal orders included in the report. You can click any of them to see the result for that order only. On the right side, you see the report result, separated into columns for actual costs, plan costs, and their variance in absolute amounts and as a percentage.

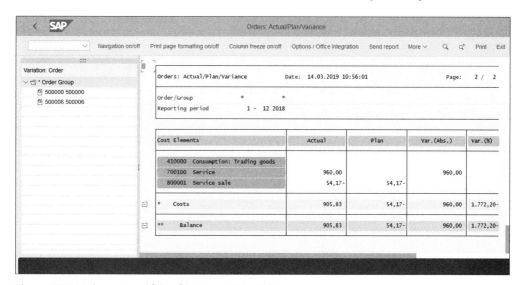

Figure 11.55 Orders: Actual/Plan/Variance Output Screen

In the different lines of the report, you can see the various cost elements posted. If you double-click a line, you can drill down to another report, as shown in Figure 11.56.

Figure 11.56 Orders Drilldown Reports

For example, selecting **Orders: Actual Periods** provides a breakdown of the actual costs per period, as shown in Figure 11.57.

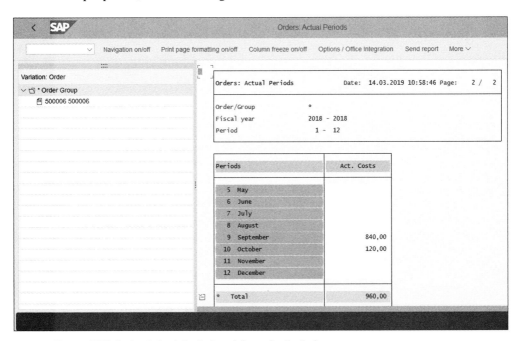

Figure 11.57 Order Actual Costs Breakdown by Periods

Another commonly used report is for order actual line items, located at application menu path **Accounting • Controlling • Internal Orders • Information System • Reports for Internal Orders • Line Items • KOB1—Orders: Actual Line Items**.

In the selection screen shown in Figure 11.58, you can select data by controlling area, from and to posting date, internal orders, and cost elements. Then, execute the report with the **Execute** button.

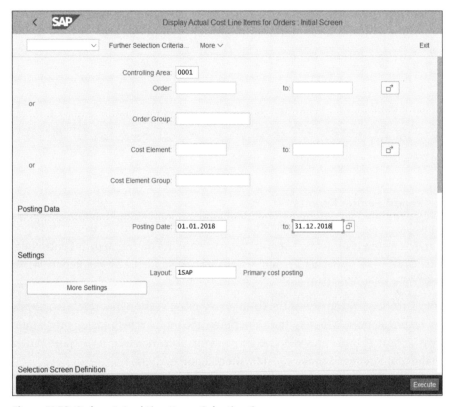

Figure 11.58 Orders Actual Line Items Selection Screen

In the report output screen, shown in Figure 11.59, the system shows the actual costs per order and per line item and provides subtotals and a total at the end. Double-clicking a line shows the original document.

There are many other standard reports that you can explore in similar fashion. These reports can provide invaluable management analysis for your internal orders.

11 Internal Orders

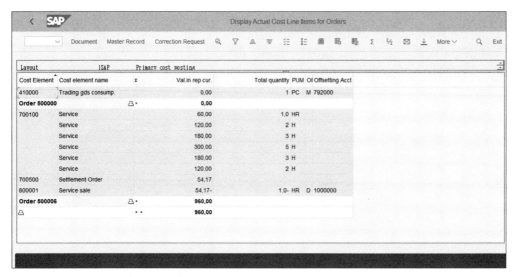

Figure 11.59 Orders Actual Line Items Output Screen

11.5.2 Report Painter Reports

You can also create powerful and flexible reports for internal orders using the Report Painter. To create a user-defined Report Painter report, follow menu path **Controlling • Internal Orders • Information System • User-Defined Reports • Define Report Writer Reports**, then select **Create Report**.

Don't be surprised by the title of the configuration transaction that refers to a *Report Writer report*. Report Writer is the underlying technology behind Report Painter, whereas Report Painter provides the convenient user interface for Report Writer. Often the tool is referred to as Report Painter/Report Writer.

On the initial screen shown in Figure 11.60, you need to specify a report name and select a library. SAP provides two standard libraries for internal orders: 601 and 602. You should rarely need to create a new library, as the standard ones already contain all the relevant fields. Select library 601 and enter a reference report to copy from—in this case, standard report 6000-001. Then click the **Create** button.

On the next screen, shown in Figure 11.61, you can modify the report according to the reporting requirements. The reference report shows actual cost values for various periods, which are totaled in the first columns.

11.5 Information System

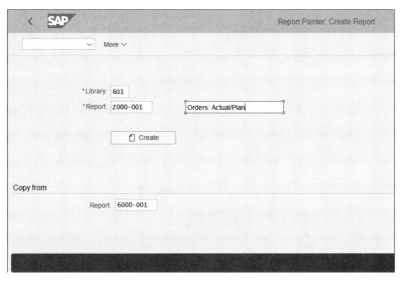

Figure 11.60 Create Internal Order Report

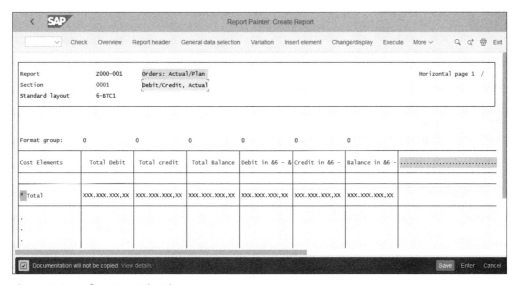

Figure 11.61 Define Internal Order Report

Let's say you want to add columns with plan cost values. Position the cursor after the last column and double-click. Then in the pop-window shown in Figure 11.62, select **Key figure with characteristics**.

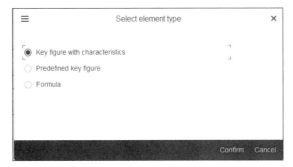

Figure 11.62 Select Element Type

This option allows you to select a value field from the library, such as plan costs, together with characteristics to determine the source. **Predefined key figure**(s) are provided by the system-standard value fields, which are already combined with certain characteristic values. **Formula** allows you to perform a mathematic operation on already defined columns or rows from the report.

Select the basic key figure **Plan Costs -> Line Items**, as shown in Figure 11.63. For the **Fiscal Year** and **Period** characteristics, check the variables and enter variable "6-GJAHL" for the **Fiscal Year** and "6-PERIK" for the **Period**. For the debit/credit indicator (**Dr/Cr ind. CO**), enter "S", which indicates a debit. Because costs are posted on the debit side, we're interested only in the debits. After confirming, the new columns appear at the end of the report.

Then you can add yet another column to provide the variance between the actual and plan values. Double-click again after the last column; this time, select **Formula**.

In the **Enter Formula** screen shown in Figure 11.64, create a formula that indicates that the plan costs should be subtracted from the total balance by selecting the **X003** ID button, then the minus sign from the calculator buttons on the right side, and then the **X007** button.

11.5 Information System

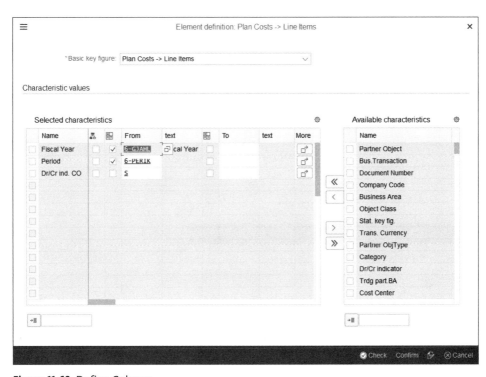

Figure 11.63 Define Column

Figure 11.64 Define Formula

11   Internal Orders

After confirming, you need to specify the column title information, as shown in Figure 11.65.

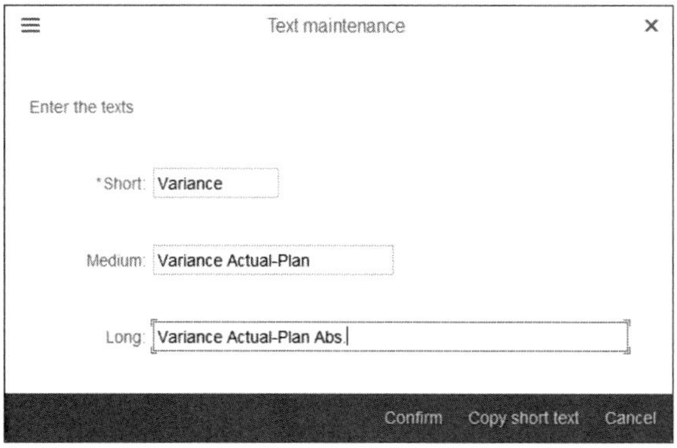

Figure 11.65 Column Title

Because this is a formula column, there's no text to be transferred from the key figure, so we manually enter short, medium, and long descriptions here.

Using these techniques, you can add further columns and rows to the report. After you're satisfied with the report layout, save the report with the **Save** button.

11.6   Summary

This was yet another extensive chapter in which we covered the functionalities and configuration of internal orders accounting inside and out.

After finishing this chapter, you should be very familiar with and know how to configure the various internal orders functionalities:

- Master data
- Budgeting
- Actual postings and periodic processing
- Planning
- Information system

With that, we complete our guide to the overhead controlling areas of the system. In the next chapter, we'll teach you how to configure profit center accounting, which is a cross section of controlling and financial accounting. Technically, it's part of controlling, but it provides comprehensive reporting not only for profit and loss accounts, but also for the balance sheet.

Chapter 12
Profit Center Accounting

This chapter gives step-by-step instructions for configuring profit center accounting. We also explain how SAP S/4HANA can provide full reporting for the whole company because users can drill down into every area of responsibility.

Profit center accounting provides full financial statement reporting on everything lower than the legal entity level. Most companies set up their separate legal entities as company codes, but they want to draw both balance sheets and profit and loss statements on a lower level, such as a product line or division. Here, profit center accounting comes to the rescue because these lower levels can be set up as profit centers easily. Also, profit centers are posted with every single financial line item, as opposed to cost objects such as cost centers or internal orders, which are posted only with profit and loss accounts. Therefore, with profit center accounting it's possible to create a complete balance sheet and profit and loss statement for a profit center or for a group of profit centers.

Profit center accounting in SAP S/4HANA can be regarded as a crossroads between financial accounting and controlling. In the application and customizing menu, profit center accounting is part of controlling. In the older SAP ERP releases, before the introduction of the new general ledger, profit center accounting used special purpose ledger 8A. However, in SAP S/4HANA, profit center accounting is fully integrated with the general ledger. Therefore technically we could argue that profit center accounting is part of financial accounting, but because it integrates information for both the balance sheet and profit and loss accounts and often can be derived from other cost objects, we decided to discuss profit center accounting within our controlling chapters, after you were familiar with overhead costing.

In this chapter, we'll guide you through the main components of profit center accounting:

12   Profit Center Accounting

- Master data
- Profit center derivation and splitting
- Information system

As always, we'll start with setting up the master data required for the module.

12.1   Master Data

Master data for profit center accounting consists of the following, each of which will be discussed in this section:

- Standard hierarchy
- Profit centers
- Profit center groups

We'll start by defining profit centers.

12.1.1   Profit Center

The *profit center* is a financial organizational unit used to structure the organization from a management point of view. It's used to report based on the various areas of the organization, as structured by the management. Therefore, one of the most fundamental design decisions in the finance area of an SAP S/4HANA implementation is how to structure the profit centers. Normally, the business needs extensive support from implementation consultants to help define the best structure for the profit centers. This is also a cross-functional effort because not only finance people should be involved, but also sales, purchasing, and production teams. Many profit centers are derived from cost centers and are normally defined by finance, but also a main derivation point for profit centers is the material master, and here the linkage should be provided by logistic teams.

Different scenarios are possible, but most commonly profit centers are defined with a sales-based approach because management is mostly interested in analyzing how the various products and divisions of the company are performing. Therefore, profit centers often represent various product lines or sales departments of the organization. The goal is that management should be able to quickly and transparently create a full set of financial statements, including costs, revenues, assets, and liabilities, for each important area of the organization.

12.1 Master Data

To create a profit center, follow application menu path **Accounting • Controlling • Profit Center Accounting • Master Data • Profit Center • Individual Processing • KE51— Create**.

On the initial screen, shown in Figure 12.1, enter a profit center to be created and a controlling area. Profit centers are always created for controlling areas. The naming convention for profit centers should be straightforward and easy to understand and recognize. Good practice is to use numeric codes for profit centers. The length depends on the number of profit centers required; four or six digits is common. There should be consistency among the profit centers; for example, profit centers from the same division should start with the same number.

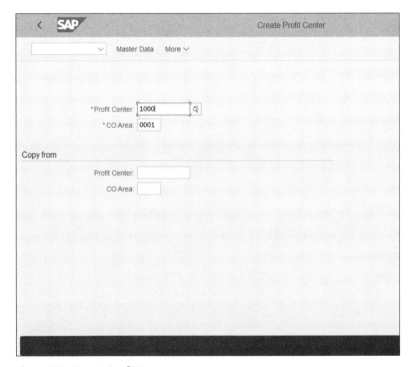

Figure 12.1 Create Profit Center

In the **Copy from** section, you can enter an existing profit center and its controlling area, in which case the values of that profit center will be proposed as defaults on the next screens.

The profit center master record is organized into tabs, which contain logically connected fields. The first tab, shown in Figure 12.2, is **Basic Data**.

479

12 Profit Center Accounting

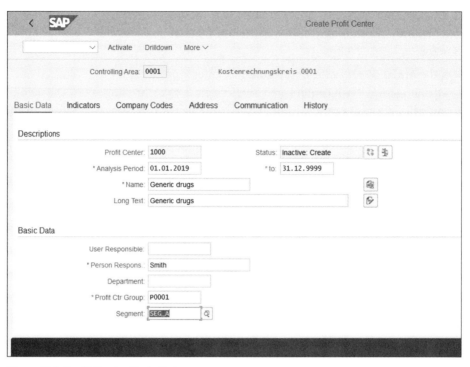

Figure 12.2 Profit Center Basic Data

On this tab, the following fields are available (as with other SAP screens, required fields are marked with a red asterisk):

- **Analysis Period**
  This is the validity period of the profit center. The profit center can be posted only within that validity period. Unless there is some specific reason (e.g., the profit center represents a product being phased out), usually the end date is set as 31.12.9999, which means it's indefinite.
- **Name**
  Short description of the profit center.
- **Long Text**
  Long description of the profit center.
- **User Responsible**
  In this field, you can enter an SAP user responsible for the profit center.
- **Person Respons.**
  This is the name of the responsible manager for the profit center.

12.1 Master Data

- **Department**

  This is the department to which the profit center belongs.

- **Profit Ctr. Group**

  Each profit center should be assigned to a group. The profit center groups are structured in a standard hierarchy, which we'll configure in the next section.

- **Segment**

  Segments are higher-level organizational units than profit centers and are used to produce financial statements and reports on the few main areas into which the organization is structured from a management point of view. If segments are used, each profit center should be assigned to a segment here. Then, when this profit center is posted, the segment is posted also. Both fields are part of table ACDOCA, the Universal Journal.

In the next tab, **Indicators**, shown in Figure 12.3, you can lock the profit center for postings. You can also enter a template for formula planning which contains functions that can be used to find plan values using formulas.

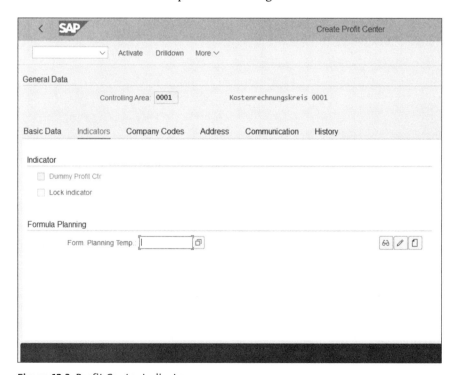

Figure 12.3 Profit Center Indicators

481

In the next tab, **Company Codes**, shown in Figure 12.4, you have to select the company codes for which the profit center is active. Here all company codes assigned to the controlling area will be in the list. By ticking the checkbox in the **Assigned** column, you can select a company code.

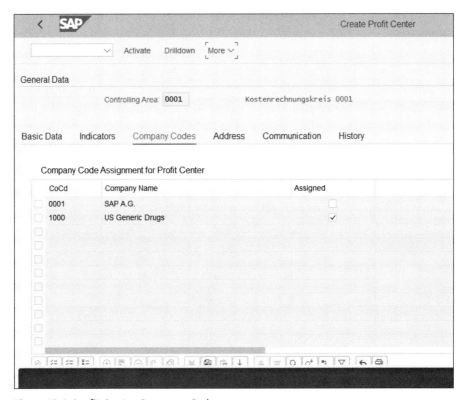

Figure 12.4 Profit Center Company Codes

On the next tab, **Address**, shown in Figure 12.5, you can enter various address information relevant for the profit center, such as street address, region, country, and so on.

On the next tab, **Communication**, shown in Figure 12.6, you can enter communication information relevant for the profit center, such as language key, telephone, fax, and so on.

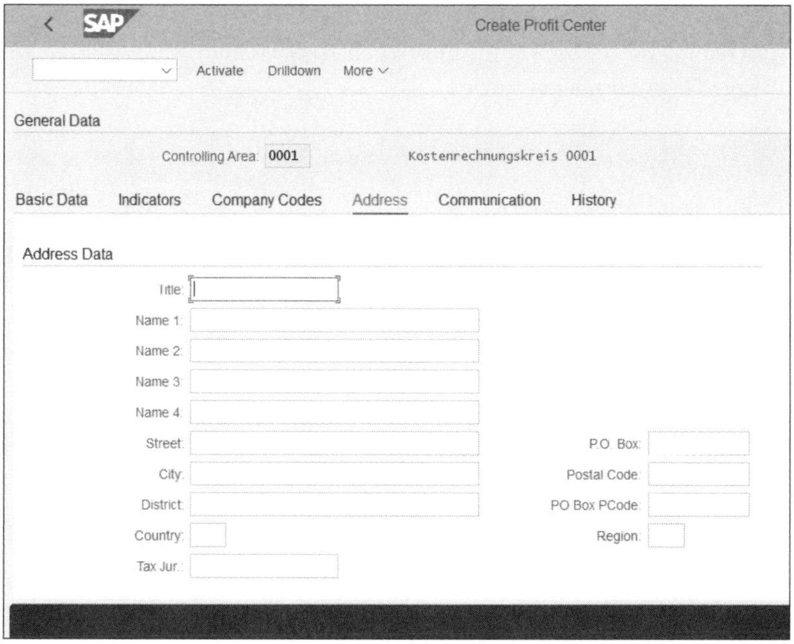

12.1 Master Data

Figure 12.5 Profit Center Address Data

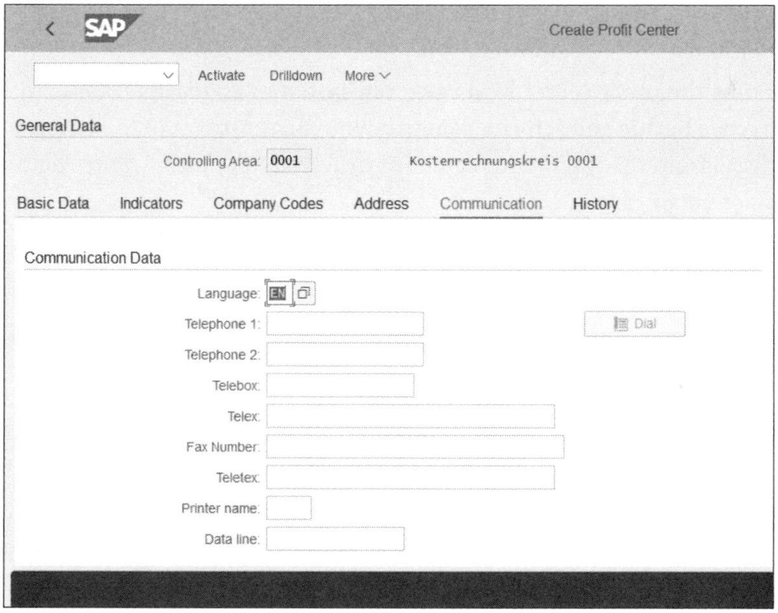

Figure 12.6 Profit Center Communication Data

Finally, on the **History** tab shown in Figure 12.7, the system shows information about the user who created the profit center and when they created it, as well as any changes to the profit center master record.

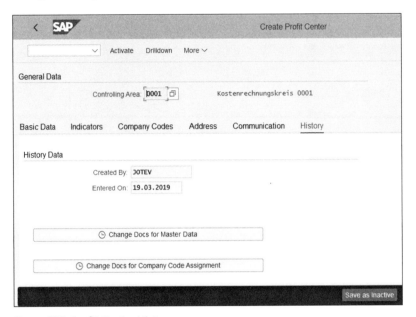

Figure 12.7 Profit Center History

After maintaining the profit center fields, you can save it in an inactive status with the **Save as Inactive** button and activate it later, or you can activate it immediately by selecting **Activate** from the top menu. You have to activate the profit center before you can use it for postings.

12.1.2 Profit Center Group

The profit center group serves to classify profit centers with similar functions. They are then used heavily for reporting.

To create a profit center group, follow application menu path **Accounting • Controlling • Profit Center Accounting • Master Data • Profit Center Group • KCH1—Create**.

On the initial screen, shown in Figure 12.8, enter a name for the **Profit Center Group**, which can consist of letters or numbers. All profit center groups should follow a consistent approach, with either numeric values or descriptive names that point to the function of the group.

12.1 Master Data

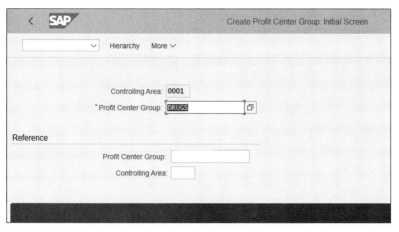

Figure 12.8 Create Profit Center Group

In the **Reference** section, you can provide an existing profit center group to use as a template for the new group. Proceed with the **Enter** button.

On the next screen, shown in Figure 12.9, initially you see just the name of the profit center group you entered in the previous screen. Enter a long description for the group and then you can start to build the lower hierarchy of the group. You can assign additional groups on the same level you're positioned on with the **Same Level** command from the top menu and on a lower level with the **Lower Level** option from the top menu. You can assign profit centers to groups by selecting **Profit Center** from the top menu.

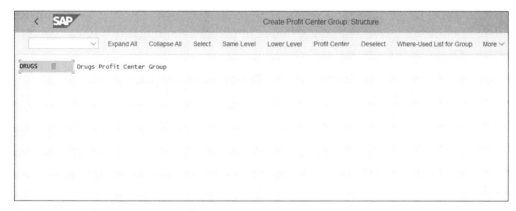

Figure 12.9 Profit Center Group Definition Screen

12   Profit Center Accounting

Figure 12.10 shows the profit center group **DRUGS** without additional groups underneath, with three profit centers assigned.

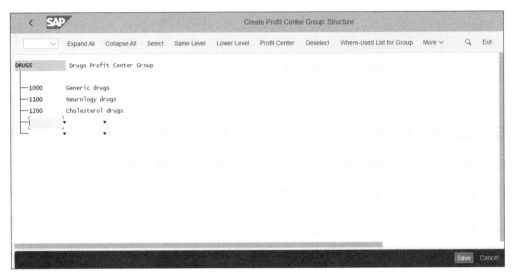

Figure 12.10 Profit Center Group Defined

When you're done, save the profit center group with the **Save** button.

12.1.3   Standard Hierarchy

A *standard hierarchy* is a structure that contains all the profit centers in your organization, grouped in logical groups, based on the purpose and the strategy of classification of the profit centers. It's similar in purpose and function to the cost center standard hierarchy we created before.

To maintain a profit center standard hierarchy, follow application menu path **Accounting • Controlling • Profit Center Accounting • Master Data • Standard Hierarchy • KCH5N—Change**.

In essence, the standard hierarchy is a profit center group at the top level, which contains the other profit center groups in your controlling area. The interface provided by this transaction, shown in Figure 12.11, is very flexible and full of options.

12.2 Profit Center Derivation

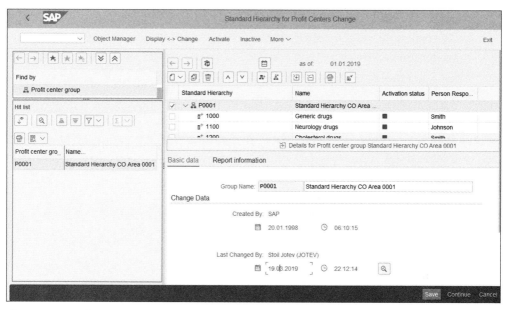

Figure 12.11 Profit Center Standard Hierarchy

The screen is separated into multiple windows for easier navigation. On the left side, you have a search function, in which you can search by profit center or profit center group. On the right side at the top, you have the profit center standard hierarchy; at the bottom, you see details about the selected node. The standard hierarchy is a tree-like structure, which you can expand using the arrows ⟩ to the left of each node. You can also expand the whole hierarchy from the node in which you're positioned with the ⊞ (**Expand Subtree**) button, and collapse the whole hierarchy from the node in which you're positioned with the ⊟ (**Collapse Subtree**) button.

You can right-click any node and create an additional profit center group at the same or a lower level or assign a profit center. Include all your profit center groups in the standard hierarchy and save it with the **Save** button.

12.2 Profit Center Derivation

A profit center is an object that's almost always derived from another object, such as a cost center or material, or derived via configuration and substitution techniques. In this section, we'll examine the various methods by which the system derives the profit center and how it splits documents to ensure that every line item has a profit center.

12 Profit Center Accounting

12.2.1 Account Assignment Objects

Let's review the various scenarios by which the SAP S/4HANA system determines the profit center and what settings you need to make to ensure correct determination.

For profit and loss accounts, the most common mechanism is to derive the profit center from the cost object posted. The profit center should be a required field on the cost center master.

As shown in Figure 12.12, enter "1000" in the **Profit Center** field to assign profit center **Generic Drugs**. Any postings to that cost center will simultaneously post to profit center 1000 also.

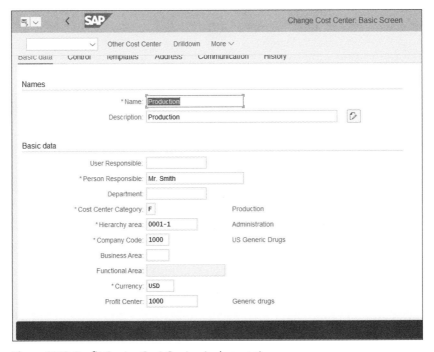

Figure 12.12 Profit Center Cost Center Assignment

Based on the same logic, the system will derive the profit center from an internal order or other cost object in which the master record is assigned to that profit center.

Another order of assignment is also possible. In Transaction OKB9 (Default Account Assignment), which we configured in Chapter 10, you can specify a default cost center and internal order.

As shown in Figure 12.13, you need to enter "3" in the **Acct assign detail** column. Then double-click **Detail per profit center** on the left side of the screen.

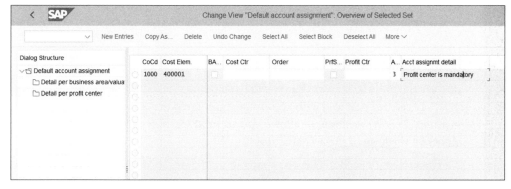

Figure 12.13 Profit Center Default Account Assignment

On the next screen, shown in Figure 12.14, you specify profit centers for the selected cost elements, which should be the default profit centers for the selected cost centers or internal orders.

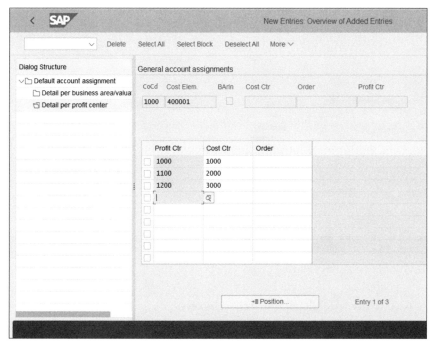

Figure 12.14 Default Account Assignments

For inventory accounts, the profit center is derived from the material master record. However, there are many balance sheet accounts for which there is no direct assignment of profit center possible, such as bank accounts, for example. That's when document splitting comes to the rescue.

12.2.2 Document Splitting

Document splitting enables the system to split line items into multiple line items so that they can be enhanced with account assignment objects such as profit center and segment. This key technique allows you to produce fully balanced financial statements on the level of these characteristics.

We already configured document splitting in Chapter 4. We'll now just activate the profit center as a splitting characteristic by following menu path **Financial Accounting** • **General Ledger Accounting** • **Business Transactions** • **Document Splitting** • **Define Document Splitting Characteristics for General Ledger Accounting**.

In this configuration transaction, shown in Figure 12.15, you enter the profit center in the **Field** column (here, enter "PRCTR") and the partner object in the **Partner Field** column (here, enter "PPRCTR").

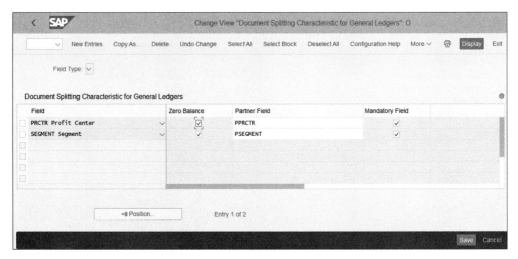

Figure 12.15 Profit Center as Splitting Characteristic

You also need to tick the following two checkboxes:

- **Zero Balance**
  This checkbox is needed so that the system during posting will check whether the

12.2 Profit Center Derivation

balance for the profit center is equal to zero on the document level. If not, it will generate additional clearing lines on clearing accounts that achieve the zero balance.

- **Mandatory Field**
  This field ensures that the profit center must be filled with a value after document splitting for each line item. If the system isn't able to fill it because of missing customizing, there will be an error message and the document won't be posted.

After making the settings for the profit center and other splitting characteristics, save your entry with the **Save** button.

12.2.3 Profit Center Substitution

The profit center in the sales order line item is derived from the material master record, where the profit center needs to be maintained (in table MARC and field PRCTR). However, in some circumstances, you need to substitute it in the sales order based on certain criteria. To define this substitution, follow menu path **Controlling • Profit Center Accounting • Assignments of Account Assignment Objects to Profit Centers • Sales Orders • Sales Order Substitutions • Define Substitution Rules**.

On the screen shown in Figure 12.16, select **Substitution** from the top menu.

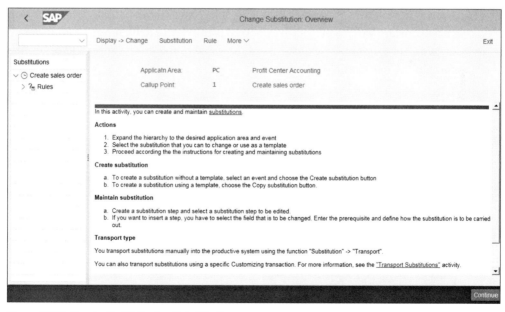

Figure 12.16 Create Profit Center Substitution

491

Enter a name and description for the substitution, and select **Step** from the top menu. The only field available to substitute is the profit center, as shown in Figure 12.17. You also can develop a user exit to program more complex logic.

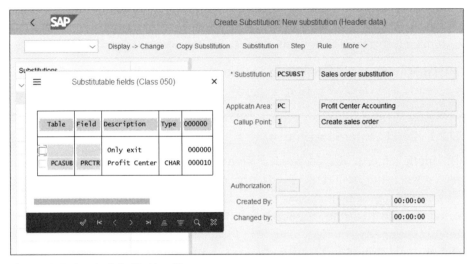

Figure 12.17 Create Substitution Step

In the pop-window shown in Figure 12.18, select whether the substitution will work with a constant value (you will enter a fixed profit center to be substituted), user exit, or field-field assignment, which means the value of the profit center will depend on the value of another field.

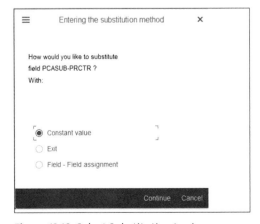

Figure 12.18 Select Substitution Logic

12.2 Profit Center Derivation

Select **Constant value**. Now in the **Prerequisite** section of the substitution step, shown in Figure 12.19, you can define the criteria that will trigger the substitution when met.

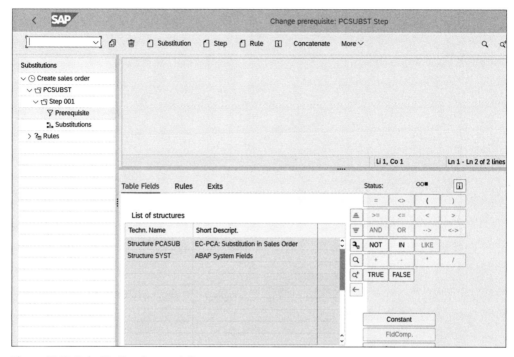

Figure 12.19 Substitution Prerequisite

You can choose from fields of **Structure PCASUB** or **Structure SYST** by double-clicking them. Then you can click fields to add them to the prerequisite definition in the top part of the screen and use the mathematical and logical operators on the right side of the screen to construct the rule. The screen shown in Figure 12.20 depicts a rule based on the sales district.

In the **Substitution** section, define the profit center to be substituted as shown in Figure 12.21.

12 Profit Center Accounting

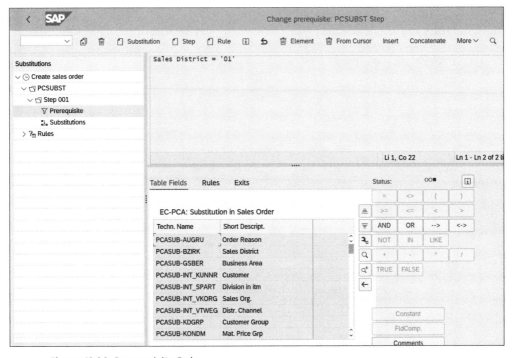

Figure 12.20 Prerequisite Rule

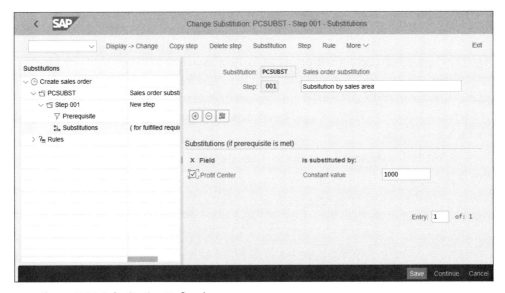

Figure 12.21 Substitution Defined

12.2 Profit Center Derivation

After that, save your entry with the **Save** button.

Then there is one more step to activate the substitution for your controlling area. Follow menu path **Controlling • Profit Center Accounting • Assignments of Account Assignment Objects to Profit Centers • Sales Orders • Sales Order Substitutions • Assign Substitution Rules**. Here, click **New Entries** from the top menu and activate the substitution as shown in Figure 12.22.

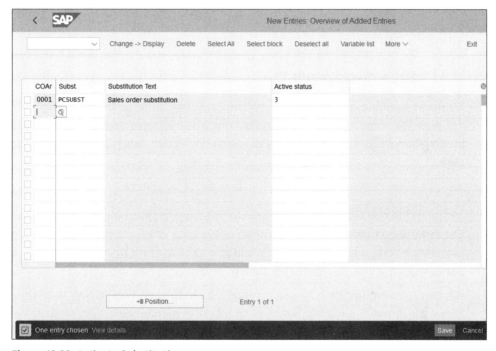

Figure 12.22 Activate Substitution

The following fields are to be configured on this screen:

- **COAr**
  The controlling area for which you activate the substitution.
- **Subst.**
  The substitution created in the previous transaction.
- **Active status**
  Determines how the substitution should be activated, with the following options:
  - 0: Not Active
  - 1: Other Transactions + Cross-Company (Billing Documents)

495

- 2: Only for Cross-Company (Billing Documents)
- 3: Other Transactions + Cross-Company (Orders + Billing Doc.)
- 4: Only for Cross-Company (Sales Orders + Billing Documents)

As you can see, you can activate the substitution only for billing documents or for orders and billing documents, and only for cross-company transactions or for all transactions.

Select the required activation status and save your entry with the **Save** button.

12.3 Information System

Let's now examine the information system for profit center accounting. First we'll examine the standard reports provided, which can meet most business requirements. Then we'll explain how to create your own reports using the drilldown technique.

12.3.1 Standard Reporting

The information system for profit center accounting offers extensive standard reports that enable you to analyze actual and plan values per profit center. However, in SAP S/4HANA, it can cause some confusion.

We first need to discuss how profit center accounting evolved in SAP over time. Before the introduction of the new general ledger, as mentioned earlier, profit center accounting was, technically speaking, a special purpose ledger with the code 8A. With the new general ledger, it became integrated with the general ledger, and now in SAP S/4HANA this integration is complete and all profit center postings are part of table ACDOCA, the Universal Journal. As such, profit center reporting is also part of the general ledger module and menu.

However, in the profit center application menu, SAP keeps the older reports based on the 8A ledger and library. These could be useful for you during a migration project for reconciling old profit center reports. They're available at application menu path **Accounting** • **Controlling** • **Profit Center Accounting** • **Information System** • **Reports for Profit Center Accounting**, as shown in Figure 12.23.

12.3 Information System

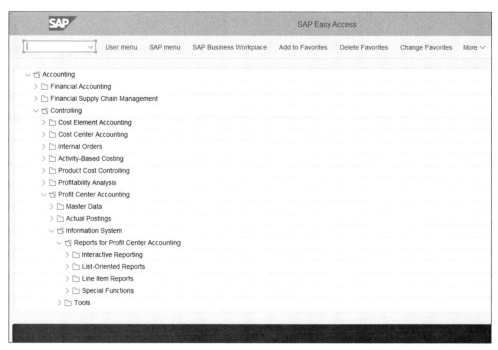

Figure 12.23 Profit Center Classic Reports

On this screen, you can see that these classic reports are grouped into the following major sections:

- **Interactive Reporting**
  The interactive reports use the drilldown reporting tool and enable you to drill down in real time to various sections of the report to obtain further relevant information.

- **List-Oriented Reports**
  The list-oriented reports are designed using Report Painter and offer valuable actual/plan comparisons, similar to the cost center reports we analyzed in Chapter 10.

- **Line Item Reports**
  The line item reports provide reporting on the lowest level, the individual line items per profit center.

- **Special Functions**
  Here you can find more rarely used reports, such as average balance reports, transfer price reports, and reports related to transfer of profit center data to the executive information system.

12 Profit Center Accounting

Let's now examine the new reports available for profit center accounting in the general ledger. First follow application menu path **Accounting • Financial Accounting • General Ledger • Information System • General Ledger Reports • Reports for Profit Center Accounting**.

As shown in Figure 12.24, you have standard reports for profit centers in the general ledger, which provide actual and plan comparison of values, key figures, and return on investments.

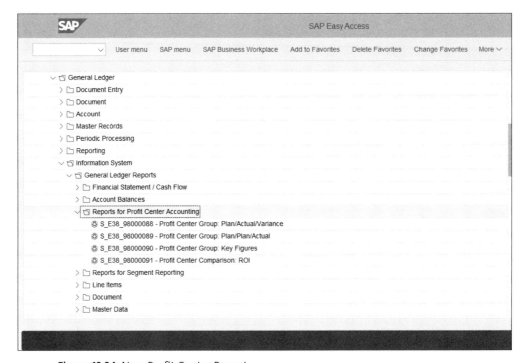

Figure 12.24 New Profit Center Reports

Enter Transaction S_E38_98000088 (Profit Center Group: Plan/Actual/Variance). After selecting profit centers/profit center groups, date parameters, a ledger, a currency, and a version, you see the output screen, shown in Figure 12.25.

This is a drilldown-based report. On the left side, you can navigate through the various available characteristics, such as profit center, segment, cost element, and so on. On the right side, you see the values for the selection in separate **Plan**, **Actual**, and **Variance** columns. At the bottom you see the key figures available in the report.

12.3 Information System

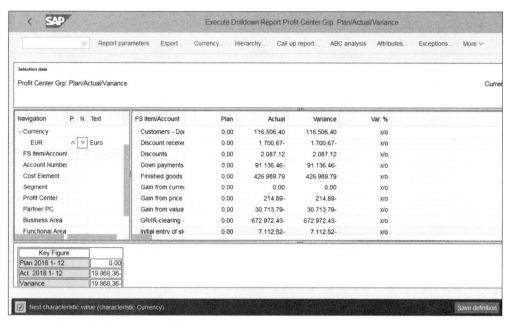

Figure 12.25 Profit Center Actual/Plan Comparison Report

This report, as well as the other profit center reports in the general ledger, was developed using the drilldown reporting tool, which in previous SAP releases was heavily used in controlling. Now in SAP S/4HANA, there are numerous standard drilldown reports available for FI as well, which provide valuable information on the profit center level.

You can get a good overview and access all of them using Transaction FGI0. It provides a tree-like structure showing the available reports by report type, as shown in Figure 12.26.

For profit centers, the following reports from report type 002 are particularly noteworthy:

- 0SAPBSPL-03: Profit Center Grp: Plan/Actual/Variance
- 0SAPBSPL-04: Profit Center Group: Plan/Plan/Actual
- 0SAPBSPL-05: Profit Center Group: Key Figures
- 0SAPBSPL-06: Profit Center Comparison: ROI

12 Profit Center Accounting

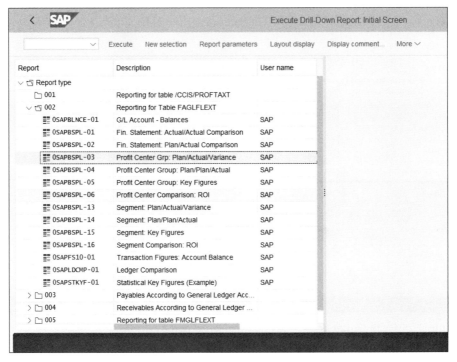

Figure 12.26 Financial Drilldown Reports

The previous report we executed was report S_E38_98000088 (Profit Center Group: Plan/Actual/Variance), which in fact is report 0SAPBSPL-03 (Profit Center Grp: Plan/Actual/Variance) defined here. You can also execute and analyze the other reports, which can provide you with plan values from more than one version, with calculation of the ROI (based on financial statement items you specify in the selections) and the key figures.

12.3.2 Drilldown Reporting

You can also define your own profit center reports, similar to the standard ones you see in Transaction FGI0. They are also drilldown reports and thus are very flexible and interactive.

Drilldown reports usually are created by referencing a form. The form is like a template that already contains the characteristics and key figures used in the report and the structure of columns and rows.

To create a new form, enter Transaction FGI4. In the initial creation screen, shown in Figure 12.27, first select the report type. Use **Reporting for Table FAGLFLEXT** for a profit center report on the general ledger level, **Payables According to General Ledger Account Assignments** for reports on the vendor level, and **Receivables According to General Ledger Account Assignments** for reports on the customer level. Then enter your form name starting with Z or Y in the custom name range and its description.

Figure 12.27 Create Form

In the **Structure** section, you have the following three options:

- **Two axes (matrix)**
  In forms with two axes, you define both columns and rows.
- **One axis with key figure**
  In one-axis forms, you define only one dimension (columns or rows). In this case, you also define the key figures within the form.
- **One axis without key figure**
  In this case, you define the key figures during the report creation.

Enter the standard form 0SAPBSPL-03 in the **Copy from** section to create a new form with it as a reference. Then click the **Create** button.

Figure 12.28 shows the definition screen of the form. In the rows, the various key figures used in the form are defined, such as plan and actual values.

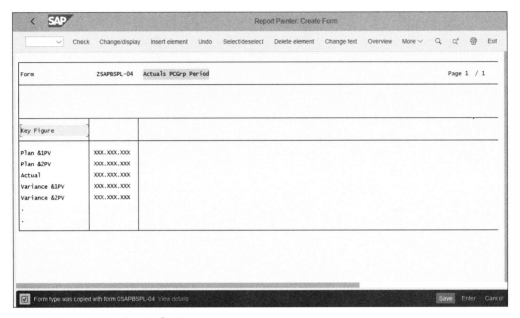

Figure 12.28 Form Definition

What we want to do in our example is display only the actuals, but per period. Therefore, delete the other elements, except the **Actual** row, by selecting **Delete element** from the top menu. First start with the **Variance** rows at the end because they're formulas; the **Plan** rows can be deleted only after that.

Then double-click the **Actual** line and a pop-up window will display the definition of the row.

Figure 12.29 shows that the **Key figure** used is the balance sheet value, which is fine. Fiscal year is variable, to be entered at report execution. **Record Type** selects actual values. **Posting period** is variable, asking the user to enter to and from periods, which will be displayed cumulatively in the same column. Instead, we want to make a report that will show the values per quarter in separate columns. So we want four different columns, showing the values for periods 1–3, 4–6, 7–9, and 10–12.

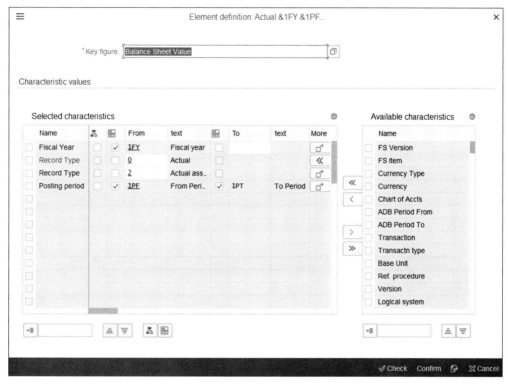

Figure 12.29 Actual Values Definition

Next, change the posting periods to 1–3, as shown in Figure 12.30.

After that, copy that row three more times and adjust the periods to represent the other quarters. To do so, select the row by right-clicking it, then choose **Select/deselect**. Then move your cursor to the next empty row and select **Copy** from the top menu. Do this until you have four rows. Then change the periods for the other rows to 4–6, 7–9, and 10–12. Also update the descriptions of the rows by clicking the (**Change Short, Middle, and Long Texts**) button. At the end, you should have a form like the one shown in Figure 12.31. To proceed, save the form with the **Save** button.

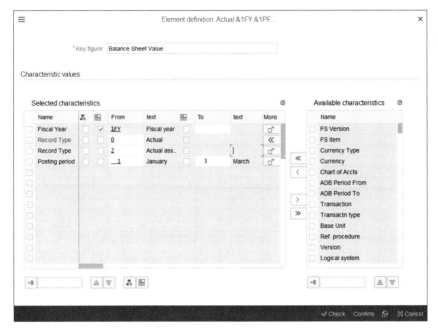

Figure 12.30 Actual Values Per Period

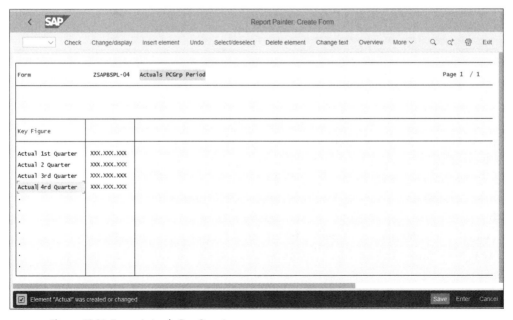

Figure 12.31 Form Actuals Per Quarter

Now you're ready to create a new report. To do so, enter Transaction FGI1. In the initial creation screen, shown in Figure 12.32, first select the report type.

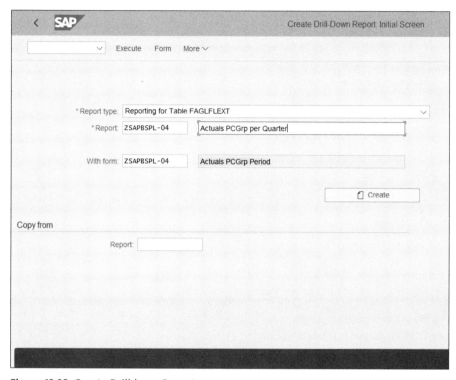

Figure 12.32 Create Drilldown Report

Use **Reporting for Table FAGLFLEXT** for a profit center report on the general ledger level, **Payables According to General Ledger Account Assignments** for reports on the vendor level, and **Receivables According to General Ledger Account Assignments** for reports on the customer level. Then enter your report name (starting with Z or Y in the custom name range) and its description. Then specify a form in the **With form** field. In this case, enter the form you created in the previous step, ZSAPBSPL-04. Finally, click the **Create** button.

We already defined the structure of the report in the form at the beginning of this section. On the screen shown in Figure 12.33, you can define default values for the characteristics if needed or enter additional characteristics.

12 Profit Center Accounting

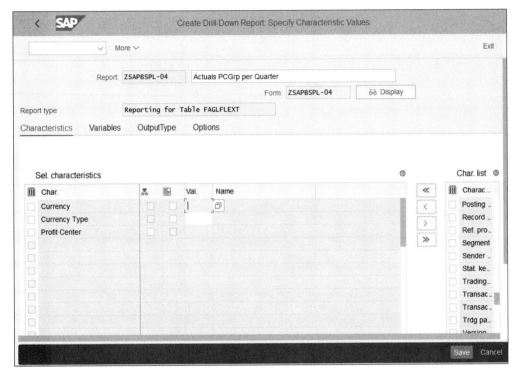

Figure 12.33 Drilldown Report Definition

Enter the profit center by selecting it on the right side, where all the available characteristics appear, and move it to the selected characteristics with the [<] (**Add Char.**) button.

In the **OutputType** tab, you can define some options for the output of the report, as shown in Figure 12.34.

You can choose the default output of the report:

- **Graphical Report Output**
- **Classic drilldown**
- **Object list (ALV)**
- **XXL (Spreadsheet)**

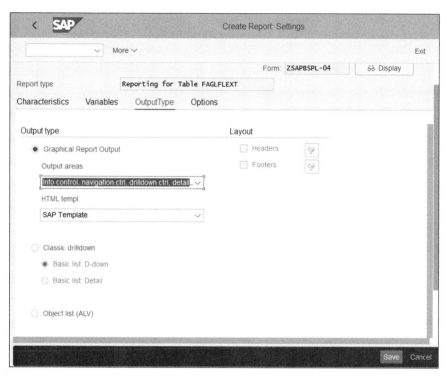

Figure 12.34 Drilldown Report Output Type

We examined how these options look in the previous controlling chapters. As with other reports, this can be the default, but the user can still select other options on the report selection screen if you tick the **Available on Selection Screen** checkbox.

After defining the report, save it with the **Save** button. Now your new report should be available in Transaction FGI0 in report type 002, reporting for table FAGLFLEXT. This way, you can create various forms and reports for profit center accounting, which can provide you with invaluable management analysis.

12.4 Summary

In this chapter, we covered the functionalities and configuration of profit center accounting in SAP S/4HANA. For users with a background in SAP ERP, this is a major change because profit center accounting is now fully integrated with the general ledger and no longer uses a separate special purpose ledger. Still, logically and functionally it's part of management accounting and therefore is positioned in the controlling application and configuration menus in SAP S/4HANA.

After reading this chapter, you should be very familiar with and ready to configure the following in your SAP S/4HANA project:

- Profit center master data
- Profit center derivation and actual postings
- Information system

Properly designed profit center master data is the foundation of correct, transparent, and effective management reporting. Equally important is to set up the derivation of profit centers using the various techniques you learned in this chapter because profit centers are almost always derived automatically through proper configuration. Finally, the drilldown reporting techniques you learned in this chapter should be very useful. You can create powerful reports using this technique, not only for profit centers but also in other controlling and financial areas.

In the next chapter we will show you how to configure profitability analysis, which is one of the functional areas that benefits the most from the power and flexibility of SAP S/4HANA.

Chapter 13
Profitability Analysis

This chapter gives step-by-step instructions for configuring profitability analysis in SAP S/4HANA. Our emphasis is on account-based profitability analysis, which is now the mandatory form of profitability analysis in SAP S/4HANA, but you'll also learn about the optional costing-based profitability analysis.

Profitability analysis is one of the most important controlling areas. It provides detailed analysis of the profitability of a company, thus enabling accurate contribution margin calculation. It can calculate the profitability on the level of numerous characteristics, which can be configured when implementing SAP S/4HANA. Therefore, the initial setup and configuration to be done is of paramount importance to the correctness and flexibility of profitability analysis. This very integrated area receives almost all its data from other areas, such as sales, purchasing, production, and finance. Manual postings in profitability analysis are very rare.

In this chapter, we'll provide an extensive guide to the global settings for profitability analysis. We'll then analyze in detail the settings and processes for the various data flows into profitability analysis, and finally we'll guide you through its information system. Before attacking this extensive workload, we need to start with some general information on how profitability analysis evolved in SAP and what the differences are between the two types of profitability analysis: account-based and costing-based.

13.1 Overview of Profitability Analysis

For a long time, profitability analysis was a core component of controlling in SAP because it provides answers to the most important questions for high-level management: How profitable is the company? How well is this segment performing? How profitable is this product? What is the contribution margin of that division? And so on. But the power and flexibility of profitability analysis mostly lies in its multidimensional structure. You can analyze profitability not in only one or two dimensions, but in as many dimensions as are needed and configured in the system. Therefore, profitability analysis is often referred to as a multidimensional cube, as shown in Figure 13.1.

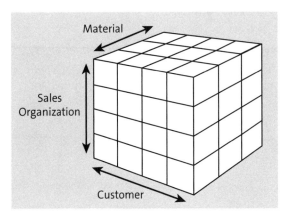

Figure 13.1 Profitability Analysis Multidimensional Cube

The dimensions are the different characteristics by which the data is sliced and diced from a profitability analysis point of view, such as material, customer, sales organization, and so on. Each posting transferred to profitability analysis contains several characteristics, which together constitute a multidimensional profitability segment.

There are two types of profitability analysis: account-based and costing-based. We'll explain these types in detail in the following sections.

13.1.1 Costing-Based Profitability Analysis

Costing-based profitability analysis analyzes profitability using value fields such as material costs, discounts, revenues, and so on. These value fields can be regarded as buckets into which similar values are grouped. Costing-based profitability analysis was the most used type of profitability analysis in previous SAP releases.

Costing-based profitability analysis is very powerful because these value fields can be defined as required in each SAP client, thus allowing for a great degree of flexibility.

From a data point of view, costing-based profitability analysis uses the following tables:

- CE1xxxx (actual line item table)
- CE2xxxx (plan line item table)
- CE3xxxx (segment level)
- CE4xxxx (segment table)

In this nomenclature, xxxx stands for the operating concern name. For example, if the operating concern is 1000, the actual line item table will be CE11000, the plan line item table will be CE21000, and so on.

There are some limitations to costing-based profitability analysis. For example, it's quite challenging to reconcile costing-based profitability analysis with financial accounting. This stems from the very essence of costing-based profitability analysis: the value fields don't match the accounts used in financial accounting. Furthermore, there are differences in the value flow to financial accounting and the value flow to profitability analysis. The basic sales process involves creating a sales order, then delivery with goods issue, and finally sending an invoice document to the customer. In financial accounting, the cost of goods sold is posted with the goods issue and the sales revenue with the invoice. In costing-based profitability analysis, however, both the sales revenue and cost of goods sold are transferred with the invoice document. This leads to mismatches between financial accounting and profitability analysis at month-end.

These considerations prompted many companies to use account-based profitability analysis.

13.1.2 Account-Based Profitability Analysis

Account-based profitability analysis uses accounts (cost elements) to collect profitability values. Therefore, by design, it's very easy to reconcile it with financial accounting. However, in SAP ERP and earlier versions, it had some considerable limitations. The cost of goods sold couldn't be split among different cost components, which is possible in costing-based profitability analysis. Also, variance analysis was only possible for totals variance, but not by variance category. That made costing-based profitability analysis the choice for most companies prior to SAP S/4HANA, and account-based profitability analysis was sometimes implemented in parallel to facilitate the accounting reconciliation process.

As we discussed in the Introduction of this book, SAP made a huge improvement to profitability analysis with SAP S/4HANA, and more specifically to account-based profitability analysis. Now in SAP S/4HANA the advantages of costing-based profitability analysis are available in account-based profitability analysis, combined with easy reconciliation with financial accounting. Cost component split is now possible in account-based profitability analysis. Also, variance analysis is made possible by variance categories. Therefore, in SAP S/4HANA it's possible to construct a profit and loss statement with a contribution margin calculation, very much like in costing-based profitability analysis.

Based on all of this, in SAP S/4HANA account-based, profitability analysis is not only the recommended but also the required approach. You can activate costing-based profitability analysis in addition if you choose. However, our recommendation for a greenfield SAP S/4HANA implementation is to use only account-based profitability analysis. You have all the benefits of the costing-based now in the account-based version. For brownfield implementations, it makes more sense to continue using the costing-based approach together with the now mandatory account-based approach.

From a technical point of view, account-based profitability analysis doesn't create any additional tables. The segment postings update table ACDOCA, the Universal Journal, thus making the profitability analysis fully integrated with financial accounting.

Now it's time to delve deep into the secrets of the profitability analysis configuration. We'll start with the global settings.

13.2 Master Data

In this section, we'll set up the master data structures required for profitability analysis. The main organizational object for profitability analysis is the operating concern. We discussed operating concerns in Chapter 3, Section 3.2.4, so we'll just briefly revisit this subject to create a new operating concern for our controlling area. We'll then create the characteristics and value fields for the operating concern. The value fields are used only in costing-based profitability analysis, whereas the characteristics are the foundation of both types of profitability analysis.

13.2.1 Operating Concern

The operating concern is an organizational object that enables you to analyze an organization from a profitability point of view. It's assigned to a controlling area. To create an operating concern, follow menu path **Enterprise Structure • Definition • Controlling • Create Operating Concern**.

Figure 13.2 shows some standard operating concern templates provided by SAP. To create a new operating concern, select **S001** by placing a check in the box to its left and select **Copy As...** from the top menu. Name the new operating concern "US01" and enter "US Operating Concern" for the description. Then save your entry with the **Save** button.

Figure 13.2 Operating Concern

13.2.2 Data Structure

After creating an operating concern, you need to define its data structure. It consists of characteristics and value fields (in the case of costing-based operating concerns), as you'll see in the following sections.

Characteristics

The profitability analysis characteristics define the criteria by which you classify the values in profitability analysis. Characteristics are required both for account- and costing-based profitability analysis.

There are standard characteristics delivered by SAP, such as customer, material, profit center, cost center, company code, and so on. You also can create additional custom characteristics specific to your organization. There is a limitation on the number of custom characteristics that can be defined: up to 50 in the older SAP ERP system and now up to 60 in SAP S/4HANA.

To maintain characteristics, follow menu path **Controlling • Profitability Analysis • Structures • Define Operating Concern • Maintain Characteristics**. On the initial screen shown in Figure 13.3, you can choose to maintain all characteristics or only those assigned to a specific operating concern or not assigned. Select **All Characteristics** and click the **Change** button.

13 Profitability Analysis

Figure 13.3 Define Characteristics

The system issues a warning that you are changing cross-client data, as shown in Figure 13.4.

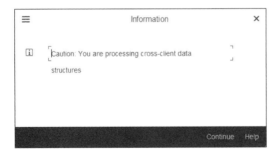

Figure 13.4 Cross-Client Data Warning

The profitability analysis structures are cross-client and can be changed only in a client that allows cross-client customizing. This should be your golden customizing client. Once changed there, the changes will be available immediately in other clients within the same instance. Proceed with the **Continue** button.

13.2 Master Data

You'll see a list of the characteristics defined in the system. All those displayed in Figure 13.5 are standard characteristics provided by SAP. The naming convention for characteristics is that custom-defined characteristics should start with WW.

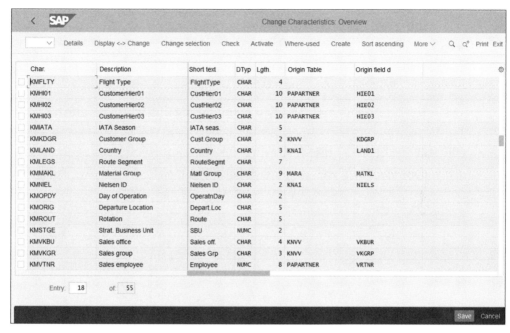

Figure 13.5 Characteristics List

You can see that characteristics are defined with some of the following settings in the columns of this configuration table:

- **Char.**
  The characteristic code. For standard characteristics, this often corresponds to the field name to which it refers. For custom-defined characteristics it should start with WW.
- **Description**
  Long description of the characteristic.
- **Short text**
  Short description of the characteristic.
- **DTyp**
  The data type of the characteristic. It will be **CHAR** for alphanumerical values and **NUMC** for numerical values.

515

- **Lgth.**
  Length of the characteristic, which defines the length of the values supplied by this characteristic.
- **Origin Table**
  The ABAP dictionary table from which the field is taken.
- **Origin field d**
  The field in the ABAP dictionary table from which the field is taken.

Let's now create a new custom characteristic. Go back to the previous screen, shown in Figure 13.3, and enter a characteristic name starting with WW in the **Create Characteristic** section. For this example, create characteristic WWGRP to represent a customer group. Then click the **Create/Change** button.

A pop-up window appears, as shown in Figure 13.6, in which you need to define how the data will be retrieved for this characteristic.

Figure 13.6 Create Characteristic

Here you need to configure the following fields:

- **Transfer from SAP table/User defined**
  There are two options: transfer the field from an SAP table, which you specify in the **Table** field, or create a custom characteristic starting with WW.

13.2 Master Data

- **With own value maintenance**
  With this option, you can create your own check table in which to maintain the possible values of the characteristic.
- **Without value maint.**
  Used for characteristics for which you don't define values.
- **With reference to existing values**
  With this option, you define a characteristic that has the same data structure as an already existing field. The characteristic contains the same technical properties and the same value set as the referenced field.

Proceed with the **Create** button. On the screen shown in Figure 13.7, you define the characteristic.

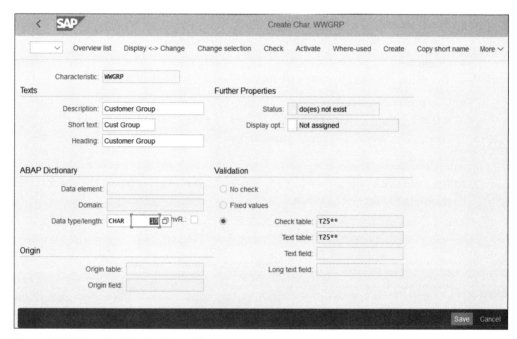

Figure 13.7 Characteristic Configuration

The following fields are available for configuration:

- **Description**
  Long description of the characteristic.
- **Short text**
  Short description of the characteristic.

- **Heading**
  A description of the characteristic used as column heading.
- **Data element**
  The characteristic can be defined with reference to an existing data element.
- **Domain**
  Data domain of the existing data element the characteristic refers to.
- **Data type**
  The data type of the characteristic. It can be **CHAR** for alphanumerical values and **NUMC** for numerical values.
- **Data length**
  Length of the characteristic, which defines the length of the values supplied by this characteristic.
- **Origin table**
  The ABAP dictionary table from which the field is taken.
- **Origin field**
  The field in the ABAP dictionary table from which the field is taken.
- **Check table/Text table**
  When creating a characteristic that doesn't refer to an existing field, the system creates check tables in which you can maintain the values of the characteristic.

After configuring the characteristic, you need to activate it. Select **Activate** from the top menu. Then a pop-up window appears with a message regarding the check tables, as shown in Figure 13.8.

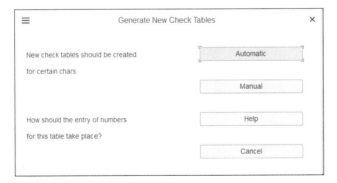

Figure 13.8 Check Table Message

The new check tables can be created automatically or manually. If you have only one SAP system in your landscape, you can choose automatic and let the system assign

13.2 Master Data

the table names. However, if you have more than one system it's advisable to create the check tables manually to avoid inconsistencies.

After that, save the characteristic with the **Save** button. The check tables are filled in automatically, as shown in Figure 13.9.

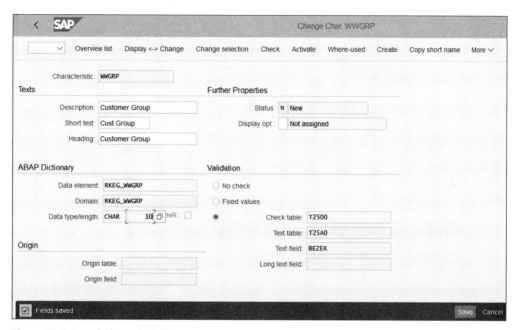

Figure 13.9 Saved Characteristic

In the next step, you need to select the characteristics to be used in the operating concern. Follow menu path **Controlling • Profitability Analysis • Structures • Define Operating Concern • Maintain Operating Concern**.

Figure 13.10 shows the screen in which you define the data structure of the operating concern.

Enter "US01" for the **Operating Concern** and click the button to switch from display to change mode. Then activate both the costing- and account-based profitability analysis by ticking both checkboxes in the **Type of Profitability Analysis** section. Click the → **Create** button to select the characteristics for the operating concern.

On the next screen, shown in Figure 13.11, you see that the system creates data structure **CE1US01** for the operating concern, which will be used for the costing-based profitability analysis.

13 Profitability Analysis

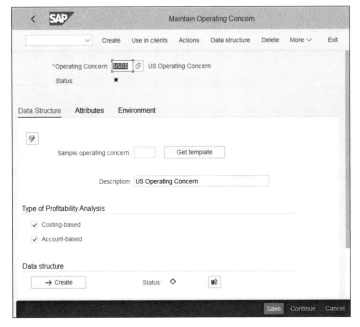

Figure 13.10 Maintain Operating Concern

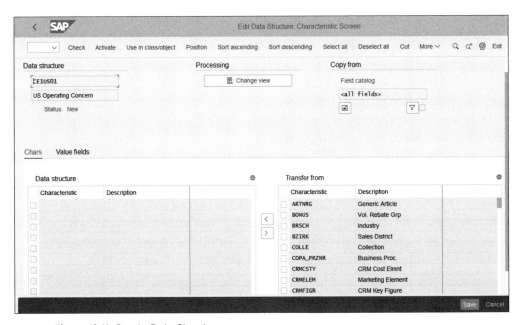

Figure 13.11 Create Data Structure

13.2 Master Data

In the lower-left section of the screen, note that as of now there are no characteristics selected for the operating concern. On the right side, all available characteristics are listed. Select the needed characteristics by ticking the checkboxes to their left and clicking the [<] button.

Now the selected characteristics are transferred to the left side, as shown in Figure 13.12.

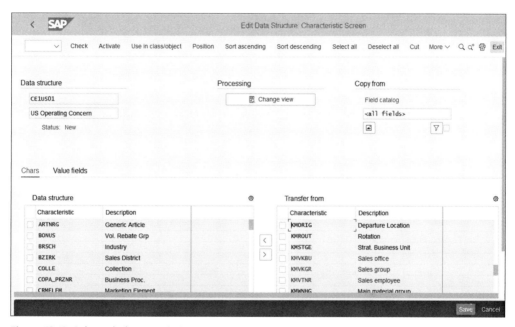

Figure 13.12 Selected Characteristics

Value Fields

The next step is to add value fields. Click the **Value fields** tab and add the value fields you want to use in your operating concern, as shown in Figure 13.13.

Double-clicking a value field shows you its definition, as shown in Figure 13.14. A value field can be defined either as an amount or quantity field. Also, its data element is specified here.

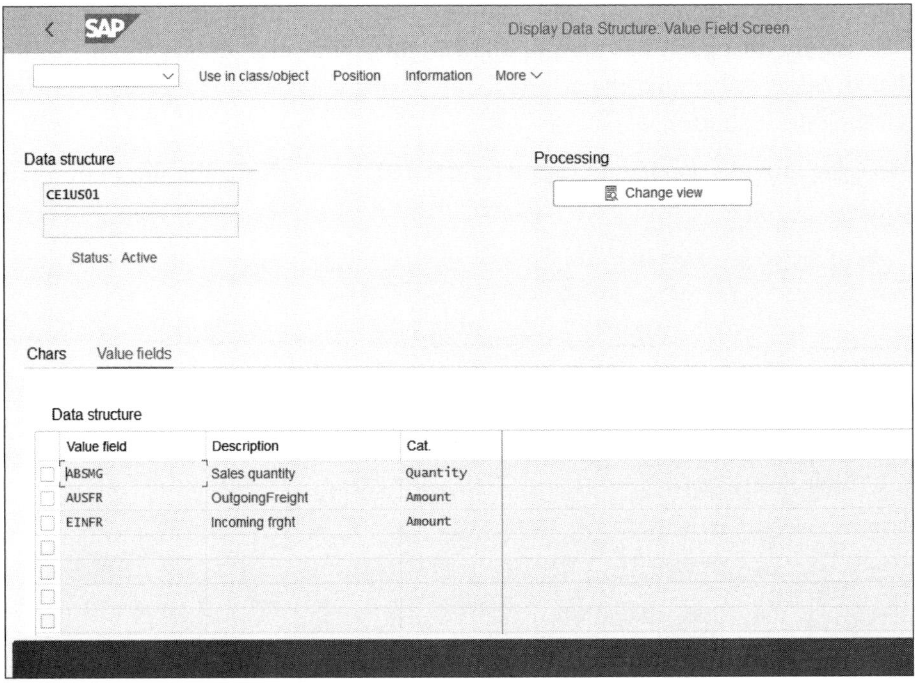

13 Profitability Analysis

Figure 13.13 Selected Value Fields

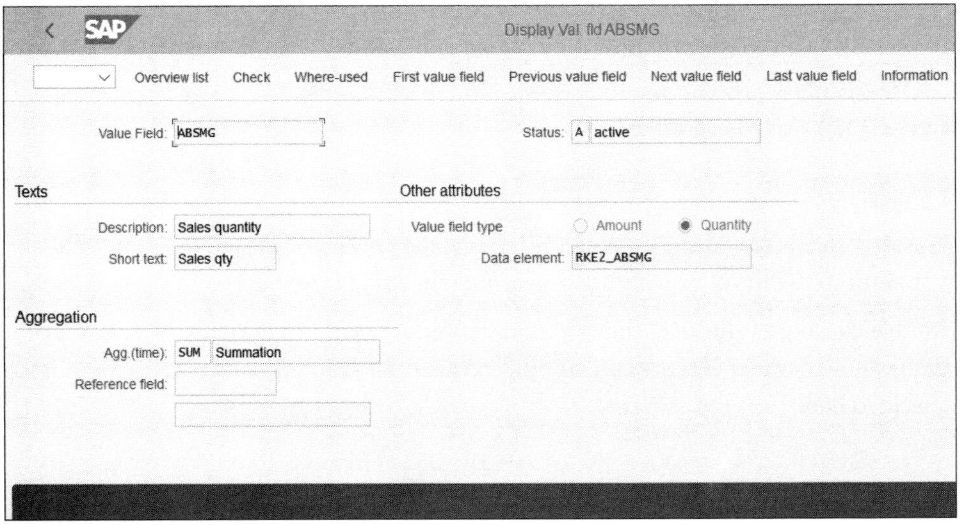

Figure 13.14 Value Field Definition

After adding the characteristics and value fields, save with the **Save** button, then activate the data structure of the operating concern by clicking **Activate** from the top menu.

13.2.3 Operating Concern Attributes

You also need to maintain the attributes of the operating concern. From the screen shown in Figure 13.10, select the **Attributes** tab.

In the screen shown in Figure 13.15, you can maintain the operating concern attributes.

Figure 13.15 Operating Concern Attributes

You need to configure the following fields:

- **Operating concern currency**
  The currency in which the data transferred to profitability analysis is converted and updated.
- **Fiscal year variant**
  The fiscal year variant used to determine the periods in the profitability analysis.

- **Currency types for costing-based Profitability Analysis**
  Here you define in which additional currency types the values should be stored in costing-based profitability analysis.

- **2nd period type—weeks**
  With these indicators, the system will store the actual and/or plan data in costing-based profitability analysis in weeks as well.

As a last step, you need to generate the profitability analysis environment. Click the **Environment** tab, which will result in the screen shown in Figure 13.16.

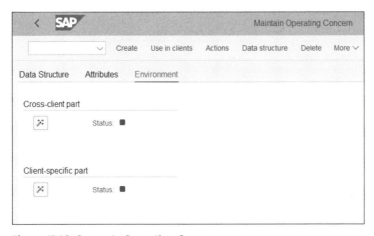

Figure 13.16 Generate Operating Concern

You need to activate both the cross-client and client-specific parts by clicking the button in both sections.

After completing the data structure of the operating concern (characteristics and value fields, in the case of costing-based profitability analysis), you now need to maintain further master data–related settings: the characteristics hierarchy and characteristic derivation.

13.2.4 Characteristics Hierarchy

You can group characteristic values in a hierarchy, which then can be used in the derivation of the values and in reporting. Follow menu path **Controlling • Profitability Analysis • Master Data • Characteristic Values • Define Characteristics Hierarchy**.

Enter your operating concern, as shown in Figure 13.17.

13.2 Master Data

Figure 13.17 Enter Operating Concern

After that, you'll see the screen shown in Figure 13.18. On the left side is a list of the characteristics for which you can create a hierarchy. Select **Customer** so that you can build a hierarchy of customers. Enter variant "001" in the **Variant** field; you can maintain multiple hierarchies under different variant names. Then click the **Create/Change** button.

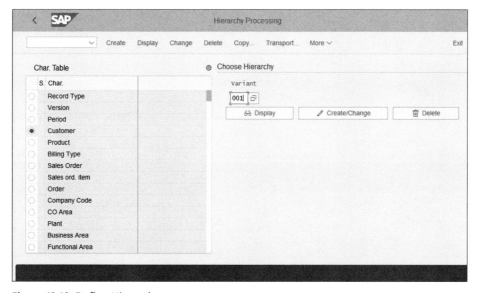

Figure 13.18 Define Hierarchy

On the next screen, shown in Figure 13.19, you need to enter a description for the variant. You also can select the **Visible system-wide** checkbox, which will make the

525

13 Profitability Analysis

hierarchy available not only in profitability analysis but also in other components of the system.

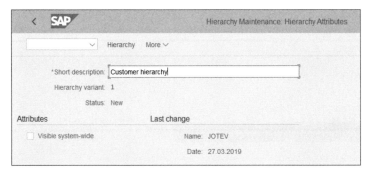

Figure 13.19 Hierarchy Attributes

After that, select **Hierarchy** from the top menu. In the next screen, shown in Figure 13.20, you can build a hierarchy in a tree-like structure with nodes and subnodes. In our example, customer number **11** is at the top, underneath which is **10000022**, the top-level customer for several other customers. You can add nodes and subnodes using the **Same Level** and **Lower Level** commands from the top menu.

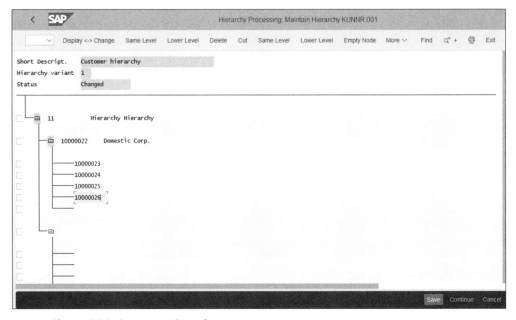

Figure 13.20 Customer Hierarchy

13.2 Master Data

After maintaining the characteristics hierarchy, save it with the **Save** button.

13.2.5 Characteristic Derivation

Each posting to profitability analysis forms a profitability segment, which is a combination of posted characteristics. A lot of these are determined by the standard logic of the system. For example, the ship-to customer from the sales order will go into the ship-to characteristic (KUNWE) in the profitability segment. But you also will have to build logic for some of the characteristics to tell the system how to derive their values. For that, you need to maintain the characteristic derivation by following menu path **Controlling • Profitability Analysis • Master Data • Define Characteristic Derivation** or entering Transaction KEDR.

In the screen shown in Figure 13.21, you can see the defined derivation steps. You can have multiple steps that derive various characteristics.

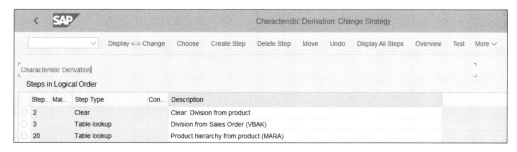

Figure 13.21 Derivation Steps

To create a new derivation step, click **Create Step** from the top menu. For this example, create a step to derive the reference item from the sales document. A pop-up window appears, as shown in Figure 13.22.

Figure 13.22 Create Derivation Step

527

Here you need to select what type of derivation rule to use from the following options:

- **Derivation rule**
  With this option, you can specify if-then logic, which combines several characteristics.
- **Table lookup**
  This rule tells the system to derive the values of the characteristics from another table/field.
- **Move**
  This option enables you to move the content of a source field or a constant to a target field.
- **Clear**
  This option enables you to delete a characteristic value (resets it to blank for CHAR fields or to 0 for NUMC fields).
- **Enhancement**
  You can also program your own logic for the derivation, developed using component 003 of customer enhancement COPA0001.

For now, select **Table lookup** and proceed with the **Continue** button. In the next pop-up window, shown in Figure 13.23, specify the table to read the values from.

Figure 13.23 Table Lookup

Enter "VBAP" in this case; table VBAP is the Sales Document: Item Data table. Proceed with the **Continue** button.

In the next screen, shown in Figure 13.24, you define the rules for the derivation.

In the **Step Description** field, enter a description for the step. Good practice is to name it with the name of the characteristic being derived.

In the **Source Fields for Table Lookup** section, the fields from the sales document item table used to derive the characteristic and their corresponding fields in profitability analysis are defined. Then in **Assignment of Table Fields to Target Fields** in the lower section of the screen, you define that from field **VBAP-VGBEL** (document number of the reference document) from the sales document line item table, a characteristic you created that represents the document number of the reference document (**WW001**) should be derived.

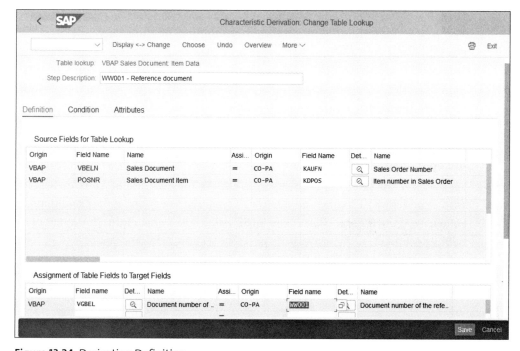

Figure 13.24 Derivation Definition

After maintaining the derivation, save it with the **Save** button.

Using these derivation techniques, you should complete the derivation steps for all your characteristics not already derived by the standard logic of the system.

13.3 Data Flow

In this section, we'll discuss how profitability analysis receives its data from the various SAP S/4HANA areas and what settings you need to make to ensure the correct data flow into profitability analysis.

13.3.1 Invoice Value Flow

The invoice value flow plays a very important role in profitability analysis because it provides the revenues and the cost of goods sold to profitability analysis. In this section, you'll learn how to transfer sales, sales deductions, and cost of goods sold to profitability analysis.

There is a difference in the value flow in account-based and costing-based profitability analysis. The account-based approach uses the general ledger accounts to transfer the values into profitability analysis, so you only need to configure the automatic account determination from sales to financial accounting using the condition technique we briefly touched upon in Chapter 4. In costing-based profitability analysis, however, the values are transferred using value fields, and you need to assign the value fields to sales and distribution (SD) conditions.

Account-Based Settings

Let's analyze the value flow, starting from an SD billing document. You can display it with Transaction VF03, then double-click a material line item. In the **Conditions** tab shown in Figure 13.25, you'll see a list of SD conditions, which form the values of the billing document, such as sales price, discounts, taxes, total, and so on.

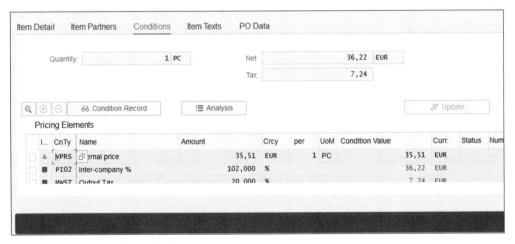

Figure 13.25 Billing Line Item

Double-click pricing condition **PI02** in the **CnTy** field and you'll see its details, as shown in Figure 13.26.

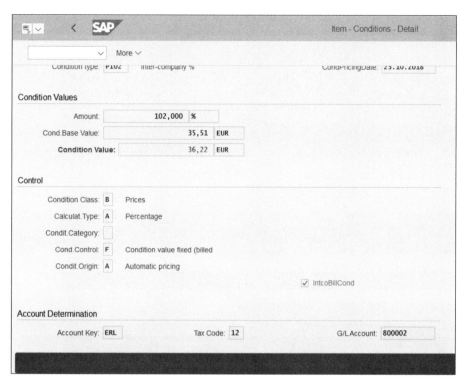

Figure 13.26 Condition Details

At the bottom of the screen is the **Account Determination** section. You'll see account key **ERL** here, which is used to determine the revenue account to be posted. Then you'll see that account **800002** was determined.

Let's configure how to determine the general ledger accounts. In a new session, start Transaction VKOA, which is the transaction to assign general ledger accounts to sales conditions based on certain criteria.

Figure 13.27 shows several condition tables, which have various combinations of characteristics. The system tries to find general ledger accounts by going through the tables from the most specific (with the highest number of characteristics) to the least specific, as defined in the pricing procedure.

Double-click **Table 003**, **Material Grp/Acct Key**, for example. In the screen shown in Figure 13.28, you assign the revenue general ledger accounts.

13 Profitability Analysis

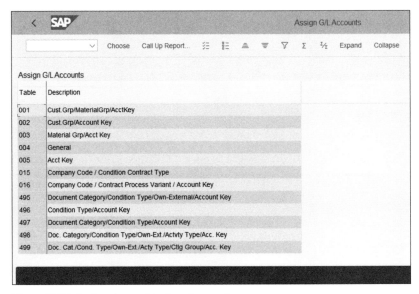

Figure 13.27 Assign General Ledger Accounts

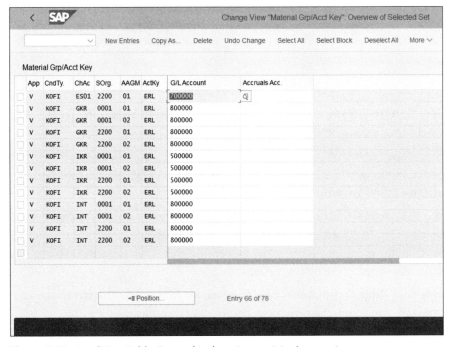

Figure 13.28 Condition Table General Ledger Account Assignments

The following fields serve as criteria:

- **App**
  The condition technique is used throughout the SAP S/4HANA system. Here, **V** is used, which means SD.
- **CndTy.**
  There are two condition types for account determination:
  - **KOFI** for billing documents without an account assignment object
  - **KOFK** for billing documents with an account assignment object
- **ChAc**
  The chart of accounts, for which you maintain the account determination.
- **SOrg.**
  The sales organization, for which you maintain the account determination.
- **AAGM**
  The account assignment group material is a group from the material master, thus enabling different accounts based on the material.
- **ActKy.**
  This is the account key being used by the condition to derive the account.
- **G/L Account**
  This is the general ledger account to which the revenue is posted. It should have cost element category 11 (revenue) or 12 (revenue reduction).
- **Accruals Acc.**
  This is entered only if there are conditions relevant for accrual—for example, rebate conditions.

Make new entries in this or the other condition tables to derive all your relevant revenue accounts.

Going back to the billing document, shown in Figure 13.25, let's review the generated profitability analysis document. Click **Accounting** in the top menu. The system shows the related accounting documents in financial accounting and controlling, as shown in Figure 13.29. Double-click the controlling document.

As shown in Figure 13.30, the controlling object is **PSG**, which means profitability segment. There are no value fields in the account-based profitability analysis, and the value is being posted to a cost element.

13　Profitability Analysis

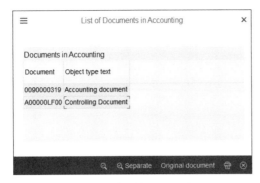

Figure 13.29 Accounting Documents

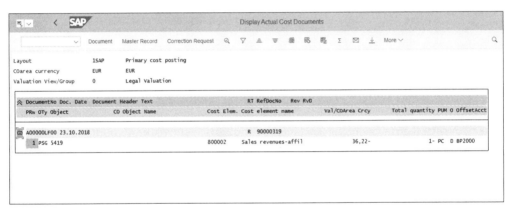

Figure 13.30 Controlling Document

Costing-Based Settings

Let's see now how costing-based profitability analysis received its values from SD. You need to configure the flow of values from SD conditions to profitability analysis value fields by following menu path **Controlling • Profitability Analysis • Flows of Actual Values • Transfer of Billing Documents • Assign Value Fields**, then selecting **Maintain Assignment of SD Conditions to Profitability Analysis Value Fields**.

Initially, the table is empty, as shown in Figure 13.31. Click **New Entries** from the top menu.

In the screen shown in Figure 13.32, assign the condition type **PR00 Price** to standard value field **ERLOS Revenue**. That ensures that values posted with condition type PR00 will update the ERLOS value field in profitability analysis.

13.3 Data Flow

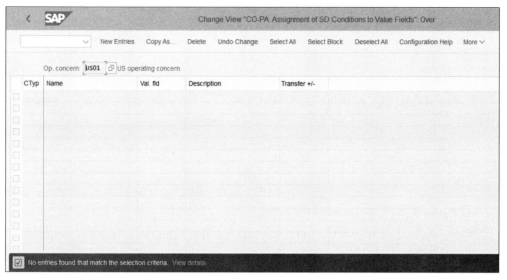

Figure 13.31 Assign Value Fields to Conditions

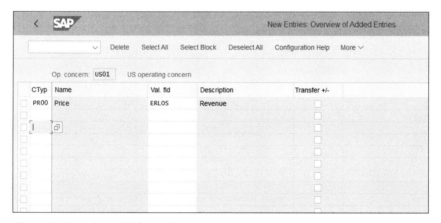

Figure 13.32 Assigned Value Fields

Assign all your relevant value fields in this table, then save with the **Save** button.

Similarly, you should assign sales quantity fields to your relevant profitability analysis quantity fields also. To do so, follow menu path **Controlling • Profitability Analysis • Flows of Actual Values • Transfer of Billing Documents • Assign Quantity Fields**.

As with value fields, initially the table is empty when you start with a new operating concern, as shown in Figure 13.33.

535

13 Profitability Analysis

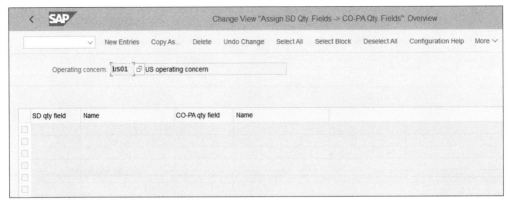

Figure 13.33 Assign Quantity Fields to Conditions

Click **New Entries** from the top menu.

In the screen shown in Figure 13.34, assign the sales quantity field **FKIMG Billed Quantity** to profitability analysis quantity field **ABSMG Sales quantity**.

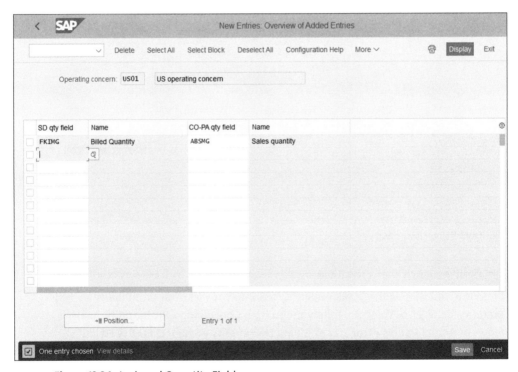

Figure 13.34 Assigned Quantity Fields

Assign all your relevant quantity fields in this table, then save with the **Save** button.

13.3.2 Overhead Costs Flow

The flow of overhead costs comes to profitability analysis from other controlling objects, such as internal orders, cost centers, or WBS elements. These are originally posted with the cost, then at month end these costs are transferred to profitability analysis to provide accurate profitability, which includes the overhead costs allocated to the proper profitability analysis characteristics. In the case of internal orders and WBS elements, these rules come from the settlement profile; in the case of cost centers, the overhead costs are transferred via assessment. Direct postings to profitability analysis profitability segments in financial accounting documents are also possible.

In account-based profitability analysis, the flow of actual values doesn't require further setup beyond correctly setting up the settlement profiles and assessment cycles in the sender cost objects, which we already analyzed in Chapter 10 and Chapter 11. In costing-based profitability analysis, however, you need to assign value fields to the relevant cost elements. This is done using a customizing object called a *transfer structure*.

To maintain a transfer structure for order settlement, follow menu path **Controlling • Profitability Analysis • Flows of Actual Values • Order and Project Settlement • Define PA Transfer Structure for Settlement**.

On the first screen, shown in Figure 13.35, you see a list of transfer structures. Select **E1**, which is the standard profitability analysis transfer structure, and select **Copy As...** from the top menu to copy it to a transfer structure specific to your project.

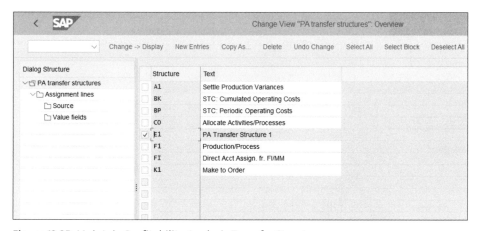

Figure 13.35 Maintain Profitability Analysis Transfer Structure

13 Profitability Analysis

Enter "Z1" as the transfer structure name (as always, it's good practice to name your project-specific objects in the custom name range, starting with Z or Y), and enter a description. Proceed by pressing the `Enter` key, and the system issues a message that it also copied the dependent objects, as shown in Figure 13.36.

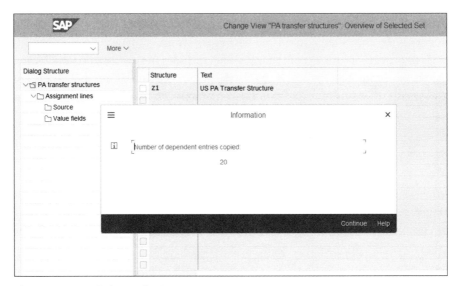

Figure 13.36 Copied Transfer Structure

Then select the **Z1** structure and click **Assignment lines** on the left side of the screen. In the example shown in Figure 13.37, there is only one line to include all cost elements. Create as many lines as needed here to segregate the various cost elements that need to go to different value fields.

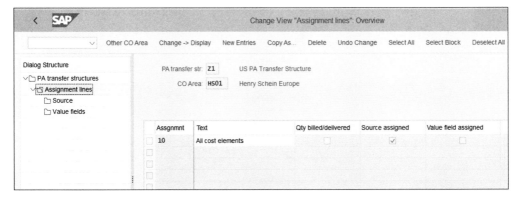

Figure 13.37 Assignment Lines

538

Click **New Entries** from the top menu and create new lines, as shown in Figure 13.38.

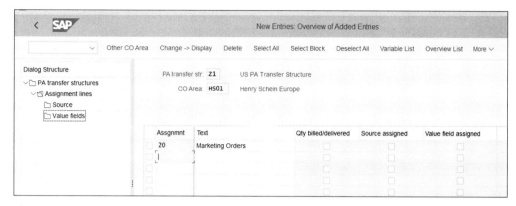

Figure 13.38 New Assignment Line

Then select a line and click **Source** on the left side of the screen. In the next screen, shown in Figure 13.39, enter a range of cost elements or a cost element group. The postings to these cost elements will be allocated to the value field we'll configure in the next screen. Enter the relevant cost element range.

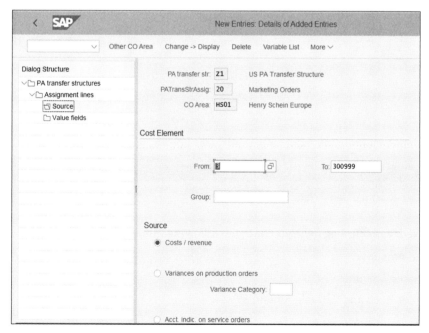

Figure 13.39 Source Cost Elements

13 Profitability Analysis

In the **Source** section of the screen, you have the following three options:

1. **Costs/revenue**
2. **Variances on production order**
3. **Acct. indic. on service orders**

Choose **Costs/revenue** for internal orders. The other options are used for production and for service orders, which we'll touch on in the next section on manufacturing costs.

Then you need to assign the **Quantity/value** field. Click **Value fields** on the left side of the screen. The screen is initially empty. Click **New Entries** from the top menu.

On the screen shown in Figure 13.40, you assign the **Quantity/value** field that should receive the costs in profitability analysis for the cost elements configured in the previous step.

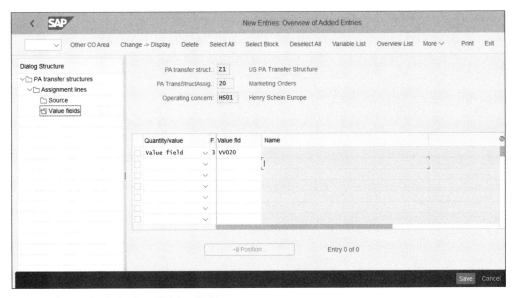

Figure 13.40 Assigned Value Field

In the **Fixed/Variable Flag** field, you have the following three options:

- **1: Fixed Amounts**
- **2: Variable Amounts**
- **3: Sum of Fixed and Variable Amounts**

Enter "3" unless you need to transfer only the fixed or the variable portion of the costs. Maintain the whole structure in a similar fashion and save it with the **Save** button.

13.3 Data Flow

In the next step, you need to assign the newly created transfer structure to the relevant order types. To do so, follow menu path **Controlling • Profitability Analysis • Flows of Actual Values • Order and Project Settlement • Assign PA Transfer Structure to Settlement Profile**. You'll see a list of the settlement profiles defined. Double-click settlement profile **20** to see its configuration screen, as shown in Figure 13.41.

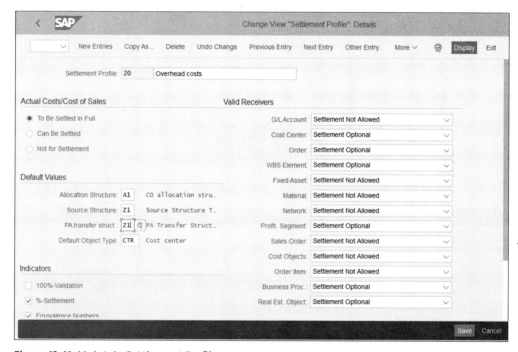

Figure 13.41 Maintain Settlement Profile

Enter "Z1" in the **PA transfer struct.** field, then save with the **Save** button.

Similarly, you need to maintain another transfer structure for direct postings from FI/MM. Follow menu path **Controlling • Profitability Analysis • Flows of Actual Values • Direct Posting from FI/MM • Maintain PA Transfer Structure for Direct Postings**.

Select transfer structure **FI Direct Acct Assign. fr. FI/MM** and click **Assignment lines** on the left side of the screen, as shown in Figure 13.42.

You'll see the various assignment lines defined, as in Figure 13.43, which correspond to the types of costs to be transferred to profitability analysis. Select them line by line and assign the relevant cost element ranges or groups using **Source** on the left side of the screen.

541

13 Profitability Analysis

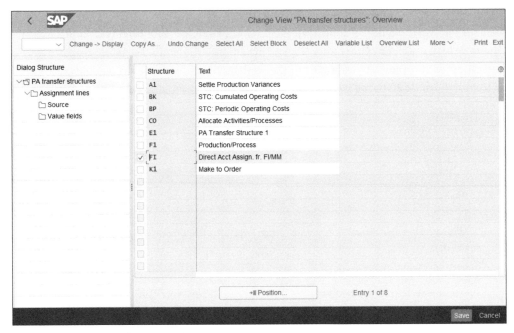

Figure 13.42 FI/MM Transfer Structure

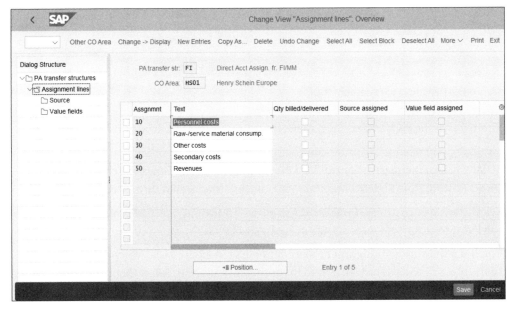

Figure 13.43 FI/MM Assignment Lines

13.3 Data Flow

In the screen shown in Figure 13.44, you can assign a range of cost elements or a cost element group.

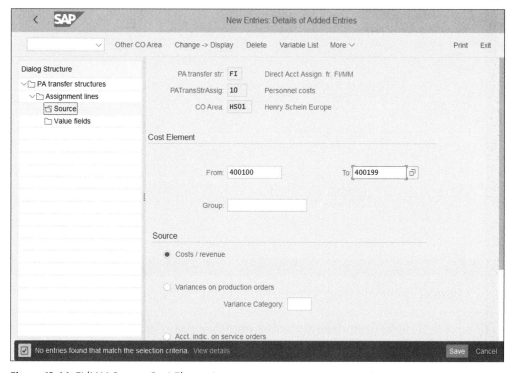

Figure 13.44 FI/MM Source Cost Elements

Now you need to assign the value/quantity field. Click **Value fields** on the left side of the screen. The screen is initially empty. Click **New Entries** from the top menu.

On the screen shown in Figure 13.45, you assign the value or quantity field that should receive the costs in profitability analysis for the cost elements configured in the previous step.

In the **Fixed/Variable Flag** field, you have the following three options:

- 1: Fixed Amounts
- 2: Variable Amounts
- 3: Sum of Fixed and Variable Amounts

543

13 Profitability Analysis

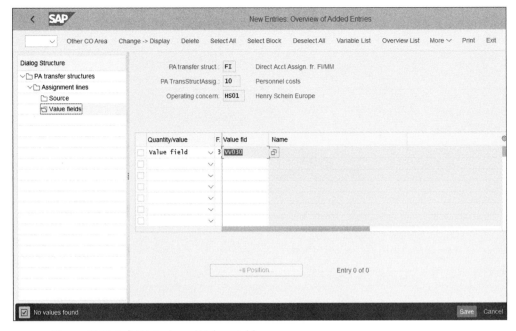

Figure 13.45 FI/MM Assigned Value Field

Enter "3" unless you need to transfer only the fixed or the variable portion of the costs. Maintain the whole structure in this way and save it with the **Save** button.

You can also configure that a profitability analysis segment is the default cost object for specific cost elements so that it can receive costs directly. To do so, follow menu path **Controlling • Profitability Analysis • Flows of Actual Values • Direct Posting from FI/MM • Automatic Account Assignment** or enter Transaction OKB9.

Select **New Entries** from the top menu to enter new cost elements, as shown in Figure 13.46.

Enter a company code and cost element and put a tick in the **Prf.Seg.** column. Then during postings in financial accounting, the profitability segment will be populated using derivation, and you also can maintain it manually.

13.3 Data Flow

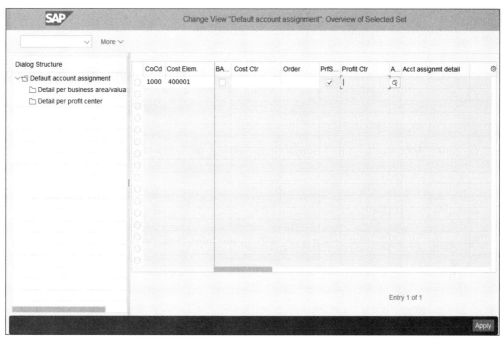

Figure 13.46 Default Account Assignment

13.3.3 Production Costs Flow

Production costs are an important part of the profitability calculation of companies in the production and process industries. How production costs are passed to profitability analysis in costing-based versus account-based profitability analysis is fundamentally different. In account-based profitability analysis, when there is a goods issue with a financial document generated, the profitability segment is posted with the production costs and the Universal Journal, table ACDOCA, contains not only the financial document but also the profitability analysis data.

In costing-based profitability analysis, at the time of goods issue there is no posting to profitability analysis. At the time of the invoice to the customer, both the revenue and the cost of goods sold are posted, which include the production costs. That leads to reconciliation problems with financial accounting, as explained in the beginning of this chapter. Therefore, account-based profitability analysis represents a significant improvement, which SAP S/4HANA offers for to be able to track and analyze production costs much more transparently and correctly.

You may wonder why companies didn't opt for account-based profitability analysis before SAP S/4HANA, when account-based profitability analysis was used mostly in addition to costing-based profitability analysis, only as a reconciliation tool with financial accounting. The main reason is that now the fast, in-memory database SAP HANA has enabled SAP to significantly improve account-based profitability analysis, thus overcoming its previous shortcomings, that made the costing-based approach the recommended option.

One of the main improvements in account-based profitability analysis is that it's possible to define different accounts for splitting the cost of goods sold and price differences. That enables posting them on different cost elements and thus analyzing them effectively in profitability analysis.

The cost of goods sold account posted during goods issue is derived using the account determination in Transaction OBYC, which we configured in Chapter 4. It's defined in account grouping code GBB-VAX. Now in SAP S/4HANA you have the opportunity to differentiate this account into multiple accounts, based on cost components of the cost component structure.

To define accounts for splitting of cost of goods sold, follow menu path **Financial Accounting • General Ledger Accounting • Periodic Processing • Integration • Materials Management • Define Accounts for Splitting the Cost of Goods Sold**, and click **New Entries** from the top menu to create a new splitting profile.

On the first screen, shown in Figure 13.47, enter a name for the new splitting profile, advisably starting with Z or Y. Enter the controlling area, from which the system also will populate the chart of accounts. You can check the **Acc Based Split** checkbox, in which case cost of goods sold always will be split when the source account is posted. After that, select a splitting profile and click **Source Accounts and Valuation Views** from the left side of the screen. Then click **New Entries** from the top menu to enter cost of goods sold source accounts.

On the screen shown in Figure 13.48, you need to enter all accounts that are defined in the GBB-VAX grouping code in Transaction OBYC. Also, you can define them separately per different type of valuation: legal, group, or profit center.

13.3 Data Flow

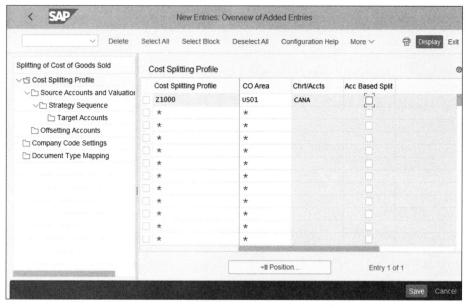

Figure 13.47 Cost-Splitting Profile

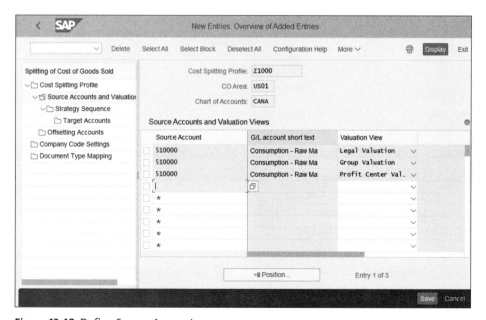

Figure 13.48 Define Source Accounts

547

13　Profitability Analysis

After that, for each source account you need to define target accounts. Select a source account and click **Strategy Sequence** on the left side of the screen. Then click **New Entries** from the top menu.

On the screen shown in Figure 13.49, you need to enter a strategy sequence. You can define separate sequences for released and upcoming cost estimates. Next, click **Target Accounts** from the left side of the screen and click **New Entries** from the top menu.

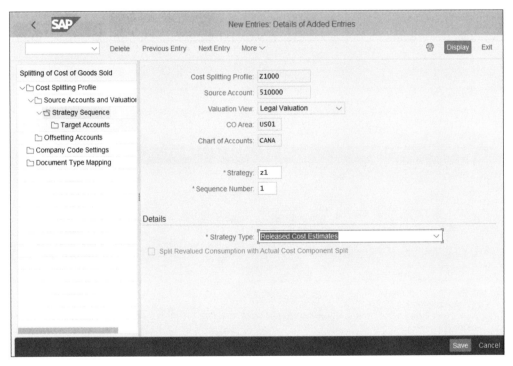

Figure 13.49 Strategy Sequence

On the screen shown in Figure 13.50, you define a target account for each component of the cost component structure.

Then click **Company Code Settings** on the left side of the screen and click **New Entries** from the top menu.

Here, in the screen shown in Figure 13.51, you need to activate the splitting structure per company code. Enter a validity start date and company code, select the created splitting structure, and save with the **Save** button.

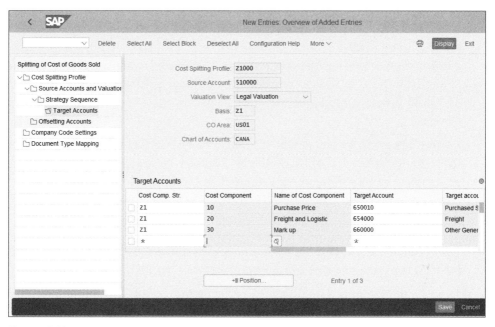

Figure 13.50 Target Accounts

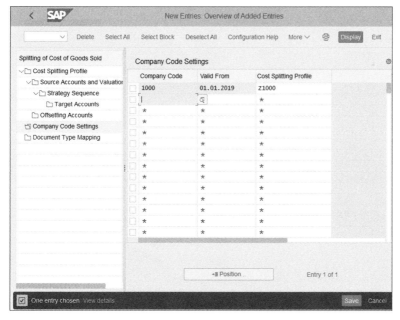

Figure 13.51 Company Code Activation

13.4 Integrated Planning

Profitability analysis is a functional area that receives almost all its data from other applications, as you saw in the previous section. Therefore it offers a very flexible and powerful integrated planning component, by which you can analyze integrated plan data from various other functional areas, such as cost center planning or logistics planning.

Now we'll examine this integrated planning process, starting with the definition of the planning framework, which combines the integrated planning parameters. Then we'll explain how to configure the various planning elements.

13.4.1 Planning Framework

The planning framework is the environment that contains the various planning elements, such as planning level, planning package, planning method, and parameter sets.

To access the planning framework, follow application menu path **Accounting • Controlling • Profitability Analysis • Planning • KEPM—Edit Planning Data**.

Initially you'll see the screen shown in Figure 13.52, where you'll define the planning elements.

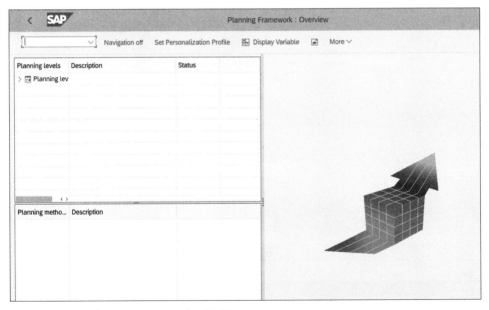

Figure 13.52 Planning Framework Initial Screen

13.4 Integrated Planning

Now you need to define the various planning elements, such as planning level, planning package, planning method, and parameter sets.

13.4.2 Planning Elements

First you need to define the planning level, which determines the level at which the planning is performed. To do so, right-click **Planning Level** and select **Create Planning Level**, as shown in Figure 13.53.

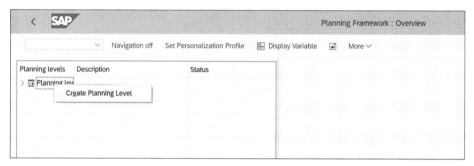

Figure 13.53 Create Planning Level

Enter a planning level name and description, as shown in Figure 13.54.

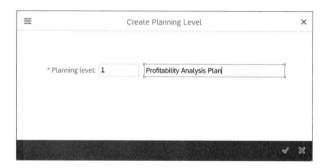

Figure 13.54 Planning Level Description

Proceed with the ✓ button.

On the next screen, shown in Figure 13.55, you'll see the list of available characteristics on the right side of the screen, and on the left side are the selected characteristics for the planning level.

13 Profitability Analysis

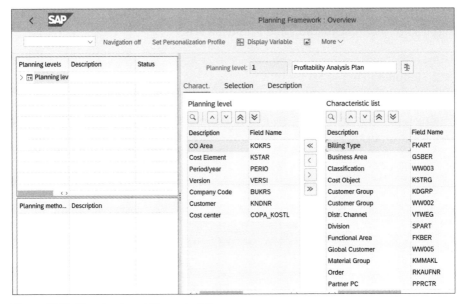

Figure 13.55 Planning Level Characteristics

You can add characteristics to the planning level with the ◁ (**Column Left**) button, and remove them with the ▷ (**Column Right**) button.

The next step is to define the characteristic values. Click the **Selection** tab and enter characteristic values as shown in Figure 13.56.

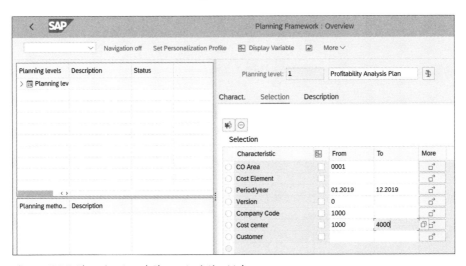

Figure 13.56 Planning Level Characteristics Values

13.4 Integrated Planning

After that, save the planning level with the **Save** button. The system automatically generates planning methods for the level, as shown in Figure 13.57.

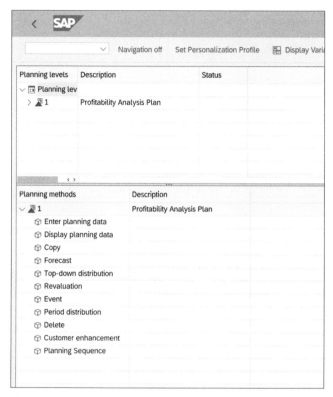

Figure 13.57 Planning Methods

The planning methods enable you to copy and evaluate a large amount of planning data.

The next step is to create a parameter set, which will contain various settings required to execute the planning. Right-click the **Enter planning data** method, then select **Create Parameter Set**, as shown in Figure 13.58.

On the next screen, shown in Figure 13.59, enter a name and description for the set.

13 Profitability Analysis

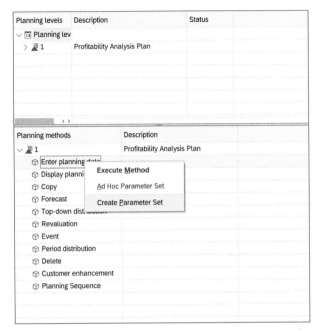

Figure 13.58 Enter Parameter Set

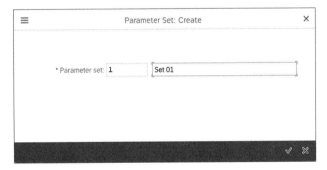

Figure 13.59 Create Parameter Set

Proceed with the ✓ button.

On the next screen, shown in Figure 13.60, enter a name for the new planning layout and a currency for the planning. Click the ☐ (**Create**) button to the right of the layout name to create it.

On the next screen, shown in Figure 13.61, you define the characterstics to be used in the planning layout.

13.4 Integrated Planning

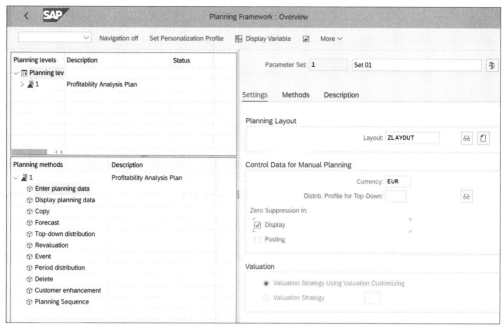

Figure 13.60 Parameter Settings

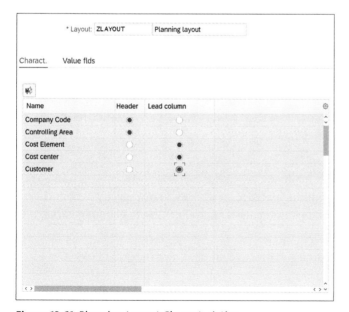

Figure 13.61 Planning Layout Characteristics

Select which characteristics should be on the header level and which should be lead columns when entering plan data. Then click the **Value flds** tab. In the screen shown in Figure 13.62, you can select value fields to be used in the layout.

Figure 13.62 Planning Layout Value Fields

Then save with the 💾 button.

You also need to create a planning package, in which you can specify default characteristic values. Right-click the planning level and select **Create Planning Package**, as shown in Figure 13.63.

Enter a planning package name and description, as shown in Figure 13.64.

13.4 Integrated Planning

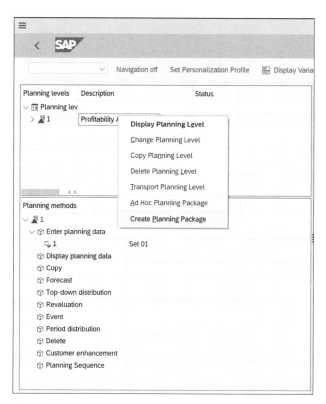

Figure 13.63 Create Planning Package

Figure 13.64 Planning Package Name

Proceed with the ✓ button. On the next screen, shown in Figure 13.65, enter default values for the characteristics.

Then save your entries with the **Save** button.

557

13 Profitability Analysis

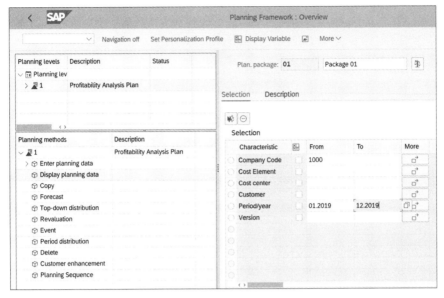

Figure 13.65 Planning Package Defaults

With that, we complete the configuration for integrated planning.

13.5 Information System

Profitability analysis almost entirely receives data. Its main strength is that when properly designed and configured, it can provide very important and powerful management reports to track and analyze the profitability of the company. Not surprisingly, the information system of profitability analysis provides myriad sophisticated and flexible reports that can meet all reporting requirements.

In addition, you can easily create your own reports. In SAP S/4HANA, the profitability analysis reports are extremely fast, benefiting from the revolutionary columnar in-memory database. Users of profitability analysis from older SAP releases are probably well aware that some profitability analysis reports had a very long runtime and had to be executed in the background, sometimes over a day or even more. Now in SAP S/4HANA you can produce huge and powerful profitability analysis reports extremely fast and on the fly.

Let's first examine the line item list reports, which enable you to analyze the data on the line item level. Then we'll discuss drilldown reporting, which is the main tool to build your own profitability analysis reports.

13.5.1 Line Item Lists

As with other controlling components, profitability analysis also provides reporting on the lowest line item level.

To execute line item reports, follow application menu path **Accounting • Controlling • Profitability Analysis • Information System • Display Line Item List**. You can view actual line items with the **KE24—Actual** transaction and view plan line items with the **KE25—Plan** transaction.

Enter Transaction **KE24—Actual**. Figure 13.66 shows the selection screen of the line item report, in which you can select line items by various criteria such as period, company code, cost element, date and user of creation, and defined profitability analysis characteristics. Then click **Execute** from the top menu.

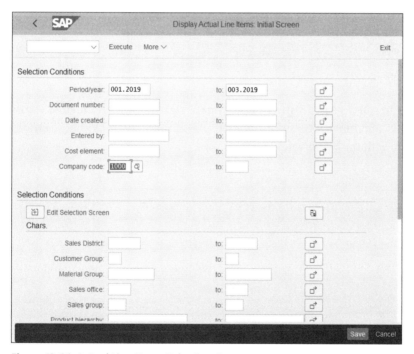

Figure 13.66 Actual Line Items Selection Screen

On the next screen, shown in Figure 13.67, you see the output of the report.

You see the profitability analysis line items based on the selections. You can add and remove fields from the report by selecting **More • Change Layout** from the top menu.

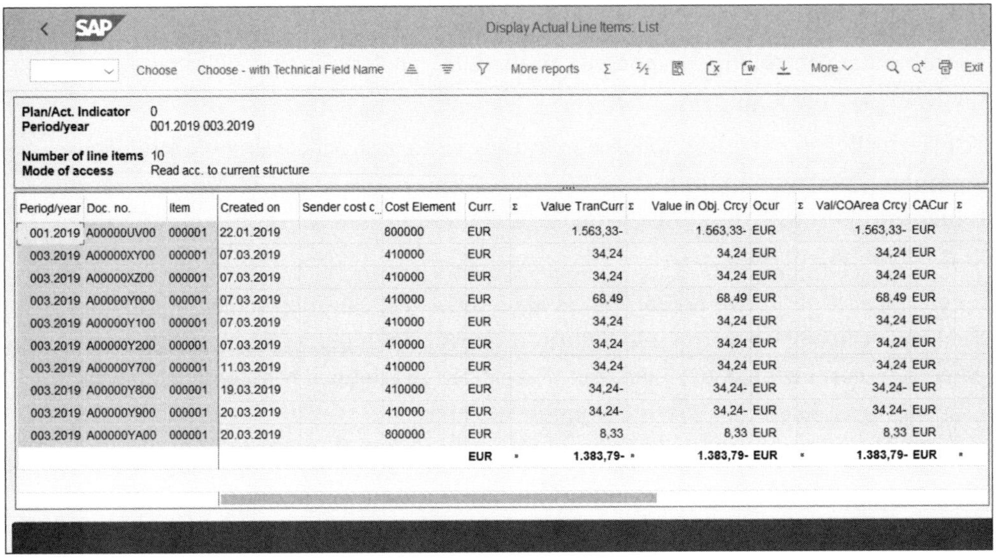

13 Profitability Analysis

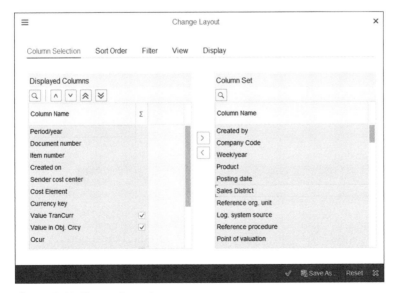

Figure 13.67 Actual Line Items Output

For example, to add the **Sales District**, mark it on the right side of the screen, as shown in Figure 13.68, and move it to the left side (**Displayed Columns**) with the ⟨ button. Then confirm with the ✓ button.

Figure 13.68 Change Layout

560

13.5 Information System

Double-clicking a line item shows the controlling document and the profitability segment, as shown in Figure 13.69.

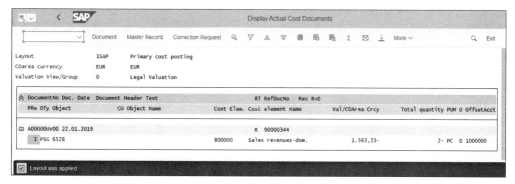

Figure 13.69 Profitability Document

13.5.2 Drilldown Reporting

Profitability analysis uses the drilldown reporting technique, which you're already familiar with from Chapter 12. This is a flexible form of reporting that enables you to slice and dice the profitability data in any dimension you need.

Drilldown reports, as you already know, can be basic or use a form, which can be reused in multiple reports. Let's create a contribution margin report using a form. To create a form, enter Transaction KE34.

In the screen shown in Figure 13.70, you define the form name and structure.

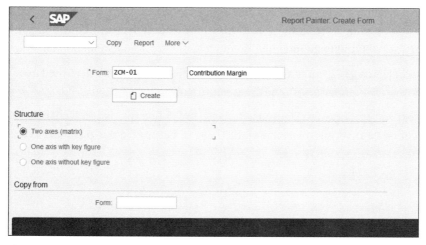

Figure 13.70 Create Form

Enter a name and description for the form and select the **Two axes (matrix)** type of structure. Then proceed with the **Create** button.

Figure 13.71 shows a sample contribution margin structure that you can define. In the rows, define various cost elements that represent the relevant categories. In the columns, select the actual amount and actual quantity.

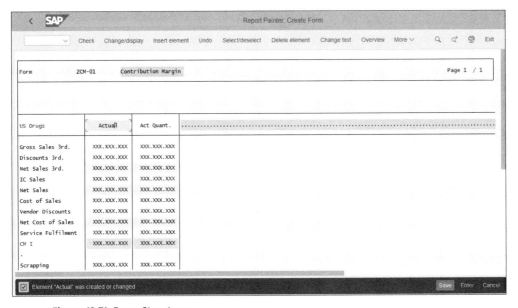

Figure 13.71 Form Structure

To create a drilldown profitability analysis report, follow application menu path **Accounting • Controlling • Profitability Analysis • Information System • Define Report • KE31—Create Profitability Report**.

As shown in Figure 13.72, enter a report name and description and select the already created form. Then proceed with the **Create** button.

On the next screen, shown in Figure 13.73, select and define the required characteristics. We define them as variables; therefore, they can be entered during report execution at the selection screen.

13.5 Information System

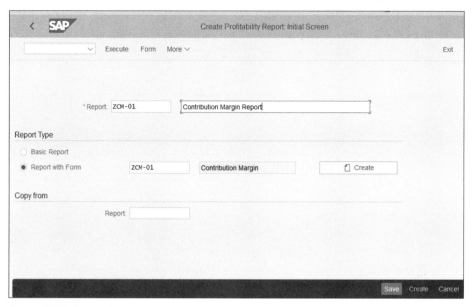

Figure 13.72 Create Report

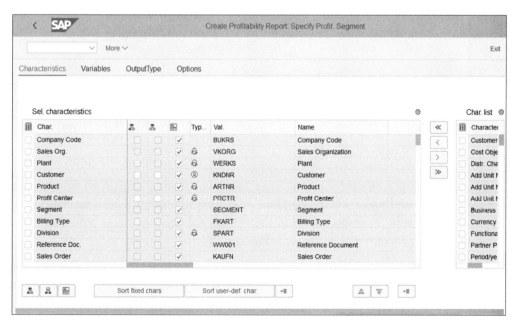

Figure 13.73 Report Definition

13 Profitability Analysis

To execute an already defined drilldown report, follow application menu path **Accounting • Controlling • Profitability Analysis • Information System • KE30—Execute Report**. Select the report you defined, **ZCM-01 Contribution Margin Report**, and execute it. Figure 13.74 shows the selection screen of the report.

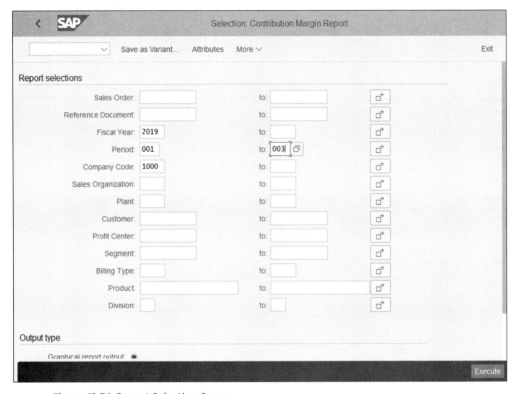

Figure 13.74 Report Selection Screen

Enter selections for the report, such as fiscal year, period, company code, sales organization, and other profitability characteristics. You can also choose the type of report output: graphical report output, classical drilldown list, or object list. Then execute the report with the **Execute** button.

Figure 13.75 shows the graphical report output. On the top-left side of the screen, you navigate between the characteristics, thus drilling down on them. On the right side, you see the calculated contribution margin as defined in the report.

13.6 Summary

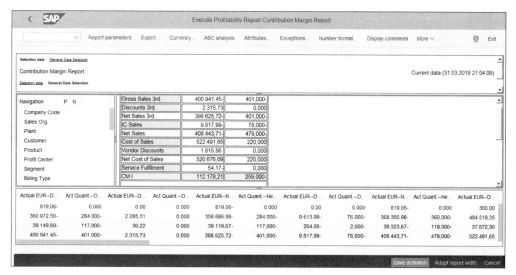

Figure 13.75 Contribution Margin Report

Using these techniques, you can define profitability reports that will help you analyze various characteristics and profitability margins.

13.6 Summary

In this chapter, we provided an extensive guide to profitability analysis. You learned the differences between and benefits of account-based and costing-based profitability analysis. We put special emphasis on the improvements SAP S/4HANA brings to account-based profitability analysis.

You've learned the global settings required for activating profitability analysis, and you also learned how the various value flows are posted to profitability analysis. Then you walked through how to configure integrated planning in profitability analysis. Finally, you learned about the powerful line item and drilldown reports that profitability analysis provides and how you can create your own reports.

In the next chapter you will learn how to configure and use product costing in SAP S/4HANA.

Chapter 14
Product Costing

This chapter gives step-by-step instructions for configuring product costing in SAP S/4HANA, including for the material ledger, which is now required in SAP S/4HANA. It discusses real-life scenarios and their implementation in the product costing model.

Product costing is a part of controlling that enables planning of the product costs and maintaining material prices. Thus it's a vital part of the calculation of the profitability of the enterprise. It's integrated with the production planning and materials management functionalities of SAP S/4HANA.

Broadly, we can segregate product costing into two parts: product cost planning, which is used to calculate the planned costs of the products and materials, and actual costing, which allocates the actual costs to them. Before analyzing these processes in detail, we'll set up the required master data for product costing. At the end of the chapter, we'll guide you through the information system for product costing.

14.1 Master Data

Master data plays an important role for product costing because it provides the backbone for the correct calculation of product prices. We'll review how to set up the following for product costing:

- Material master
- Bill of materials (BOM)
- Work center
- Routing

The material master is the foundation for the product costing. In product cost planning, you create so-called material cost estimates, which provide the costing on the material level. The BOM and routing create the so-called quantity structure, which is used in costing estimates.

Let's start with the material master.

14.1.1 Material Master

The material master is a foundational master data object for all of logistics in SAP S/4HANA. All raw materials, semifinished products, and finished products are created as material masters. Furthermore, services also can be set up as materials.

The material master is a complex structure that consists of numerous views organized into tabs related to sales, purchasing, accounting, product costing, and so on. We'll focus on the setup required from a costing point of view.

We'll start with some basic configuration that's required for the materials to enter the costing information. Each material is created for a material type, which is used to classify materials with similar functions. For material types relevant for costing, you need to activate the costing and accounting views.

To define the settings for material types, follow menu path **Logistics—General • Material Master • Basic Settings • Material Types • Define Attributes of Material Types**.

On the screen shown in Figure 14.1, you see the material types defined in the system. The standard material types provided by SAP are shown here, and you can see that SAP provides myriad standard material types to represent different kinds of materials. When creating custom material types, they should start with Z or Y in the custom name range.

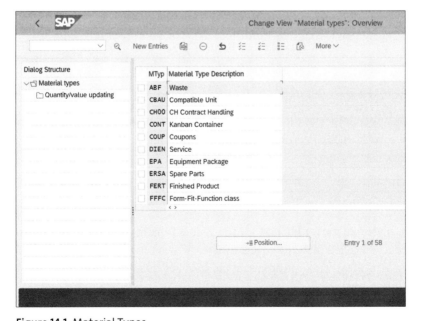

Figure 14.1 Material Types

Let's create our own material type for raw materials. Locate material type **ROH**, which is a standard SAP material type for raw materials, select it, and click the 🗐 (**Copy As...**) button from the top menu.

Figure 14.2 shows the configuration of the material type. Enter "ZRAW" as the name for the new material type and enter "Raw materials" for the long description.

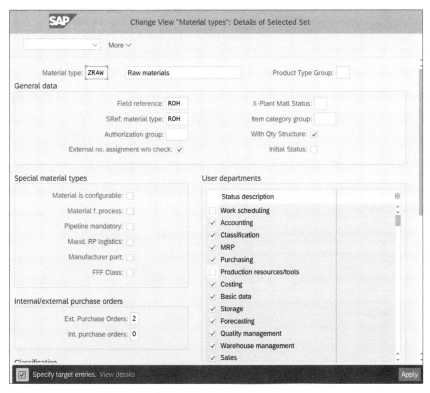

Figure 14.2 Material Type Definition

In the **User departments** area, you'll see all the possible views to be maintained on the material master. Make sure that **Accounting** and **Costing** are checked because they provide the required data for financial accounting and product costing. Then click the **Apply** button to adopt the settings.

The system issues a message that it copied the dependent entries for the material type also, which you can confirm with the ✅ button, as shown in Figure 14.3. After that, save the new material type with the **Save** button.

14 Product Costing

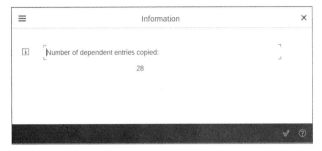

Figure 14.3 Copied Dependent Entries

To create a new material, follow application menu path **Logistics • Materials Management • Material Master • Material • Create (General) • Create (General)**.

Figure 14.4 shows the initial screen of material creation. Enter a material number, which can consist of numbers or letters, depending on the number range configuration. You also need to select **Industry Sector** and **Material type**. For this example, select **Pharmaceuticals** and the new material type, **Raw materials**. Then proceed with the **Continue** button.

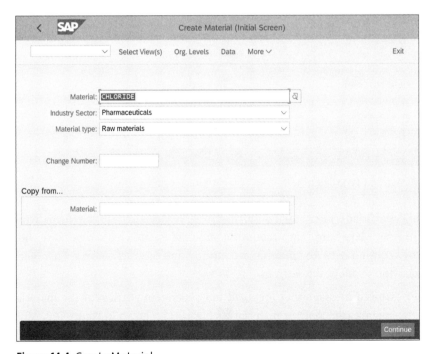

Figure 14.4 Create Material

On the next screen, shown in Figure 14.5, check the views you're going to maintain in the material master. Make sure the accounting and costing views are selected, then proceed with the ✓ button.

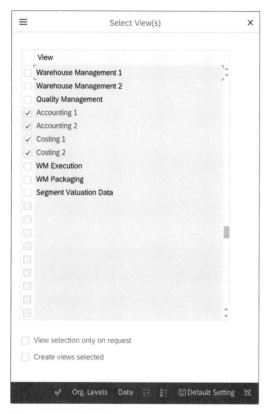

Figure 14.5 Select Views

We won't detail all the material views, which contain logistic information. We'll concentrate instead on the accounting and costing views. Figure 14.6 shows the **Accounting 1** view, which contains some important settings from a product costing point of view. Most important is the **Prc. Ctrl.** (price control) field, for which you have two options:

- **S**: Standard price
- **V**: Moving average price/periodic unit price

14   Product Costing

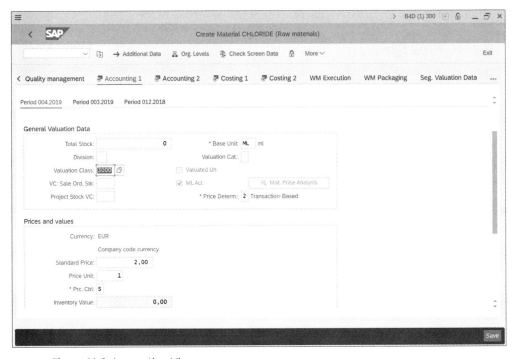

Figure 14.6 Accounting View

If material has a standard price (**S**), the value of the material is calculated at this standard price. If goods movements or invoice receipts have a price that is different from the standard price, those differences are posted to the price difference account. The variance isn't taken into account in inventory valuation. If, however, the material has a moving average price, each goods movement and invoice receipt updates its price. When goods movements and invoice receipts are posted with a price that's different from the current moving average price of the material, the differences are posted to the stock account itself. Therefore, the moving average price and the inventory value change.

We recommend using standard prices for finished and semifinished products. The moving average price can be used for raw materials and external purchases. Especially when the price could vary significantly, the moving average price is recommended. Also, in some countries where prices can fluctuate significantly, especially in an inflation environment, the moving average price is recommended—and sometimes even legally required.

14.1 Master Data

This is a major decision point and important strategy in each SAP S/4HANA implementation that affects how to value the inventory. Each type of material should be carefully considered and the business should make a well-informed decision with the help of its integration partner. It's important to make good decisions for all materials because the valuation of the materials provides the backbone for proper product costing. To do that, you need to understand what types of materials are used as raw, semifinished, and finished materials. You also need to understand the specifics of these materials, how're they affected by market price movements, and how stable their prices are. It's important to consider the economy in which the company operates and whether there are inflationary forces in play.

Next, navigate to the **Costing 2** view of the material, as shown in Figure 14.7.

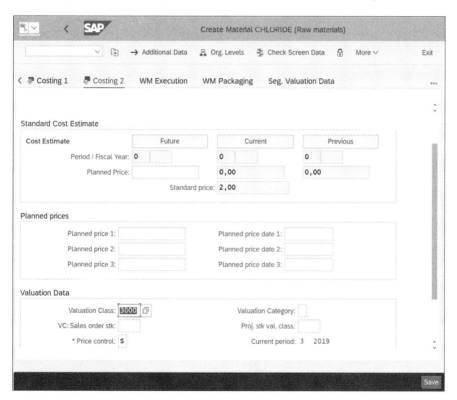

Figure 14.7 Costing View

Here in the **Standard Cost Estimate** section, you'll maintain the standard cost estimate, which is used to valuate materials with standard price control. There are sections for

573

future, current, and previous prices, which currently are empty because we haven't costed the material yet. We will do that in Section 14.2, but for now we'll just save after maintaining the other views.

14.1.2 Bill of Materials

The BOM is a quantity structure that provides the list of components and activities to manufacture a product. It's a master data object from production planning, but from a costing point of view a costing BOM is needed also. It's created with Transaction CS01.

In the definition screen, shown in Figure 14.8, enter the components that are needed for the manufacturing a finished or semifinished product.

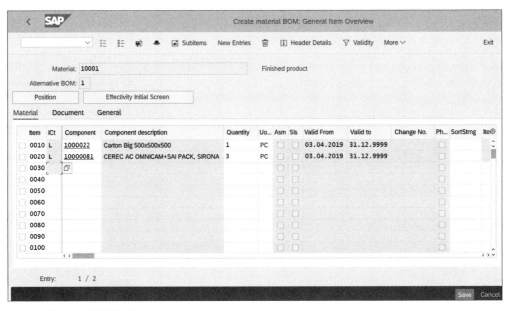

Figure 14.8 Bill of Materials

From a costing point of view, to cost the finished product, its components also need to be costed. The BOM may consist of just one or a few materials or other components, but in some industries, like the automotive industry, it may consist of hundreds of items.

14.1.3 Work Center

A *work center* is a logistic organizational unit that defines where and by whom operations are performed. It's important for product costing because through its assignment of a cost center it provides the costs occurring during the production process to the products.

Let's see how to establish a link with costing. To create a work center, enter Transaction CR01.

On the initial screen shown in Figure 14.9, enter the **Plant** and **Work center** to be created and a **Work Center Category**, which indicates the purpose of the work center. You can also provide an existing work center to copy from in the **Copy from** section.

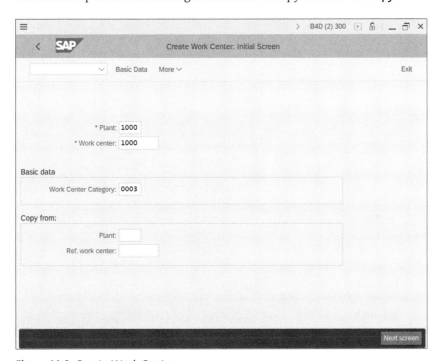

Figure 14.9 Create Work Center

An important part of the product costing view is the **Costing** tab shown in Figure 14.10.

Here, you assign the cost center that will bear the costs related to production processes in this work center. You also assign activity types to the activities to be performed in this work center. (We discussed the activity types in Chapter 10.)

14 Product Costing

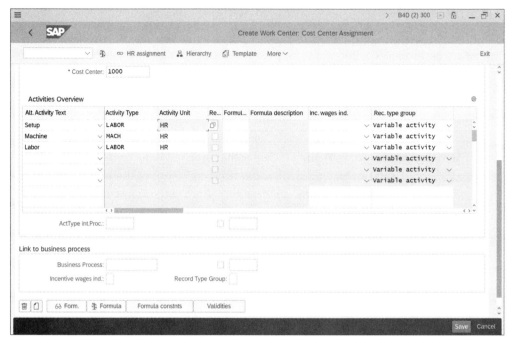

Figure 14.10 Work Center Costing Data

After establishing the costing link in the work center, production confirmations posted in production planning will pass costs to calculate the actual costs of the involved products.

14.1.4 Routing

Routing is another production planning object that affects the product costing. It's a description of which operations have to be carried out during the production process. It also defines the sequence of the activities that need to be carried out in work centers. The work center is linked with the cost center, and thus the costs incurred during these operations pass through the product costing and are included in calculating the cost of the products.

To create a routing, enter Transaction CA01. On the initial screen shown in Figure 14.11, enter a material and plant for which you'll create the routing. This is the finished or semifinished product that will be manufactured using the operations specified in the routing.

14.1 Master Data

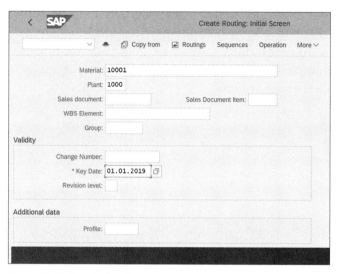

Figure 14.11 Create Routing

On the screen shown in Figure 14.12, click **Operation** from the top menu to define the operation steps.

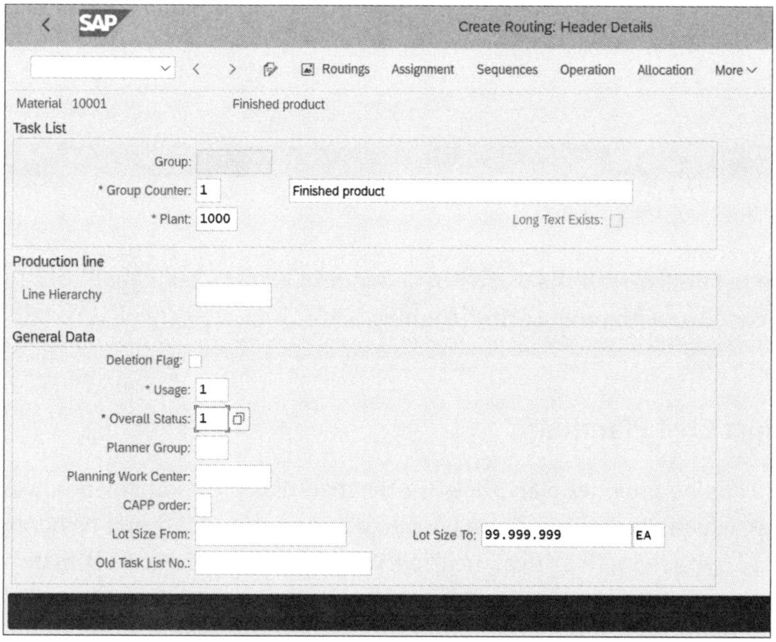

Figure 14.12 Routing Definition

On the screen shown in Figure 14.13, define the operation steps in the lines and assign work centers to each operation. Because each work center has a cost center, when the operation steps of the routing are executed, costs are accumulated in the respective cost centers. Therefore when you cost the finished product, it includes the costs associated with the activities in the work centers.

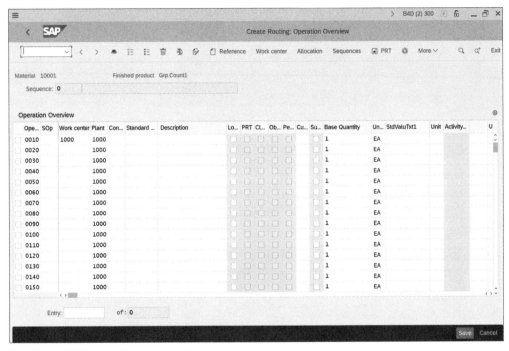

Figure 14.13 Operations Definition

Now that you're familiar with the master data that provides the building blocks for costing, let's continue with product cost planning.

14.2 Product Cost Planning

Product cost planning provides plan prices for the materials the company produces and purchases, which are then used as a basis to value the inventory and recognize price differences with the actual prices incurred. We create a material cost estimate to provide the planned standard price.

Product costing planning is based on a costing variant that determines what costs are included. We'll start by configuring the various components that need to be included in a costing variant. Then we'll configure the costing variant. After that, we'll guide you through how to configure a cost component structure, which groups the cost elements into cost components. Then we'll configure a costing sheet, which defines how overhead costing is performed. Finally, you'll learn how to perform a material cost estimate, which costs the materials.

14.2.1 Costing Variant Components

The costing variant determines how the manufacturing costs are valuated and which master data objects are included for calculation of the standard price. The costing variant combines the following configuration components, as we'll discuss in this section:

- Costing type
- Valuation variant
- Quantity structure control
- Transfer control
- Reference variant

Costing Type

The costing type defines the use of the calculation. To define a calculation type, follow menu path **Controlling** • **Product Cost Controlling** • **Product Cost Planning** • **Material Cost Estimate with Quantity Structure** • **Costing Variant: Components** • **Define Costing Types**.

Figure 14.14 shows the defined costing types. We can use the standard ones for this example. Double-click costing type **01**, **Standard Cost Est. (Mat.)**.

The settings are divided into three tabs. Figure 14.15 shows the first tab, **Price Update**. The **Price Update** field lists the field in which the calculated price will be updated in the material master. In addition to the **Standard Price** field, there are also other price fields in the material master that can be used, such as tax price and commercial price.

14 Product Costing

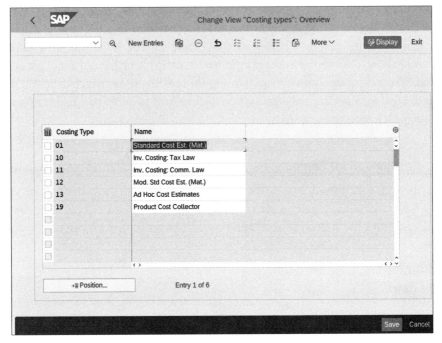

Figure 14.14 Costing Types

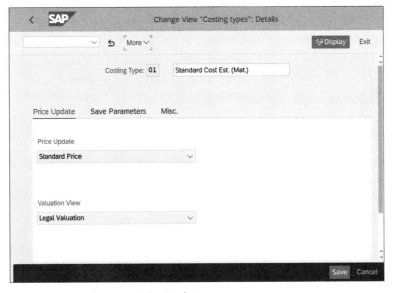

Figure 14.15 Costing Type Price Update

14.2 Product Cost Planning

In the **Valuation View** field, select the type of valuation to be used in case of parallel valuation.

Next, click the **Save Parameters** tab, as shown in Figure 14.16.

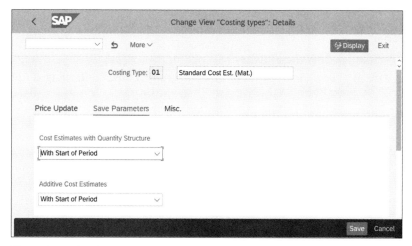

Figure 14.16 Costing Type Save Parameters

Here, there are two fields to configure. You define whether to save the date information related to the calculation in the **Cost Estimates with Quantity Structure** and **Additive Cost Estimates** fields. We recommend setting both options with date information. Finally, click the **Misc.** tab, as shown in Figure 14.17.

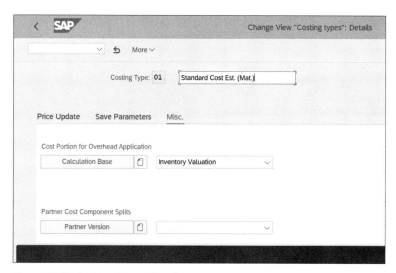

Figure 14.17 Costing Type Miscellaneous

581

14   Product Costing

In the **Cost Portion for Overhead Application** field, you define which cost elements serve as a basis for the calculation of the overhead costs. In the **Partner Cost Component Splits** field, you define which characteristics, such as company code or profit center, will be saved in the cost estimate for further analysis.

Valuation Variant

The valuation variant defines the valuation settings for the cost calculation. To define a calculation variant, follow menu path **Controlling • Product Cost Controlling • Product Cost Planning • Material Cost Estimate with Quantity Structure • Costing Variant: Components • Define Valuation Variants**.

On the first screen, shown in Figure 14.18, you'll see a list of the defined valuation variants. Plenty of standard valuation variants are defined, which cover different types of valuation concepts. You can use them or create your own in the custom name range.

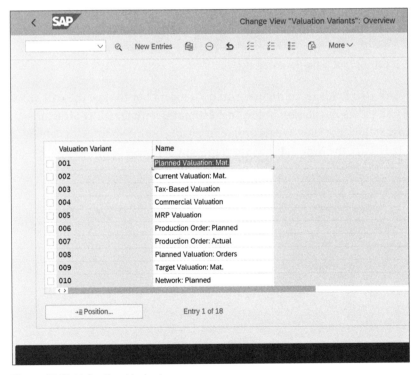

Figure 14.18 Valuation Variants

14.2 Product Cost Planning

Let's review the settings behind the valuation variant. Double-click variant **001**, **Planned Valuation: Mat**. The settings are structured in six tabs. In the first tab, **Material Val.**, shown in Figure 14.19, you define which prices to use for material valuation. You can enter up to five types of prices and thus build a strategy.

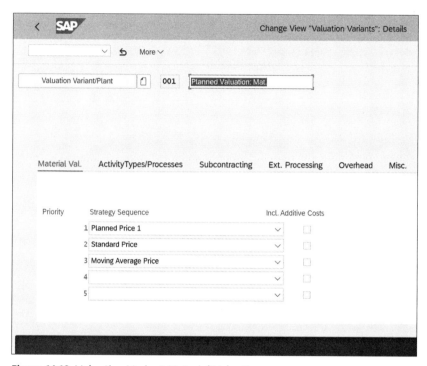

Figure 14.19 Valuation Variant Material Valuation

The system will first try to find the price specified in the **Priority 1** field—**Planned Price 1** in our example. If it isn't maintained in the material master, it will look at the second priority—**Standard Price** in our example. Then **Priority 3**, and so on. If it doesn't find a price at all, there will be an error in the costing calculation.

Next, click the **ActivityTypes/Processes** tab, as shown in Figure 14.20.

Here you define the strategy sequence for price determination of the activity types. You can provide up to three priorities for pricing. In the **CO Version Plan/Actual** field, specify the version in which the activity prices are stored.

Now click the **Subcontracting** tab, as shown in Figure 14.21.

583

14 Product Costing

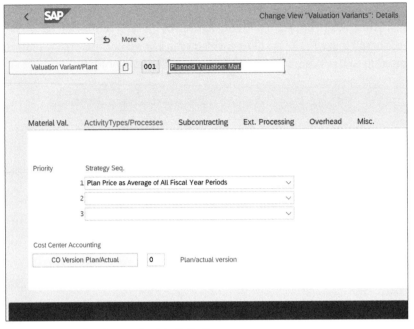

Figure 14.20 Valuation Variant Activity Types

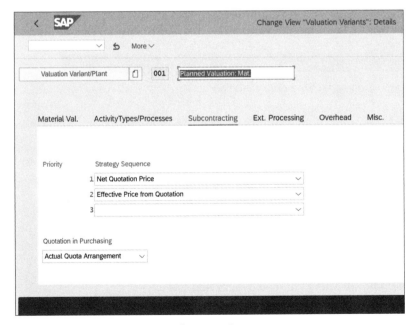

Figure 14.21 Valuation Variant Subcontracting

14.2 Product Cost Planning

Here you define the strategy sequence for price determination for subcontracting. You can provide up to three priorities, such as net quotation price, gross purchase order price, and so on. *Subcontracting* is a process in which you deliver one or more components to a subcontractor, who manufactures the final product and returns it to you. The materials are your property during the subcontracting process. There are costs incurred during this process, which need to be included in the costing.

Next, click the **Ext. Processing** tab, as shown in Figure 14.22.

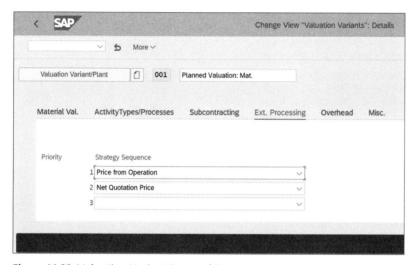

Figure 14.22 Valuation Variant External Processing

In this tab, you define the strategy sequence for price determination for the valuation of externally processed operations. To calculate the costs for external processing, the externally-processed operations are valued with the prices specified here. Various options are available to determine the price, such as:

- Net quotation price from the purchasing info record
- Net order price from the purchase order
- Price from operation

Now move to the **Overhead** tab, as shown in Figure 14.23.

On this tab, you define how overhead rates apply to the cost estimate by entering a costing sheet in the valuation variant. In Figure 14.23 we have the costing sheet **PP-PC Standard** assigned for the finished and semifinished materials. We'll configure a costing sheet later in this chapter in Section 14.2.4.

585

14 Product Costing

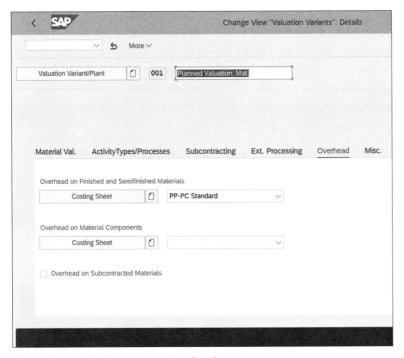

Figure 14.23 Valuation Variant Overhead

Finally, click the **Misc.** tab, as shown in Figure 14.24.

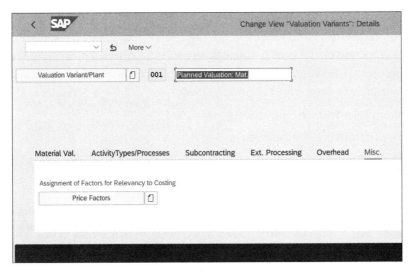

Figure 14.24 Valuation Variant Miscellaneous

14.2 Product Cost Planning

Here you can assign price factors. Price factors are used to multiply single line items by a certain price factor. This could be used, for example, for inventory valuation for tax purposes.

Quantity Structure Control

The quantity structure control defines how to search for the BOM and the routing included in the cost calculation.

To define the quantity structure control, follow menu path **Controlling • Product Cost Controlling • Product Cost Planning • Material Cost Estimate with Quantity Structure • Costing Variant: Components • Define Quantity Structure Control**.

On the first screen, shown in Figure 14.25, you'll see a list of the controls defined. You can create a new one with the **New Entries** option from the top menu or copy an existing one with the 🗐 button from the top menu.

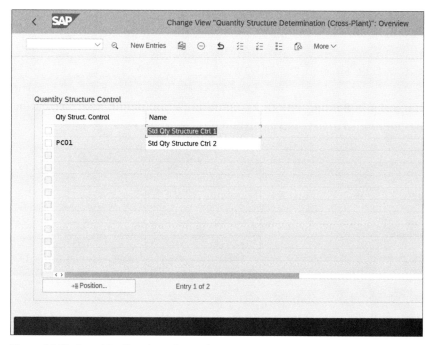

Figure 14.25 Quantity Structure Control

Let's review the possible settings. Double-click **PC01**. On the first tab, **BOM**, shown in Figure 14.26, you define the BOM determination. There are several BOM applications—

for example, BOM for production, BOM for sales, or costing BOM. Here you select which one should be used for the costing calculation.

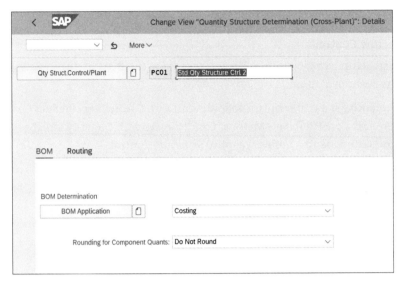

Figure 14.26 Quantity Structure Control BOM

On the second tab, **Routing**, shown in Figure 14.27, you define how the routings should be determined in the cost calculation.

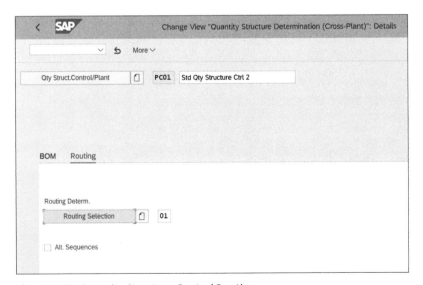

Figure 14.27 Quantity Structure Control Routing

14.2 Product Cost Planning

Transfer Control

Transfer control ensures that when a cost estimate already exists, it will be transferred rather than creating a new cost estimate.

To define transfer control, follow menu path **Controlling • Product Cost Controlling • Product Cost Planning • Material Cost Estimate with Quantity Structure • Costing Variant: Components • Define Transfer Control**.

On the first screen, shown in Figure 14.28, you'll see a list of the transfer controls defined. You can create a new one with the **New Entries** option from the top menu or copy an existing one with the 🗐 button from the top menu.

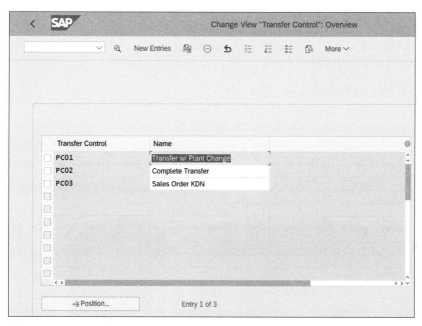

Figure 14.28 Transfer Control

Let's review the possible settings. Double-click **PC01 Transfer w/ Plant Change**. On the **Single-Plant** tab, shown in Figure 14.29, you can set up to three strategies. The system will search for the first appropriate calculation. If no cost estimate is found, it will check the next priority, and so on. If you activate the **Fiscal Year** indicator, it will search only for cost estimates in the current fiscal year. The **Periods** field allows you to specify how many months in the past the system should check for a suitable calculation.

589

14 Product Costing

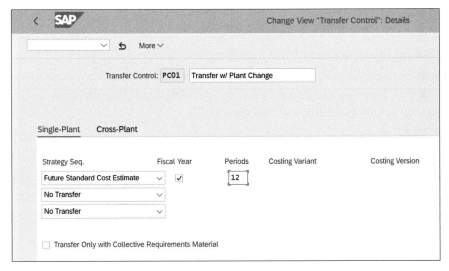

Figure 14.29 Transfer Control Single Plant

On the **Cross-Plant** tab, shown in Figure 14.30, you can maintain a strategy sequence if you work with special procurement keys. The system will search for allocations from other sources according to the special procurement key. The special procurement key is part of the material master record and determines whether material is procured externally or produced in house.

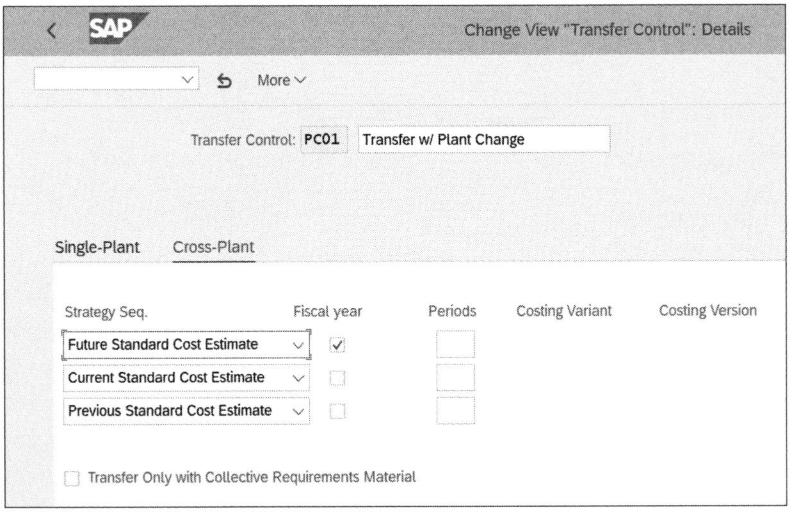

Figure 14.30 Transfer Control Cross-Plant

14.2 Product Cost Planning

Reference Variant

The reference variant enables you to create cost estimates based on the same quantity structure. To define a reference variant, follow menu path **Controlling • Product Cost Controlling • Product Cost Planning • Material Cost Estimate with Quantity Structure • Costing Variant: Components • Define Reference Variants**.

No variant exists initially, so click **New Entries** from the top menu of the initial screen shown in Figure 14.31.

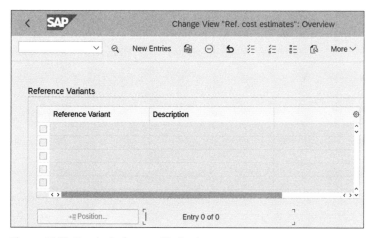

Figure 14.31 Create Reference Variant

Provide a name and description for the variant, as shown in Figure 14.32. In the first tab, **Cost Estimate Ref.**, you define the strategy for how to transfer an existing cost estimate.

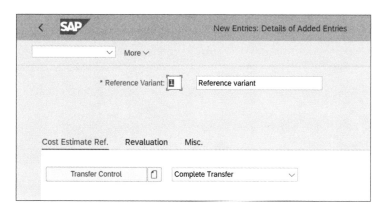

Figure 14.32 Reference Variant Cost Estimate

591

On the **Revaluation** tab, shown in Figure 14.33, you define which components should be recalculated and which should remain the same from the original cost estimate. Those that are checked here will be revaluated.

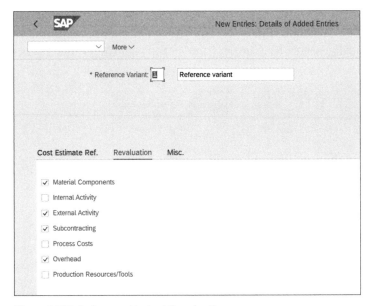

Figure 14.33 Reference Variant Revaluation

On the **Misc.** tab, shown in Figure 14.34, if you check the **Transfer Additive Costs** checkbox you can define that additive costs also should be transferred when creating a cost estimate with a reference.

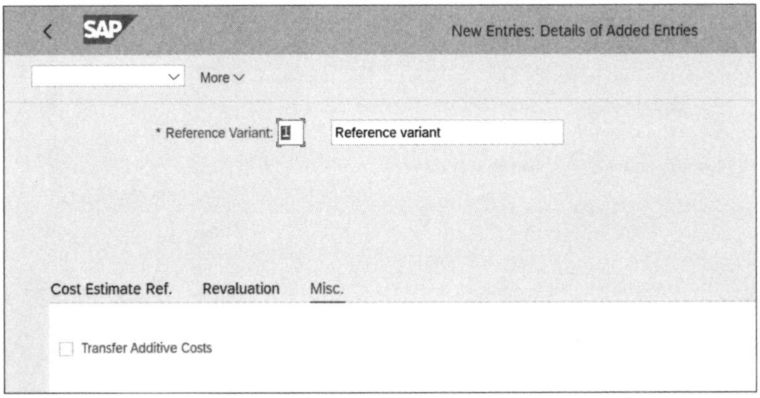

Figure 14.34 Reference Variant Miscellaneous

After completing all the tabs, save the reference variant with the **Save** button.

14.2.2 Creating the Costing Variant

In the previous section, we configured the components that are part of the costing variant. Now we're ready to create the costing variant itself. Follow menu path **Controlling • Product Cost Controlling • Product Cost Planning • Material Cost Estimate with Quantity Structure • Define Costing Variants**.

On the initial screen shown in Figure 14.35, you see the defined costing variants. You can create new variants with the **New Entries** option from the top menu or copy an existing one with the ⧉ button from the top menu.

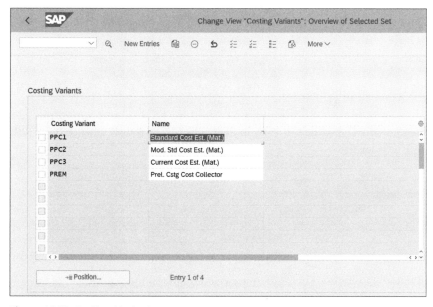

Figure 14.35 Costing Variants

Let's review the settings. Double-click **PPC1**, **Standard Cost Est. (Mat.)**, which is a standard costing variant provided by SAP.

In the first tab, **Control**, shown in Figure 14.36, you assign the various costing variant components, which we defined in the previous steps:

- Costing Type
- Valuation Variant
- Date Control
- Qty Struct. Control
- Transfer Control
- Reference Variant

593

14 Product Costing

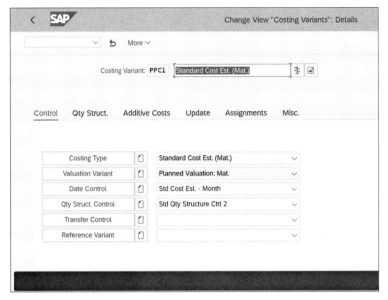

Figure 14.36 Costing Variant Control

On the second tab, **Qty Struct.**, shown in Figure 14.37, you define settings related to the quantity structure.

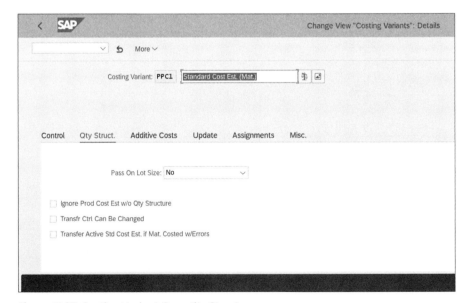

Figure 14.37 Costing Variant Quantity Structure

14.2 Product Cost Planning

These settings are as follows:

- **Pass on Lot Size**
  Defines whether the system will determine the costing lot size using the lot size of the highest material in the BOM.

- **Ignore Prod Cost Est w/o Qty Structure**
  If you tick this checkbox, all cost estimates without quantity structures will be ignored.

- **Transfr Ctrl Can Be Changed**
  If you tick this checkbox, you can manually enter the transfer control parameters for existing costing data.

- **Transfer Active Std Cost Est. if Mat. Costed w/Errors**
  Defines whether the active standard cost estimate is used to continue costing when errors occur.

On the next screen, shown in Figure 14.38, you define whether additive costs should be considered in the cost estimate. If you tick the **Include Additive Costs with Stock Transfers** checkbox, additive costs also will be considered during transfer of a material from a different plant.

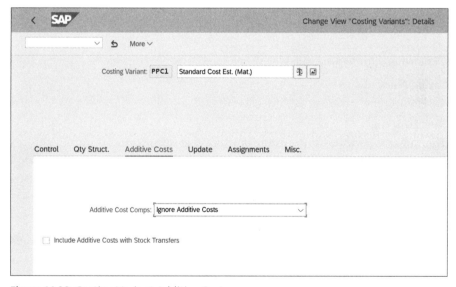

Figure 14.38 Costing Variant Additive Costs

595

On the next tab, shown in Figure 14.39, you define settings related to saving the cost estimate.

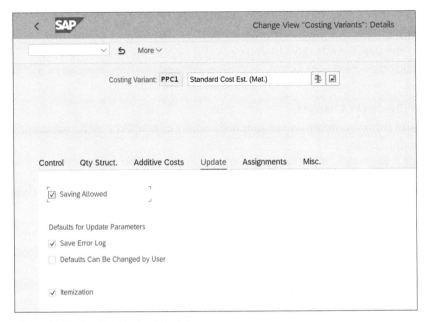

Figure 14.39 Costing Variant Update

The following options are available:

- **Saving Allowed**
  This indicator enables saving the cost estimate.
- **Save Error Log**
  This indicator enables saving the error log during calculation.
- **Defaults Can Be Changed by User**
  If you tick this checkbox, you can change the save parameters during the calculation of the cost estimate.
- **Itemization**
  If you tick this checkbox, the itemization also is saved to the cost component split. We recommend ticking this indicator for better information for reporting.

In the next tab, **Assignments**, shown in Figure 14.40, you define several more settings.

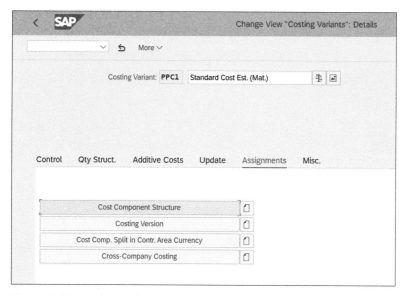

Figure 14.40 Costing Variant Assignments

These settings are as follows:

- **Cost Component Structure**
  The cost component structure defines the costing structure. It structures the separate components under which costing is divided, such as labor, materials, and so on. Double-clicking this button will let you see/modify the assigned cost component structures per company code.

- **Costing Version**
  With the costing version, you can perform different costing calculations for the same material.

- **Cost. Comp. Split in Contr. Area Currency**
  With this option, you can perform the calculation in the controlling area currency in addition to the company code currency.

- **Cross-Company Costing**
  Activate this option if you want to perform cross-company code cost calculation.

In the last tab, **Misc.** (see Figure 14.41), you define whether the error log should be saved when running the cost estimate and whether it should be sent automatically to existing SAP users.

14 Product Costing

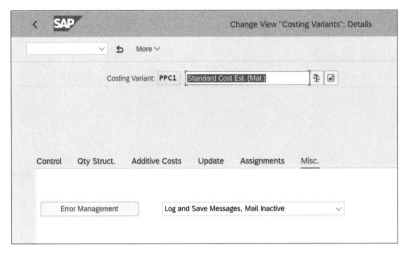

Figure 14.41 Costing Variant Miscellaneous

With that, we finish the configuration of the costing variant.

14.2.3 Cost Component Structure

We mentioned the cost component structure briefly when we assigned it to the costing variant. The cost component structure defines how the results of material cost estimates are updated. It structures the costs for the individual materials into cost components (such as material costs, overhead, internal activities, external activities, and so on).

To define a cost component structure, follow menu path **Controlling • Product Cost Controlling • Product Cost Planning • Basic Settings for Material Costing • Define Cost Component Structure**.

On the initial screen shown in Figure 14.42, on the right side you see the defined cost component structures (in this example, only **01**). On the left side of the screen, you can configure various functions related to the cost component structure.

Normally you'd create your own cost component structure in the Z/Y custom name range. Let's examine the settings by selecting structure **01** and clicking **Cost Components with Attributes** from the left side of the screen.

14.2 Product Cost Planning

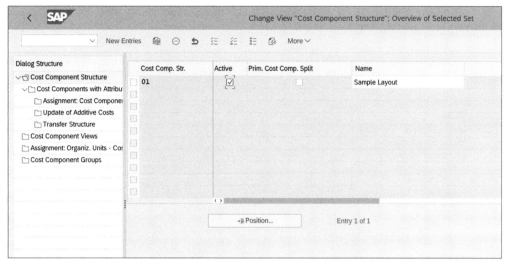

Figure 14.42 Define Cost Component Structure

On the screen shown in Figure 14.43, you see the defined cost components for cost component structure **01**. Nine components are defined, which represent the main types of costs related to manufacturing and selling a product. There is a limit to the number of cost components that can be defined. It used to be 40 in older SAP releases, but in SAP S/4HANA it's been significantly increased to 120.

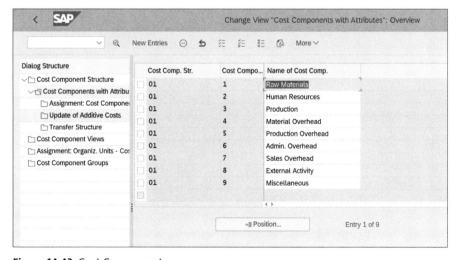

Figure 14.43 Cost Components

Double-clicking any element brings you to the detailed configuration screen for it, as shown in Figure 14.44.

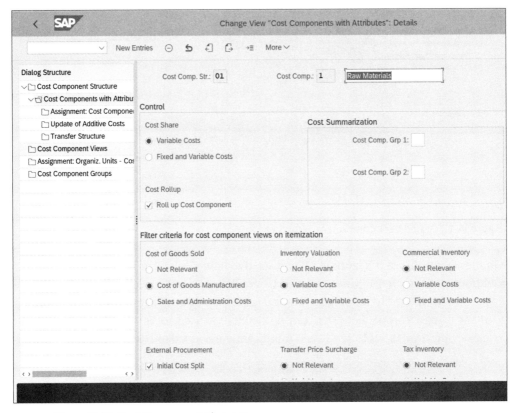

Figure 14.44 Cost Component Element

Here, you configure which types of costs should be included, the relevance for inventory valuation, transfer price surcharge, tax inventory, and other settings. If you check the **Roll up Cost Component** indicator, the costing results of a cost component will be rolled up into the next-highest costing level.

Click **Assignment: Cost Component—Cost Element Interval** from the left side of the screen. On the screen shown in Figure 14.45, you assign the cost elements to the cost component. These assignments are made on the chart of accounts level. It's also possible to use the **Origin group** as a restriction criterion so that the same cost elements are assigned to different cost components based on a different origin group. This enables you to post to the same cost elements but to allocate certain costs in a different part of the cost component structure.

14.2 Product Cost Planning

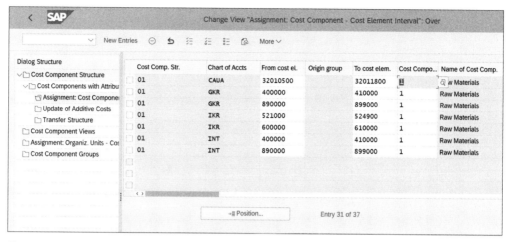

Figure 14.45 Cost Component Cost Element Assignment

Click **Update of Additive Costs** from the left side of the screen. Figure 14.46 shows the configuration for additive costs.

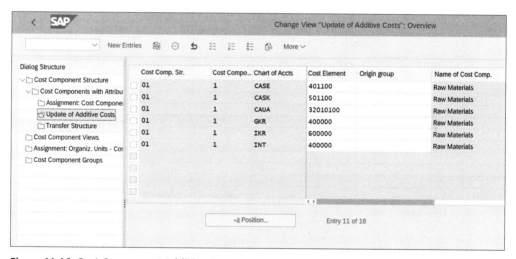

Figure 14.46 Cost Component Additive Costs

Here, you assign the relevant cost elements to the cost components in case your calculation uses additive costs. The assignment is on the chart of accounts level.

Click **Cost Component Views** from the left side of the screen. You can build cost component views here, which can include one or more of the cost components (see Figure 14.47).

601

14   Product Costing

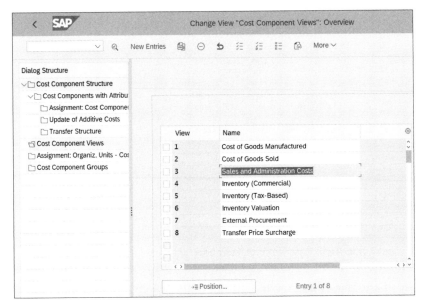

Figure 14.47 Cost Component Views

Double-click a view to see/maintain its components, as shown in Figure 14.48.

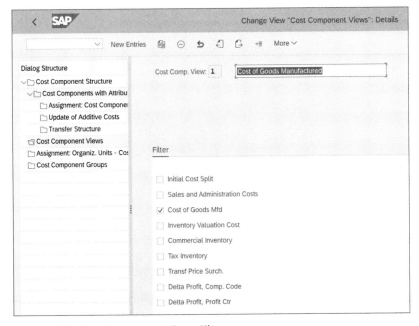

Figure 14.48 Cost Component Views Filter

602

14.2 Product Cost Planning

Click **Assignment: Organiz. Units—Cost Component Struct** from the left side of the screen. On the screen shown in Figure 14.49, you assign a cost component structure to a company code and specify its validity dates. You also can refine the assignments per plant costing variant.

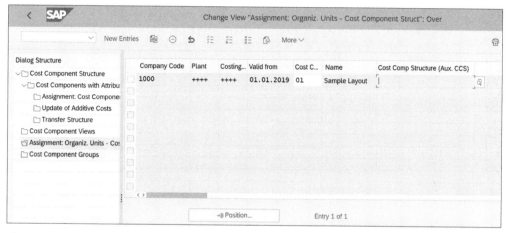

Figure 14.49 Assignment of Organizational Units

With that, we complete the settings for the cost component structure. Save your entries with the **Save** button.

14.2.4 Costing Sheet

For the allocation of overhead costs in the product costing, you need to define a costing sheet. The costing sheet consists of three elements: calculation bases, overhead rates, and credits. Using these elements, you define how the overhead costs should be allocated to products. In this section, we'll first look at each of the elements before describing how to define the costing sheet itself.

Calculation Bases

The calculation bases define the cost elements to which overhead costs are posted. They can be defined as individual cost elements, a cost element interval, or a cost element group.

To define calculation bases, follow menu path **Controlling • Product Cost Controlling • Product Cost Planning • Basic Settings for Material Costing • Overhead • Costing Sheet: Components • Define Calculation Bases**.

14 Product Costing

On the first screen, shown in Figure 14.50, you see the defined standard bases provided by SAP. The bases correspond to different types of overhead costs.

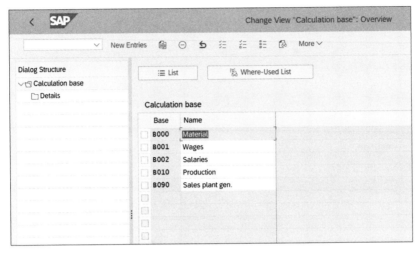

Figure 14.50 Calculation Bases

Let's modify the settings for base **B001**, **Wages**. Select it and click **Details** from the left side of the screen. After entering the **Controlling Area**, you'll see the screen in Figure 14.51.

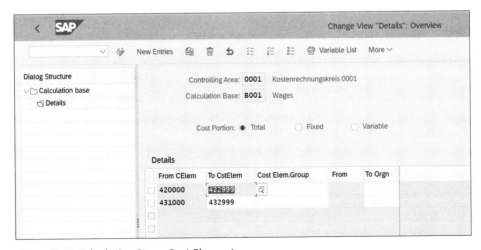

Figure 14.51 Calculation Bases Cost Elements

14.2 Product Cost Planning

Here, you enter ranges of cost elements on which the overhead costs for wages are posted. Thus they'll be included in this base in the costing sheet calculation.

Using this logic, define your overhead bases and assign the respective cost elements.

Overhead Rates

Overhead rates define the conditions based on which overhead is applied. There are two types of overhead rates: percentage rates, which are defined as percentages to be applied, and quantity-based rates, which depend on the underlying quantity.

To define percentage overhead rates, follow menu path **Controlling • Product Cost Controlling • Product Cost Planning • Basic Settings for Material Costing • Overhead • Costing Sheet: Components • Define Percentage Overhead Rates**.

On the screen shown in Figure 14.52, you see the defined overhead rates along with their dependencies. The dependency defines which characteristics and organizational elements are available for the definition of the surcharge.

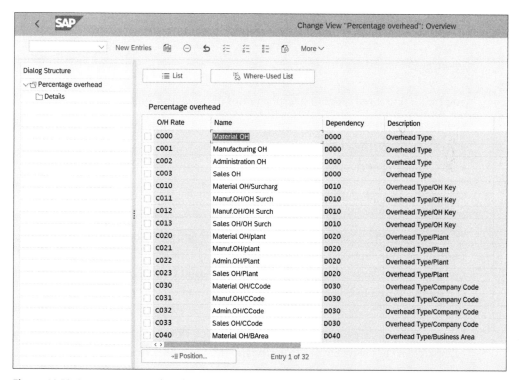

Figure 14.52 Percentage Overhead Rates

14 Product Costing

These overhead rates are standard-delivered by SAP, and you can define your own if needed. Select overhead rate **C000 Material OH** and click **Details** from the left side of the screen.

On the screen shown in Figure 14.53, you define the overhead percentage rates and their validity dates.

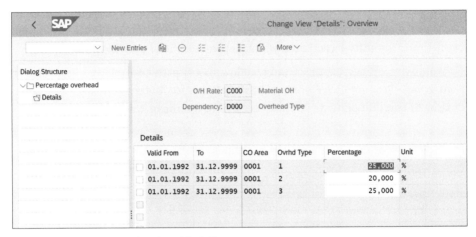

Figure 14.53 Percentage Overhead Details

You can define three types of overhead rates in the **Ovrhd Type** field:

- 1: Actual Overhead Rate
- 2: Planned Overhead Rate
- 3: Commitment Overhead Rate

The overhead type defines the types of costs that are being allocated. Often, new overhead rates are maintained each year, with a new validity period.

To define quantity-based overhead rates, follow menu path **Controlling • Product Cost Controlling • Product Cost Planning • Basic Settings for Material Costing • Overhead • Costing Sheet: Components • Define Quantity-Based Overhead Rates**. On the screen shown in Figure 14.54, you see the defined overhead quantity rates along with their dependencies.

Select overhead rate **C100 Material OH** and click **Details** from the left side of the screen. Then click **New Entries** from the top menu and maintain the rates.

14.2 Product Cost Planning

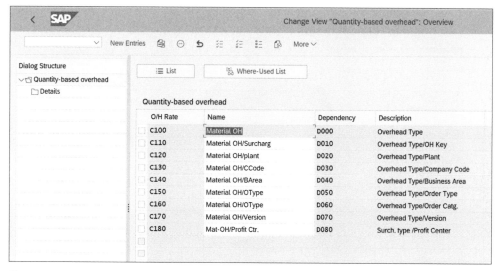

Figure 14.54 Quantity Overhead Rates

As you can see in Figure 14.55, here you enter amounts that depend on quantities. In our example, an overhead of 10 USD per 1000 pieces of the material is defined.

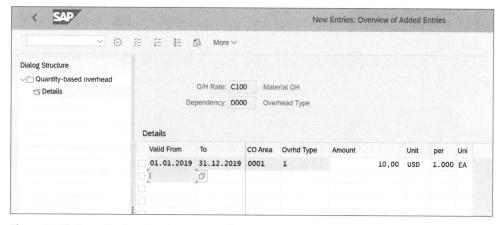

Figure 14.55 Quantity Overhead Rates Details

Credits

Credits define account assignment objects such as cost elements and cost centers, which are posted during the costing calculation.

14 Product Costing

To define credits, follow menu path **Controlling** • **Product Cost Controlling** • **Product Cost Planning** • **Basic Settings for Material Costing** • **Overhead** • **Costing Sheet: Components** • **Define Credits**. On the screen shown in Figure 14.56, you see the defined credits.

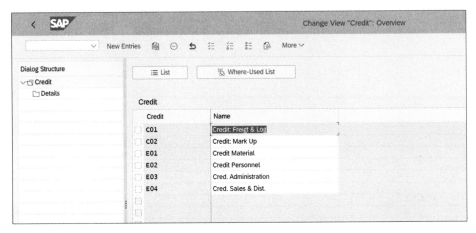

Figure 14.56 Define Credits

Select credit **C01**, **Credit: Freigt & Log** and click **Details** from the left side of the screen. Then click **New Entries** from the top menu, enter a controlling area, and maintain the assignments.

Figure 14.57 shows an example of how you can assign cost elements and cost objects to the credit key. The cost element should be a secondary cost element with category 41.

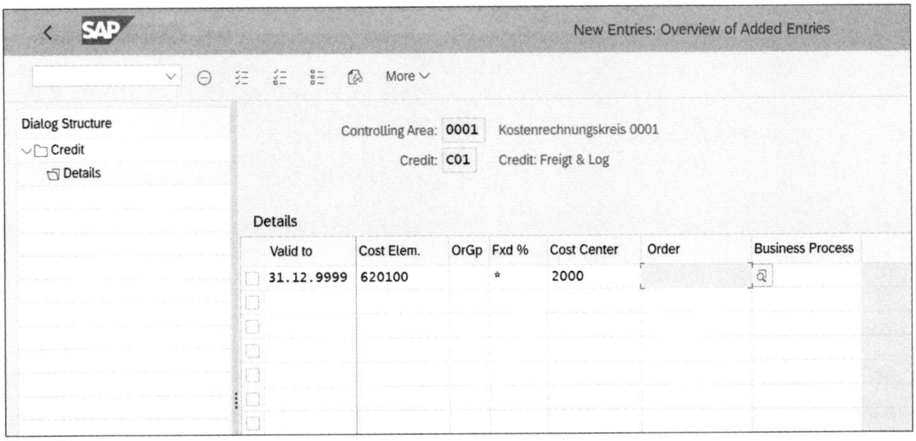

Figure 14.57 Credit Details

14.2 Product Cost Planning

Define Costing Sheet

Finally, after defining the bases, overhead rates, and credits, you need to define the costing sheet itself, which combines these elements. To do so, follow menu path **Controlling • Product Cost Controlling • Product Cost Planning • Basic Settings for Material Costing • Overhead • Define Costing Sheets**. On the screen shown in Figure 14.58, you'll see the costing sheets.

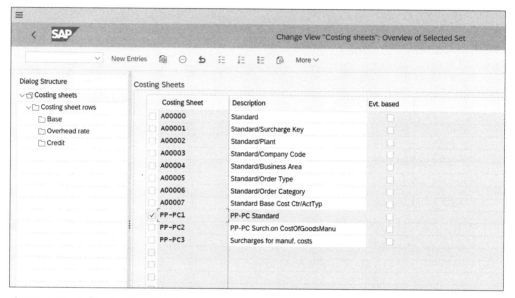

Figure 14.58 Define Costing Sheet

Select costing sheet **PP-PC1 PP-PC Standard** and click **Costing sheet rows** from the left side of the screen.

Figure 14.59 shows the details of the costing sheet. In the different rows, the bases you defined earlier are assigned. The overhead rates and credits refer to the rows with the bases. In the **Base**, **Overhead rate**, and **Credit** sections from the left side of the screen, you can see/modify the relevant details.

With that, we finish the configuration for the overhead calculation through the costing sheet.

609

14 Product Costing

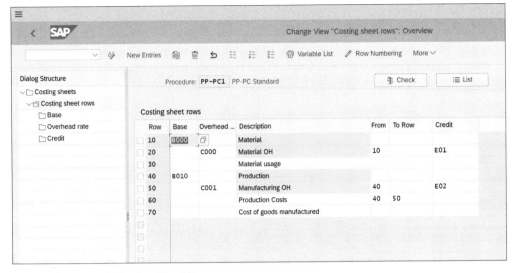

Figure 14.59 Costing Sheet Rows

14.2.5 Material Cost Estimate

The product cost calculation created a material cost estimate, which calculates the cost per material. You can create the material cost estimate individually per material or for multiple materials in a costing run.

To create a material cost estimate for one material, follow application menu path **Accounting • Controlling • Product Cost Controlling • Product Cost Planning • Material Costing • Cost Estimate with Quantity Structure • CK11N—Create**.

As shown in Figure 14.60, you need to enter a material and plant to be costed. You also need to select a costing variant, which we configured in Section 14.2.2. Optionally, you can specify a costing lot size, which serves as a basis to determine the costing quantity. Also optionally, you can enter a transfer control, which we defined in Section 14.1.1 and controls how the system should search for existing cost estimates.

In the **Dates** tab, you specify a costing date, quantity structure, and valuation date. On the **Qty Struct.** tab, you can enter BOM and routing data related to the cost estimate.

You also can create costing estimates for multiple materials through a costing run. Usually in the beginning of the year, companies run a costing run to cost all relevant materials and determine their standard prices.

610

14.2 Product Cost Planning

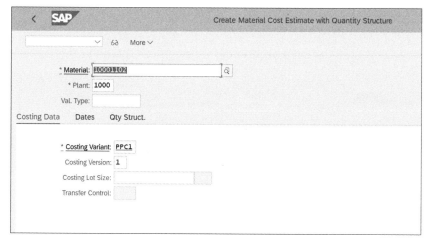

Figure 14.60 Material Cost Estimate

To create a costing run, follow application menu path **Accounting • Controlling • Product Cost Controlling • Product Cost Planning • Material Costing • Costing Run • CK40N—Edit Costing Run**. Click the ☐ button to create a new costing run. On the screen shown in Figure 14.61, enter a name, date, and parameters for the costing run.

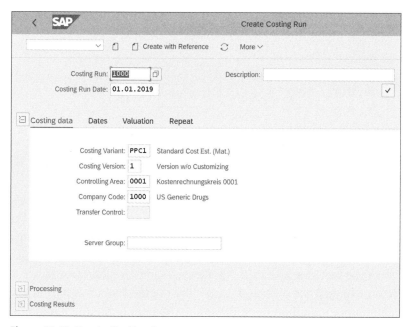

Figure 14.61 Create Costing Run

Then in the **Costing data** tab, select the costing variant you created earlier, along with costing version, controlling area, company code, and transfer control if necessary.

After saving the run, in the lower section of the screen you can enter the material selections.

On the screen shown in Figure 14.62, click the ![icon] button in the **Selection** row.

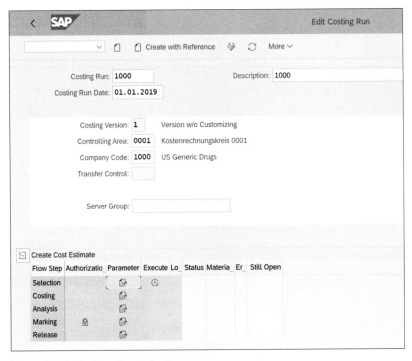

Figure 14.62 Costing Run Selections

This opens a selection screen in which you can select materials and make further restrictions, as shown in Figure 14.63.

The next row, **Costing**, creates the material cost estimates for the selected materials. Then in the **Analysis** row, you can review them. The next row, **Marking**, updates the calculated cost estimate in the material master as a future standard cost. The last row, **Release**, makes the future standard cost a current cost estimate, and the current cost estimate becomes the previous cost estimate. Then existing inventory is revalued at the new standard cost, and the difference caused by the revaluation is posted as a finance document.

14.3 Actual Costing and Material Ledger

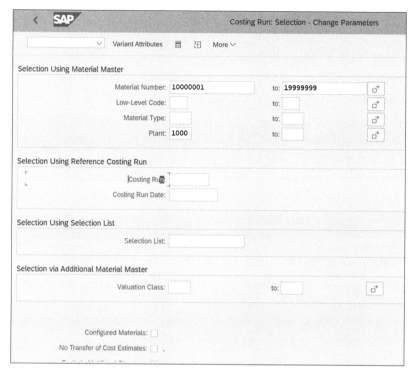

Figure 14.63 Material Selections

With that, we finish our guide to product cost planning and move to actual costing and the material ledger.

14.3 Actual Costing and Material Ledger

Actual costing calculates the actual prices of inventory based on the goods movements that occurred. The actual price calculated is called the periodic unit price and can be used to revaluate inventory.

The material ledger stores on the line item level the changes in stock and prices with each material movement in multiple currencies. Material movements related to goods receipt, invoice receipt, and so on are recorded in the material ledger with the price and exchange rate differences.

In the next section, we'll provide an overview of the material ledger and how to activate it. Then we'll guide you through how to configure the material ledger to be used

14 Product Costing

for multiple valuations and currencies. In the next section, we'll configure the material ledger update. In the last section, we'll configure the actual costing and provide guidance how to use the actual costing cockpit to perform actual costing.

14.3.1 Overview and Material Ledger Activation

With SAP S/4HANA, the use of the material ledger becomes mandatory, but actual costing does not. Actual costing is particularly important for companies that operate in countries with high-inflation environments and unstable price levels. In some countries, it's even required to use it to value inventory, such as in Brazil and Russia.

With SAP S/4HANA, actual costing is one of the functions that the material ledger can enable, but it also provides material valuation in parallel currencies and valuation principles.

With SAP S/4HANA, the tables of the material ledger are integrated into table ACDOCA.

For a brownfield implementation of SAP S/4HANA, if you're not using the material ledger, it's mandatory to activate it. This is because the inventory valuation tables (EBEW, EBEWH, MBEW, MBEWH, OBEW, and so on) no longer update transactional data, which is retrieved from the table ACDOCA line items table in real time.

To activate the material ledger, follow menu path **Controlling • Product Cost Controlling • Actual Costing/Material Ledger • Activate Material Ledger for Valuation Areas**, then select **Activate Material Ledger**.

Figure 14.64 shows the plants for which the material ledger should be active.

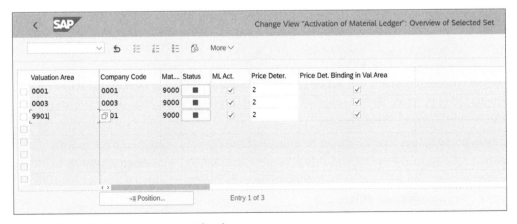

Figure 14.64 Activate Material Ledger

14.3 Actual Costing and Material Ledger

You need to tick the **ML Act.** checkbox. In the **Price Deter.** field, select **2** (transaction-based). This means that both materials with the price control V (moving average price) and materials with the price control S (standard price) will be evaluated.

14.3.2 Multiple Currencies and Valuations

One of the main purposes of actual costing is to valuate materials in multiple currencies and valuations. The material ledger enables this process because it stores each inventory transaction on the line item level in multiple currencies. Then, using the actual costing cockpit, inventory can be revaluated based on these multiple valuations.

We start the configuration for multiple currencies and valuations by defining material ledger types, which control in which currencies the materials will be valuated. Follow menu path **Controlling** • **Product Cost Controlling** • **Actual Costing/Material Ledger** • **Assign Currency Types and Define Material Ledger Types**.

On the screen shown in Figure 14.65, you'll see the defined material ledger types.

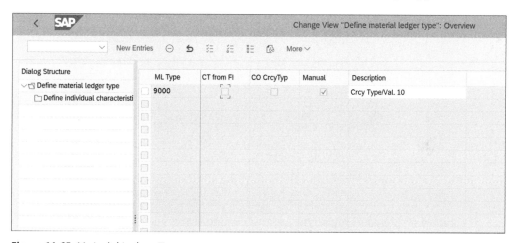

Figure 14.65 Material Ledger Types

Select **9000**, a standard type in the company code currency. Click on **Define individual characteristics** on the left side of the screen to review its settings. As you see in Figure 14.66, for this type the company code currency is selected. You can also define other types for group currency, hard currency, and so on.

14 Product Costing

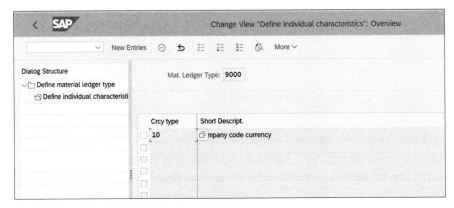

Figure 14.66 Material Ledger Type Details

In the next step, we need to assign the material ledger type to a valuation area. Follow menu path **Controlling • Product Cost Controlling • Actual Costing/Material Ledger • Assign Material Ledger Types to Valuation Area**.

Figure 14.67 shows the assignment of material ledger types for valuation areas.

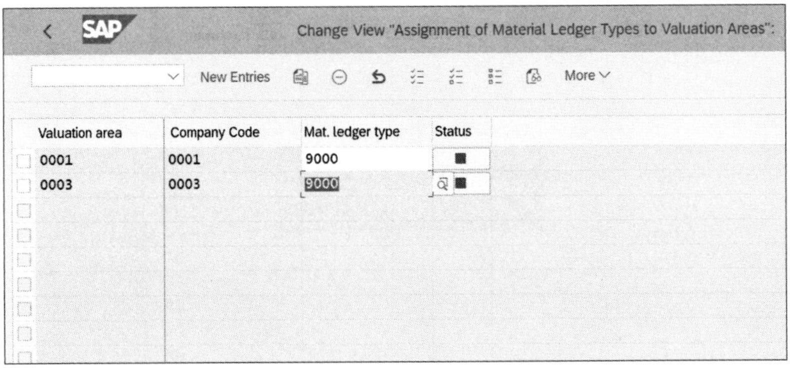

Figure 14.67 Material Ledger Type Assign Valuation Area

Click **New Entries** from the top menu and assign your valuation area to the defined material ledger type by entering a valuation area and a material ledger type.

14.3.3 Material Ledger Update

You also need to configure how the material ledger is updated from the various inventory movements.

14.3 Actual Costing and Material Ledger

First, you need to define movement type groups for the material ledger. To create a movement type group, follow menu path **Controlling • Product Cost Controlling • Actual Costing/Material Ledger • Material Update • Define Movement Type Groups of Material Ledger**.

Initially the configuration screen is blank, as shown in Figure 14.68.

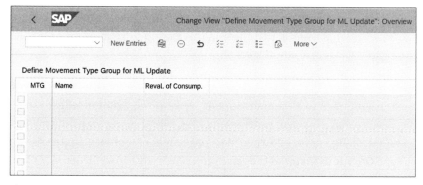

Figure 14.68 Movement Type Group for Material Ledger Update

Let's create a movement type group for revaluation of consumption on the general ledger account level. Select **New Entries** from the top menu and create a new movement type group, as shown in Figure 14.69.

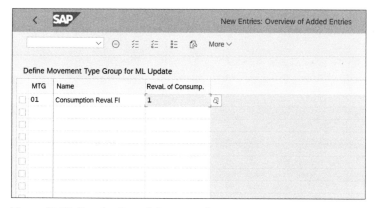

Figure 14.69 Revaluation of Consumption Movement Type Group

The following fields should be configured:

- **MTG**
  Code for the movement type group.

14 Product Costing

- **Name**
  Description of the movement type group.

- **Reval. of Consump.**
  Indicator that determines how to revalue consumption. With option **1** or **2**, the account assignment objects (general ledger account and/or controlling account assignment) are stored when consumption occurs for movement types that are assigned to the movement type group. This enables the account assignment objects to be revaluated using actual costs. In this example, we select **1** for revaluation on the general ledger account level only.

Save your entry with the **Save** button.

In the next step, you need to assign movement types to the material ledger movement type group created. To assign movement types, follow menu path **Controlling • Product Cost Controlling • Actual Costing/Material Ledger • Material Update • Assign Movement Type Groups of Material Ledger**. In the screen shown in Figure 14.70, you see how management type groups are assigned to movement types.

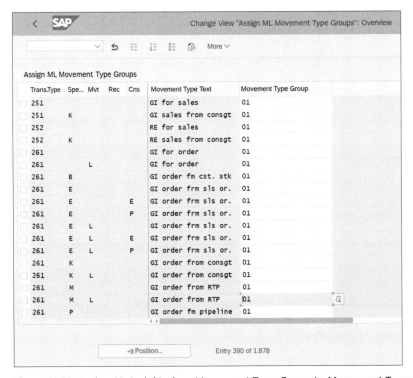

Figure 14.70 Assign Material Ledger Movement Type Group to Movement Types

618

14.3 Actual Costing and Material Ledger

Assign the movement type group **01** created for revaluation of consumption on the financial accounting level to movement types **251** and **261**, then save with the **Save** button. You also need to define a material update structure, which defines how the values from valuation-relevant transactions are stored in the material ledger.

To define a material update structure, follow menu path **Controlling • Product Cost Controlling • Actual Costing/Material Ledger • Material Update • Define Material Update Structure**.

Figure 14.71 shows the standard material update structure **0001**.

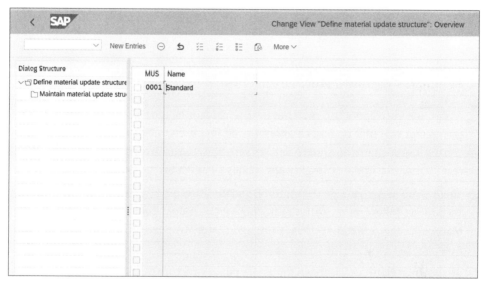

Figure 14.71 Material Update Structure

In most cases, the standard structure **0001** should suit your needs. With the standard logic, the material stocks in the closed posting period will be valuated with the weighted average price.

In our example, we'll use standard structure **0001**. Select it by ticking the checkbox to its left, then click **Maintain material update structure** on the left side of the screen.

Figure 14.72 shows the assignment of material update categories to process categories.

Select **B+** to indicate procurement and **V+** to indicate consumption process categories. Procurement process categories lead to inventory receipts, such as purchase orders or production. Consumption process categories lead to goods withdrawal, such as withdrawals from production orders or cost centers.

14　Product Costing

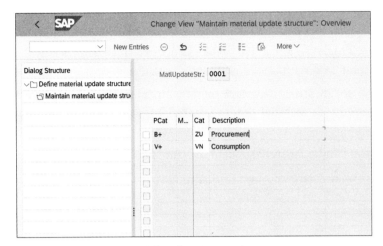

Figure 14.72 Assign Material Update Categories

Assign the corresponding material update categories to the process categories (i.e., the receipt material update category to the procurement process category and the consumption material update category to the consumption process category). A receipt material update category always has an effect on the valuation price.

Finally, you need to assign the material update structure to your valuation categories. To do that, follow menu path **Controlling • Product Cost Controlling • Actual Costing/ Material Ledger • Material Update • Assign Material Update Structure to a Valuation Area**.

As shown in Figure 14.73, assign material update structure **0001** to the valuation areas.

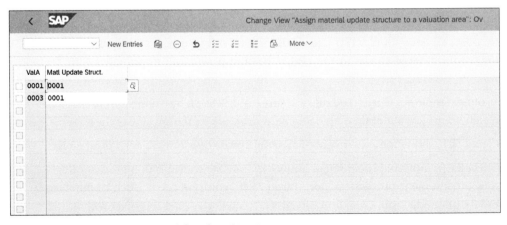

Figure 14.73 Assign Material Update Structure

14.3.4 Actual Costing

To use actual costing, you need to activate it on the plant level. Follow menu path **Controlling • Product Cost Controlling • Actual Costing/Material Ledger • Actual Costing • Activate Actual Costing**, then select **Activate Actual Costing**. Figure 14.74 shows the activation of actual costing per plant.

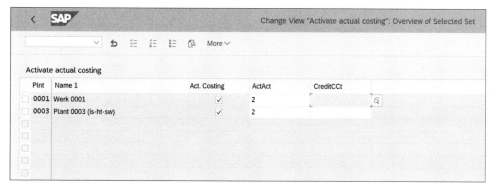

Figure 14.74 Activate Actual Costing

Tick the **Act. Costing** checkbox for the plants for which you want to activate actual costing. In the **ActAct** field, you define settings for updating the consumption of activities in the actual quantity structure. Three settings are possible:

- 0

  Update is not active.

- 1

  Update is active but isn't relevant to price determination. In this case, consumption is updated in the quantity structure but not included in the price determination.

- 2

  Update is active and included in price determination.

The next step is to activate the actual cost component split. The actual cost component split is used to analyze actual costs over multiple production lines. Profitability analysis uses the actual cost component split to revaluate the manufacturing costs at the end of the period.

To activate the actual cost component split, follow menu path **Controlling • Product Cost Controlling • Actual Costing/Material Ledger • Actual Costing • Activate Actual Cost Component Split**.

14 Product Costing

In the screen shown in Figure 14.75, you can activate the actual costing per valuation area and company code.

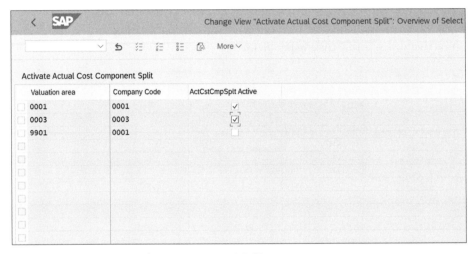

Figure 14.75 Activate Actual Cost Component Split

Tick the checkbox in the **ActCstCmpSplt Active** column, and save your entries with the **Save** button.

14.3.5 Actual Costing Cockpit

The actual costing program is known as the actual costing cockpit. The actual costing cockpit enables you to perform balance sheet valuation by revaluing your inventory based on the periodic unit price. It also allows you to revalue your cost of goods sold, thus providing more accurate valuation for your profit margin calculation.

In SAP S/4HANA, the actual costing cockpit was redesigned and improved with a simplified data structure and calculation logic. In the new program, there's one settlement step that replaces the single-level price determination, multilevel price determination, revaluation of consumption, and WIP revaluation steps. Therefore the program is simplified and benefits from the new SAP HANA database.

To run actual costing, follow application menu path **Accounting • Controlling • Product Cost Controlling • Actual Costing/Material Ledger • Actual Costing • CKMLCP—Edit Costing Run**.

Figure 14.76 shows the definition screen of the actual costing run.

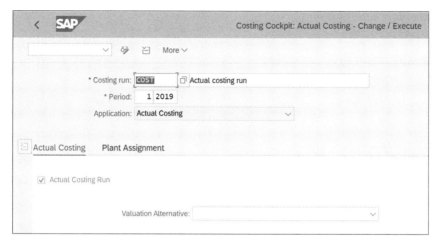

Figure 14.76 Actual Costing Run

To create a new actual costing run, click the ☐ button.

Similar to the cost estimate run, you need to enter an actual costing run name and description. You also need to enter a period. Normally, an actual costing run is run every month as part of the month-end closing procedures. Then in the **Plant Assignment** tab, you select the plants for which you want to run actual costing.

After you run actual costing, it calculates the periodic unit price for the materials. It then generates financial accounting and material ledger documents. It's optional to use the calculated periodic unit price as a valuation price for the inventory, which is an option used mostly in countries with unstable prices.

With that, we finish our guide to actual costing and the material ledger.

14.4 Information System

The information system for product costing provides numerous reports to help you analyze the material cost estimates, the actual costs incurred, and the variances between them. This data helps management analyze the production costs and how they compare to the plan.

Let's examine the information system for product cost planning and for actual costing and the material ledger.

14.4.1 Product Cost Planning

Product cost planning reports are available at application menu path **Accounting • Controlling • Product Cost Controlling • Product Cost Planning • Information System**. They're grouped into the following types of reports:

- Summarized analysis
- Object lists
- Detailed reports
- Object comparisons

For example, open the useful report S_P99_41000111 (Analyze/Compare Material Cost Estimates), available at application menu path **Accounting • Controlling • Product Cost Controlling • Product Cost Planning • Information System • Object List • For Material**.

On the selection screen shown in Figure 14.77, enter the plant and material numbers, and you also can restrict by costing variant, version, and costing date. You also need to select the cost component view to display.

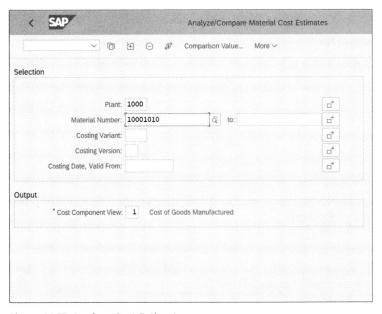

Figure 14.77 Analyze Cost Estimates

Figure 14.78 shows the result of the cost estimate. The costing result is provided per material and costing lot size.

14.4 Information System

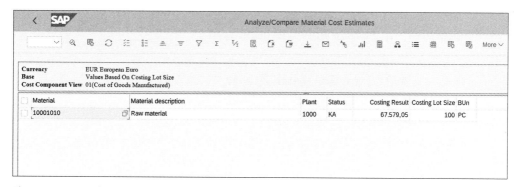

Figure 14.78 Analyze Cost Estimates Output

Double-clicking the material lets you display the cost estimate. Figure 14.79 shows the material cost estimate.

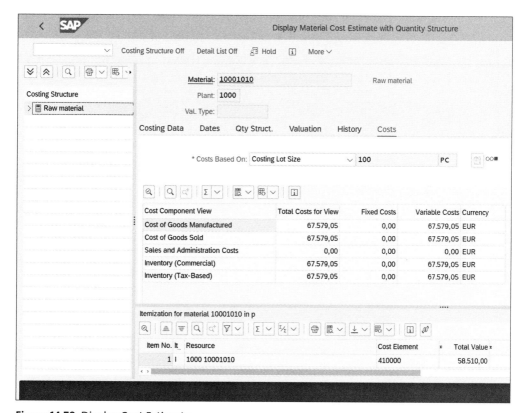

Figure 14.79 Display Cost Estimate

Using this report, you can get an overview of cost estimates per material and drill down to each material to analyze its details.

14.4.2 Actual Costing and Material Ledger

Actual costing/material ledger reports are available at application menu path **Accounting • Controlling • Product Cost Controlling • Actual Costing/Material Ledger • Information System**. They're grouped into the following types of reports:

- Object lists
- Detailed reports
- Document reports

For example, access report S_P99_41000062 (Prices and Inventory Values), available at application menu path **Accounting • Controlling • Product Cost Controlling • Actual Costing/Material Ledger • Information System • Object List**.

Figure 14.80 shows the selection screen of the report. Enter the plant, period, and fiscal year, and select a currency valuation. You can also restrict by material number; if not entered, all relevant materials per plant will be included. Execute with the ⏲ button.

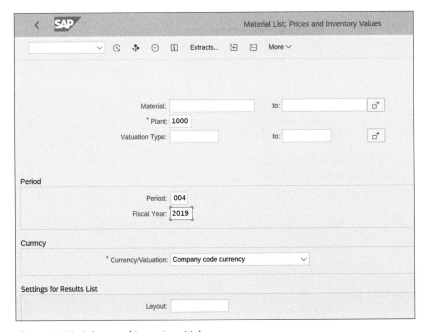

Figure 14.80 Prices and Inventory Values

14.4 Information System

Figure 14.81 shows the resulting output list. Per material, it provides the total stock, total inventory value, standard price, and per unit price. Double-clicking a material provides further information and all goods movements.

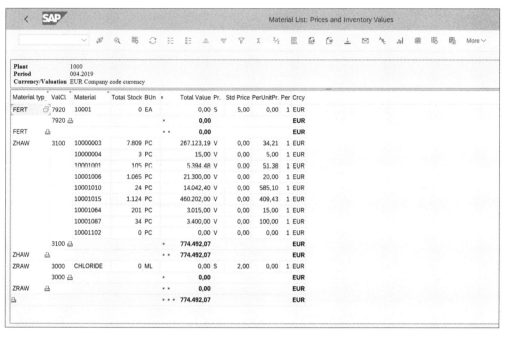

Figure 14.81 Prices and Inventory Values Output

14.4.3 Drilldown Reporting

As with other functional areas, such as profitability analysis and general ledger, drilldown reports also are available for product costing. There are many standard reports that enable you to analyze and compare planned and actual costs.

To access the drilldown reports, enter Transaction KKO0 (KK, followed by letter O and number 0). As shown in Figure 14.82, all the reports are conveniently organized in a tree-like structure, and you can run any of them with the (**Execute**) button from the top menu.

You need to run a summarization of the data first, though. To do this, enter Transaction KKRV. Figure 14.83 shows the selection screen of the program.

14 Product Costing

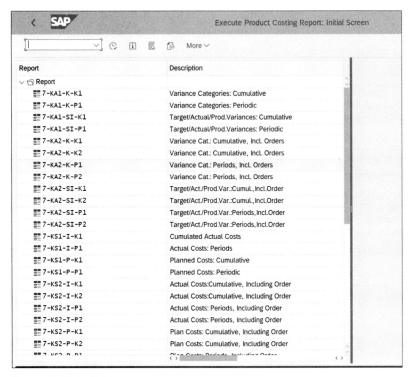

Figure 14.82 Drilldown Reports

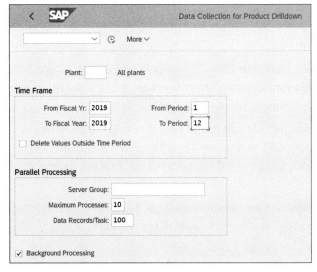

Figure 14.83 Data Collection for Drilldown Reporting

You can enter a specific plant in the **Plant** field, or leave it empty so the data collection can run for all plants. Specify from and to fiscal years and a period, leave the **Background Processing** checkbox ticked for better performance, and run the program with the ⓖ (**Execute**) button.

After that, the drilldown reports can be run based on the summarized data.

14.5 Summary

In this chapter, you learned how to configure product costing. You learned how to create a cost estimate and how to configure actual costing and the material ledger. You're now aware of the guidelines and best practices for selecting appropriate standard or moving average prices for material valuation. You learned that the material ledger is now required in SAP S/4HANA and is fully integrated with the Universal Journal, table ACDOCA, but actual costing is still optionally available. You learned how to configure both the material ledger and actual costing. Finally, you received an overview of the information system for product costing, which provides powerful reports for product cost planning and for actual costing and the material ledger.

In the next chapter, you'll learn about the new solution for group reporting that first became available with the SAP S/4HANA 1809 release: SAP S/4HANA Finance for group reporting.

Chapter 15
Group Reporting

This chapter will teach you how to configure SAP S/4HANA Finance for group reporting. Because this solution is newly introduced with the latest SAP S/4HANA 1809 release, the chapter provides an overview of the consolidation topic and previous SAP solutions before presenting step-by-step instructions for configuration.

Most companies that run SAP are global companies with multiple legal entities that operate in multiple regions and countries. As such, most SAP clients need to provide group reporting to create consolidated financial statements for the whole group that includes all legal entities and countries. But consolidation is a complicated process because profit between the group legal entities (known as intercompany profit) needs to be eliminated, as well as intercompany transactions. Therefore, there's always been a separate group reporting solution to provide group reporting. SAP has a long history of providing different modules to cover the consolidation group reporting requirement, which deserves a detailed explanation—which we'll provide in this chapter.

And now with the latest release (as of the time of writing of this book), SAP S/4HANA 1809, which came out in September 2019, SAP has provided a wholly new, revamped, and significantly enhanced solution based on the impressive capabilities of the SAP HANA database to meet the consolidation requirements: SAP S/4HANA Finance for group reporting.

In this chapter, we'll provide an overview of the previous consolidation options and how they compare with SAP S/4HANA Finance for group reporting. Then we'll go over the main configuration activities for the new solution.

15.1 Group Reporting Basics

We'll start this section by first outlining the need for group reporting and describing what the objectives of group reporting are from a business point of view. We'll then

15 Group Reporting

briefly cover the history of group reporting in SAP before discussing some of the key benefits of the new SAP S/4HANA Finance for group reporting solution.

15.1.1 What Is Group Reporting?

Group reporting is accounting reporting that aims to provide consolidated financial statements. *Consolidated financial statements* are financial statements in which the information related to a parent company and its subsidiaries is presented as if they're a single entity. The accounting framework that defines the consolidated financial statements is defined in International Accounting Standard 27: Consolidated and Separate Financial Statements, and International Financial Reporting Standard 10: Consolidated Financial Statements.

The main challenge in creating group reporting is that multiple eliminations of transactions and results should be performed between the various entities, such as the following:

- Intercompany debt
- Intercompany revenue and expenses
- Intercompany stock ownership

The consolidation process requires a lot of work and should be automated as much as possible. Therefore companies have long been using consolidation software to help with these tasks.

15.1.2 Historical Group Reporting in SAP

Because SAP provides software to record and manage all accounting data, it's only natural for companies that run SAP also to use SAP to create their group reporting. The SAP offerings in that area have a long history, with multiple modules provided with varying degrees of success.

Twenty years ago, the SAP solution for group reporting in SAP R/3 was called Legal Consolidation (LC). This was module closely integrated with the other financial modules of SAP, but this solution couldn't provide management consolidation. *Management consolidation* is used to analyze different scenarios for management and to perform parallel consolidations with different levels of data using different accounting principles. Therefore, this module wasn't very popular, and in fact many companies chose to use separate consolidation software from other vendors, such as Hyperion, and interface it with their SAP data for consolidation purposes.

The next solution from SAP was the Enterprise Controlling Consolidation (EC-CS) module in SAP ERP. This was a good solution that could provide group reporting from both legal and management points of view. It was closely integrated both with FI and CO.

With the introduction of the SAP Business Warehouse (SAP BW) system that could combine data from both SAP and non-SAP systems, SAP made the next step forward with its SAP Strategic Management Business Consolidation (SEMC-BCS). It was released in 2002, and it was a very useful solution, offering a high degree of automation in the consolidation process. SEMC-BCS operates fully on top of SAP BW and thus is an online analytical processing (OLAP) system, as opposed to earlier solutions, which were online transactional processing (OLTP) systems.

The next step in the evolution and optimization of the consolidation process was the SAP Business Planning and Consolidation (SAP BPC) solution. It's similar to SEMC-BCS, but it also offers sophisticated planning functions on top of its consolidation capabilities.

And now we come to SAP S/4HANA. Because the power of the new SAP HANA database is colossal, it's only natural that SAP would provide a new SAP HANA-based solution that can streamline the consolidation processes within SAP. This solution is called SAP S/4HANA Finance for group reporting and it's fresh off the shelf—first introduced with SAP S/4HANA 1809.

15.1.3 Key Benefits

SAP S/4HANA Finance for group reporting is the latest offering in the group reporting space and is indeed vastly superior to its predecessors. It's entirely based on the SAP HANA database, which is particularly well optimized for consolidation processes and group reporting. The user interface is mostly based on the new SAP Fiori applications, which provide users with an efficient and beautiful interface to perform group reporting.

SAP provides best practices for consolidation, which can be copied and modified as appropriate. This is a great benefit, since they can serve as a foundation to be built upon with best practices from the years of consolidation experience SAP brings into the new solution.

The new solution provides both legal and management consolidation and in terms of functionalities is similar to the well-known EC-CS solution, but it's SAP HANA and SAP Fiori based and benefits from the fast and optimized new data structure.

15 Group Reporting

Consolidation postings are stored in the new table ACDOCU (Universal Consolidation Journal Entries). Table ACDOCU is similar to table ACDOCA, but it's used to store consolidation postings (see Figure 15.1).

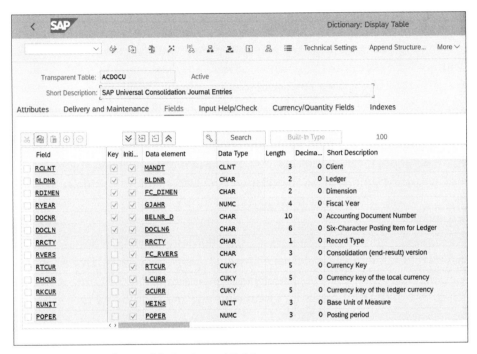

Figure 15.1 Universal Consolidation Journal Entries

Let's examine in detail how to configure SAP S/4HANA Finance for group reporting.

15.2 Global Settings

We'll now delve into the configuration of the global settings for the new group reporting solution. We'll start with the prerequisites to install SAP Best Practices configuration content. Then we'll configure configuration ledgers, which store the documents generated by the group reporting solution. After that, we'll configure the consolidation versions, which define dedicated data areas in the database. Finally, we'll configure dimensions, which define the basis for consolidation, such as companies and profit centers.

15.2 Global Settings

15.2.1 Prerequisites

A mandatory prerequisite before configuring SAP S/4HANA Finance for group reporting is to install SAP Best Practices configuration content, as shown in Figure 15.2.

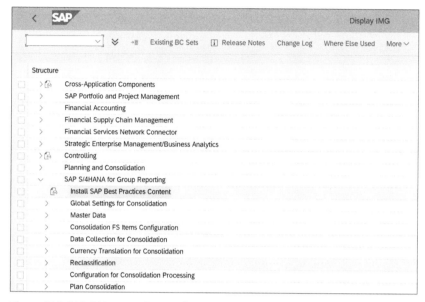

Figure 15.2 SAP S/4HANA Finance for Group Reporting Configuration

This task normally would be performed by your Basis team through Transaction /SMB/BBI, which is called the Building Block Builder. It's delivered with a separate add-on called BP-INSTASS. The scope item in the Building Block Builder for SAP S/4HANA Finance for group reporting is XX_1SG_OP.

15.2.2 Consolidation Ledger

The consolidation ledger is a ledger that stores the documents generated by the group reporting module.

To configure the consolidation ledger, follow menu path **SAP S/4HANA for Group Reporting • Master Data • Define Consolidation Ledgers**, then select **Create Ledger**.

On the screen shown in Figure 15.3, enter a ledger name, starting with Z or Y. You can also copy from an existing consolidation ledger using the **Reference ledger** field.

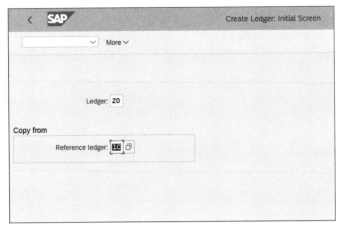

Figure 15.3 Create Ledger Initial Screen

Proceed by pressing [Enter] and you'll see the configuration screen of the ledger, as shown in Figure 15.4.

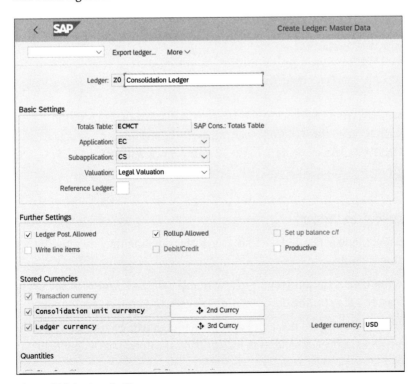

Figure 15.4 Ledger Settings

In Figure 15.4, you see and can modify the following details of the ledger:

- **Valuation**
  Here, you select the type of consolidation valuation to perform in this ledger: legal, group, or profit center. You need separate ledgers for the different types of valuation.

- **Ledger Post. Allowed**
  Defines that postings can be made to the ledger.

- **Rollup Allowed**
  Defines that rollup can be made in the ledger. Users of the special purpose ledger will be familiar with this concept. To *roll up* means to summarize data from other ledgers into the target ledger.

- **Write line items**
  Defines whether line items should to be written during the update.

- **Productive**
  When you set this indicator, the transaction data cannot be deleted.

- **Ledger currency**
  The currency of the ledger, which should be the currency in which you need to perform the group reporting.

After configuring the ledger, save it with the **Save** button.

15.2.3 Consolidation Version

The *consolidation version* identifies a dedicated data area in the database. Versions are used to consolidate various sets of financial data. It's possible to create separate versions for actual and plan data.

To configure a consolidation version, follow menu path **SAP S/4HANA for Group Reporting • Master Data • Define Versions**. On the initial screen shown in Figure 15.5, you'll see the defined versions.

Select the standard actual version 100 and click (**Copy As...**) from the top menu to create your own actual version.

15 Group Reporting

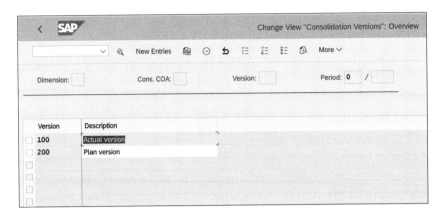

Figure 15.5 Defined Versions

On the next screen, shown in Figure 15.6, enter a consolidation version name starting with Z or Y and a description.

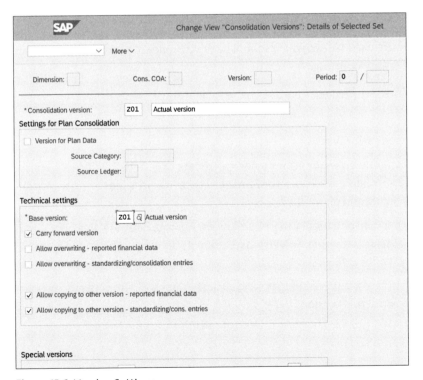

Figure 15.6 Version Settings

You can define that the version is used for plan data if you tick the **Version for Plan Data** checkbox. Also assign a base version in the **Base version** field, which can be the same or a different version. If you have a different base version, you can include the totals records and journal entries of both the current consolidation version and the base version in reports.

If you tick the **Carry forward version** checkbox, the balances will be carried forward into the new fiscal year.

You can also allow data from the consolidation version to be copied to other versions and vice versa with the **Allow overwriting** and **Allow copying** checkboxes.

Scroll down to define the settings for special versions. *Special versions* enable you to use different methods or rules for specific consolidation goals. As shown in Figure 15.7, define your own special version. After that, click the **Apply** button.

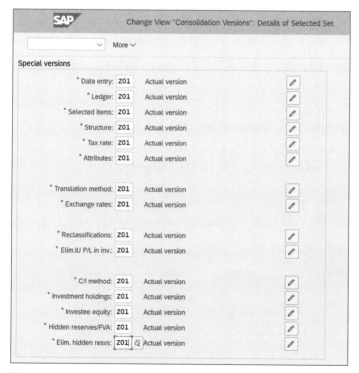

Figure 15.7 Special Versions

Then a pop-up window opens, in which you need to enter descriptions for the special versions, as shown in Figure 15.8. Then save with the **Save** button.

15 Group Reporting

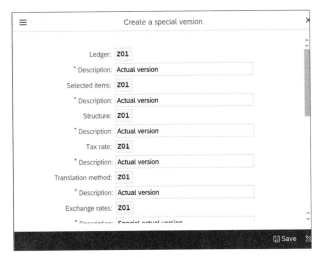

Figure 15.8 Create Special Versions

15.2.4 Dimensions

Dimensions define the basis for consolidation, such as companies, profit centers, and so on. To define the settings on the dimension level, follow menu path **SAP S/4HANA for Group Reporting • Master Data • Display Dimension**.

On the initial screen shown in Figure 15.9, you can choose the consolidation dimension: **Companies**, **Business areas**, **Profit center**, or **User-defined**.

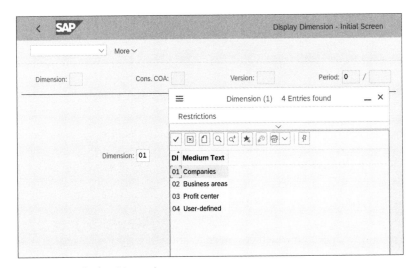

Figure 15.9 Display Dimension

15.3 Data Collection and Consolidation Configuration

On the next screen, shown in Figure 15.10, you need to enter the global parameters for consolidation.

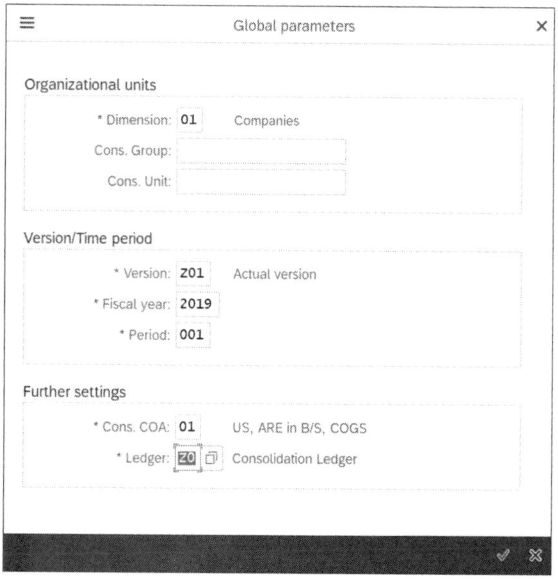

```
☰                    Global parameters                    ✕

Organizational units
              * Dimension: 01      Companies
              Cons. Group:
              Cons. Unit:

Version/Time period
               * Version: Z01     Actual version
            * Fiscal year: 2019
                * Period: 001

Further settings
              * Cons. COA: 01     US, ARE in B/S, COGS
                 * Ledger: Z0     Consolidation Ledger


                                                    ✓  ✕
```

Figure 15.10 Global Parameters

Enter the dimension, version, fiscal year and period, consolidation chart of accounts, and consolidation ledger for which you will make consolidation settings.

Proceed with the ✓ button. This sets the environment for which you will make the following settings.

15.3 Data Collection and Consolidation Configuration

Having configured the global configuration settings, we'll continue with the various data collection and consolidation settings.

First, let's configure financial statement items, which are linked to the operating accounts and serve to classify the items that need to be reported on in the group financial statements. Then we'll configure subitems, which further categorize the financial statement items in conjunction with the subitem categories. We'll also configure document types for posting consolidation transactions and the associated number ranges. Then we'll configure the data collection tasks, which are used to collect the data that will be consolidated.

15 Group Reporting

After that, we'll configure the consolidation methods, which define how to consolidate your data and perform interunit eliminations. Finally, we'll configure the task groups, which group together the various consolidation tasks to be performed.

15.3.1 Financial Statement Items

Financial statement items are the foundation of the group reporting settings. They're linked to the operating accounts and serve to classify the items that need to be reported on in the group financial statements. SAP delivers lots of standard financial statement items, but you also can create your own. Figure 15.11 shows some of the standard SAP-delivered financial statement items.

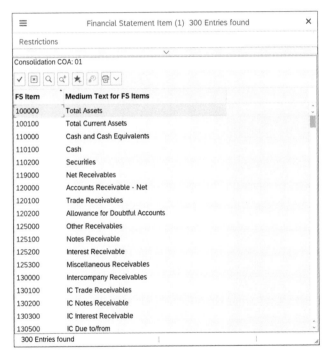

Figure 15.11 Financial Statement Items

Financial statement items are grouped into reporting items, which are grouped into reporting hierarchies. Maintenance of the financial statement items and the hierarchies is handled via the Manage Global Accounting Hierarchies SAP Fiori app (app ID F2918).

15.3 Data Collection and Consolidation Configuration

As shown in Figure 15.12, you can review the app details and implementation information on the SAP Fiori apps reference library, available at *http://s-prs.co/v485701*.

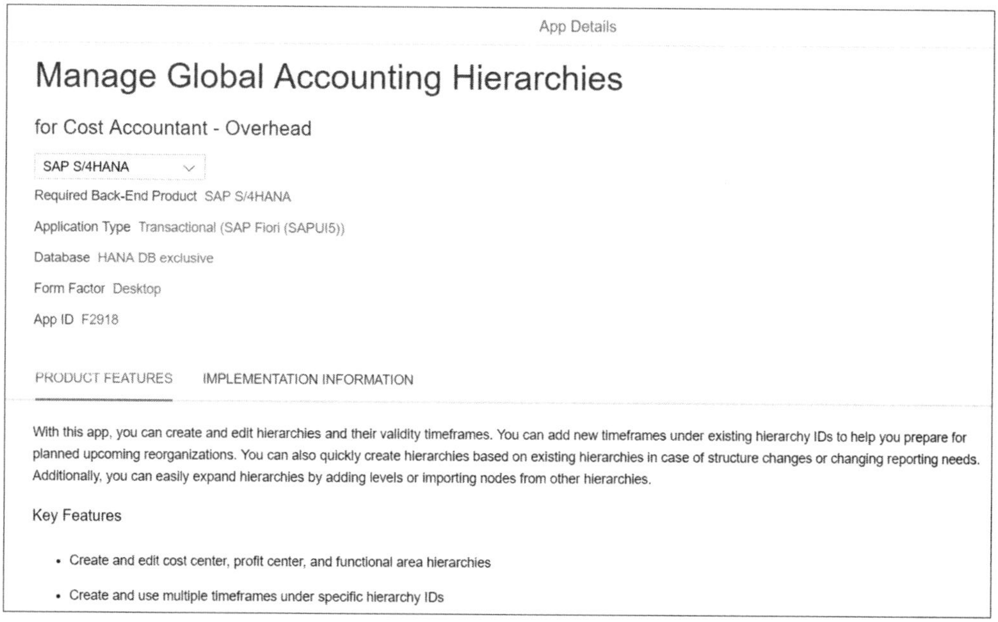

App Details

Manage Global Accounting Hierarchies
for Cost Accountant - Overhead

SAP S/4HANA

Required Back-End Product SAP S/4HANA
Application Type Transactional (SAP Fiori (SAPUI5))
Database HANA DB exclusive
Form Factor Desktop
App ID F2918

PRODUCT FEATURES   IMPLEMENTATION INFORMATION

With this app, you can create and edit hierarchies and their validity timeframes. You can add new timeframes under existing hierarchy IDs to help you prepare for planned upcoming reorganizations. You can also quickly create hierarchies based on existing hierarchies in case of structure changes or changing reporting needs. Additionally, you can easily expand hierarchies by adding levels or importing nodes from other hierarchies.

Key Features

- Create and edit cost center, profit center, and functional area hierarchies
- Create and use multiple timeframes under specific hierarchy IDs

Figure 15.12 Manage Global Accounting Hierarchies: SAP Fiori App Details

15.3.2 Subitem Categories and Subitems

Subitems further categorize the financial statement items in conjunction with the subitem categories. Typical subitem categories include the consolidation transaction type and the functional area. With their help, you can categorize financial statement items. Subitems are dependent on the subitem categories. For example, for the subitem category transaction type, you may create the subitems opening balance, acquisitions, retirements, closing balance, and so on.

To configure subitems and subitem categories, follow menu path **SAP S/4HANA for Group Reporting • Master Data • Define Subitem Categories and Subitems**.

Figure 15.13 shows a tree-like structure with the defined subitem categories delivered by SAP, such as transaction type, region, or functional area. You can create further subitem categories with the ⬜ (**Create**) button from the top menu.

15 Group Reporting

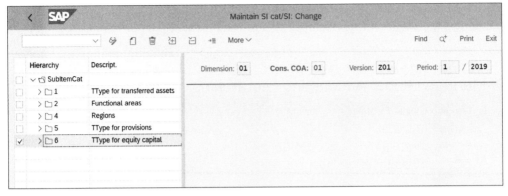

Figure 15.13 Subitem Categories

You can expand a subitem category with the > button to its left side. Let's expand subitem category **1**, **TType for transferred assets**. In Figure 15.14 on the left side, you can see the subitems created for this subitem category, such as acquisitions, retirements, transfers, and so on.

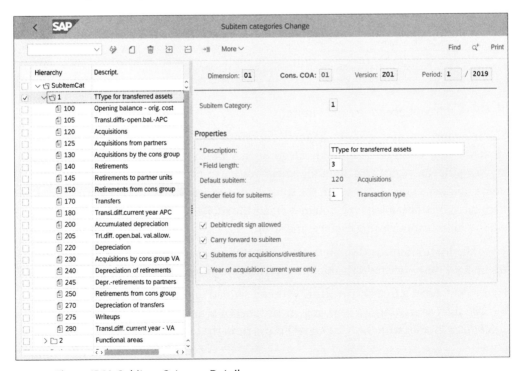

Figure 15.14 Subitem Category Details

644

On the right side, you see the configuration of the subitem category itself, where you can configure the following fields:

- **Description**
  Long text of the subitem category.
- **Field length**
  Defines the length of the subitems for this subitem category.
- **Sender field for subitems**
  Establishes the link with the field that's providing data for consolidation.
- **Debit/credit sign allowed**
  When this is checked, you're allowed to specify a sign for the subitems.
- **Carry forward to subitem**
  When you check this, you can enter a separate subitem for carrying forward balances.
- **Subitems for acquisitions/divestitures**
  Controls the screen sequence for subitems. If you tick it, you can enter a retirement subitem.
- **Year of acquisition: current year only**
  If you check this indicator, you can specify that only acquisitions for the current year can be entered for the subitem.

Now double-click subitem **120**. Figure 15.15 shows its configuration screen.

Here you can configure the following fields:

- **Medium Text**
  Description of the subitem.
- **Debit/credit sign (+/-)**
  In the subitem category, you defined that you can enter a debit/credit sign for the subitems. The system will multiply the value entered by the debit/credit sign of the subitem.
- **Carry forward to subitem**
  The subitem to which the value of the current subitem will be carried forward during the balance carryforward.
- **Retire./divest. subitem**
  Indicates that this is a subitem for retirements/divestitures.
- **Acquisition subitem**
  Indicates that this is a subitem for acquisitions.

- **No Posting/Entry**
  Defines that there can be no postings or entries for this subitem.

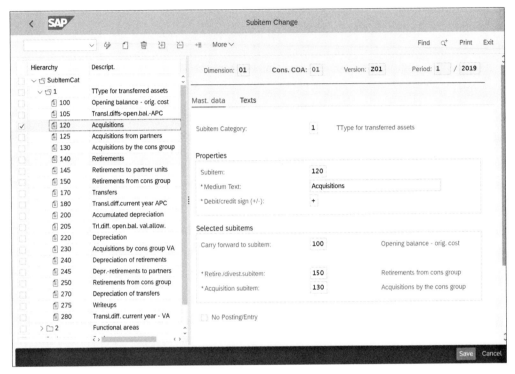

Figure 15.15 Subitem Settings

You can also define default assignments. In Figure 15.14, the default subitem is **120**, which is a grayed-out field. To define default assignments, follow menu path **SAP S/4HANA for Group Reporting • Master Data • Define Default Values for Subassignments**.

On the initial screen shown in Figure 15.16, you see the characteristics for which you can define default assignments. Click the → **Dflt values** button on the **Subitem** line.

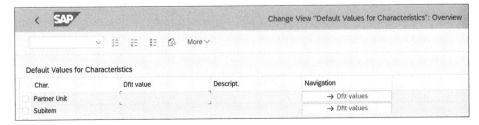

Figure 15.16 Default Assignments

15.3 Data Collection and Consolidation Configuration

Figure 15.17 shows the default assignments for the various subitem categories. Here is where subitem **120** is assigned as a default for subitem category **1**.

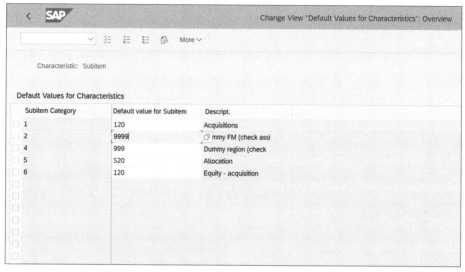

Figure 15.17 Subitem Default Assignments

Modify as appropriate and save with the **Save** button.

15.3.3 Document Type

As you've seen in other financial areas, in group reporting you need to define document types for posting various consolidation transactions. The configuration transactions are located at menu path **SAP S/4HANA for Group Reporting • Master Data**, and there are five separate transactions there to maintain different document types, as follows:

- Define Document Types for Reported Financial Data
- Define Document Types for Manual Posting in Data Monitor
- Define Document Types for Reclassification in Data Monitor
- Define Document Types for Manual Posting in Consolidation Monitor
- Define Document Types for Reclassification in Consolidation Monitor

The data monitor and consolidation monitor should be familiar tools for consultants and users with experience in EC-CS consolidation. The data monitor is used to run

15 Group Reporting

the activities for collecting and preparing the financial data reported by the consolidation units, which are called *tasks*. The consolidation monitor presents a graphic overview of the consolidation units and groups and an interface for executing tasks for collecting and consolidating the reported financial data and for monitoring the progress of these tasks. Now in SAP S/4HANA Finance for group reporting, these transactions are still used, available both through SAP Fiori and the SAP GUI interface.

Let's review the settings for manual postings in the data monitor. The available settings for the other document types are similar. Follow menu path **SAP S/4HANA for Group Reporting • Master Data • Define Document Types for Manual Posting in Data Monitor**.

Figure 15.18 shows the defined standard document types for manual postings in the data monitor. Double-click document type **05, Adjustments to reported data** to review its settings.

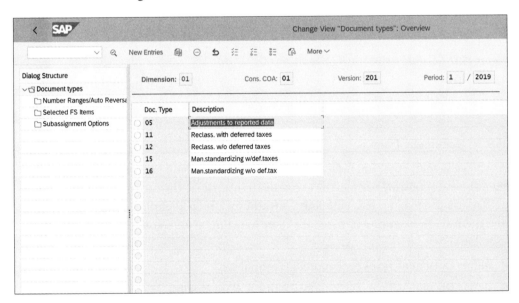

Figure 15.18 Document Types

Figure 15.19 shows the fields you can configure on the document type level.

The fields are as follows:

- **Posting Level**
  Classifies the consolidation entry among the following options:

648

- Adjustments to reported financial data
- Standardizing entries
- Reconciliation entries
- Elimination entries
- Consolidation entries
- Divestitures

- **Balance check**
Defines what the system should do when checking the balance when posting to statistical items:
  - Error message when balance not equal to zero
  - Warning message when balance not equal to zero
  - No balance check

- **Bus.application**
Classifies the document types according to their business application, such as elimination of intercompany receivables and payables, consolidation of investments, and so on.

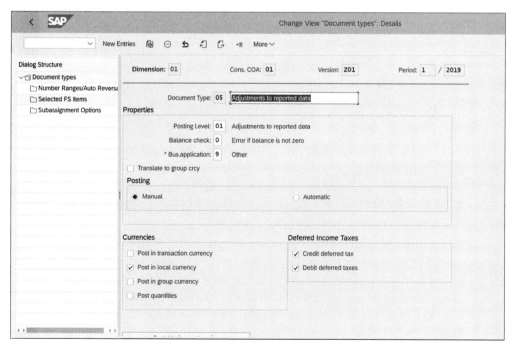

Figure 15.19 Document Type Settings

- **Translate to group crcy**
  If you check this flag, the local and transaction currency values are translated into group currency.
- **Posting**
  In this section, you choose whether the document type is used for manual or automatic consolidation postings.
- **Currencies**
  In this section, you choose in which type of currencies the document type can post and whether it can post quantities.
- **Deferred Income Taxes**
  In this section, you define whether the document type can credit and/or debit deferred taxes. Deferred taxes align the tax expenses of the individual financial statements with the group's consolidated earnings.

Now click **Number Ranges/Auto Reversal** from the left side of the screen. On the screen shown in Figure 15.20, you see the number ranges assigned to the document type per consolidation version.

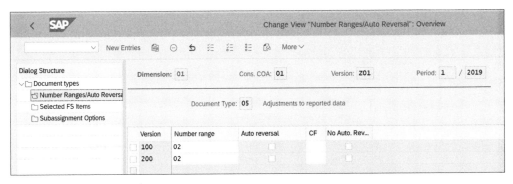

Figure 15.20 Number Ranges

Select **New Entries** from the top menu to assign a number range to the new version Z01 you created previously, as shown in Figure 15.21.

Assign number range **02** for consolidation version **Z01**. If you tick the **Auto reversal** checkbox, the system will automatically create a reversal entry in the subsequent period for documents that are only valid in one period. In the **CF** field, you can define the consolidation frequency for determining the reversal period for automatic reversals, such as annually, semiannually, quarterly, or monthly. If you tick the **No Auto. Reversal in Foll. Year** checkbox, the system will post automatic reversal entries only in the current fiscal year, not in the following one.

15.3 Data Collection and Consolidation Configuration

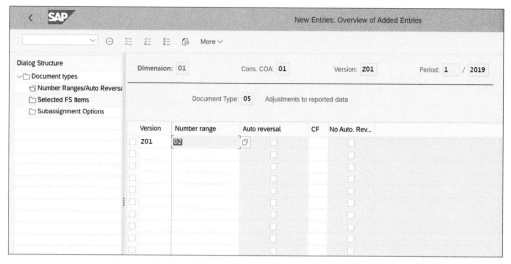

Figure 15.21 Adding New Version

In the **Selected FS Items** section from the left side of the screen, you can specify financial statement items for the deferred taxes. In the **Subassignment Options** section, you can hide certain subassignments and determine whether subassignments can be displayed and changed during posting with this document type, as shown in Figure 15.22.

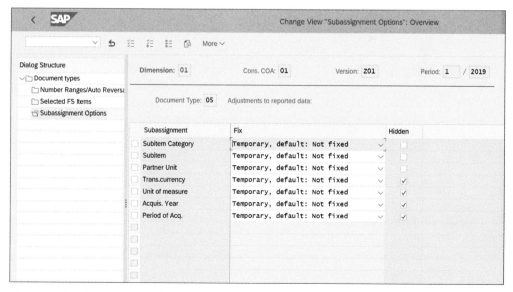

Figure 15.22 Subassignment Options

651

Once you're done modifying the settings of the document type, save it with the **Save** button.

15.3.4 Number Ranges

You also need to define the number ranges for consolidation documents. We already assigned number range intervals to the document type per consolidation version. These number range intervals need to be valid for the year in which postings are to be made.

To maintain number ranges, follow menu path **SAP S/4HANA for Group Reporting • Master Data • Edit Number Range Intervals for Posting**.

On the initial screen shown in Figure 15.23, select the correct **Dimension**. Then click the ⁄ Intervals (**Change Intervals**) button. Then with the ⊕ (**Insert Line**) button from the top menu, enter new lines to extend the number ranges for the required year(s) of validity.

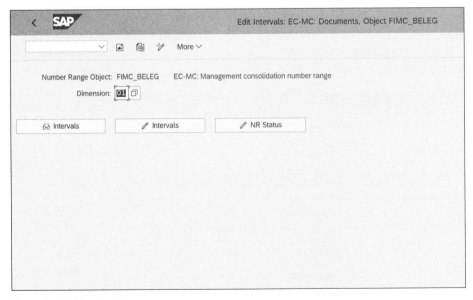

Figure 15.23 Number Ranges

Enter the number ranges and years of validity, as shown in Figure 15.24. Entering "9999" enables the number ranges indefinitely. Then save with the **Save** button.

15.3 Data Collection and Consolidation Configuration

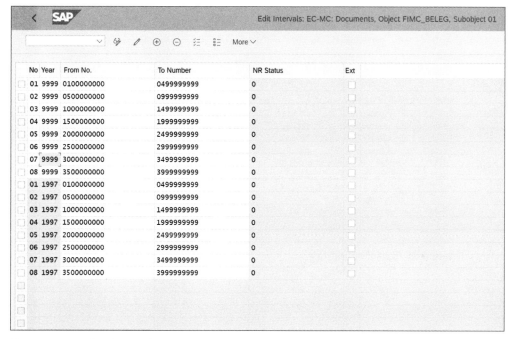

Figure 15.24 Maintain Number Ranges

15.3.5 Data Collection Tasks

As already discussed, the group reporting process from a user point of view has two main parts: data consolidation, which is done in the data monitor, and consolidation, which is done in the consolidation monitor. In the data monitor, data collection tasks are performed. To define these tasks, follow menu path **SAP S/4HANA for Group Reporting • Data Collection for Consolidation • Define Task**.

Figure 15.25 shows the standard tasks provided by SAP. If you need to create other tasks, you can do so with the **New Entries** option from the top menu, but these should suffice for most standard requirements.

In a separate step, you define tasks for manual postings by following menu path **SAP S/4HANA for Group Reporting • Data Collection for Consolidation • Define Tasks for Manual Posting**. Figure 15.26 shows the standard manual tasks provided. You can create new tasks with the **New Entries** option from the top menu.

653

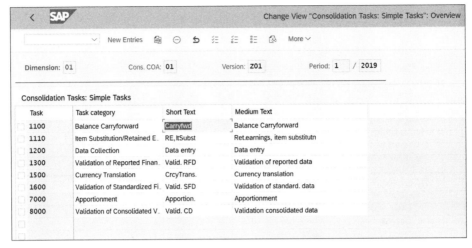

Figure 15.25 Consolidation Tasks

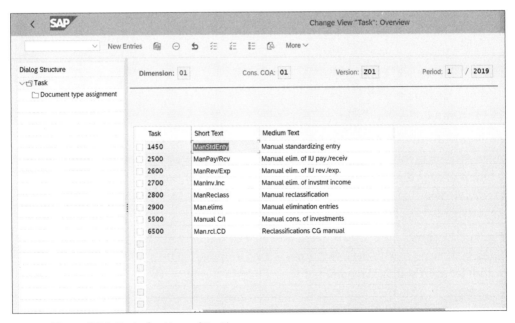

Figure 15.26 Tasks for Manual Postings

Select task **1450**, **ManStdEnty** and click **Document type assignment** from the left side of the screen. On the screen shown in Figure 15.27, you select a document type and from year and period from which the assignment is valid.

15.3 Data Collection and Consolidation Configuration

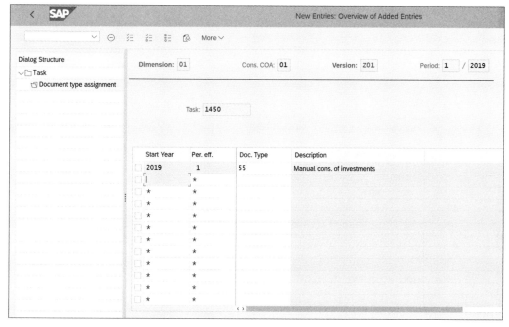

Figure 15.27 Document Type Assignment

Enter the document type to be used with the task in the field **Doc. Type**.

Maintain the document types for the other tasks as well and save with the **Save** button.

15.3.6 Consolidation of Investments Methods

Consolidation of investments, often abbreviated as C/I, is a fundamental group accounting task. It determines how goodwill, negative goodwill, fair value adjustments, and other investment valuation topics that arise in the consolidation process are treated.

Consolidation of investments methods are used to carry out this task, and there are several standard methods provided by SAP. To define consolidation methods, follow menu path **SAP S/4HANA for Group Reporting • Configuration for Consolidation Processing • Define Methods**.

Figure 15.28 shows the standard methods provided by SAP. Multiple methods according to US GAAP and HGB (Handelsgesetzbuch; German commercial laws) are provided. You also can create your own methods, but for now let's review the settings.

15 Group Reporting

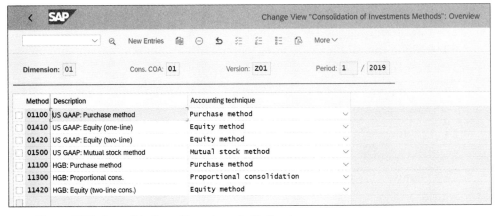

Figure 15.28 Consolidation of Investments Methods

Double-click method **01100**, **US GAAP: Purchase method**. You'll see the definition of the method, as shown in Figure 15.29.

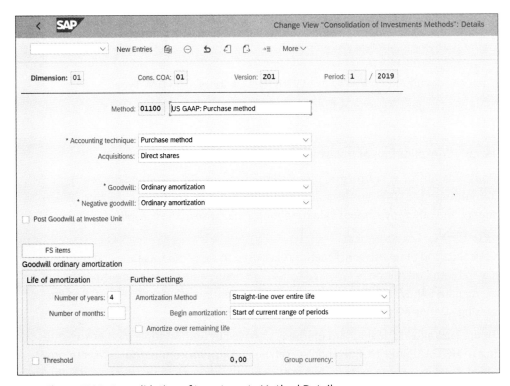

Figure 15.29 Consolidation of Investments Method Details

You can configure the following fields:

- **Accounting technique**
  The technique for consolidating investments in investee companies. You can choose from the following:
    - Purchase method
    - Proportional consolidation
    - Equity method
    - Mutual stock method
    - Cost method
- **Acquisitions**
  Defines how the minority interest is calculated. There are two options:
    - Direct shares
      In this case, the minority interest in equity changes based on the increase of the investor's direct share in the investee.
    - Group shares
      In this case, the minority interest in equity changes according to the increase of the group share of the investee.
- **Goodwill**
  Defines how goodwill will be treated.
- **Negative goodwill**
  Defines how negative goodwill will be treated.
- **Post Goodwill at Investee Unit**
  With this indicator, you can post goodwill at the investee when this consolidation of investments method is assigned.
- **Goodwill ordinary amortization**
  In this section, you make settings for the amortization of goodwill, such as the life of the amortization in years and months, the start date and method of amortization, and the amortization threshold amount.
- **Negative goodwill ordinary amortization**
  In this section, you make settings for the amortization of negative goodwill, such as the life of the amortization in years and months, the start date and method of amortization, and the amortization threshold amount.

15.3.7 Task Group

Task groups group the various tasks that are performed in the data monitor and the consolidation monitor. As with the other configuration objects we analyzed, there are standard task groups provided by SAP, and you can create your own.

To define a task group, follow menu path **SAP S/4HANA for Group Reporting • Configuration for Consolidation Processing • Define Task Group**.

Figure 15.30 shows the standard task groups provided by SAP. In the same transaction, task groups for the data monitor and the consolidation monitor are maintained. You can create additional task groups by selecting **New Entries** from the top menu or copy from an existing task group by selecting it and clicking the (**Copy As...**) button from the top menu.

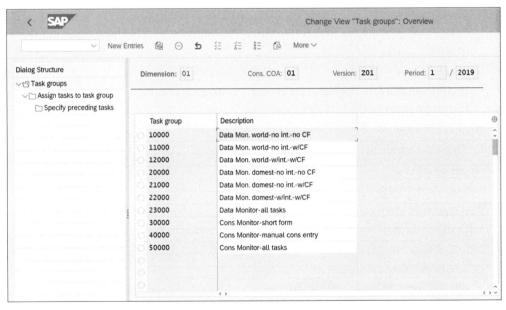

Figure 15.30 Task Groups

Let's review the settings for task group **40000**, **Cons Monitor-manual cons entry**. Select it and click **Assign tasks to task group** from the left side of the screen. In this section, you can assign the individual tasks to the task group.

Figure 15.31 shows all the tasks assigned to the task group. If you tick the **Block auto.** checkbox, the relevant task will be blocked automatically after successful execution. If not, you will have to block it manually.

15.3 Data Collection and Consolidation Configuration

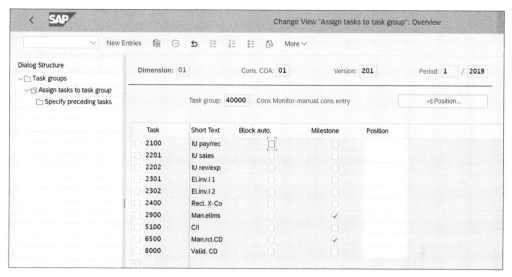

Figure 15.31 Assign Tasks to Task Groups

The checkbox in the **Milestone** field marks the task as a milestone task. During the automatic execution, the consolidation process stops after a task marked as a milestone, and the user has to continue the process manually. In the **Position** field, you can define the sequence order of execution of equal-ranking tasks.

By selecting a task and clicking **Specify preceding tasks** from the left side of the screen, you can define a preceding task for a given task. When a task has a preceding task, that preceding task should be successfully executed and blocked before its successor task can be executed, as shown in Figure 15.32.

In the last step, you need to assign the task groups to the dimension and version by following menu path **SAP S/4HANA for Group Reporting • Configuration for Consolidation Processing • Assign Task Group to Dimension**.

In this transaction, you assign the task groups for both the data monitor and the consolidation monitor for a combination of dimension, version, and consolidation chart of accounts.

As shown in Figure 15.33, you assign the task groups and specify the start year and period from which the assignment is valid. In the **Period cat.** field, you define the period category, which specifies the periods in which the tasks are executed. Period categories group together periods that have the same consolidation requirements, such as annual, quarterly, or some other frequency.

15 Group Reporting

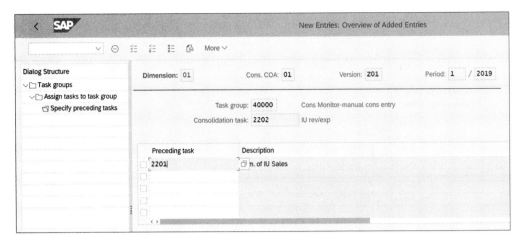

Figure 15.32 Specify Preceding Task

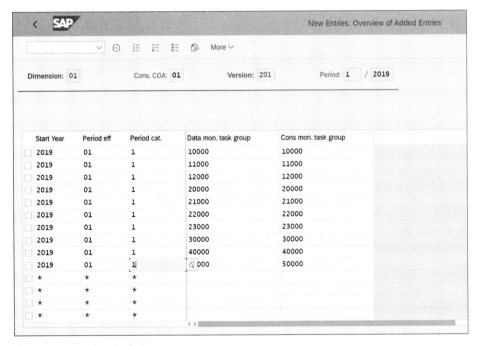

Figure 15.33 Assign Task Groups

With that, we finish our guide to configuring SAP S/4HANA Finance for group reporting.

15.4 Summary

In this chapter, you learned how to configure the new consolidation solution from SAP: SAP S/4HANA Finance for group reporting. You received a general overview of the group reporting process and requirements and a look at the benefits SAP S/4HANA Finance for group reporting provides. Then we took a deep dive and provided a detailed guide to configuring the SAP S/4HANA Finance for group reporting solution.

This solution is a major step forward for SAP. It brings the group reporting process into SAP S/4HANA, fully benefiting from the power and flexibility of the SAP HANA database and effectively eliminating the need to use third-party solutions for consolidation. SAP S/4HANA Finance for group reporting technically is similar to the EC-CS consolidation solution, but with heavy use of SAP Fiori apps for the end user, which streamlines the data collection and consolidation tasks and enhances reporting.

In the next chapter, we will cover migrating your data in SAP S/4HANA.

Chapter 16
Data Migration

An important part of every implementation is to migrate legacy data into the new system. This chapter teaches you how to perform finance migration for both greenfield implementations and for brownfield implementations, for which SAP S/4HANA provides special tools and programs to facilitate the migration process.

Data migration is a crucial part of every SAP S/4HANA implementation. No matter how well the system is configured, no matter how precise the business processes are portrayed in the new SAP S/4HANA solution, without a properly performed migration, there can be no successful implementation. Therefore, we've dedicated an entire chapter to the data migration topic, in which we'll explain in detail how to plan and execute the migration and what the best practices are that will help you avoid dangerous pitfalls.

The migration process involves numerous objects. We'll concentrate on the finance objects because other logistics objects, such as materials, lie outside the scope of the book. From the finance point of view, we need to migrate general ledger accounts, cost centers, profit centers, internal orders, customers, vendors, assets, banks, and related balances and open items.

Because the migration strategy is different for greenfield and brownfield implementations, we'll dedicate separate sections to each path. After that, we'll discuss in detail the most important finance migration objects and what should be considered when migrating them.

16.1 Brownfield Implementation Migration

As noted previously, converting an existing SAP ERP system to SAP S/4HANA is called a *brownfield implementation*. You need to make sure that your system meets various technical requirements before you can make the conversion, and you need to

16 Data Migration

make sure that all finance settings and objects are correctly migrated to the new SAP S/4HANA landscape.

Before you start, you should be familiar with the Conversion Guide for SAP S/4HANA 1809, available at *http://s-prs.co/v485702*. A lot useful information also is available in the conversion document for accounting, attached to SAP Note 2332030. To search for SAP Notes, visit *https://support.sap.com/en/index.html*. An SAP S-user ID is required to access SAP Notes, which can be provided by your project manager. SAP delivers various checks and migration tools and programs to help you in these tasks.

We'll first teach you how to use the check programs to make sure your system is ready for conversion to SAP S/4HANA. Then we'll guide you through how to execute the migration.

16.1.1 Check Programs for SAP S/4HANA Readiness

SAP delivers multiple programs to help you identify areas that aren't compatible with the future SAP S/4HANA solution.

Program RASFIN_MIGR_PRECHECK checks the readiness of fixed assets for SAP S/4HANA. To execute programs in SAP, enter Transaction SE38.

Figure 16.1 shows the initial screen to run programs. Enter the program name, "RASFIN_MIGR_PRECHECK", in the **Program** field, and execute it with the button from the top menu.

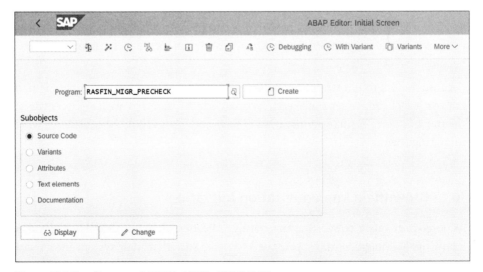

Figure 16.1 Run Program RASFIN_MIGR_PRECHECK

On the first screen, shown in Figure 16.2, there are two options:

- SAP Simple Finance Add-On
- SAP Simple Finance/SAP S/4HANA

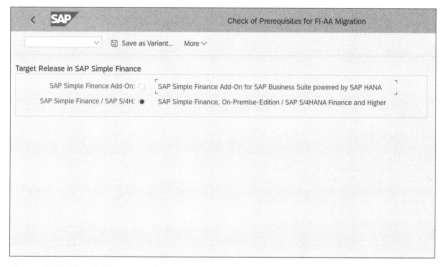

Figure 16.2 Check Fixed Assets

As explained in the Introduction, SAP Simple Finance (now called SAP S/4HANA Finance) was the first application from SAP to be provided for the SAP HANA database, and it was delivered as an add-on solution for SAP ERP. This program can be used to carry out checks for compliance with this solution, which are in general less stringent than the checks for SAP S/4HANA.

Select checks for SAP S/4HANA (the second radio button) and proceed with the **Execute** button on the lower-right side of the screen.

Figure 16.3 shows the errors that the program identified related to fixed assets. Double-clicking an error message provides additional information about it. These errors need to be resolved before you continue with the migration activities for fixed assets.

To check the overall financial customizing settings, follow menu path **Conversion of Accounting to SAP S/4HANA • Preparations and Migration of Customizing • Check Customizing Settings Prior to Migration**. This transaction checks whether the customizing settings are ready for migration to SAP S/4HANA Finance. It checks whether your ledger, company code, and controlling area settings meet the prerequisites for migration.

665

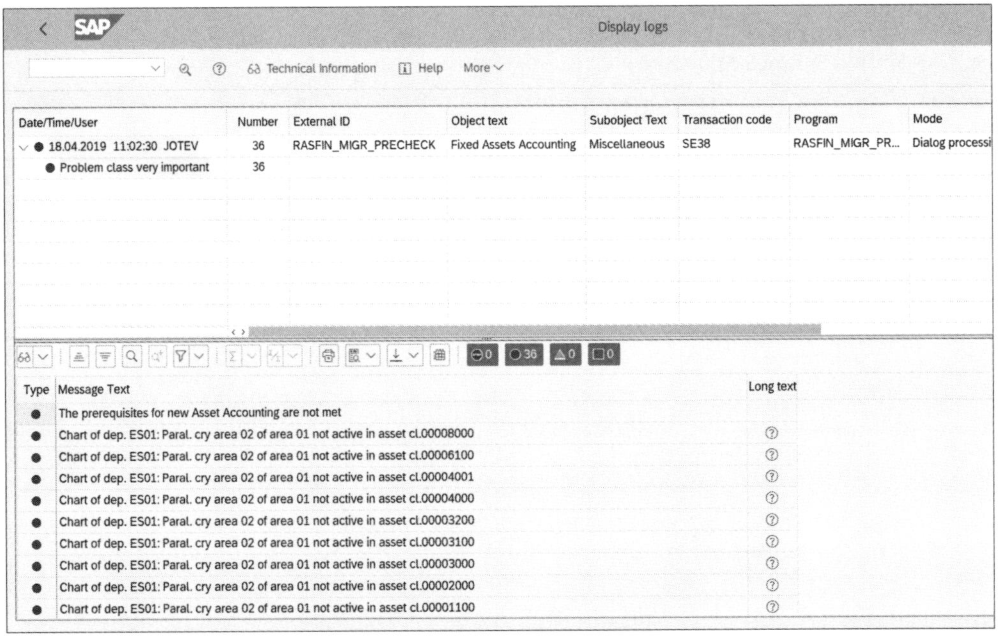

16 Data Migration

Figure 16.3 Check Fixed Assets Output List

Starting with menu path **Conversion of Accounting to SAP S/4HANA • Preparations and Migration of Customizing**, you can the various transactions to check the configuration objects for the different financial modules that need to be checked prior to migration, as shown in Figure 16.4:

- Preparations and Migration of Customizing for General Ledger
- Preparations and Migration of Customizing for Accrual Engine
- Preparations and Migration of Customizing for Asset Accounting
- Preparations and Migration of Customizing for Controlling
- Preparations and Migration of Customizing for Material Ledger
- Preparations for Migration of House Bank Accounts
- Preparations for Migration of Financial Documents to Trade Finance
- Preparatory Activities and Migration of Customizing for Credit Management

16.1 Brownfield Implementation Migration

Figure 16.4 Check Customizing Settings

There is also a readiness program to check the general ledger settings, available at **Conversion of Accounting to SAP S/4HANA • Preparations and Migration of Customizing • Preparations and Migration of Customizing for General Ledger • Execute Consistency Check of General Ledger Settings**. It checks the customizing settings for the ledgers. This check needs to be performed without error messages before you can continue with the migration.

Figure 16.5 shows the result of the checks performed by the program.

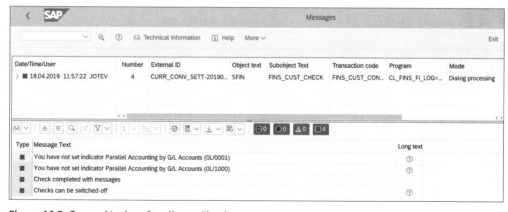

Figure 16.5 General Ledger Readiness Check

16 Data Migration

16.1.2 Migration to SAP S/4HANA

After successfully eliminating all errors in the readiness check programs, you can continue with the migration of the existing SAP system data to SAP S/4HANA.

In this section, we'll examine the steps in the migration process. We'll start by checking that the system is converted to Unicode. Then we'll analyze the transactional data that needs to be converted. After that, we'll teach you how to start and monitor the migration process. Then we'll migrate the general ledger allocations. We'll finish off with the technical step to complete the migration and the activities to be performed after the migration.

Check for Unicode Conversion

As a prerequisite to the migration, you need to make sure your existing SAP system is already converted to Unicode. Unicode is a standard for representing text symbols, which allows the use of a huge number of characters not available in older standards. Unicode is a mandatory prerequisite for SAP S/4HANA. To check whether your system is Unicode-based, select **More • System • Status** from the menu, as shown in Figure 16.6.

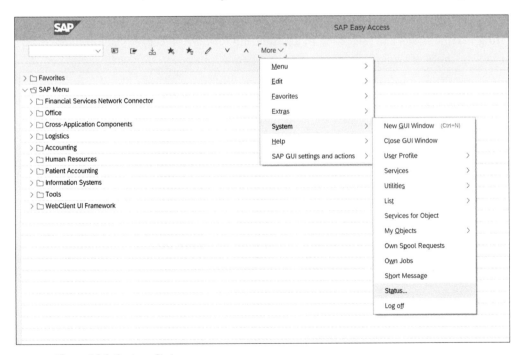

Figure 16.6 System Status

16.1 Brownfield Implementation Migration

Then in the **Unicode System** field, a **Yes** or **No** indicates whether the system is Unicode-based. If not, your Basis team needs to upgrade it to Unicode.

Now you're ready to start the migration from the existing SAP system to S/4HANA. You need to plan downtime so that there are no users in the system during the migration. The migration should be done over a weekend when there will be enough downtime to not disrupt the business.

Analysis of Transactional Data

The first step in the data migration process is to analyze the existing transactional data.

On the initial screen shown in Figure 16.7, you can restrict the number of background jobs in the **Number of batch jobs** field, which makes sense in systems with a very high volume of transactional data. You can specify a server group name in the **Server Group Name** field.

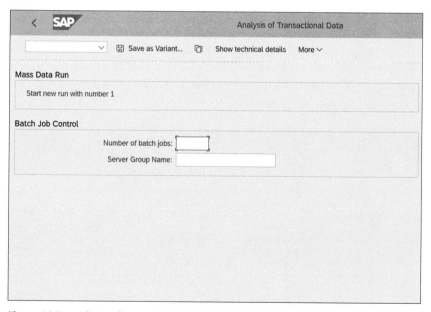

Figure 16.7 Analysis of Transactional Data

Then proceed with the **Execute** button on the lower-right side of the screen. The system displays a log of the scheduled jobs, as shown in Figure 16.8.

16   Data Migration

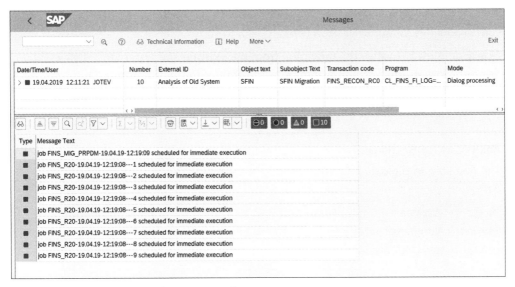

Figure 16.8 Analysis of Transactional Data Log

Go back to the customizing menu. To display the results, follow menu path **Conversion of Accounting to SAP S/4HANA • Display Status of Analysis of Transactional Data**. As shown in Figure 16.9, the system shows the number of errors found—in this case, **1,751**. Double-click the number of errors.

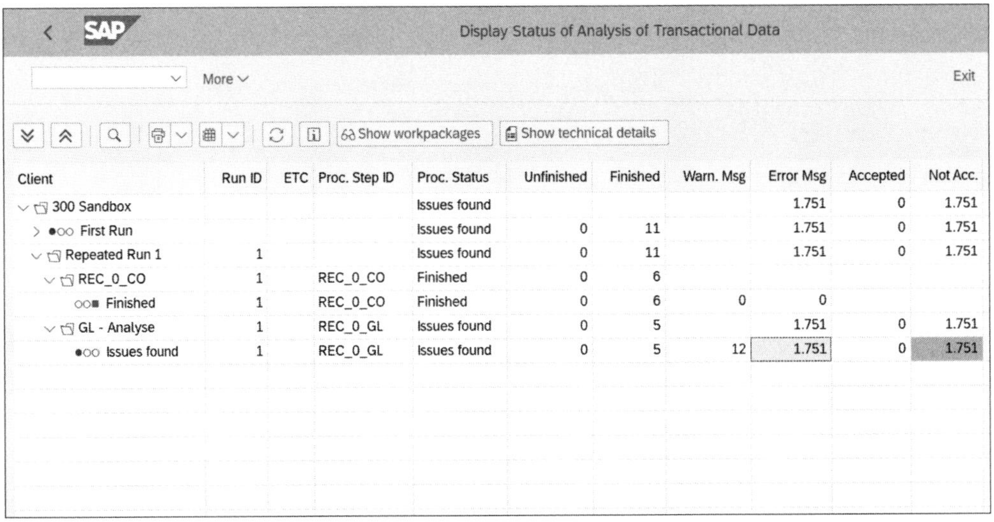

Figure 16.9 Display Analysis of Transactional Data

16.1 Brownfield Implementation Migration

Figure 16.10 shows an overview of the errors. In each line, errors related to specific combinations of company code and year are shown, known as package keys for mass data processing.

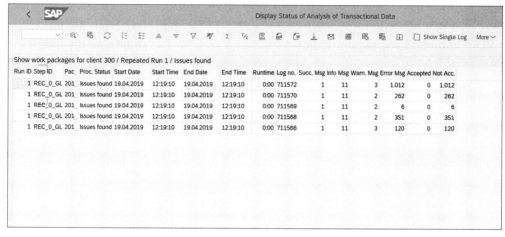

Figure 16.10 Display Overview of Transactional Data

Double-click a line to see the individual documents with errors, as shown in Figure 16.11.

5000000215 2018 004: Field DMBE2 of BSEG does have a non initial Amount (15.00) ⓘ		00:00:00
5000000215 2018 003: Field DMBE2 of BSEG does have a non initial Amount (15.00) ⓘ		00:00:00
5000000215 2018 001: Field DMBE2 of BSEG does have a non initial Amount (15.00) ⓘ		00:00:00
5000000213 2018 004: Field DMBE2 of BSEG does have a non initial Amount (75.00) ⓘ		00:00:00
5000000213 2018 003: Field DMBE2 of BSEG does have a non initial Amount (75.00) ⓘ		00:00:00
5000000207 2018 004: Field DMBE2 of BSEG does have a non initial Amount (711.00) ⓘ		00:00:00
5000000207 2018 003: Field DMBE2 of BSEG does have a non initial Amount (711.00) ⓘ		00:00:00
5000000207 2018 001: Field DMBE2 of BSEG does have a non initial Amount (711.00) ⓘ		00:00:00
5000000172 2018 004: Field DMBE2 of BSEG does have a non initial Amount (703.89) ⓘ		00:00:00
5000000172 2018 003: Field DMBE2 of BSEG does have a non initial Amount (703.89) ⓘ		00:00:00
5000000172 2018 001: Field DMBE2 of BSEG does have a non initial Amount (703.89) ⓘ		00:00:00
5000000171 2018 004: Field DMBE2 of BSEG does have a non initial Amount (500.00) ⓘ		00:00:00
5000000218 2018 001: Field DMBE2 of BSEG does have a non initial Amount (75.00) ⓘ		00:00:00
5000000218 2018 003: Field DMBE2 of BSEG does have a non initial Amount (75.00) ⓘ		00:00:00
5000000218 2018 004: Field DMBE2 of BSEG does have a non initial Amount (75.00) ⓘ		00:00:00
5000000220 2018 001: Field DMBE2 of BSEG does have a non initial Amount (30.00) ⓘ		00:00:00
5000000220 2018 003: Field DMBE2 of BSEG does have a non initial Amount (30.00) ⓘ		00:00:00

Figure 16.11 Detailed Errors

You need to analyze the errors. It's possible to accept some errors as not critical by selecting them and clicking the **Accept error** button.

Start and Monitor Migration

To perform and monitor the data migration, SAP provides the SAP S/4HANA migration cockpit, available at menu path **Conversion of Accounting to SAP S/4HANA • Data Migration • Start and Monitor Data Migration**.

On the screen shown in Figure 16.12, you'll see the migration runs. The runs will be numbered according to their start date and time. The first run will be numbered **1**.

Figure 16.12 Start and Monitor Migration Overview

In the **Tables** tab, you can see the number of entries in the involved tables, as shown in Figure 16.13.

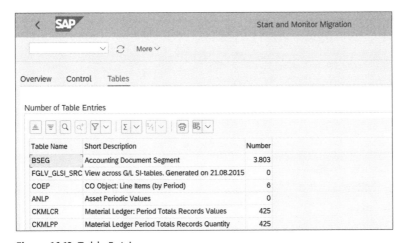

Figure 16.13 Table Entries

16.1 Brownfield Implementation Migration

Migrate General Ledger Allocations

In the next step, you need to migrate general ledger allocations into the new SAP HANA database tables. Follow menu path **Conversion of Accounting to SAP S/4HANA • Data Migration • Migrate General Ledger Allocations**.

With this step, you move all fields and allocations from the new general ledger totals table FAGLFLEXT to the new view ACDOCT. Figure 16.14 shows the selection screen of the program.

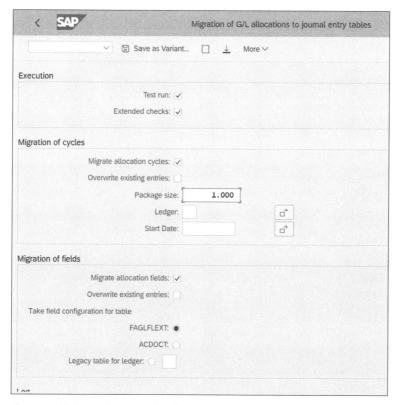

Figure 16.14 Migrate General Ledger Allocations

Completing the Migration

After all data migration steps are completed, you should complete the migration. Under menu path **Conversion of Accounting to SAP S/4HANA • Data Migration • Complete Migration • Reconcile and Compare Migrated Data**, there's extensive documentation available that explains which programs are available to reconcile and compare

16 Data Migration

migrated data. You can review customizing documentation by clicking the button from the left side of the customizing node.

After reconciling and comparing the data using the listed programs, set the migration to completed by following menu path **Conversion of Accounting to SAP S/4HANA • Data Migration • Complete Migration • Set Migration to Completed**. The status indicators all should now be green, as shown in Figure 16.15.

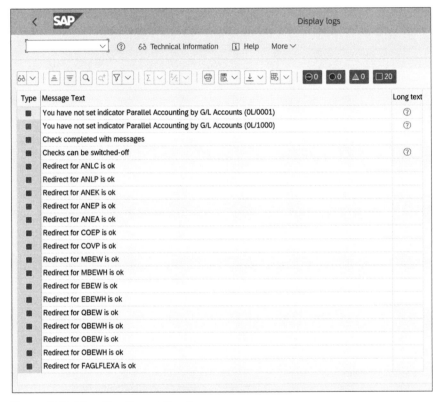

Figure 16.15 Migration Completion Log

Activities after Migration

After the migration is completed, you can open the system for users. There are several activities to be done after the migration, but they can be performed in a productive SAP S/4HANA system.

Figure 16.16 shows the various technical steps that should be performed after the system is converted to SAP S/4HANA from a finance point of view.

16.1 Brownfield Implementation Migration

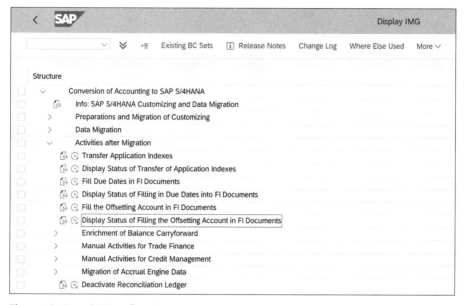

Figure 16.16 Activities after Migration

These activities include the following:

- Transfer Application Indexes
- Display Status of Transfer of Application Indexes
- Fill Due Dates in FI Documents
- Display Status of Filling in Due Dates into FI Documents
- Fill the Offsetting Account in FI Documents
- Display Status of Filling the Offsetting Account in FI Documents
- Enrichment of Balance Carryforward
- Manual Activities for Trade Finance
- Manual Activities for Credit Management
- Migration of Accrual Engine Data
- Deactivate Reconciliation Ledger

The migration to SAP S/4HANA migrates the documents from the general ledger, accounts payable, accounts receivable, fixed assets, and material ledger to the new Universal Journal, table ACDOCA. You also may need to migrate data also from other tables in separate steps if you use certain components, such as the following:

- Cost of sales accounting, table GLFUNCT
- Classic profit center accounting, table GLPCT
- Costing-based profitability analysis, tables CE1*

With that, we finish our guide to data migration for brownfield implementation. In the next section, we'll discuss how to perform migration for a brand-new SAP implementation or for cases in which existing SAP customers choose the greenfield implementation option.

16.2 Greenfield Implementation Migration

In a greenfield SAP S/4HANA implementation, the migration process needs to bring in migration objects from other systems, which may be SAP or non-SAP systems. From a migration point of view, the same concepts, methods, and processes apply whether the system is for a new SAP customer or a legacy SAP system is being retired and replaced with a new SAP S/4HANA system.

We'll start by discussing the migration options you have for a greenfield implementation. Then we'll teach you how to use the migration cockpit and the SAP S/4HANA migration object modeler, which are the SAP-provided tools to migrate your legacy data. Finally, we'll discuss how to perform the data load.

16.2.1 Migration Options

Traditionally, to migrate legacy data in SAP, a certain amount of programming or at least technical skill was needed. Different tools were used for the migration, such as the Legacy System Migration Workbench (LSMW), which was very popular not only for migration but also for various mass data maintenance tasks. In fact, many consultants are very familiar with LSMW because it's very flexible and has lots of capabilities.

In SAP S/4HANA, however, SAP provides new tools specifically tailored for migration: the migration cockpit and the migration object modeler. The migration cockpit allows you to work with various data migration objects and generate Excel templates for them, which then can be populated with legacy data and migrated to SAP S/4HANA. The migration object modeler is the brain behind the migration cockpit, and it enables you to modify the structure of the data objects to be migrated or to create new ones.

Although LSMW is still a possible option for migration, it's not recommended. SAP Note 2287723 states the following:

16.2 Greenfield Implementation Migration

The Legacy System Migration Workbench (LSMW) should only be considered as a migration tool for SAP S/4HANA for objects that do not have interfaces or content available after carefully testing for each and every object. The use of LSMW for data load to SAP S/4HANA is not recommended and at the customer's own risk.

Therefore, unless there is some very specific, very strong reason to use something else, the migration tool for greenfield SAP S/4HANA implementations should be the migration cockpit in combination with the migration object modeler.

16.2.2 Migration Cockpit and Migration Object Modeler

The *migration cockpit* is a web-based tool that's used to migrate legacy data to SAP S/4HANA. It helps generate predefined Excel templates in which to populate the legacy data and then helps upload it.

To start the migration cockpit, enter Transaction LTMC. Authorization role SAP_CA_DMC_MC_USER is required. The transaction opens a web browser, as shown in Figure 16.17.

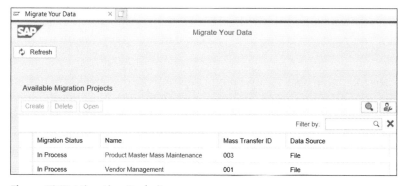

Figure 16.17 Migration Cockpit

You'll see a list of defined migration projects. You'll probably have multiple finance projects that cover the various areas, such as banks, cost centers, general ledger balances, and so on.

Some basic logic is provided by SAP in the migration cockpit, but normally you have to enhance it or create new migration objects using the migration object modeler. The *migration object modeler* is an SAP GUI-based tool that enables you to define the structure of migration objects.

16 Data Migration

Start the migration object modeler by entering Transaction LTMOM. Authorization role SAP_CA_DMC_MC_DEVELOPER is required. In the initial screen shown in Figure 16.18, you can choose to view/modify either migration objects or projects.

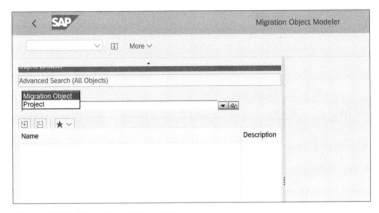

Figure 16.18 Migration Object Modeler

You can create a new user-defined migration object or create one from a template. SAP provides many templates, which should cover most objects that need to be migrated, but if you need something else, you can define it yourself with the user-defined option.

Let's create a new migration object from a template. From the menu, select **Migration Object Modeler • Create Migration Object • From Template**, as shown in Figure 16.19. You can also create a new migration object as a copy from an existing one with the **Copy Migration Object** option.

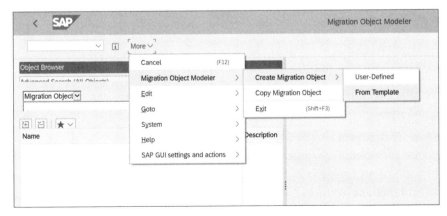

Figure 16.19 Create New Migration Object

678

On the next screen, shown in Figure 16.20, enter the migration project name. As discussed, you should have separate migration objects that group together similar migration objects. The project name should start with Z or Y in the custom name range. These are the projects that you can see in the migration cockpit, visible in Figure 16.17.

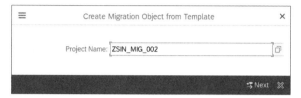

Figure 16.20 Project Name

Continue with the **Next** button. On the next screen, you need to select a migration object template. As shown in Figure 16.21, SAP provides templates for the main migration tasks—for example, accounts receivables open items, accounts payable open items, general ledger accounts, and so on.

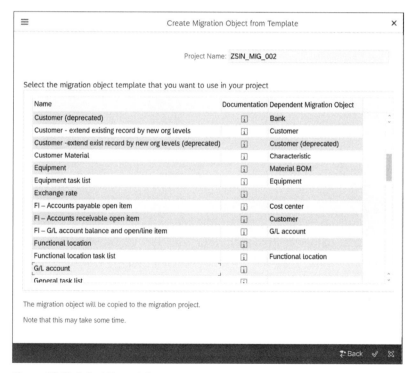

Figure 16.21 Select Template

Let's create a migration object for migration of general ledger accounts. Select the **G/L account** template and proceed with the ✓ button. On the next screen, you see an overview of the created migration object. Initially it's shown in display mode; click the 🖉 button from the top menu to switch to change mode.

As shown in Figure 16.22, the migration object has a tree-like structure, in which you can define the global data, source structure, target structure, structure mapping, and field mapping.

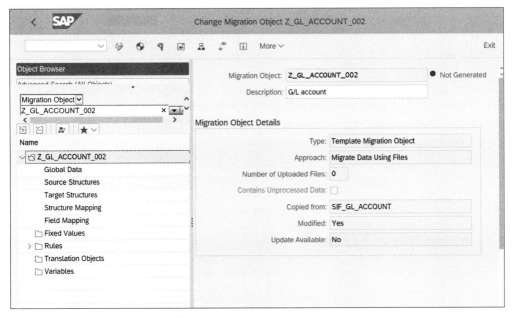

Figure 16.22 Migration Object Overview

Click **Global Data** on the left side of the screen. As shown in Figure 16.23, you'll see the technical information related to the migration object. Here you can see that the migration is being done using function module GL_ACCT_MASTER_SAVE, which is a standard SAP function module provided for the migration of general ledger master data.

You also can see that this migration template was first provided for release 1809, which means that if you're implementing an older SAP S/4HANA release, this template won't be available. SAP continuously expands the technical capabilities in this and other areas. If you're implementing an older SAP S/4HANA release and you can't find the template you need, search to see if the relevant function module is available; if so, create a user-defined object based on this function module. If you deal with some very specific, rarely used migration object your development team may need to create a custom function module.

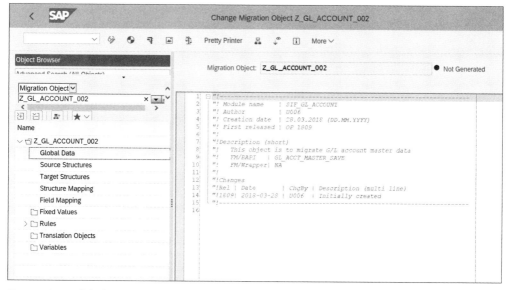

Figure 16.23 Global Data

Click **Source Structures** on the left side of the screen. As shown in Figure 16.24, you'll see the source structures for the migration object. A source structure contains several related fields and often corresponds to a database table in the SAP HANA database. In this case, because you created the migration object from a template, the source structures are already predefined; if you create a user-defined migration object, you'll have to define your own structures.

In this example for general ledger accounts, there are separate structures that represent the general data for general ledger accounts (fields from table SKA1), the company code data (fields from table SKB1), and descriptions of general ledger accounts (fields from table SKAT).

Let's review the fields on the company code level. Click the **S_COMPANY** structure. Figure 16.25 shows the fields that are included in the structure for company code data. It's like looking at table SKB1 in the data browser (Transaction SE11). The system displays the field **Name**, **Data Type**, **Length** of field, and **Description**. Key fields are marked with a checkmark in the **Key Field** column. The key fields are the primary keys of a table and provide unique identification for a table row.

16 Data Migration

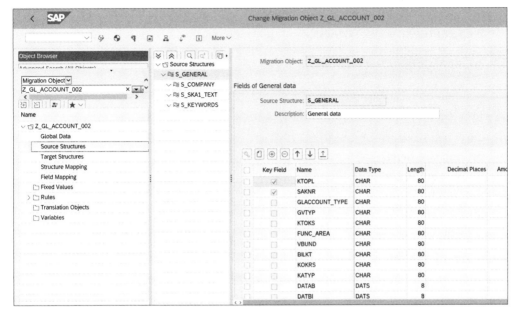

Figure 16.24 Source Structures

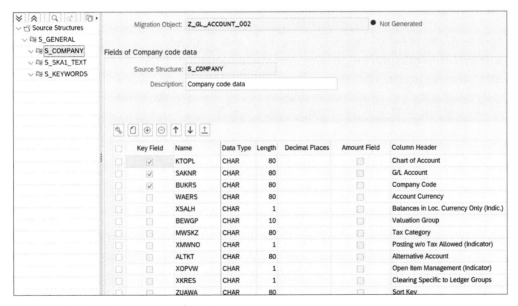

Figure 16.25 Company Code Data

16.2 Greenfield Implementation Migration

You can also modify structures. With the [▢] button, you can add an additional field at the end of the structure. With the [⊕] button, you can add field in between the existing field. With the [⊖] button, you can remove existing field from the structutre. With the [↑] and [∨] buttons, you can rearrange the order of the fields.

To insert additional structures or delete structures, right-click on the structure to be changed.

You can append a structure at a lower, a higher, or the same level as the selected level, and you can also delete a selected structure as shown in Figure 16.26. This way, you can rearranage the structures needed for the migration object, and even add additional structures if you determine that more complexity is needed after creating the migration object. This may save you from the need to delete the migration object and create it again.

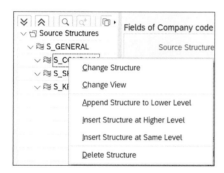

Figure 16.26 Modify Structure

Click **Target Structures** on the left side of the screen. The target structure, shown in Figure 16.27, defines the fields to be included in the template file to be generated. It's predefined by the function module used.

Click **Structure Mapping** on the left side of the screen. In this step, shown in Figure 16.28, you map each source structure with a target structure. Again, in this case it's predefined by the function module. With drag and drop, you can map source and target structures.

16 Data Migration

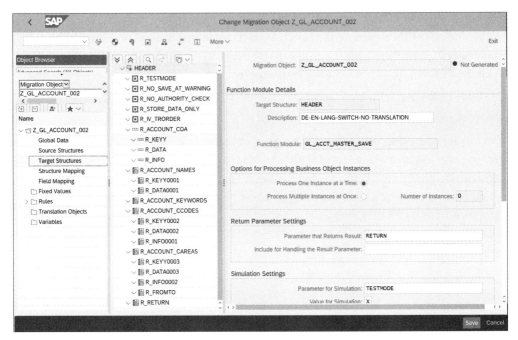

Figure 16.27 Target Structures

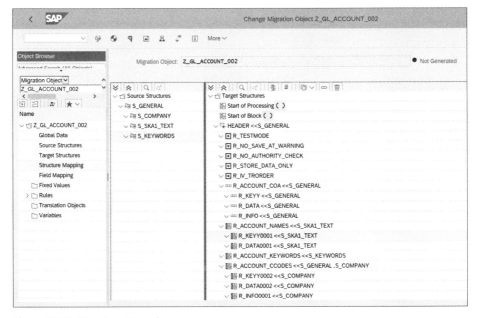

Figure 16.28 Structure Mapping

684

16.2 Greenfield Implementation Migration

Click **Field Mapping** on the left side of the screen. Figure 16.29 shows the field mapping between the source and target structures. Standard mapping is provided by the template, which you can modify.

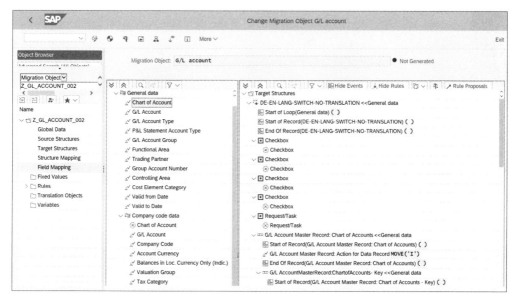

Figure 16.29 Field Mapping

It's convenient when working with fields to be able to view their technical names. To do that, select **More • Settings • Technical Names On/Off** from the top menu, as shown in Figure 16.30.

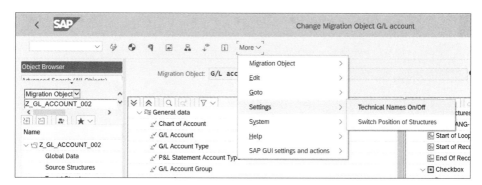

Figure 16.30 Technical Names On

Then the view changes, as shown in Figure 16.31.

685

16 Data Migration

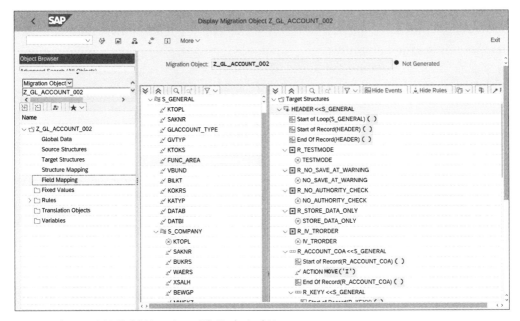

Figure 16.31 Field Mapping with Technical Names

To change the field mapping, you can drag and drop fields from the source structures to the target structures. When defining field mapping, you need to specify rules for the mapping. If you're changing a defined rule mapping, the system shows a pop-up window, as shown in Figure 16.32.

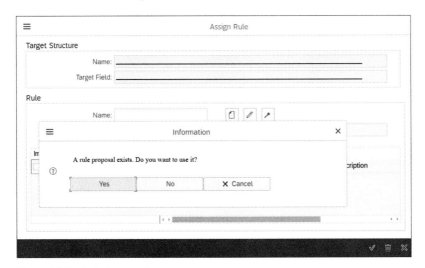

Figure 16.32 Rule Proposal

16.2 Greenfield Implementation Migration

To use an existing rule, click the **Yes** button. Then on the next screen, shown in Figure 16.33, you can choose an existing rule.

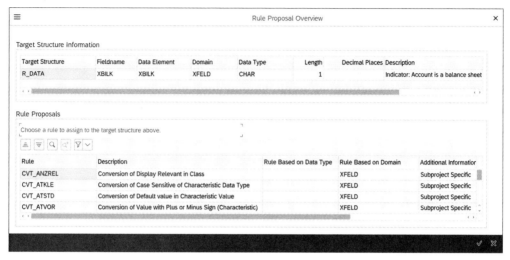

Figure 16.33 Rule Proposal Overview

Otherwise, you can create your own rule after clicking **No** on the screen shown in Figure 16.32. Then on the screen shown on Figure 16.34, enter a name for your rule (again, starting in the custom name range with Z or Y), and click the button.

Figure 16.34 Create Rule

You need to define input and output parameters, as shown in Figure 16.35. View the ABAP code for the rule by clicking the **Open Editor** button at the bottom of the screen. Normally, this will be a task for your development team.

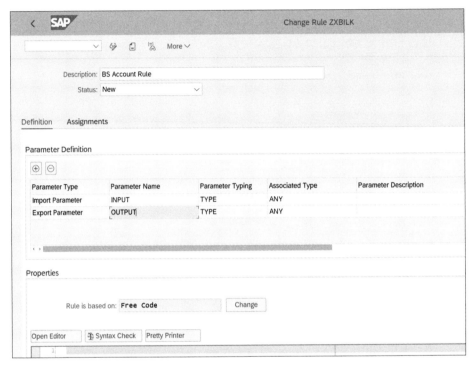

Figure 16.35 Rule Definition

After modifying all relevant settings of the migration object, save it with the **Save** button in the lower-right part of the screen, and generate it with the ● button from the top menu.

16.2.3 Legacy Data Load

After generating the migration objects, you generate Excel-based data templates through the migration cockpit, which need to be loaded with legacy data.

Figure 16.36 shows a sample upload sheet for the general data of general ledger account master records (table SKA1). The key fields have green column headers.

16.2 Greenfield Implementation Migration

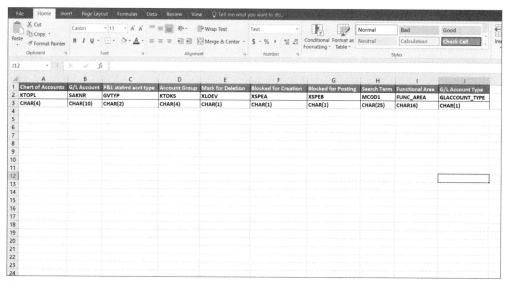

Figure 16.36 Upload Sheet for General Ledger General Data

Another example is shown in Figure 16.37. Here you need to populate the fields to be migrated on the general ledger company code level (table SKB1).

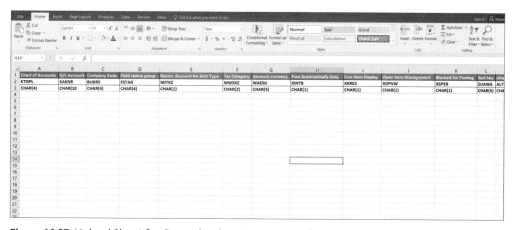

Figure 16.37 Upload Sheet for General Ledger Company Code Data

Filling out the upload sheets normally would be a joint responsibility of consultants and client users with a more technical background. It isn't an easy task, and it requires time and effort. Normally the client should provide a set of raw data from their system, which the consultants then should clean, enhance, and validate. Database tools such as Microsoft Access are quite helpful for the process of cleaning and enhancing the legacy data. The final result should be correct upload files for all relevant migration objects, which can be uploaded using the migration cockpit without any issues. Normally, this process requires multiple iterations in test systems.

16.3 Financial Migration Objects

In this section, we'll discuss the specifics of the most important and commonly migrated financial objects. You already learned how to perform migrations in both brownfield and greenfield implementations, and we provided many examples of how to prepare and load the data. In this section, we'll concentrate on the specifics of each object, what data and fields need to be migrated, and which tables are filled in SAP S/4HANA.

In the following sections, we'll examine the general ledger data, the accounts payable/accounts receivable data, the fixed assets data, and the controlling-related data.

16.3.1 General Ledger Data

In the general ledger, you need to migrate the chart of accounts with all general ledger accounts and cost elements, and general ledger account balances and open items.

The complexity of the general ledger data migration depends on the number of charts of accounts used. At a minimum, you must migrate the operational chart of accounts, which is the main chart of accounts to post all the financial documents. There are three related tables in SAP S/4HANA that are filled with data:

- Table SKA1
  Contains the general level data that's relevant for all company codes
- Table SKB1
  Contains the company code level data, which is migrated for each company code
- Table SKAT
  Contains the names and descriptions of the general ledger accounts in multiple languages

Alternative charts of accounts often are used to depict accounting requirements in various countries. In this case, alternative accounts are mapped to the operational chart of accounts. From a migration point of view, this means that you need to migrate each alternative chart of accounts, which usually are one per country for the countries that have this requirement to have an alternative chart of accounts. The migration of alternative charts of accounts is simpler than that of the operational chart of accounts because you only need to migrate data in table SKAT to load the account descriptions. Then, of course, these alternative accounts need to be mapped to the operational chart of accounts.

Figure 16.38 shows the mapping of the alternative account in the **Alternative Account No.** field.

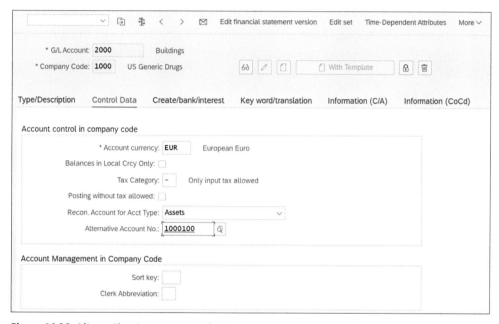

Figure 16.38 Alternative Account Mapping

The group chart of accounts is also used sometimes to depict the group accounts, which also need to be migrated. Its source is the current consolidation system used.

After migrating the general ledger master data, you also need to migrate the general ledger account balances. You should receive the final balances as of the migration date, as close as possible to the production go-live. These balances will become the initial balances in your new SAP S/4HANA system.

General ledger accounts that are managed on an open item basis are more complicated because you need to migrate not just the total balance but all open items. It's good practice to reduce these accounts as much as possible before go-live.

16.3.2 Accounts Payable and Accounts Receivable Data

In accounts payable/accounts receivable, you need to migrate the business partner master records and their open items.

Because customers and vendors in SAP S/4HANA are managed as business partners, which share the same general data, it's important to carefully check the legacy data files and avoid redundancies. Make sure that the general data is created only once for a business partner that serves as both customer and vendor.

One of the key areas to pay attention to is the bank data for business partners. Correctly migrated bank accounts and IBAN numbers ensure proper payment processes in the new SAP S/4HANA system. The main tables that are being updated when migrating the business partners on the customer and vendor levels remain the same as in previous SAP releases:

- **Table KNA1**
  Customer general data
- **Table KNB1**
  Customer company code data
- **Table LFA1**
  Vendor general data
- **Table LFB1**
  Vendor company code data

After migrating the business partner master records, you need to migrate the customer and vendor open items as well. They form the overall accounts receivable and payable balances. As with general ledger account open items, it makes sense to try to minimize them before go-live. This is a good time for the accounts receivable department to try to collect as many outstanding receivables as possible or to write off uncollectable receivables. Similarly, the accounts payable department should try to close as many open payables as possible.

16.3.3 Fixed Assets Data

Fixed assets usually are one of the more complex migration objects. In SAP S/4HANA, the process of migrating fixed assets has been significantly improved. We'll start by reviewing the fixed asset migration process prior to SAP S/4HANA. The migration was done in multiple separate steps:

1. Migrate the fixed assets master records and the fixed assets values
2. Migrate the corresponding general ledger account balance values using journal entries and reconcile them with the individual fixed assets values

Now in SAP S/4HANA, because fixed assets and the general ledger are fully integrated, there's no separate load of fixed asset values and general ledger entries. Entering the asset values automatically populates the general ledger. You can do the migration automatically or manually. For the automatic upload (which will be used in most companies, unless the volume is low enough to consider manual entry), you should use the Business Application Programming Interface (BAPI) BAPI_FIXEDASSET_OVRTAKE_CREATE for carrying over the asset data, which both creates the assets and posts the carried over values.

The manual process in SAP S/4HANA starts with the creation of the fixed assets master record using Transaction AS91 (Create Legacy Data), which was used in SAP ERP also, but is no longer used to enter asset values.

Figure 16.39 shows the initial screen of the Create Legacy Data transaction.

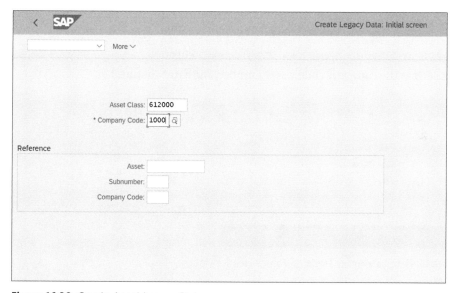

Figure 16.39 Create Asset Legacy Data

Proceed by pressing the [Enter] key. On the next screen, enter the fixed asset master data information in the different tabs, as we did in Chapter 7, Section 7.2. The difference is that now you also need to enter the capitalization date of the asset, as shown in Figure 16.40.

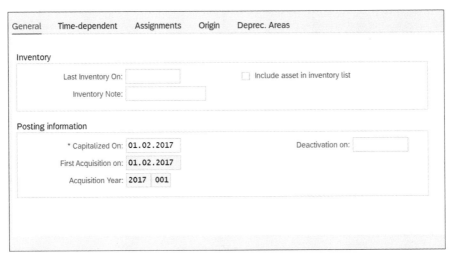

Figure 16.40 Legacy Asset Posting Information

In this transaction, you don't enter any asset values. You do that in the next step, which will post the values in the general ledger at the same time. For this step in SAP S/4HANA, you use the new Transaction ABLDT.

Figure 16.41 shows the first screen of Transaction ABLDT, in which you enter header data such as the company code, asset number, and asset subnumber.

Figure 16.41 Legacy Asset Transfer Values Header Data

Proceed with the (**Continue**) button.

In the next screen, shown in Figure 16.42, you enter the carried over values per depreciation area.

16.3 Financial Migration Objects

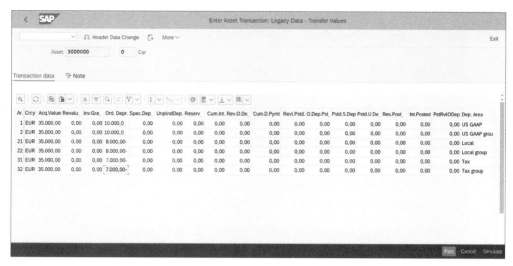

Figure 16.42 Legacy Asset Transfer: Enter Values

Typically, you enter the acquisition value and ordinary depreciation here, and the system will calculate the net book value. You can also enter other types of values, such as revaluation or unplanned depreciation. Then save the entry with the **Post** button.

In this way, both the asset master data and the values are migrated, either manually or automatically. During the migration, the system populates the general ledger tables automatically also, so there's no need for manual reconciliation between fixed assets and general ledgers as there was in SAP ERP.

16.3.4 Controlling-Related Data

In controlling, you need to migrate all the controlling objects that you're going to use and have analogs in the legacy system that you're migrating from. Cost centers and profit centers can be migrated using data from the legacy system if their structure is to remain similar. If during the implementation of SAP S/4HANA you come up with a completely new structure for your cost centers and profit centers, their data load likely won't be based on legacy system data.

If there are open internal orders and production orders in the legacy system, they also need to be migrated. This typically is required if the legacy system is an older SAP ERP system.

Another very important object to migrate that has a significant impact on controlling is the material master. The material master migration is a joint effort of almost all functional areas in the project because it combines data relevant for most functional teams. For finance, the correct migration of the material master is of utmost importance because it determines the correct costing of materials and the correct account determination and tax determination. Therefore the finance team should be heavily involved from the very beginning in the material master migration process.

Particularly important from a controlling point of view is table MBEW (Material Valuation). Table 16.1 shows the important fields from table MBEW, which should be included in the migration of the material master from a controlling valuation point of view.

Field Name	Field Format	Field Type	Field Length	Field Description
MBEW	MATNR	CHAR	40	Material
MBEW	BWKEY	CHAR	4	Valuation area
MBEW	BWTAR	CHAR	10	Valuation Type
MBEW	VPRSV	CHAR	1	Price control
MBEW	VERPR	CURR	11	Moving price
MBEW	PEINH	DEC	5	Price unit
MBEW	BKLAS	CHAR	4	Valuation Class

Table 16.1 Material Master Valuation Fields for Migration

16.4 Summary

In this chapter, you learned the processes and best practices for migrating legacy data into your new SAP S/4HANA system. We first discussed various migration options. Fundamentally, there are two different migration processes, depending on whether you're performing a brownfield or greenfield SAP S/4HANA implementation, so we dedicated a separate section to each of these migration paths, which each comes with its own tools and transactions.

You learned that during a brownfield implementation, there are number of programs and transactions provided by SAP to help you validate your customizing and data

16.4 Summary

from the existing legacy SAP system. You also learned how to perform the migration and what subsequent steps are required.

For greenfield implementations, regardless of whether the legacy system is a non-SAP system or an older SAP system that isn't going to be converted to SAP S/4HANA, SAP provides very powerful and flexible migration tools: the migration cockpit and migration object modeler. You learned that the migration cockpit is a web-based tool used to generate Excel-based data-upload templates to be populated with legacy data and to uploaded into SAP S/4HANA. You also learned how to use the migration object modeler to modify the standard-provided migration templates or to create your own.

Finally, you learned about the main finance objects that need to be migrated, along with best practices and advice on how to properly migrate them into SAP S/4HANA. In the next chapter, we will discuss testing your new SAP S/4HANA system.

Chapter 17
Testing

This chapter explains how to properly organize and conduct various stages of testing when implementing SAP S/4HANA.

In the previous chapters, we configured the finance and controlling areas of SAP S/4HANA. We also covered important integration topics related to logistics areas. You should now have a fully configured, robust solution that meets your business requirements.

The next phase in the project is the testing phase. It's hard to overstate how important testing is. No matter how well the business requirements are defined, no matter how well the system is configured, it's properly planned and executed testing that ensures the success of the project. Even the best configured system will have some glitches that only well-performed testing can track and resolve. In fact, in our experience all major issues that occurred after go-live could have been avoided with better test execution.

In this chapter, we'll teach you how to plan, organize, and execute the testing in your SAP S/4HANA implementation. We aren't going to focus on specific testing tools, of which there are many, including SAP Solution Manager and many third-party tools. A lot of them are very good, and we have successfully used a few of them. However, what's most important is to learn the processes and best practices you need to follow. At that point, most tools available on the market for test management will suffice to help you deliver very good test results.

17.1 The Testing Process

Testing is a process, and as such you need to define clearly what needs to be accomplished, in what timeframe, and what the expected results are. In this section, we'll discuss testing strategies, which will be helpful for your SAP S/4HANA implementation, but also for any other ERP implementation because the testing concepts and best practices we'll discuss are valid also for many other IT projects.

Before you start with testing, you need to plan your testing carefully, choose your testing tools, and define the requirements for testing documentation, as you'll see in the following sections.

17.1.1 Test Plan

Your test plan should clearly define the various testing phases and their subphases, if relevant; the milestones that mark the successful completion of each phase and subphase; the resources from the client and consulting sides that will participate; and the deadlines according to the project plan.

The phases of a test plan should be well integrated within the overall project plan. Typical phases in SAP S/4HANA testing include the following:

- **Unit testing**
  Unit testing is performed during the realization phase of the project, while consultants and developers are building the system. In this testing, they test the functionalities being configured and developed, usually in a sandbox client and unit testing client.

- **Integration testing**
  Integration testing is performed after the build is complete and is part of the testing phase in the overall project plan. Integration testing is usually performed in a specially designated test client. It involves experts from the whole project team, who test the integrated functionalities, focusing on end-to-end processes.

- **User acceptance testing**
  User acceptance testing (UAT) is done by the future users of the system with the help of the consulting team. Usually every process area has one or more key users designated, whose job is to thoroughly test the new system and functionalities and sign off on the testing results, thus giving a green light for the productive start of the system.

There are different ways to plan the test cycles and and the involvement of the resources. Figure 17.1 shows a possible test plan across the phases.

The testing should be organized into test cycles. Each cycle has a different scope, as you can see in the example test plan in Figure 17.1. The first test cycle starts with unit testing, in which the consulting team tests the functionalities of the system in their individual units. The execution is done by the functional consultants with only limited support from key users. After completing the unit testing within that first testing cycle, limited integration tests can be executed that test the integrated processes, still mainly by the functional consultants with some involvement of the key users.

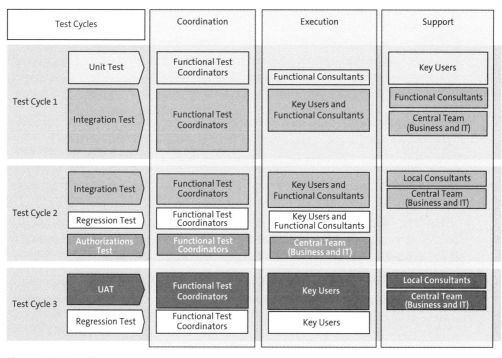

Figure 17.1 Test Plan

The second test cycle should be very expansive, testing all functionalities of the integrated system. It's a considerable effort, which should start after the system is fully developed. It includes not only testing all integrated functionalities, but also regression testing, which tests existing functionalities in the new environment. This is needed in brownfield implementations, in which even functionalities that haven't changed should be regression-tested for possible problems. It also includes authorization testing, which tests the security built in the new system and the various user profiles and roles. This is needed in both brownfield and greenfield implementations and involves checking the access to the various transactions that the users should and should not be able to access. This second test iteration should involve most project resources, not only consultants and key users, but also business and IT resources from the client side, which are not engaged by the project on a regular basis. That enables high-level confidence that the system that is being built meets the business requirements and there are no technical glitches.

Finally, the third test cycle in this sample test plan is when the users actively test the fully-integrated system This user acceptance testing is especially important because users can best attest that the new system is performing well and according to the business requirements. The role of the consultants in this cycle is supportive: to guide the users and resolve defects, but not to test directly. It also involves regression testing performed by the users to make sure the existing functionalities (for a brownfield implementation) still work correctly in the new environment.

17.1.2 Testing Tools

We touched on the testing tools topic at the beginning of the chapter. Without a doubt, your testing effort will benefit tremendously from using specialized testing tools to record the test results, manage the defects, and track the progress. Of course, it's also possible to do these tasks without specialized software, just by using Excel sheets and Word documents, but a central testing tool is much more efficient because it's easier to check and audit and can provide numerous helpful reports on the test progress.

We won't recommend a specific tool. In our experience, SAP Solution Manager does a very good job as a testing tool, and it's needed for many other functions within the project as well, so it makes perfect sense to utilize it as a testing tool. But many companies decide to adopt specialized testing software options, which often have great functionalities tailored to the testing process. We'll provide a brief overview and comparison of some of the popular testing tools that customers use when implementing SAP S/4HANA.

Quality Center

Quality Center is one of the most used testing tools for SAP implementations. It's been acquired a few times by different companies. We used it in SAP implementation projects in the early 2000s when it was known as Mercury Quality Center, with its components Test Director and LoadRunner. It was developed by the company Mercury Interactive, which was then acquired by Hewlett Packard, and the testing tool was rebranded as HP Quality Center. Since 2017, it's been owned by Micro Focus and is called Quality Center Enterprise (QC). You can find it at the following link, and a free trial version is available: *http://s-prs.co/v485703*.

QC has many useful features and is a well-organized tool. You can build a complex test plan that incorporates the various testing cycles and further subdivides them by process area and functionality. This enables you to create test scripts, which can be automatically executed in sequence in test sets. Once you encounter an issue, you can raise a defect from within the test execution screen. Defect management is another major component of the tool. You can track and manage defects, record all relevant information in a defect, and monitor the progress of the resolution. QC also offers a rich report selection, which enables you to efficiently track the testing progress and the defect management process.

Ranorex Studio

Ranorex Studio is a sophisticated testing tool tailored to the needs of testing SAP systems. It's provided by the Austrian company Ranorex GmbH. It offers reusable components, which enables faster test case development and execution. Among its useful features is video reporting of test execution, which enables you to see what happened in a test run without rerunning it.

The Ranorex Studio SAP testing offering is available at *http://s-prs.co/v485704*.

Selenium

Selenium is a web-based and portable solution, which is especially useful when testing web-based applications such as the SAP Fiori apps and portal applications. Selenium offerings are available at *https://www.seleniumhq.org/*.

Worksoft

Worksoft is a US company with extensive experience in providing sophisticated testing technology. Its Worksoft tool has proven itself in countless SAP implementations. It offers good test automation and is especially good for testing end-to-end processes. More information is available at *http://s-prs.co/v485705*.

Whatever the tool, its main tasks are to provide a comprehensive and clear structure for all the test phases and areas, record the test results, manage defects arising from testing, and provide useful and flexible reporting. Therefore, consider the following requirements for the testing tool you are going to use in your test planning.

When using a testing tool, normally there should be a hierarchical structure that contains all the test cases, organized by testing phase and functional area. It should have functions to run the test case, record test results, and record the status (passed versus failed). For a failed test case, it should provide a convenient interface to create defects, in which you can describe the expected outcome, the actual outcome, and the possible causes. The defect should be linked with the relevant test case. After resolving the defect, the test case execution can continue until it passes.

17 Testing

At any moment, the testing tool should be able to provide detailed reporting that shows the actual test results in comparison with the testing plan. This is one of the main benefits of having a dedicated test tool, versus recording the test results in separate documents.

17.1.3 Testing Documentation

Testing documentation is one of the most important deliverables of the testing process. It keeps track of the results of the testing and serves as support for the decision to sign off on the testing and approve the productive start of the system. Testing is also subject to IT audits, and the testing documentation should be kept according to the internal control standards of the organization.

The testing documentation should consist of test cases that include all the relevant functional steps that were tested. It should clearly indicate the test result in each iteration, and for failed tests it should provide detailed information about what went wrong and how it was corrected. It should also contain information about the user IDs used for testing and the date and time of execution.

There should be standardized templates used by the project for the test cases, the defects, and any other relevant testing documentation. They can be generated from the testing tool or can be set up in in Word or Excel.

Let's review sample test document designed in Excel. Figure 17.2 shows a sample test document, which has two separate sheets. The first sheet keeps track of the version control. The creation of and every change in the document is recorded here, with information regarding who made the change, when, and any relevant comments.

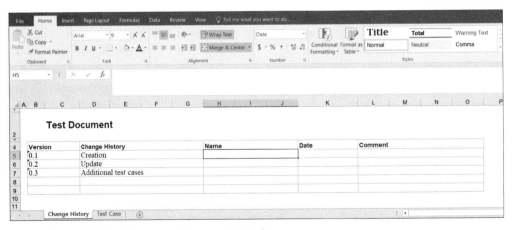

Figure 17.2 Test Document Version Control

On the second sheet, the test steps within the test case are recorded. The example shown in Figure 17.3 is for a test case for fixed asset depreciation. The header section records administrative information about the test case, such as the author of the test case, what the prerequisites are, who approved the test case, who the tester is, and the date of execution.

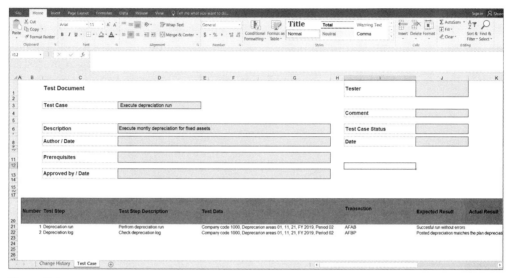

Figure 17.3 Test Case Document

This is a relatively simple test case with two test steps: run depreciation and check the depreciation log. The test document contains information about the transactions that should be executed and the required input data. It has columns for the expected and actual results.

Depending on the testing phase, test cases can be short as this one, which is a unit test, or very long, which typically is an end-to-end test case, aiming to test full business processes in their logical entirety across the areas of SAP.

Now let's deep dive into the various testing phases and their specifics.

17.2 Unit Testing

Unit testing is the process of testing newly configured and developed functionalities on their own, rather than in integration with the whole end-to-end process.

Unit testing is performed during the realization phase of the project. The realization phase consists of configuration of the various functionalities required to meet the business requirements, and the development of reports, interfaces, conversions, enhancements, forms, and workflows (RICEFW) objects, which comprise all the objects that require custom programming.

All the configuration and development objects need to be tested first by the consultants responsible for their development in unit testing. Unit testing is very important because it is the first test of the newly configured or developed functionality. Proper unit testing saves a lot of time for the next iterations of integration testing because it catches bugs and issues early in the project lifecycle.

Let's now examine the specifics of unit testing in a system sandbox client and in a unit test client.

17.2.1 Sandbox Client Testing

When you need to configure new functionality, normally first you do it in a so-called sandbox client. A sandbox client is a client in your development system that's open for customizing and is intended for research and experimenting. Normally it doesn't have very clean data because many consultants use it as a proof of concept area and to research various topics. Nothing from this client is transported to other clients. From time to time, such a client needs to be refreshed to maintain some level of quality for the test data.

Let's examine a specific example to see how the testing process should flow. Suppose you need to configure withholding tax functionality for vendors. Withholding tax is a tax that's withheld from the payment remitted to the vendor, usually according to some government mandate. You receive a requirement from the accounts payable department that it needs to submit report to the tax authorities in the system for the rollout in France for the amounts that should be withheld from the relevant vendors, but the amounts shouldn't actually be withheld from the vendor payments but just reported to the government. This is a France-specific requirement called *DAS2 reporting*.

Your first step should be to build a prototype of the solution in the sandbox client, which involves setting up the relevant withholding tax code and global settings for withholding tax calculation, then maintaining a business partner to have that withholding tax code. Then you need to test the process in the sandbox. You should construct a simple unit test case, which includes the following:

- Post the vendor outgoing invoice to a vendor with the relevant withholding tax code.
- Check withholding tax data on the invoice.
- Process a payment run that includes the vendor.
- Run report S_P00_07000134 (Generic Withholding Tax Reporting) to check the results and generate a reporting file for the authorities.

After you're satisfied with the results, you can configure the solution in a clean configuration client.

17.2.2 Unit Testing Client Testing

You should have a client that doesn't contain any transactional data, which is used for configuration. After configuring your solution there, it should be transported to a client dedicated to unit testing. In there, you can post the test data that you used in the sandbox client also. Here, however, the testing should be more formal, and you should keep test documentation proving that the test was passed successfully. It should include screenshots from the posted document, payment, and executed report.

Figure 17.4 shows the output of the withholding tax report proving the correctness of the test case.

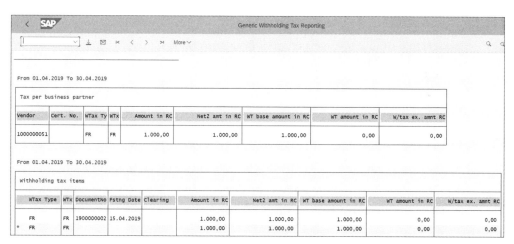

Figure 17.4 Test Case Support Documentation

Part of this test case is also to generate a DAS2 reporting file that needs to be submitted to tax authorities in France. The unit testing should include a step to check the correctness of this file to make sure the format works.

After successful completion of the unit test, the new functionality is ready to be moved to a special integration testing client, where it can tested in relation to the whole end-to-end process into which it fits.

17.3 Integration Testing

Integration testing is part of the testing phase in the overall project plan, along with the user acceptance testing, and it starts after the realization phase is complete. This means the whole build of the system should be completed, including all configuration activities and all development RICEFW objects. Any changes to configuration or development beyond that should be considered change requests.

Integration testing is performed in a quality assurance client, which should contain valid test data that closely resembles the expected productive data.

Let's discuss how to plan integration testing. Then we'll examine the various phases of the integration testing and the documentation that should be produced during the testing.

17.3.1 Planning

The project planning of the integration testing is very important because the integration testing should involve a big part of the project resources. The plan for integration testing should clearly define its objectives. These objectives could be any or all of the following:

- The defined solution that had been built is working correctly from a technical point of view.
- The integration of all SAP S/4HANA areas works correctly.
- The integration of all SAP S/4HANA business processes works correctly.
- The SAP S/4HANA configuration and development objects are well integrated.
- All interfaces with other systems work correctly.

The plan should also identify all the needed resources. There should be a clearly defined integration test plan, which lists all end-to-end scenarios to be tested and the steps that need to be executed. Each step should have a deadline to complete and assigned resources. One end-to-end scenario should be executed by multiple teams, such sales, purchasing, and finance, depending on the process.

Another part of the planning process for integration testing is to set up the testing environment. The following steps should be considered:

1. Move all configuration and workbench transports to the integration testing client. Pay special attention to the sequence of moving transports because moving them in the wrong order often poses problems. You should be using a tool for managing transports like Change Request Management (ChaRM) in SAP Solution Manager, rather than keeping track of them in Excel sheets.
2. Check the configuration before starting integration testing.
3. Complete manual configuration steps such as setting up number ranges and other nontransportable objects.
4. Set up the required master data, such as material masters, business partners, and so on.
5. Set up test user IDs and communicate them to the relevant resources.

17.3.2 Phases

The integration phase is the most extensive part of the testing life cycle and therefore should be well structured in separate phases with clearly defined milestones. Broadly, it consists of a preparation phase and an execution phase, which we'll now examine in detail.

Preparation Phase

The preparation phase includes all the steps that need to be performed prior to starting testing. Part of it is to set up the testing environment, which we explained in the previous section when we talked about planning the integration testing.

In addition, the preparation phase includes the following activities:

- Defining the scope of testing. The testing scope should include all business scenarios, end-to-end processes, interfaces to other systems, and period-end closing activities that are planned to go live in the productive system.
- Defining the relevant organizational structures and master data for testing.
- Creating test scripts, either manually or in the testing tool that's going to be used.
- Assigning testers to the testing scripts and defining the timeline of each testing step.

Also in the preparation phase, you need to define the defect management process. For each error identified during integration testing, a defect should be created that explains the issues, includes screenshots if relevant, and is assigned to the relevant technical expert that can solve it. Testing tools are especially handy in the defect management process because they enable transparent and smooth tracking and management of the defect. When using a testing tool, each defect will have its own number assigned, it will be easy to assign it to the correct resource from the correct team, and you'll be able to track the status of all relevant defects easily. You'll be able to set priorities for defects as well. The most important issues that prevent further test execution would have a very high priority, and you can assign high, medium, and low priorities to other defects as appropriate.

Execution Phase

The execution phase is when all the integration tests are executed and defects are created for the various issues that arise during testing. More specifically, the following flow of activities can be considered:

1. The testers run their assigned test cases and record the test results.
2. Testers raise defects against the issues they discover.
3. Project management tracks the timely resolution of defects and escalates when needed.
4. Project management organizes daily testing status meetings in which issues can be discussed, progress can be monitored, and any changes to the testing schedule can be communicated to the testing team.
5. Project management follows up on any reported test blockers that need to be removed promptly.
6. Test reports are generated and communicated to all relevant stakeholders.
7. The testing documentation is completed after successful execution of integration tests.

The execution of the tests when using many test tools is done from within the tool. In the tool, test scripts will be set up that contain the specific transactions to be executed in the system and the specific steps to take within these transactions. The tool should have a start and stop button, which allows you to start and end the test. At the same time, it will record a log of the execution time and duration.

When you encounter an error in the system, the testing tools normally provide a convenient interface to create a defect from within the test script.

Defects are central objects within the testing documentation, which aim to track issues in detail and record how they were resolved. These are some of their functions:

- Provide information regarding the impact of the issue in the system: low, medium, high, or very high. Low-impact issues should be able to be resolved quickly because they have lower complexity, such as a missing number range, for example. On the other hand, very high impact defects are very complex issues that require time and effort and sometimes work from multiple teams.

- Provide information regarding the priority of the issue: low, medium, high, very high. Low-priority defects do not stop further test execution, whereas without resolving the high-priority and very high priority defects further execution will be blocked, thus resulting in loss of time and resources.

- Provide information regarding the classification of issues. Using the reports on defects, you should be able to locate all defects related to product costing quickly, for example, or all very high priority defects in finance.

- Track and improve the progress of issue resolution. One of the main tasks of a proper defect management process is to speed up resolution, which is done by utilizing the reporting capabilities of the testing tool and applying management pressure where needed.

- Manage the escalation progress and keep track of the escalation logs. If a high-priority defect hasn't been resolved for a long time, project management should be able to analyze why and what happened and use these analytics to improve the process for the future.

The execution phase can consist of a couple of different iteration cycles. The first iteration may start before the build is fully complete to get better confidence in the system setup ahead of time. The more iteration cycles of integration testing are performed, the better the system will be tested and the more error-free the setup will be. But of course the benefit of having such extensive integration testing needs to be compared with the extended cost and time investments.

17.3.3 Documentation

As you can imagine, integration testing requires a lot more extensive documentation than unit testing. Whereas unit test cases can be prepared by the functional consultant, and it's not that important to use test tools for unit testing, integration test cases should be defined in strong collaboration with the business.

Figure 17.5 shows an integration test case template for controlling operating expenses (OPEX). As opposed to the unit test case that included only transactions related to the asset accounting depreciation, which was tested on its own, here the

17 Testing

process steps include transactions from controlling and the project system, along with various logistic transactions.

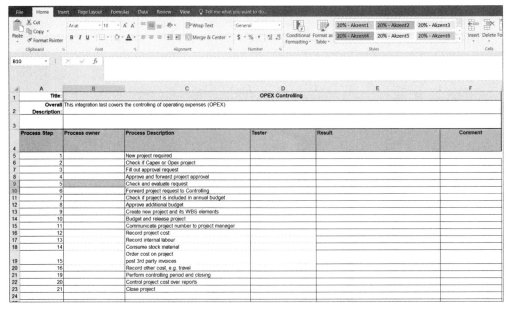

Figure 17.5 Integration Test Case

Therefore, to document integration testing is a more demanding task, and it requires the collaboration of team of testers from different process areas.

Also, the integration testing documentation is very important because it serves as a basis for approval of the test results and for moving to the next phase of user acceptance testing. Therefore it's subject to IT system audits and should comply with established internal control standards.

Once all iterations of the integration testing are complete, the testing phase of the project continues with the UAT.

17.4 User Acceptance Testing

UAT is the final testing phase, in which the customers, the future users of the system, test the system. It is after UAT that a company should finally provide sign off that the system is functioning correctly and in line with business requirements.

17.4 User Acceptance Testing

In this section, you'll learn how to perform the planning that's the foundation of successful UAT. Then we'll delve into the execution of the UAT, which has a lot of specifics compared to the execution of the unit and integration testing.

17.4.1 Planning

User acceptance testing is very important and very sensitive. In this phase, users should get the level of comfort needed to agree that the new system is built correctly and that they can use it in their day-to-day business activities. Because the most important factor in the success of an implementation is to have a good relationship and trust between the customer and the implementation team, the planning for the user acceptance testing should be perfect so that there are no unexpected issues that could have a negative impact on the user perception of the new system.

User acceptance testing should be performed in a specially designated quality assurance client. The data should be very clean and similar to the real productive data of the company. Users should be able to run through the real business processes they're going to perform in the productive system. Test cases therefore should be carefully selected sets of integration test scripts, tailored to the most important business processes.

To determine that the system is ready for user acceptance testing, the following prerequisites should be met:

- Business requirements are fully agreed upon and signed off on.
- Custom developments are fully developed.
- Unit testing, integration testing, and system testing are completed.
- There are no outstanding high- and medium-level defects in the unit and integration testing.
- Only a few cosmetic errors are acceptable before entering user acceptance testing.
- Authorization testing should be completed.
- Regression testing should be completed.
- The user acceptance testing system environment should be fully set up.

17.4.2 Execution

The execution of the user acceptance testing takes a lot collaboration because it's done exclusively by business users. The consultants take a backseat advisory role, helping users if they need advice on how to execute a test and working on any possible defects.

Logistically, it makes sense to organize the business users that are going to perform the user acceptance testing together during the weeks of the testing in an onsite location. This makes the testing process much more efficient because most business processes require a lot of collaboration between the various departments and users. Also, when working in a common onsite location, there can be efficient collaboration between the users and the consultants, the consultants can support the users much more effectively, and working on defects will be much faster.

It isn't an easy task to organize a big group of business users together at a specific time, especially because some users may have high-level positions with a lot of important obligations. Therefore the UAT should be planned well in advance so that the availability of the resources is assured in the planned weeks of testing.

It's good to organize the user acceptance testing in a central location close to a major airport. Travelling to remote locations will limit the onsite availability of resources, and executing test cases remotely and connecting with consultants via remote chats and conference calls isn't optimal for the UAT phase.

As specific steps, the UAT should include the following:

1. Development of UAT test plan
2. Defining test scenarios
3. Development of UAT test cases
4. Setting up test data, which is similar to production data
5. Execution of the test cases by the users
6. Recording the test results
7. Confirming the results and signing off on the UAT

The test scenarios should be as similar to the real business processes as possible. The users should run through test cases that represent their real productive work. Because extensive test cases are already created for integration testing, normally you should be able to reuse a lot of them, with some minor adjustments so that they include only the user-relevant steps. Therefore, the UAT should be part of the integration testing test scripts scope.

The best way to execute UAT is again using a testing tool, which allows the user to run test scripts from within the tool, record the results, and raise defects when needed. Then, because the timeframe for UAT is usually much shorter than for integration testing, these defects should be managed well and resolved in a timely manner. UAT can be organized over one, two, or three weeks, depending on the complexity, and also in a few different iterations because it will be very difficult to book three consecutive weeks of full dedication of the business users. Therefore in the weeks when you have these users onsite for UAT, there should be good project organization in place so that all technical resources are available and can collaborate efficiently to resolve defects.

Often the defects reported during user acceptance testing are not in fact system defects but are caused by misunderstandings from users or are due to not knowing the new system setup well enough. Of course, users should have already had training sessions on the new system before starting UAT, but it's still very important for the relevant functional teachers to be there with the business users and "hold their hands" along the way if needed. Some users need more attention than others because they all have different levels of SAP experience, from nonexistent to very extensive.

More specifically in the area of finance, the users performing UAT will be well experienced in accounting and controlling and may already have very good SAP experience in case of a brownfield or greenfield implementation that replaces an existing SAP ERP system. That can make the UAT execution process much easier, but still, as you've learned throughout the course of this book, there are a lot of new functionalities in SAP S/4HANA. Make sure that you educated the finance users extensively about the new finance integration in SAP S/4HANA and the Universal Journal. Controlling and fixed asset users especially will need a lot of support to learn how all data is integrated in the Universal Journal and posted in real time. For example, legacy SAP users would expect ledger-specific posting in fixed assets to be completed with a separate month-end procedure, whereas now in SAP S/4HANA it happens in real time.

17.4.3 Documentation

After execution of the various iterations of UAT and resolving the high- and medium-priority defects, explicit sign offs should be obtained for every test case from the users. This is very important because it's the UAT sign off that serves as one of the most important factors for the decision to go live with the new system. Therefore the documentation that's prepared during UAT is key and should be stored carefully for future IT audits.

The defects that are being worked on should have a clear log that shows what the causes of the issues were, how they were resolved, and the timeframe of resolution.

Sometimes defects are not truly defects but process questions or misunderstandings from the business users. Still, they need to be tracked as defects so that you can analyze what caused them and how the training and communication processes can be improved.

The test scripts for UAT should be organized by process area in the testing tool being used. An archive of these scripts should also be downloaded and made available in a collaboration platform such as SharePoint folder for future checks and audits.

UAT test scripts should contain more details about how to execute test cases because business users need more extensive instructions.

Figure 17.6 shows a sample UAT test script. As you can see, it contains more specific instructions on how to execute transactions and what the expected outcome is. The tester should record the test results, such as posted documents, and indicate whether a test passed or failed.

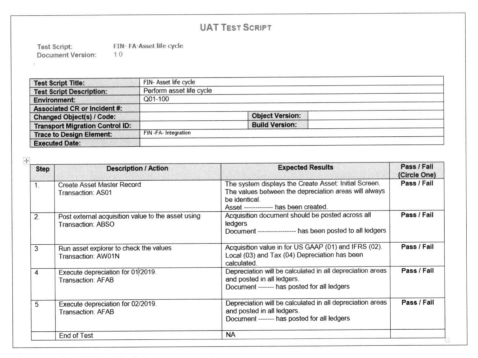

Figure 17.6 UAT Test Script

It's also good practice to have formal UAT sign-off forms that users can sign to confirm successful user acceptance testing. Figure 17.7 shows a sample user acceptance form to be completed after completing UAT. In it, the user should sign off to indicate that she's confident that the testing performed represents the business processes well and the test cases passed.

> **USER ACCEPTANCE TESTING – ACCEPTANCE SIGN-OFF FORM**
>
> Name: _____
> Position: _____
> Date: _____
>
> I confirm the testing performed represents the business process that will be performed in the productive system. As user acceptance tester, I represent the business area of .... and I confirm:
>
> - I am familiar with the features and functions of the system.
> - I perform all expected job function activities using the new features and functions.
> - I am confident that the goals of all job functions can be achieved with the features and functions of the new system.
> - I completed the provided test scripts and passed all of them.
> - I formally sign off acceptance of the ability of the system to perform all mine job function tasks and activities with the new features and functions in the business area of........................
>
> Signature/Date: _____

Figure 17.7 UAT Acceptance Form

Obtaining such forms from each tester ensures that the business users are confident in the new system.

17.5 Summary

In this chapter, you learned how to organize testing in your SAP S/4HANA implementation. You learned fundamental techniques and best practices, which are as valid for SAP S/4HANA implementations as they are for other ERP implementation projects.

We explained how to plan the various phases of the testing and how to optimize resources. We covered all the main phases of the testing process, including unit testing, integration testing, and UAT. We discussed what guidelines to follow to choose a testing tool and what functions it should provide.

You also learned how to manage the defect process in order to speed up the resolution of tickets and to track the progress of defect resolution efficiently.

We paid special attention to the testing documentation and how important it is to obtain sign off from the users on the test results. We provided specific examples of testing documents from the various testing phases.

Now that we've covered the testing phase of the SAP S/4HANA project, we can finally discuss the most exciting time: the go-live and the support phase!

Chapter 18
Go-Live and Support

This chapter will teach you how to prepare for go-live, calling attention to the most important areas and where potential pitfalls may arise. We'll also help you manage the initial production hypercare support period.

Finally we've come to the most exciting time in an SAP S/4HANA implementation project: the go-live! After months or even more than a year of hard work and dedication, the project management and the business have decided to give a green light to the new system to go live. This is indeed the most important moment in the whole project, and it needs careful planning and management. In this chapter, we'll guide you through the process of going live with SAP S/4HANA, and then we'll discuss the initial support period, which is often called hypercare support.

18.1 Preparation for the Go-Live

As with the other project phases, preparation is crucial for the success of the go-live. Here preparation plays an even bigger role: the actual go-live is just a date, so there's much more planning and preparation than execution involved. The preparation for the go-live in fact starts very early in the project with planning the go-live date. We'll discuss the various considerations for choosing a go-live date. Then we'll guide you through preparing a cutover plan for the go-live and a backup plan in case something goes wrong.

18.1.1 Choosing a Go-Live Date

Choosing a go-live date should be done well in advance. Most companies that implement SAP S/4HANA are global companies that operate in multiple regions and countries. As such, from the beginning of the project, wave planning that sets go-live dates for various countries should have been part of the workflow. It's good practice to cluster countries that are similar geographically and/or from a business point of view so that in one go-live a couple of countries will become productive. For a pilot country or a few pilot countries, it's good to choose ones that aren't the most complicated.

Some companies go for the *big bang* approach, in which many or even all countries in scope go live together. Unless the company is relatively small and doesn't operate in many countries, this approach poses more risks than benefits in our view. Especially for the first go-live of SAP S/4HANA, it's a good idea to start the productive use of the system in one or two not very complex countries.

In terms of planning, even in the early stages of the project the project management should define a wave plan and communicate it to the project and the business. Of course, such a plan can change, and in fact more often than not it *will* change. But the goal is not to slip too much in time from the initial objectives, which would involve increased cost and effort for all parties.

From our experience working in SAP S/4HANA implementations in big global companies, it's realistic to plan the go-live of a greenfield implementation for the first pilot country around one year after the start of the project. Then in the next two to three years, the system should be deployed in the main markets of the company.

In terms of brownfield implementations, that timeframe varies greatly depending on the complexity. In general, brownfield implementations should take less time than greenfield implementations, but in some companies the existing SAP system is so complex and business processes and custom developments so difficult to adjust in the new system environment that it may take as much time or even more time to prepare the go-live.

Now let's talk about the actual go-live. It's always planned on a weekend to minimize disruptions for the business and to have more time to react to unexpected issues. In general, especially from a finance point of view, it's good for the go-live to coincide with the beginning of a new fiscal year. That makes the migration of legacy data much easier and is better from an audit point of view. However, depending on the timeline, that might not be practical. But if your project is contemplating going live in the last one or two months before the end of the fiscal year, we strongly advise setting the go-live for the beginning of the new fiscal year.

18.1.2 Defining a Cutover Plan and Responsibilities

For the go-live, you need to prepare a formal project plan called a *cutover plan*, named because the process of converting to a new system is called a *cutover*.

The cutover plan defines clearly the responsibilities of the various teams and resources related to the go-live and indicates the sequence of tasks that need to be performed.

Most project team members and a lot of business resources are involved in the go-live, so the cutover plan is very important for making everyone's responsibilities clear.

The cutover plan can vary a lot from project to project. It should include both pre-go-live tasks and post-go-live tasks.

Here are some mandatory tasks that should be included in an SAP S/4HANA cutover plan:

- Make sure all transport requests are transported in the productive system.
- Load master data in the productive system.
- Load balances and open items in the productive system.
- Make sure number ranges are defined in the productive system (because they're not transported).
- Make sure tax codes are defined in the productive system (because there's a manual import step).
- Make sure the operating concern environment is properly generated.
- Ensure cost estimates are marked and released.

18.1.3 Preparing Back-Up Plan

After months of hard work and dedication, nobody wants to even consider that the go-live could be a failure. But sometimes, very rarely, it happens. Among the myriad, mostly successful SAP implementations, there are a couple that have failed. The reasons can differ: the system couldn't meet some important business requirements by design, perhaps, or important functionality doesn't work because of some technical glitches and this stops important processes such as placing customer sales orders.

In any case, you have to be prepared for problems once the system goes live, even if you need to switch it off and go back to the legacy system. These very unlikely scenarios should be clearly defined in a back-up and restore plan.

When going live with any system, you should have the technical option if needed to go back to the legacy system. The roles and responsibilities for this should be written down in the back-up plan with clearly defined, short, and specific timelines to ensure a quick return to a productive system.

Practically, when going live with SAP, it's most important that day-to-day business operations can continue smoothly and without interruption. For example, one potential issue could be that the system can't process sales or purchase orders because of issues with material masters, prices, or interfaces with other systems. These technical details therefore should be tested many, many times in multiple iterations of the testing phase, which we described in detail in the previous chapter. Still, to err is human, so you should hope for the best but plan for the worst. With a good, clear back-up plan, such issues shouldn't be damaging to the business or to the new system. Even if you have to stop operations with the new system, if you organize timely resolution of the critical issues, you should be up and running on the new system again very quickly. It's important to maintain the trust in the new system for both the high-level management and the end users that will use the system in their day-to-day operations.

18.2 Activities during the Go-Live

As mentioned earlier, the go-live should be done over a weekend. Most of the activities related to the go-live will be performed during that go-live weekend. After that, there will be some validation activities.

In this section, we'll discuss the activities that should be performed during the go-live from a general project point of view, and more specifically for finance.

18.2.1 Technical Activities

The go-live includes a lot of technical tasks in which the Basis team plays most important role. SAP also offers extensive support to make sure the system is ready from a technical point of view with its SAP GoingLive Check service. This proactive service reduces risks and ensures the go-live will be technically sound. We strongly recommend using this service, and you can find more information at *http://s-prs.co/v485706*.

Another important SAP service is SAP EarlyWatch Check. This is used to audit and review your new SAP S/4HANA system just after the go-live and periodically thereafter. More information is available at *http://s-prs.co/v485707*.

Another important project activity is to ensure all user IDs have been created, passwords communicated, and relevant authorization roles assigned. SAP users can access the system using the traditional SAP GUI known from previous releases and the new SAP Fiori web-based interface. We strongly encourage users to use more of SAP Fiori and less of SAP GUI. SAP Fiori is the user interface of the future, and more and more user transactions and reports will become available over time as SAP Fiori apps. Users should have access to SAP Fiori and be well trained in using it and navigating it.

As with previous SAP releases, your new SAP S/4HANA system will run a lot of regularly scheduled jobs. An example is the asset depreciation program, which we covered in detail in Chapter 7. It's part of the technical activities during the go-live to make sure all jobs are scheduled.

Another technical step is to lock down the system. During the go-live, no users should be allowed in the system except those that have to perform specific tasks at specific times.

18.2.2 Financial Accounting Activities

There are several financial accounting activities that need to be performed during the go-live.

Financial master data such as general ledger accounts, cost centers, profit centers, and segments usually are migrated to the productive system ahead of the go-live because they're stable objects that are unlikely to change in the last minutes. However, it's during the go-live that general ledger account balances and customer, vendor, and general ledger open items are migrated to the productive system with the most up-to-date numbers. Of course, this process should have been tested already multiple times in the test and quality assurance systems. Fixed asset values also are migrated to production during the go-live with the most up-to date values.

We configured tax codes in Chapter 3. The tax codes are not automatically included in transports. There's a function to export and import tax codes, and the import needs to be included in your cutover activities. To export tax codes, follow menu path **Financial Accounting • Financial Accounting Global Settings • Tax on Sales/Purchases • Calculation • Define Tax Codes for Sales and Purchases**, or enter Transaction FTXP. After selecting the relevant country in the **Country Key** field in the screen shown in Figure 18.1, select **More • Transport • Export** from the top menu.

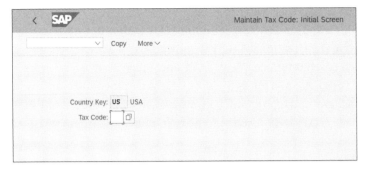

Figure 18.1 Export Tax Codes

The system will ask you for transport request in which to include the tax codes, as shown in Figure 18.2.

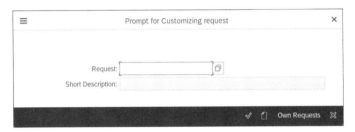

Figure 18.2 Prompt for Customizing Request

You can add the tax codes to an existing transport request or create a new one using the ▣ (**Create Request**) button.

Similarly, during go-live you need to import the tax codes. In Transaction FTXP, select **More • Tax code • Transport • Import** from the top menu, as shown in Figure 18.3.

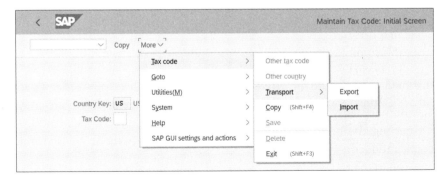

Figure 18.3 Import Tax Codes

Then select your transport with tax codes in the screen shown in Figure 18.4.

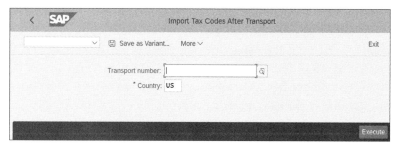

Figure 18.4 Select Import Tax Codes Transport

Process the request with the **Execute** button.

This generates a batch input session, which is executed automatically and which you can monitor in Transaction SM35, as shown in Figure 18.5.

Figure 18.5 Import Tax Codes Batch Input Session

The number of errors is shown in the column with the ⚡ heading, whereas the number of succesfully processed transactions is shown in the column with the ✓ heading. If you double-click the line, you'll see further details in the batch input log.

Another important task is setting the number ranges. You may remember that when we defined number ranges, we didn't get transport numbers. There's good reason for that: transporting number ranges can cause inconsistencies across clients, and especially in the production system. Therefore you should keep track of the number ranges defined in the test and quality assurance systems and have a checklist ready to maintain the required number ranges in the production system.

18 Go-Live and Support

18.2.3 Controlling Activities

In controlling, there are also some activities that need to be included in the cutover tasks to be performed during the go-live.

In profitability analysis, you need to make sure that the data structure is active and that both the cross-client part and client-specific part of the operating concern are activated. To do so, follow menu path **Controlling • Profitability Analysis • Structures • Define Operating Concern • Maintain Operating Concern**.

As shown in Figure 18.6, the **Data structure** should have a green status. If it doesn't, you should activate it with the → **Activate** button.

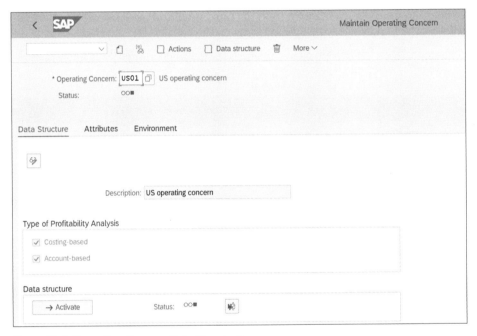

Figure 18.6 Operating Concern Data Structure

Next, click the **Environment** tab. As shown in Figure 18.7, both the **Cross-client part** and **Client-specific part** of the operating concern should have a green status. If they don't, you should activate them with the ⚙ (**Activate**) button.

In product costing, you need to make sure that the cost estimates are marked and released. There's a convenient way to check the standard prices by looking in the table content of table MBEW, which is the material valuation table. You can review the data structure of the table using Transaction SE11.

18.2 Activities during the Go-Live

Figure 18.7 Operating Concern Environment

Standard prices are stored in field STPRS, as shown in Figure 18.8. In field VPRSV, the price control of the material is stored. **S** indicates a standard price material, whereas **V** denotes a moving average price.

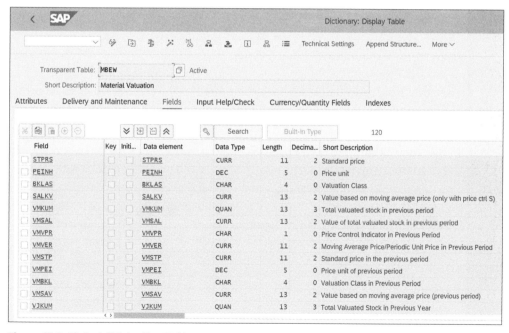

Figure 18.8 Material Valuation Table

18   Go-Live and Support

To browse the table content, enter Transaction SE16N. Enter "MBEW" in the **Table** field, as shown in Figure 18.9.

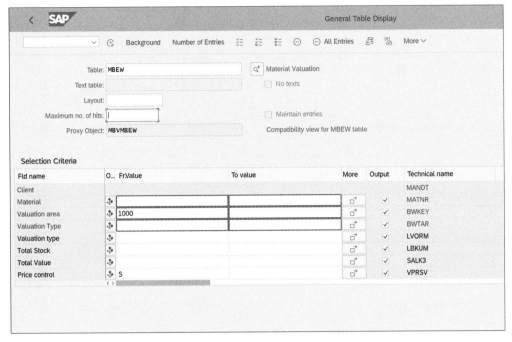

Figure 18.9 Material Valuation Table Browser

You can restrict by valuation area, which normally is the plant, and by price control S to check the standard price materials. Then proceed with the (**Online**) button, which leads to the table contents screen shown in Figure 18.10.

In the **Std price** column, you can see the standard prices. All **S** materials should have a standard price maintained. The release of the standard prices is done with Transaction CKME, which should be part of the go-live activities for controlling. As shown in Figure 18.11, you can run the transaction per plant and/or by range of materials. Normally, it will be scheduled as a background job.

18.2 Activities during the Go-Live

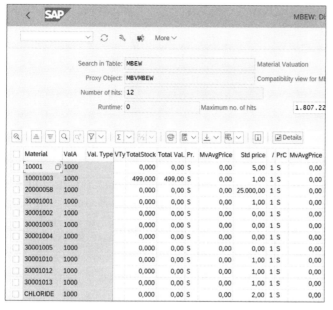

Figure 18.10 Material Valuation Table Content

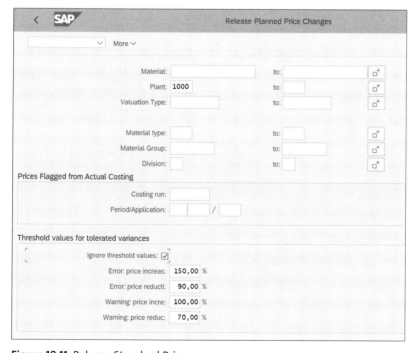

Figure 18.11 Release Standard Prices

18.3 Validation of the Go-Live

At the end of the go-live weekend, before officially opening the system to users, it needs to be validated by the project team and the key subject matter experts. Their availability should be aligned so that during the weekend they can execute some checks to make sure there will be no surprises when the users start using the system the next day.

18.3.1 Project Team Validation

The project team is the first to check the productive system. Each functional team should carefully check its process area. In terms of finance and controlling, that involves running some reports and checking some tables, which we covered in the previous sections. But once again you need to make sure that the relevant master data is properly created, number ranges are set up, tax codes are correctly set up, controlling structures and environments are generated, and materials have the correct pricing.

There should be a formal process in which each functional team goes over certain predefined key transactions and reports on and records the results. At the end, each functional team lead should sign off that the go-live is successful and the system is ready for business use.

The same is true for the Basis team, which needs to make sure that from everything is working correctly a technical point of view, and for the development team, which needs to validate the interfaces, the custom programs, and other RICEFW objects.

This is a lot of work to be done in a very short time, so good project management and coordination are extremely important. It's the responsibility of the project management to coordinate all the project resources and to be able to react quickly if issues are discovered.

Once the project team validates the system, next the business subject matter experts should check the system from their end.

18.3.2 Subject Matter Expert Validation

Several subject matter experts from the business should be nominated to be involved in the go-live weekend, executing critical checks and reports to validate the system from a business point of view. These normally would be the same key users that worked with the consultants during the implementation. They should know well what to expect from the new system and the key areas that need to be checked.

In the areas of finance and controlling, there should be separate resources to check the general ledger, fixed assets, accounts payable and accounts receivable areas, product costing, overhead accounting, and profitability analysis. They should be asked to be available during the later stages of the go-live and to execute the reports that will assure them that the system is set up correctly. For example, their activities should include checking account balances and open items, checking material prices, making sure accounting periods are open as needed, and so on.

At the end, a formal written sign-off should be required from each subject matter expert validating the system. And again, if critical issues are discovered, the project management should have a strategy in place to react quickly. Sometimes the difficult decision of whether to go live as planned or not may need to be made, depending on the criticality and the significance of the issues.

But hopefully there will be no major issues, and the project management will officially announce the go-live of the new SAP S/4HANA system! The system will be open to the business users, and the next exciting period will start: hypercare production support.

18.4 Hypercare Production Support

So finally, you've made it! The SAP S/4HANA system is productive from its first day, but the work isn't quite done yet. After the system is live, production support is very important, especially in the first weeks, which usually is called the *hypercare production support* period. This is the period starting immediately after the go-live and covering usually a month or so. During that period, the production support is provided by the project team itself. At the same time, the project team is performing knowledge transfers to hand over the production support to the long-term resources that will handle it after the hypercare production support period is over.

We'll start with a discussion of what to expect on the first day after the go-live. Then we'll cover how to schedule background jobs in the productive environment. In the next section, we'll teach you how to manage critical support incidents. Finally, we'll discuss how to transfer support from the project team that provides the hypercare support to the long-term support team.

18.4.1 The First Day

The first day after the go-live, which usually is on a Monday, is very important, so it's critical to organize the support so that the project team is onsite together with the customers, supporting the business users in their day-to-day tasks. Of course, at this point it's expected that the users are well trained and can work with the new SAP S/4HANA system by themselves, but it's still good for them to know that the consultants are there if needed.

It's vital to have a good process to manage issues. A defect management system should be used in which users can report defects found in the system. As during the testing phase, these defects should be created with the correct priority and assigned to the relevant project resources.

It's normal on the first day to battle with simple issues such as lost user passwords, not being able to log on, and so on. Although these are easy to resolve, sometimes they take unnecessary time due to lack of proper organization, which causes frustration for users. Therefore, proper organization should be in place to react immediately and be able to resolve such low-hanging-fruit types of issues quickly.

It's important during the first day to make sure there are no major issues that could stop the business or the proper financial reporting. The project team needs to explicitly ensure that all the main business processes run smoothly, such as placing sales orders, posting goods issue, invoicing customers, and purchasing materials. Particularly in finance, the invoicing process and the payment process are very critical.

Some of the specific functionalities that should be checked from a finance standpoint during the first day are as follows:

- Material prices
- Customer invoices
- Vendor invoices
- Automatic payment program
- Correct tax determination
- Account determination

18.4.2 Background Jobs

It's the responsibility of the project team to schedule the background jobs that should be performed in the live system. These vary from project to project, but some

18.4 Hypercare Production Support

typical finance jobs that should be scheduled as background jobs include the asset deprecation program, the automatic clearing program for open items, internal order settlement, and so on.

Sometimes scheduled background jobs fail due to various issues. This could happen, for example, if accounting periods are closed and a program such as asset depreciation that needs to post in financial accounting can't do so. Therefore an important part of the support activities is to monitor the scheduled jobs for errors.

Good practice is to set up email notifications so that if a background job is cancelled due to an error, certain resources will receive an email. Then the responsible experts can analyze the issue and take action.

You can monitor background jobs with Transaction SM37. If you check only the **Canceled** checkbox, as shown in Figure 18.12, you'll see the canceled jobs for the given time period. You can search by job name here. Entering "*" in the **Job Name** field uses the asterisk wild card to search all job names. The same is true for the **User Name** field, in which you can enter * for all user names or you can enter a specific user name that's used to run certain background jobs. Normally, there will be dedicated user names used to run background jobs only.

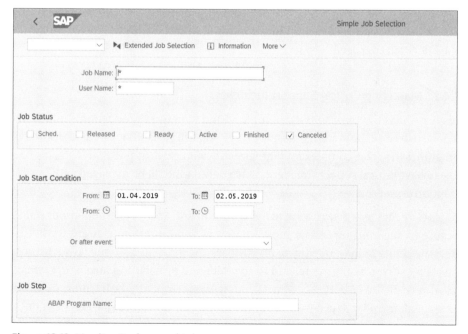

Figure 18.12 Monitor Background Jobs

Execute with the **Execute** button. On the resulting list, shown in Figure 18.13, you can see further information by selecting a job and clicking the ⓧ Spool (**Display Spool List**) button.

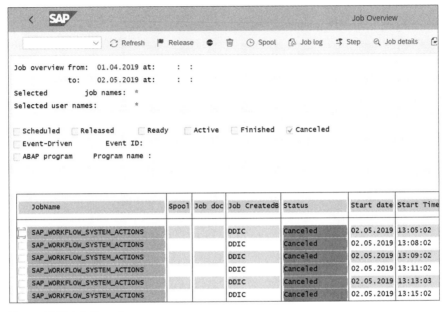

Figure 18.13 Job Overview

18.4.3 Managing Critical Support Incidents

This is perhaps the most important topic within hypercare production support. We all hope that once the system goes live everything will go as smoothly as possible, but sometimes in such a complex and integrated system such as SAP S/4HANA, it's possible to have critical support incidents. Such defects should be managed with utmost attention and speed.

We already discussed some cases in which major SAP implementations failed because of such incidents. The most important and critical defects are those that affect the ability of the company to do business and generate revenue. As such, those defects most commonly are related to the sales and distribution area. Purely financial issues can also be very problematic, but without stopping the business itself. For example, an inability to generate tax declarations may result in some fines, but nothing can compare with the inability to process sales orders, leading to missed revenue and tarnishing relationships with customers. Supply chain issues also can be very bad because they can disrupt the supply flows that are needed to generate sales and can affect relationships with vendors.

But sometimes such logistics problems are in fact rooted in finance. Missing account determinations, missing costings of materials, and the wrong tax codes are just some of the bugs in the finance area that could disrupt the whole logistics chain.

Therefore it's very important to manage critical support incidents in an integrated manner. After such a very high priority defect is opened, the work on it should be started immediately by multiple teams. The project should organize daily integration meetings to discuss and monitor the progress on high-priority tickets. Sometimes it's not that easy to determine the root cause of an issue and teams might play the blame game, stating that the issue isn't in their area. Here, project management will play a crucial role, facilitating the integration work across the teams to ensure timely resolution.

18.4.4 Organizing Long-Term Support

The hypercare support is limited because project resources are very expensive. Sooner or later, each system needs to go into regular maintenance mode. The project resources should gradually be phased out of the project, and the support work should be handed to long-term support resources.

There are various strategies to organize the long-term support of the new SAP S/4HANA system. Some companies opt in to organizing the support internally with their own SAP support teams. Such a strategy has the advantage that the company can build vast SAP knowledge internally.

But SAP support is a huge task, and some companies don't want to manage so many IT resources internally, preferring to focus on their core business. Therefore, more common currently is the approach to outsource the SAP support to companies that specialize in that area. Many consulting companies are available that have established many SAP support centers, typically in lower-cost locations.

Each approach has advantages and disadvantages, and normally this is a high-level management decision based on the needs of the company. Whatever the approach, the process of handing over the long-term support is the responsibility of the project implementation team. The project team should prepare very good documentation that includes all customizing being done, all functional specifications and development objects, and business process user guides. Such documentation should be available for reference, both for users and for future support consultants.

Also, there should be an organized formal knowledge transfer process in which experienced project resources can share their knowledge with the future support resources, which will take on their role in supporting the system after the hypercare period is finished.

The management of the long-term support also should be done using a defect management system. Users will continue to raise defects for the various issues they encounter, indicate their priority, and the long-term support team will provide resolution. The main difference is that such long-term support normally will be remote.

Some big companies will employ a long-term support strategy in which on a local level there are some SAP support specialists, who are typically called level 1 (L1) support and aren't highly experienced. Whatever issues they aren't able to resolve they'll pass to level 2 (L2) support consultants, who are more experienced and are responsible for clusters of countries. Such a hierarchy could be expanded even further with level 3 (L3) consultants, who are very experienced and work on the most difficult defects globally across regions.

18.5 Summary

In this chapter, we covered the go-live of the SAP S/4HANA system and the production support, paying special attention to the initial hypercare support period.

Now you know good practices for choosing the go-live date for your S/4HANA implementation wisely and for using the wave approach when implementing the system across multiple countries and regions. You know how to prepare a cutover project plan that clearly defines the responsibilities and timeline for the go-live activities.

We've covered the various tasks during the go-live in detail, both from a general technical perspective and more specifically for finance and controlling. You also learned how the system should be validated by project team members and subject matter experts during the go-live weekend.

Last but not least, you learned how to organize hypercare production support and how to transfer the support activities to the long-term support organization.

Congratulations! You've completed the journey to implement your new SAP S/4HANA system. Hopefully this was a very rewarding experience, and your business now has a powerful, state-of-the-art system that will enable it to streamline its operations and achieve tremendous value in all business process areas.

Appendix A
Obsolete and New Transaction Codes and Tables in SAP S/4HANA

SAP S/4HANA is the biggest change for SAP in the last 25 years. The whole data model has changed, especially in the area of finance. There are new tables and obsolete tables, and many transaction codes are obsolete either because they were moved to SAP Fiori apps or because of changes in functionality.

Because many of the readers of this book have experience with the older SAP releases, in this appendix we'll list the obsolete and new transaction codes and tables in SAP S/4HANA compared to previous releases.

Table A.1 lists obsolete and new financial transaction codes.

Old Transaction	New Transaction	Description
FS01	FS00	Create G/L Account
FS02	FS00	Change G/L Account
FS03	FS00	Display G/L Account
FK01	BP	Create Vendor
FK02	BP	Change Vendor
FK03	BP	Display Vendor
FD01	BP	Create Customer
FD02	BP	Change Customer
FD03	BP	Display Customer
KP06	FCOM_IP_CC_COSTELEM01	Change Cost and Activity Inputs
KP07	FCOM_IP_CC_COSTELEM01	Display Cost and Activity Inputs

Table A.1 Obsolete and New Transaction Codes

Old Transaction	New Transaction	Description
KP65	FCOM_IP_CC_COSTELEM01	Create Cost Planning Layout
KP66	FCOM_IP_CC_COSTELEM01	Change Cost Planning Layout
KP67	FCOM_IP_CC_COSTELEM01	Display Cost Planning Layout
CK11	CK11N	Create Product Cost Estimate
CK13	CK13N	Display Product Cost Estimate
CK41	CK40N	Create Costing Run
CK42	CK40N	Change Costing Run
CK43	CK40N	Display Costing Run
CK60	CK40N	Preselection for Material
CK62	CK40N	Find Structure: BOM Explosion
CK64	CK40N	Run: Cost Estimate of Objects
CK66	CK40N	Mark Run for Release
CK68	CK40N	Release Costing Run
CK74	CK74N	Create Additive Costs
KB11	KB11N	Enter Reposting of Primary Costs
KB21	KB21N	Enter Activity Allocation
KB31	KB31N	Enter Statistical Key Figures
KB33	KB33N	Display Statistical Key Figures
KB34	KB34N	Reverse Statistical Key Figures
KB51	KB51N	Enter Activity Posting
KE21	KE21N	Create CO-PA Line Item
KE23	KE24	Display CO-PA Line Item
KKE1	CKUC	Add Base Planning Object
KKE2	CKUC	Change Base Planning Object

Table A.1 Obsolete and New Transaction Codes (Cont.)

A Obsolete and New Transaction Codes and Tables in SAP S/4HANA

Old Transaction	New Transaction	Description
KKE3	CKUC	Display Base Planning Object
KKEC	CKUC	Compare Base Object—Unit Cost Est
KKED	CKUC	BOM for Base Planning Objects
KKB4	CKUC	Itemization for Base Planning Obj.
KKBF	KKR0	Order Selection (Classification)
F.05	FAGL_FCV	Foreign Currency Valuation
F.24	FINT	A/R: Interest for Days Overdue
F.2A	FINT	A/R Overdue Int.: Post (without OI)
F.2B	FINT	A/R Overdue Int.: Post (with OI)
F.2C	FINT	Calc.cust.int.on arr.: w/o Postings
F.4A	FINTAP	Calc.vend.int.on arr.: Post (w/o OI)
F.4B	FINTIAP	Calc.vend.int.on arr.: Post (with OI)
F.4C	FINTAP	Calc.vend.int.on arr.: w/o Postings
FA39	Obsolete	Vendors: calc.of Interest on Arrears
F.47	FINTAP	A/R: Interest for Days Overdue

Table A.1 Obsolete and New Transaction Codes (Cont.)

As we discussed many times, the data model in SAP S/4HANA has changed significantly, and the index and totals tables are obsolete. They've been replaced by compatibility views to make sure that older custom programs will continue to work. Technically, they're called CDS views. Table A.2 lists obsolete finance tables and the corresponding compatibility and data definition language (DDL) source.

Obsolete Table	Compatibility View	DDL Source
BSAD	BSAD	BSAD_DDL
BSAK	BSAK	BSAK_DDL

Table A.2 Obsolete Tables and Compatibility Views

739

Obsolete Table	Compatibility View	DDL Source
BSAS	BSAS	BSAS_DDL
BSID	BSID	BSID_DDL
BSIK	BSIK	BSIK_DDL
BSIS	BSIS	BSIS_DDL
FAGLBSAS	FAGLBSAS	FAGLBSAS_DDL
FAGLBSIS	FAGLBSIS	FAGLBSIS_DDL
GLT0	GLT0	GLT0_DDL
KNC1	KNC1	KNC1_DDL
KNC3	KNC3	KNC3_DDL
LFC1	LFC1	LFC1_DDL
LFC3	LFC3	LFC3_DDL
COSP	COSP	V_COSP_DDL
COSS	COSS	V_COSS_DDL
FAGLFLEXT	FAGLFLEXT	V_FAGLFLEXT_DDL
ANEA	FAAV_ANEA	FAA_ANEA
ANEK	FAAV_ANEK	FAA_ANEK
ANEP	FAAV_ANEP	FAA_ANEP
ANLC	FAAV_ANLC	FAA_ANLC
ANLP	FAAV_ANLP	FAA_ANLP
BSIM	V_BSIM	BSIM_DDL
CKMI1	V_CKMI1	V_CKMI1_DDL
COEP	V_COEP	V_COEP
FAGLFLEXA	FGLV_FAGLFLEXA	FGL_FAGLFLEXA
MLCD	V_MLCD	V_MLCD_DDL

Table A.2 Obsolete Tables and Compatibility Views (Cont.)

A Obsolete and New Transaction Codes and Tables in SAP S/4HANA

Obsolete Table	Compatibility View	DDL Source
MLCR	V_MLCR	V_MLCR_DDL
MLHD	V_MLHD	V_MLHD_DDL
MLIT	V_MLIT	V_MLIT_DDL
MLPP	V_MLPP	V_MLPP_DDL
T012K	V_T012K_BAM	V_T012K_BAM_DDL
T012T	V_T012T_BAM	V_T012T_DDL
FMGLFLEXA	FGLV_FMGLFLEXA	FGL_FMGLFLEXA
FMGLFLEXT	FGLV_FMGLFLEXT	FGL_FMGLFLEXT
PSGLFLEXA	FGLV_PSGLFLEXA	FGL_PSGLFLEXA
PSGLFLEXT	FGLV_PSGLFLEXT	FGL_PSGLFLEXT
JVGLFLEXA	FGLV_JVGLFLEXA	FGL_JVGLFLEXA
JVGLFLEXT	FGLV_JVGLFLEXT	FGL_JVGLFLEXT

Table A.2 Obsolete Tables and Compatibility Views (Cont.)

Appendix B
The Author

Stoil Jotev is an SAP S/4HANA FI/CO solution architect with 20 years of consulting, implementation, training, and project management experience. He is an accomplished digital transformation leader in finance. Stoil has delivered many complex SAP financials projects in the United States and Europe in various business sectors, such as manufacturing, pharmaceuticals, chemicals, medical devices, financial services, fast-moving consumer goods (FMCG), IT, public sector, automotive parts, commodity trading, and retail.

Index

A

Access sequence 107–108, 227–228
Account assignment 143, 387, 545
   objects 488
Account clearing 144
Account determination 131, 255
   select 257
Account group 117, 203–204
   copy 205
   field status 205
Account numbers 202
Account symbols 333–334
Account-based profitability analysis 37
Accounting clerks 324
Accounting principles 93
Accounts payable 157
   information system 185
Accounts receivable 197
   business transactions 213
   information system 231
   line item reports 236
   master data reports 232
Accrual calculation 391
Accrual cost center 391
Accrual order 392
ACDOCA 34, 74, 240
Acquisitions 268
   documents 270
Activity allocation 360, 406
   cycle 407
   settings 362
Activity type 379
   columns 414
   create 380
   manual planning 410
   per cost center 381
   prices 382
   settings 380
Actual cost component split 621
Actual costing 37, 613–614, 621
   cockpit 622
   reports 626

Actual costs 420
   report 470
Actual Plan Comparison report 417
Actual postings 384, 449
Additive costs 595
Aging of open items 236
Allocation structure 402
Allocations 466
Alternative reconciliation account 168
Analysis pricing 221
Application Link Enabling (ALE) 118
Assessment 400
   cycle 400
   fields 401
   segment 402
Asset accounting 38, 277
   history sheet 295
Asset balances report 290
Asset class 248
   create 249
   define 248
   SAP-delivered 249
Asset explorer 288
   comparison 289
Asset history sheet 292
Asset History Sheet app 296
Asset master record 254
Asset numbers 265
Asset subnumber 266
Asset supernumber 262, 265
Asset transaction type 272
Asset transfer 270
   variant 272
Assets under construction (AUC) 250, 268
Assignment lines 538
   FI/MM 542
   new 539
Automatic account assignment 385
Automatic payment program 308
   common issues 323
   configuration 311
   global settings 311
   parameters 309

Automatic postings 131, 134, 142
   configuration 134
   criteria 136
   group 135
   materials management 136
Availability controls 445, 447

B

Background jobs 732
Back-up plan 721
Balance carryforward 145, 152, 286
Balance check 649
Balance reports 147, 187, 233
Bank account 306
   create 308
   details 307
   determination 321
   IDs 322
Bank Account Management 306
Bank Account Management app 322
Bank Account Management Lite 306
Bank accounting 299
Bank determination 320
Bank directory
   data 300
   file formats 301
Bank key 300
   control fields 305
   details 302
   manual creation 301
Bank selection 319
Big bang approach 720
Bill of materials (BOM) 574
Billing 530
   document 533
Brownfield implementation 49, 331, 614
   migration 663
Budget profile 429, 443
   number range 446
   settings 446
Budgeting 443
Business blueprint 59
Business partner 157, 197
   company code 160
   configuration 160, 204

Business partner (Cont.)
   customer link 211
   general data 158, 198
   groups 160, 209
   groups number ranges 209
   number ranges 161
   role 159
   roles 211
   search 158
   synchronization control 164
   vendor link 164–165

C

Calculation bases 603
   cost elements 604
Calculation methods 281
CDS views 75
Change Request Management (ChaRM) 709
Characteristic 513
   create 516
   custom 516
   define 514
   list 515
   settings 517
Characteristic derivation 527
Characteristic hierarchy 524
Chart of accounts 83, 87, 114, 333, 690
   assigment 117
   country 114
   settings 115
Chart of depreciation 240–241
   copy 243
   depreciation area 245
   description 244
   settings 242
Check programs 664
Check tables 518
Checks 319
Classification 430
Closing 276
Cluster wave approach 53
Column store 30
Common Global Implementation (CGI)
   DMEE XML Format 330
   initiative 327

Index

Company 77
　confirmation of copy 81
　create 78
Company code 78, 159, 200, 303, 311, 320, 482
　activation 549
　copy 80
　create 79
　cross-company 140
　currency 615
　customer level 200
　details 82
　forms 313
　global settings 83
　selection 82
Condition records 107, 175, 225, 229
　destination country 229
　settings 175
Condition tables 219, 227, 532
　access fields 229
Condition type 105, 173, 225
　fields 174, 227
　for sales 226
Consolidated financial statements 632
Consolidation 634
　configuration 641
　data collection tasks 653
　document type 647
　investment method details 656
　investments methods 655
　number ranges 652
　task group 658
　tasks 654
Consolidation ledger 635
　create 636
　settings 637
Consolidation version 637
　copy 639
　settings 638
Controlling 65, 341
　actual postings 357
　change groups 347
　data migration 695
　general settings 341
　groups 348
　master data 352

Controlling (Cont.)
　number ranges 345
　obsolete tables 75
　partner update 430
　real-time integration 76
　requirements 65
　transactions 346
　version 462
Controlling area 81, 84, 342, 426
　activities 343
　company code assignment 87
　components 344
　copy 85
　maintain 85, 343
　settings 86, 345
Conversion guide 664
Correspondence type 214, 216
　copy 216
　cusomter invoice 217
　form names 218
　programs 217
Cost center 359, 366
　address 372
　basic data 369
　category 366
　collective processing 374
　communication data 373
　control fields 370
　create 367
　expanded hierarchy 377
　history 373
　manual planning 410
　numbering 368
　settings 366
　standard hierarchy 376
　templates 370
Cost center accounting 365
　information system 415
　periodic allocations 391
　reports 416
Cost center group 376
　maintain 378
　structure 379
Cost component 546
　cost elements 601
　define structure 599

747

Cost component (Cont.)
　　element 600
　　structure 546, 548, 597–598
　　views 602
Cost element 352, 448
　　accounting 341, 360
　　category 354
　　exempt 447
Cost element group 355
　　create 355
　　definition 356
　　structure 356
Cost object 385–386
Cost of goods sold 511
Costing run 611
　　selection 612
Costing sheet 603, 609
　　credits 607
　　define 609
　　details 609
Costing type 579–580
　　parameters 581
Costing variant 579, 593
　　additive costs 595
　　assigments 597
　　control 593
　　quantity structure 594
　　update 596
Costing view 573
Credit memos 166, 213, 218
　　incoming 170
Credits 607
　　details 608
Cross-client table 215
Cross-company code postings 345
Currencies 92, 100, 440, 459, 650
　　configuration 101
　　exchange rate 102
　　exchange rate type 101
　　multiple 615
　　types 87, 100
Customer 198
　　account balances 234
　　account management 200
　　address data 199
　　balance report 234
　　business partner link 212

Customer (Cont.)
　　correspondence 201
　　due date analysis 236
　　fields 198
　　groups 204
　　groups number ranges 207
　　invoice printing 217
　　list selections 232
　　master data report 232
　　number ranges 209
　　sales data 202
Customer-vendor integration (CVI) ... 163, 204, 211
Cutover plan 720
Cycle definition 467

D

DAS2 reporting 706
Data collection 641
Data migration 663
　　checks 665
　　preparations 666
Data monitor 648
Data structure 520, 726
Default assignments 646
Depreciation accounts 259
Depreciation area 244, 258, 278, 694
　　create 246
　　revaluation 283
Depreciation comparison 290
Depreciation key 279
　　configuration 280
Depreciation program 282
Depreciation run 281
Dimension 640
　　display 640
　　parameters 641
Distribution 392
Distribution cycle 393
　　create 393
　　segment 395
　　settings 394
Document splitting 123, 490
　　characteristics 124, 128
　　document types 127
　　example 125

Index

Document splitting (Cont.)
  general ledger accounts 125
  item categories 126
Document type 94, 648
  alternative 270
  ledgers 98
  settings 95–96, 649
Drilldown report 151, 417, 498, 561
  default output 506
  definition 506
  form 501
  posting periods 503
  selection 564
Dunning 202

E

Electronic bank statement (EBS) 331
  account symbols 333
  configuration 333
  posting rule 335
  transaction type 336
Enterprise Controlling Consolidation
  (EC-CS) 633
Evaluation group 264
Excel 235, 688
Exchange rates 101, 142
  maintain 103
Execution profile 428

F

Field status groups 206
Finance data model 73
Financial accounting 60
  configuration documents 64
  global settings 73
  obsolete tables 75
  process areas 61
  real-time integration 76
  requirements 64
Financial closing 137
Financial statement items (FSIs) 642
Financial statement version (FSV) 148
  assign accounts 150
  copy 149
  structure 149

Financial statements 148
Fiscal year 350
  settings 410
Fiscal year variant 84
Fit/gap analysis 67
Fixed assets 239
  business transactions 268
  check program 665
  data migration 692
  information system 288
  master data 251
Foreign currency valuation 141

G

General ledger 113, 200, 247, 496
  data migration 690
  master data 113
  migration 673
  readiness check 667
General ledger account ... 76, 118, 219, 258, 353
  assign 532
  control data 121, 353
  controlling settings 354
  copy 80
  languages 123
  maintain 119
  master record 354
  settings 120
Go-live 719
  activities 722
  controlling activities 726
  date 719
  day 1 732
  FI activities 723
  validation 730
Goodwill 657
GR/IR clearing 166, 176
  document type 176
  number range 178
Greenfield implementation 49
  go-live 720
  migration 676
  migration options 676
  vs. brownfield 50–51
Group asset 267
  numbers 266

749

Index

Group chart of accounts 114, 691
Group numbers 265
Group reporting 39, 631
   basics 631
   global settings 634
   history 632
Groups 437

H

History sheet 294
House banks 303, 320
   details 304–305
   ranking order 321
Hypercare production support 731

I

IBAN 308
Incoming payments 221
Information system 147, 185, 231, 288, 415, 468, 496
Integrated planning 430, 550
Integration testing 708
   documentation 711
Integration with logistics 65
Interactive reports 497
Intercompany reconciliation 140
Intercompany transfer variant 273
Internal order 425, 438
   assignments 438
   budgeting 443
   change 442
   control data 440
   cost center 439
   create 438
   number range 434
   plan allocations 466
   planning 461, 463
   reports 473
   set up 426
   settlement 449, 455
   settlement profile 450
International Bank Account Number (IBAN) 306

International Financial Reporting Standards (IFRS) 68, 90
   IFRS 15 52
Investment management 442
Invoice value flow 530
Invoices 170
   verification 177

K

Key figures 414, 474, 502

L

Layout definition 465
Ledger 90
   accounting principle 93
   company code 91
   controlling postings 352
   define 91
   groups 94
   leading ledger 90
   nonleading 99
   nonleading ledger 90
   settings 92
Legacy data load 688
Legacy System Migration Workbench (LSMW) 676
Legal consolidation (LC) 632
Line item lists 559
Line item reports 189, 236, 497
List-oriented reports 497
Local accounting standards 68
Local tax requirements 69
Localization 67
   fit-gap analysis 66
Logistics Invoice Verification (LIV) 170
Long-term support 735

M

Manage Bank Accounts app ... 306–307
Manage Global Accounting Hierarchies app 642
Management consolidation 632
Manual planning 410

Index

Manual postings 648, 653
   tasks 654
Manual reposting 357
   settings 359
Manual value correction 285
Mass maintenance 375
   execution 376
Master data 113, 251, 299, 352, 365, 426, 478, 567
   reports 186, 232
Material 570
   views 571
Material cost estimate 610–611
Material ledger 37, 71, 613
   activation 614
   reports 626
   types 615
   update 616
Material master 568, 696
Material type 568
   definition 569
Material update categories 619
Material update structure 619–620
Material valuation 727
   table 729
Migration 247, 668
   AP/AR data 692
   check program 277
   completion 673
   controlling data 695
   errors 671
   field mapping 685
   financial objects 690
   fixed assets data 692
   general ledger allocations 673
   general ledger data 690
   object 678, 680
   post-migration activities 674
   rules 687
   start and monitor 672
   structure mapping 683
   target structures 683
   transactional data 669
   transactional data log 670
Migration object modeler 676–678
Movement type group 617
Moving average price 572
Multiple valuation principles 277

N

Negative postings 97
Negative tracing 396, 404
New asset accounting 276
Number range 97, 162–163, 179, 207, 260, 347, 434–435, 455, 650
   assignment 436
   define 207
   intervals 98
   maintain 447
   transport 725

O

Online analytical processing (OLAP) 30
Online transactional processing (OLTP) 30
Operating concern 88, 350, 512, 524, 726
   attributes 523
   copy 89
   data struture 519
   define 89
Operational chart of accounts 114
Optical character recognition (OCR) 184
Order category 440
Order management 426
Order type 427
   archiving 430
   define 429
   description 428
   master data 431
   status 431
Organizational structure 77, 131, 240
   company 77
   company code 78
   controlling area 84
   operating concern 88
Outgoing invoices 213, 215, 218
Outgoing payments 179
Overhead costs 365, 537
Overhead rates 605
   percentage 605
   quantity 606–607

751

P

Parallel currencies 37
Parallel valuations 37
Payment block 182
   reasons 183–184
   settings 183
Payment file 326
Payment format 328–329
Payment Medium Workbench (PMW) 316
Payment method 310, 314
   checks 315
   company codes 317
   country settings 314
   currencies 316
   medium 328
   settings 318
Payment proposal 324
   errors 325
Payment run 309
Payment term 179–180, 221–222
   baseline date 182
   create 181
   maintain 222
   settings 181
Payment transactions 207
Period category 659
Period control 279
Period-end closing 406
Periodic allocations 391, 449
Periodic processing 137
Periodic reposting 457
   create 458
   cycle 461
   segment definition 460
   settings 459
Planning 408, 461
   basic settings 409
   elements 551
   framework 550
   level 551
   level characteristics 552
   methods 553
   number ranges 463
   package 556
   reposting 467

Planning (Cont.)
   settings 462
   versions 462
Planning layout 412, 465
   characteristics 555
   create 413
   definition 465
Planning profile 428
Post General Journal Entries app 154
Posting period 137
   account types 139
   define 139
   variant 138
Posting rules 335
Price determination 585
Pricing procedure 220
Product cost planning 578
   reports 624
Product costing 567, 726
   drilldown reporting 627
   information system 623
Production costs 545
Profit center 478
   actual/plan comparison 499
   address 483
   communications 483
   company codes 482
   copy 479
   cost center 488
   create 479
   default account 489
   derivation 487
   history 484
   indicators 481
   master record 479
   multiple assignments 386
   reports 497, 499
   settings 480
   splitting 490
   standard hierarchy 486–487
   standard reports 498
   substitution 491
Profit center accounting 477
   drilldown reporting 500
   information system 496
   master data 478
   reporting 496

Index

Profit center group 484
   create 485
   define 485
   level 485
Profitability analysis 37, 509
   account-based 511–512
   characteristics 513
   costing-based 510, 534
   data flow 529
   data structure 513
   drilldown report 562
   information system 558
   segment 544
   transfer structure 537
Project
   implementation by country 54
   objectives 48
   organizational structure 56
   planning 708
   preparation 47
   rollout 56
   scope 49, 51
   team 54
   team structure 55
   technical readiness 51
   timeline 53
Purchasing flows 132

Q

Quality Center 702
Quantity field 535
Quantity structure control 587
   BOM 588

R

Ranorex Studio 703
Rates 382
Receiver rules 403
Receiver tracing 398
   factors 396
Receiver weighting factors 406
Reconciliation accounts 122
Reference variant 591
   cost estimate 591

Reference variant (Cont.)
   create 591
   revaluation 592
Relational database management system
   (RDMS) 29
Report Painter 414, 418, 465
   reports 472
Report Writer 472
Reports 468
   definition 421
   layout 419
   output 469, 471
Requirements analysis 52, 59
   local 70
   templates 60
Requirements document 60
   configuration 62
   structure 61
Retirement 274
   account 258, 275
   general ledger accounts 274
   transaction types 275
Revaluation 283
RICEFW objects 61, 708
Routing 576
   create 577
   definition 577

S

Sales and distribution (SD) 218
Sales and use tax 150
Sales area data 203
Sales data 202
Sales flows 133
Sales output tax code 224
Sales price 220
Sandbox client testing 706
SAP Business Planning and Consolidation
   (SAP BPC) 633
SAP Cash Management 306
SAP Collections and Dispute
   Management 202
SAP EarlyWatch Check 722
SAP Fiori 33, 39
   apps reference library 306
   general ledger apps 153

753

SAP Fiori launchpad 41, 306
SAP Globalization Services 68
SAP GoingLive Check 722
SAP HANA database 29
SAP Invoice Management by OpenText 185
SAP Reference IMG 43, 63, 66
SAP Revenue Accounting and Reporting 52
SAP S/4HANA Cloud 33, 48
SAP S/4HANA Finance 665
SAP S/4HANA Finance for group
   reporting 631
   benefits 633
   prerequisites 635
SAP S/4HANA migration cockpit 676–677
   migration templates 679
   project name 679
SAP Solution Manager 62, 702
SAP Strategic Management Business
   Consolidation (SEMC-BCS) 633
Screen layout 251, 432, 560
   copy 432
   fields 252
   general data fields 254
   tabs definition 433
Screen variants 358
Segments 481
Selection variant 456
Selenium 703
Sender/receiver settings 397, 404
SEPA 312, 326
   credit transfer 327
   direct debit 223, 327
   mandate 222
Settlement 449
   collective 456
   number ranges 454
   rule 450
   type 450
Settlement profile 428, 451
   definition 452
   maintain 541
SII (Suministro Inmediato de Información
   del IVA) 52, 69
Software as a service (SaaS) 33
Software Update Manager (SUM) 247
Source accounts 547

Source cost elements 539
   FI/MM 543
Special depreciation 260
Special general ledger indicators 168–169
Special procurement keys 590
Special purpose ledger (SPL) 90
Special version 639
   create 640
Splitting profile 546
Standard price 572, 727–728
Statistical key figure 383, 464, 466
   column 466
   create 383
   manual planning 412
   settings 384
Subassignments 651
Subcontracting 585
Subitem 643
   default assigment 647
   settings 646
Subitem category 643–644
   settings 645
Subnumbers 254, 265
Substitution 387, 491
   activate 495
   activation criteria 495
   cost center 388
   create 492
   logic 492
   methods 389
   prerequisite rule 494
   prerequisites 388
   rules 491
Support incidents 734
SWIFT format 300

T

Task group 658
   assign tasks 659
   assignment 659
   preceding task 660
   settings 658
Tasks 648
Tax classification 231
Tax code 109, 167, 172, 213
   configuration 110

Index

Tax code (Cont.)
    export 723
    import 724
    invoices and credit memos 167
Taxes 103, 223
    condition types 104
    determination 172, 224–225
    financial accounting invoices 223
    jurisdiction code 224
    reports 150
    tax code 109
    tax procedure 103
Technical clearing account 269
Testing 699
    cycles 700
    documentation 704
    execution 710
    integration testing 700
    phases 709
    plan 700
    preparation phase 709
    process flow 706
    sandbox client 706
    test case 705
    tools 702
    unit testing 700, 707
    user acceptance testing (UAT) 700
Three-way match 133
Tolerance keys 170
Tolerance limits 170–171, 447
Transaction 79, 737
    /SMB/BBI 635
    /UI2/FLP 40
    ABLDT 694
    AFAB 281
    AFBP 283
    AJRW 287
    AS91 693
    ASKBN 247
    BP 158, 198
    CA01 576
    CKME 728
    CR01 575
    DMEE 329–330
    EC01 79
    FAGLVTR 145

Transaction (Cont.)
    FB60 166–167
    FB70 .... 213, 224
    FBKP 134
    FBZP 311–312
    FF_5 332
    FGI0 499
    FGI1 505
    FGI4 501
    FI01 301
    FI12_HBANK 303
    FS00 .... 119, 353
    FTXP 109, 723–724
    GGB1 387
    KAH1 355
    KB11N 358
    KB21N 361
    KE24 559
    KE34 561
    KKO0 627
    KKRV 627
    KP06 410
    KP26 410
    KP46 412
    KS12N 374
    KSC5 407
    KSH2 378
    KSU5 406
    KSV5 398
    KSW5 457
    KSW7 467
    LTMC 677
    LTMOM 678
    MEK1 174
    MR11 176
    OB13 114
    OB27 182
    OB62 116
    OBA1 141
    OBB8 180
    OBD4 117
    OBYC 135, 546
    OKB9 .... 385, 488, 544
    OKEON 378
    OKKS 378
    S_ALR_87012993 468

Transaction (Cont.)
 S_E38_98000088 498
 SE11 726
 SE16N 728
 SE38 664
 SM35 725
 SM37 733
 SPRO 43
 VF03 530
 VK11 229
 VKOA 531
Transaction types 292, 337
 bank account assignment 338
 external 337
Transfer control 589
 single plant 590
Transfer structure 537
 copy 538
 FI/MM 542
Transfers 270
Transport request 724
Two-way match 133

U

Unicode conversion 668
Unit testing 705, 707
Universal Journal 34–35, 74–75, 240, 289
Unplanned depreciation 260
Upload General Journal Entries app 155
US Generally Accepted Accounting
 Principles (US GAAP) 68, 90
User acceptance testing (UAT) 55, 700, 712,
 714–715
 documentation 715
 execution 713
 forms 717
 planning 713
 test script 716
User-defined fields 262, 265
User-defined reports 418

V

Valid receivers 452
Valuation 276, 615
Valuation variant 582
 activity types 584
 material valuation 583
 overhead 586
 subcontracting 584
Value fields 521–522, 540
 assign 535
Value-added tax (VAT) 69, 150
Variances 421, 453
Vendor groups 162, 165
Vendor invoice 166
Vendor invoice management (VIM) 166, 184
Vendors 164
 account balances 188
 due date analysis 189
 line item browser 192
 line item display report 194
 line item report 191
 line items 190
 reports 186, 191
Version 348, 651
 fiscal year settings 350
 general ledger 351
 settings 349

W

Weighting factors 399
Work center 575
Workflows 61
Worksoft 703

Y

Year-end closing 286

Z

Zero-balance clearing account 129–130

- Examine the top differences between SAP Business Suite and SAP S/4HANA Finance

- Tap the potential of the Universal Journal, SAP Fiori, and more

- Understand how key features change your finance processes

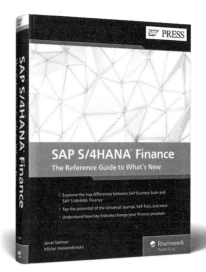

Janet Salmon, Michel Haesendonckx

SAP S/4HANA Finance: The Reference Guide to What's New

SAP S/4HANA has changed finance—but are you ready for change? Consult the experts' list of the most important innovations in SAP S/4HANA Finance. Evaluate their impact on your business users' daily routines, and learn how to make SAP S/4HANA work for you.

505 pages, pub. 04/2019
E-Book: $69.99 | **Print:** $79.95 | **Bundle:** $89.99

www.sap-press.com/4838

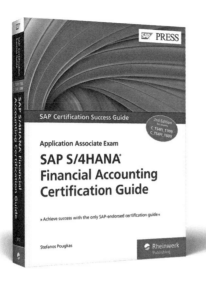

- Learn about the SAP S/4HANA Finance certification test structure and how to prepare
- Review the key topics covered in each portion of your exam
- Test your knowledge with practice questions and answers

Stefanos Pougkas

SAP S/4HANA Financial Accounting Certification Guide

Application Associate Exam

Preparing for your financial accounting exam? Make the grade with this SAP S/4HANA 1709 and 1809 certification study guide! From general ledger accounting to financial close, this guide will review the key technical and functional knowledge you need to pass with flying colors. Explore test methodology, key concepts for each topic area, and practice questions and answers. Your path to financial accounting certification begins here!

507 pages, 2nd edition, pub. 05/2019
E-Book: $69.99 | **Print:** $79.95 | **Bundle:** $89.99

www.sap-press.com/4856

www.sap-press.com

- Learn about the SAP S/4HANA certification test structure and how to prepare
- Review the key topics covered in each portion of your exam
- Test your knowledge with practice questions and answers

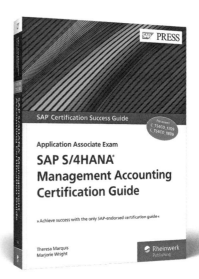

Theresa Marquis, Marjorie Wright

SAP S/4HANA Management Accounting Certification Guide

Application Associate Exam

Studying for the SAP S/4HANA Management Accounting exam? Get the tools you need to succeed with this CO certification study guide for exams C_TS4CO_1709 and C_TS4CO_1809. Understand the test structure and what to expect; then walk through each topic area, from product cost planning to profit center accounting and beyond. Quiz yourself with practice questions and answers, and ensure you're ready to make the grade!

461 pages, pub. 08/2019
E-Book: $69.99 | **Print:** $79.95 | **Bundle:** $89.99

www.sap-press.com/4886

- Configure and run actual costing in the Material Ledger
- Master group valuation, profit center valuation, balance sheet valuation, and more
- Analyze your results with SAP S/4HANA reporting tools

Paul Ovigele

Material Ledger in SAP S/4HANA

Functionality and Configuration

Unlock the potential of the Material Ledger in SAP S/4HANA with this comprehensive guide. Move beyond the basics and get the step-by-step instructions you need to configure and run actual costing, group valuation, profit center valuation, and more. Consult detailed screenshots and expert guidance as you dive deep into the major processes, specialized scenarios, and reporting and analytics. Master the Material Ledger from end to end!

540 pages, pub. 06/2019
E-Book: $79.99 | **Print:** $89.95 | **Bundle:** $99.99

www.sap-press.com/4863

www.sap-press.com

Interested in reading more?

Please visit our website for all new book
and e-book releases from SAP PRESS.

www.sap-press.com